EXPLAINING AND EXPLORING DIVERSITY IN AGRICULTURAL TECHNOLOGY

edited by

Annelou van Gijn, John C. Whittaker and Patricia C. Anderson

Oxbow Books
Oxford & Philadelphia

Other Volumes in this Series

Plants and People: Choices and Diversity through Time
(EARTH Volume 1)

Agricultural and Pastoral Landscapes in Preindustrial Society: Choices, Stability and Change
(EARTH Volume 3)

Early Agricultural Remnants and Technical Heritage (EARTH): 8,000 Years of Resilience and Innovation

Volume 2

Series Editors

Patricia C. Anderson and Leonor Peña-Chocarro

Coordinating Editor

Andreas G. Heiss

Published in the United Kingdom in 2014 by
OXBOW BOOKS
10 Hythe Bridge Street, Oxford OX1 2EW

and in the United States by
OXBOW BOOKS
908 Darby Road, Havertown, PA 19083

Hardcover Edition: ISBN 978-1-84217-515-6
Digital Edition: ISBN 978-1-78297-021-7

A CIP record for this book is available from the British Library

Library of Congress Cataloging-in-Publication Data

Explaining and exploring diversity in agricultural technology / edited by Annelou van Gijn, John C. Whittaker and Patricia C. Anderson. -- Hardcover edition.
pages cm. -- (Early agricultural remnants and technical heritage (EARTH) : 8,000 years of resilience and innovation ; volume 3)
Summary: "This volume is the outcome of collaborative European research among archaeologists, archaeobotanists, ethnographers, historians and agronomists, and frequently uses experiments in archaeology. It aims to establish new common ground for integrating different approaches and for viewing agriculture from the standpoint of the human actors involved. Each chapter provides an interdisciplinary overview of the skills used and the social context of the pursuit of agriculture, highlighting examples of tools, technologies and processes from land clearance to cereal processing and food preparation"-- Provided by publisher.
Includes bibliographical references and index.
ISBN 978-1-84217-515-6 (hardcover) -- ISBN 978-1-78297-021-7 (digital)
1. Agriculture--History. 2. Agriculture--Social aspects--History. 3. Agricultural innovations--History. 4. Agricultural implements--History. 5. Agropastoral systems--History. 6. Social change--History. 7. Ethnology. 8. Agriculture, Prehistoric. 9. Ethnoarchaelogy. 10. Landscape archaelogy. I. Gijn, Annelou L. van, 1954- II. Whittaker, John C. (John Charles), 1953- III. Anderson, Patricia C.
S419.E96 2014
630.9--dc23

2014030957

Printed in the United Kingdom by Berforts Information Press

For a complete list of Oxbow titles, please contact:

UNITED KINGDOM
Oxbow Books
Telephone (01865) 241249, Fax (01865) 794449
Email: oxbow@oxbowbooks.com
www.oxbowbooks.com

UNITED STATES OF AMERICA
Oxbow Books
Telephone (800) 791-9354, Fax (610) 853-9146
Email: queries@casemateacademic.com
www.casemateacademic.com/oxbow

Oxbow Books is part of the Casemate Group

Cover illustrations
Farmer with his daughter riding a *jarousha* (threshing sledge) to process durum wheat in the village of Ain Salem (Northwest Tunisia), 2005. Image: Patricia C. Anderson
Farmer winnowing durum wheat grain and chaff into the wind, throwing with a fork, in the village of Al Asidiya (Northwest Tunisia), 2005. Image: Patricia C. Anderson
A farmer using the wooden *mesorias* tool to harvest spelt wheat heads by pulling, in the village of Grado (Asturias, Spain), 2006. Image: Lydia Zapata

We would like to dedicate this volume to two friends and colleagues whose long-term encouragement contributed to making the interdisciplinary approach used in this book a reality. They were both members of an interdisciplinary agrarian archaeology group carrying out experiments in the 1970s and 80s in Europe, which led years later to the idea for the Earth Program.

François Sigaut

agronomist and ethno-historian specialised in the history and the anthropology of techniques of agriculture in Europe and worldwide, received his doctorate from and taught at l'École des Hautes Études en Sciences Sociales in Paris. He was a Founding Member of the Earth Program and published in Books 1 and 2 in this series. Skeptical about received wisdom on agricultural techniques, he encouraged ethnographic and historical approaches to be combined with archaeological research. His broad knowledge and passion for understanding techniques as well as their actors was legendary, and he opened new paths of research for many. He headed a network of agricultural museums worldwide. His work concerned grain storage, grinding stones, the history of soil working and field preparation, including the fallow, and harvesting and threshing tools, to name only a few of the many topics. He passed away suddenly in 2012. He is responsible for inspiring research on ethno-agrarian history and ethnoarchaeology, prevalent in this monograph.

His friend

Peter Reynolds

who studied both classics and archaeology, receiving his doctorate from Leicester University, opened up the area of experimental archaeology (or rather, 'Experiments in Archaeology', as he preferred to say), in agriculture, as founder and director of Butser Ancient Iron Age Farm in Hampshire, England, beginning in the 1970s. Also a skeptic concerning fashionable interpretations of archaeological data that were unsubstantiated by scientific evidence, he used field experiments to test hypotheses, for example about the functioning of grain storage pits, grain drying structures, the thatched roof of a roundhouse, of the ard, and of various field and harvesting methods. He encouraged researchers to work at his experimental farm harvesting the vast quantities of primitive cereals grown there, and also helped people to begin their own experiments or start their own experimental centres elsewhere. He was also a Founding Member of the Earth Program and had just hosted a preliminary meeting to draw up this project when he died suddenly in 2001. The influence of experiments in the archaeology of agriculture can be felt throughout this book.

ESF MISSION

The European Science Foundation (ESF) is an independent, non-governmental organisation, the members of which are eighty national funding agencies, research-performing agencies, academies and learned societies from thirty countries.
The strength of the ESF lies in the influential membership and in its ability to bring together the different domains of European science in order to meet the challenges of the future.
Since its establishment in 1974, the ESF, which has its headquarters in Strasbourg with offices in Brussels and Ostend, has assembled a host of organisations that span all disciplines of science, to create a common platform for cross-border cooperation in Europe.
The ESF is dedicated to promoting collaboration in scientific research, funding of research and science policy across Europe. Through its activities and instruments the ESF has made major contributions to science in a global context. The ESF covers the following scientific domains:

- Humanities
- Life, Earth and Environmental Sciences
- Medical Sciences
- Physical and Engineering Sciences
- Social Sciences
- Marine Sciences

ESF MEMBER ORGANISATIONS FUNDING THE EARTH NETWORKING PROGRAMME AND PUBLICATIONS

Fonds National de la Recherche Scientifique (FNRS) – National Fund for Scientific Research, Belgium

Fonds voor Wetenschappelijk Onderzoek Vlaanderen (FWO) – Research Foundation Flanders, Belgium

Grantová agentura České republiky (GAČR) – Czech Science Foundation, Czech Republic

Akademie věd České republiky (ASCR) – Academy of Sciences of the Czech Republic, Czech Republic

Det Frie Forskningsråd – Kultur og Kommunikation (FKK) – The Danish Council for Independent Research – Humanities, Demark

Eesti Teadusfond (ETF) – Estonian Science Foundation, Estonia

Suomen Akatemia/Finlands Akademi – Academy of Finland, Finland

Centre National de la Recherche Scientifique (CNRS) – National Centre for Scientific Research, France

Deutsche Forschungsgemeinschaft (DFG) – German Research Foundation, Germany

An Chomhairle um Thaighde sna Dána agus sna hEolaíochtaí Sóisialta (IRCHSS) – Irish Research Council for the Humanities and Social Sciences, Ireland

Nederlandse Organisatie voor Wetenschappelijk Onderzoek (NWO) – Netherlands Organisation for Scientific Research, The Netherlands

Norges Forskningsråd – Research Council of Norway, Norway

Consejo Superior de Investigaciones Científicas (CSIC) – Council for Scientific Research, Spain

Vetenskapsrådet (VR) – Swedish Research Council, Sweden

Schweizerischer Nationalfonds (SNF) – Swiss National Science Foundation, Switzerland

Arts and Humanities Research Council (AHRC), United Kingdom

Contents

SECTION 3: THE SOCIAL CONTEXT OF AGRICULTURAL TECHNOLOGY

Annexes

Foreword

As students of the human past, archaeologists typically follow two complementary avenues of research. One approach is to explore the big questions, those revolutionary episodes that set people on dramatically, and in many cases irreversibly, different paths of evolutionary development. Among these critical transformations were the initial production of stone tools, the controlled use of fire, and the creative explosion that marks the beginning of art. Arguably, none of these changes was more important than the domestication of plants and animals in the early Holocene, exemplified by Childe's term Neolithic Revolution, even though modern scholars view the shift as gradual, with some important changes beginning before the Neolithic. The complexities of the modern world are based on the Urban Revolution, which in turn was founded on the origins of agriculture. Agro-pastoralism required not only a reshaping of subsistence activities, but also a reformulation of kinship structure, political organisation, patterns of mobility, material culture, and belief systems. Its repercussions were deep and long-lasting. In the effort to understand major changes such as the Neolithic Revolution, the key question is why. What motivated people to shift livelihood from the hunting and gathering or foraging pattern that defined some four million years of hominid development to exploitation of a somewhat limited range of cultivated plants and husbanded animals? This question immediately raises the issue of human nature. While we humans are certainly an inquisitive and innovative lot, we also exhibit a wary tendency when it comes to how we feed ourselves. A key to survival is the need to sustain a group while minimising the risks to which we are exposed. As is the case with most animals, people are risk-averse, so they must weigh the costs against the benefits of adopting a new food source or process of food acquisition. This, of course, is where the element of choice enters the picture. The editors rightfully stress the importance of decision-making in agricultural activities, and the various authors demonstrate how it plays out in the things that people do.

The second approach that scholars pursue is concerned with reconstructing past lifeways, *i.e.*, gaining an understanding of the myriad activities that make up the daily lives of people, which is the focus of this book. If the first approach is concerned with the *why* questions, then this avenue deals with the *how* questions. In terms of agriculture, such questions center on issues such as the specific changes that human interference wrought on particular plants and animals as they became domesticated; the precise nature of the diet in particular periods, including the degree to which groups continued to exploit wild species; the types of tools used to prepare the soil, harvest and process crops, and convert raw plant material into edible form. The various analytical techniques reflected in the contributions cover a wide range, from the study of microwear striations and polishes on flint pieces that help to distinguish between items used in sickles (harvesting tools) and threshing sledges (for separating grains from their stems), to the identification of plants through examination of macrobotanical remains and microscopic phytoliths. These methods provide the data for building an understanding of agriculture from the ground up. The fragmentary nature of the archaeological record makes it difficult to gain a full picture of past agricultural activity, so the contributors also make use of an array of other methods. Roman historical literature provides key insights about agricultural practices in the ancient Mediterranean region, while ethnographic sources paint important images of farming activities from the past several centuries. Ethnoarchaeological research, often used in these case studies, in which archaeologists become ethnographers with a focus on how the material record is formed, and experimental work that involves recreating past practices, artifacts, and features, have expanded greatly the interpretive possibilities for archaeologists. These endeavors are all part of the effort to help archaeologists comprehend the dynamic social system that created the static archaeological record we observe, as Binford stated decades ago.

Since the goal of the book is to examine the 'culture of agriculture', it is important that the contributors provide information on a vast array of practices that leave little if any imprint in the material record. Harvesting plants may involve the use of stone blades set in bone or wooden handles, metal sickles, and scythes. However, plants could also be pulled out by hand. Threshing could be accomplished by the use of sledges, tramping by animals, and flailing; the timing of the event could vary considerably, depending on the local conditions, with the process outside over a short period immediately after the harvest in hot, dry regions, but is done indoors over a longer period in other locations, such as northern and northwestern Europe. The ethnographic studies also provide a window into the social matrix of agricultural societies. Various chapters deal with crucial issues such as how necessary skills are acquired and then transmitted through the generations; the influence of religion on techniques and tools; and the legal component.

This volume offers a rich panoply of the objects and activities that constituted preindustrial agriculture. By bringing together the work of archaeologists, historians, and others, we get to view the intricacies of traditional agricultural systems primarily in Europe and the Mediterranean region, although some studies are from further afield, such as the Far East and North and South America, all with fascinating details about specific implements and practices that demonstrate the inventive genius of those who work the land. The minutiae are balanced by consideration of agriculture as an integrative factor that weaves together the various threads of society into a tight, coherent, system. But the long view taken here shows that 'tradition' is not a monolithic thing immune to change. Linked as they are to the world near and far, farmers adopt and adapt as conditions warrant, a practice that is still as true today as throughout the preceding millennia. It is this lesson that comes through most clearly in this fine compilation.

P. Nick Kardulias
Wooster, Ohio, May 2014

Preface

Patricia C. Anderson and Leonor Peña-Chocarro

Scientific Overview

This book takes an interdisciplinary look at European preindustrial agriculture, including its origins and its diffusion outside Europe. Agriculture and its origins have long been lively and innovative subjects of research, involving people working in a variety of disciplines. Initial impetus to this area came from several quite separate disciplines. **Archaeologists** working in the 1940s and 1950s, studied **flint** blades found on archaeological sites in the Middle East and in Europe, which they hypothesised had been inserts forming the working parts of agricultural tools. Such identifications were based upon intimate knowledge of ethnographic tools (Steensberg 1943), and by examining the microscopic traces of use on their edges (Semenov 1964). **Agronomists**, in turn, interested in the origins of agriculture studied the behaviour of **wild cereals** still growing in their natural habitats and explored the way human activities may have influenced crop evolution (Harlan 1975). This area of study, pioneered by a few researchers like O. Heer (1865), lead agronomists and botanists towards the beginnings of **archaeobotanical research** (*e.g.* Hopf 1954, 1955). Over time, however, the study of agriculture evolved towards an interdisciplinary approach which tried to fully understand its multiple facets: plants, techniques, soil types, environment, agronomic practices, the attendant animal husbandry, impact on landscape, traditions, etc. In other words, the cross-disciplinary character of the subject implied the integration of the various historical, anthropological, archaeological and scientific records in order to fully appreciate the complex and interrelated issues involved.

Thus, **archaeobotanists** have looked at prehistoric **seeds** of ancient crops throughout the world to determine the presence of domesticated plants and explore the possible uses of the species identified. Other studies combined analysis of archaeological plant remains with field and genetic studies of equivalent crops still surviving today (Zohary and Hopf 2012; Harlan 1999), and documented the origin and spread of the domesticated plants from the Near East throughout Europe. Weeds have also been a subject of interest, used to explore crop husbandry practices (Charles *et al.* 1997, Jones *et al.* 2010). In particular, crop processing has been a major topic of interest in the study of agriculture. **Ethnographic observations** of the various steps within the crop processing sequence have allowed researchers to produce models (Hillman 1981, 1984; taken up later by *e.g.* Peña-Chocarro 1999) and statistical methods (*e.g.* Jones 1984, 1987; Pearsall 1988) to identify these from crop remains, which in dry sites are usually preserved by charring. This experimental approach was applied by archaeobotanists using reconstructions of specific archaeological tools preserved in European lake sites, again to see the effect on the crops (Meurers-Balke and Lüning 1999). Waterlogged archaeological sites with exceptional preservation of plant remains have allowed detailed studies of plant use with interesting insights into processing activities, among other things (Herbig 2009, Jacomet 2009) as have rare finds from permafrost (Rollo 2002; Heiss and Oeggl 2009; Dickson 2011).

Pollen analysis has also contributed to treating the question of cereal domestication (Bottema 1999), reconstructing archaeological fields and characterising agricultural practices such as the identification of the slash-and-burn technique in the Nordic areas (Vuorela 1986). And, on a more general level, palynologists have shown the role of both the natural environment and prehistoric human activities in the shaping of landscapes, both based on pollen evidence itself (Kalis *et al* 2003; Carrión *et al.* 2007) and on other microfossils in pollen records, the so-called NPPs ('non-pollen palynomorphs', such as fungal and fern spores, remains of green algae and cyanobacteria, or invertebrate eggs; *e.g.* Van Geel *et al.* 2003).

Experimental archaeology has also proven to be a powerful approach to the study of agriculture, with a myriad of examples focusing on specific processes. Agricultural field experiments were carried out in northern Europe by archaeologists, for example using reconstructed ploughs and ards to work fields (Lerche and Steensberg 1980). This led to precise reconstructions of marks in the soil, as well as of archaeological finds of ploughs and ards (Lerche 1994). Other experiments focused on soil, climate and yield of ancient cereal types (Reynolds 1981; Steensberg 1943). By combining soil micromorphology and archaeobotany with an experimental approach (Boissinot and Brochier 1997), it was possible to show field use patterns from as early as the Neolithic in southern Europe (Rösch *et al.* 2002).

Historical records and agronomy were combined to evaluate the cultural influence on concepts of yield (Sigaut 1999) while historians interested in archaeology carried out research into the effect on the soil and the crops of certain farming techniques in large-scale field experiments (Reynolds 1974, 1979, 1981, 1999). Storage pits, for instance, were reproduced and tested using historic and ethnographic references, and were shown to be effective for storing grain similar to ancient varieties (Reynolds 1974). This allowed new interpretations of such archaeological features (Gast and Sigaut 1979–85).

Some **field experiments** were also carried out using reconstructions of harvesting tools based on prehistoric finds which involved measuring the harvesting yield with different kinds of tools, and then studying the experimental tools for traces of use (Korobkova 1981). High-power microscopy, combined with the experiments, allowed finer distinctions of different tool uses and their contact with plant material (Anderson 1999).

New field experiments were designed on a regional basis to test specific tools on ancient crops grown under climatic conditions similar to the past, or on wild cereals (Anderson *et al.* 1991; Anderson 1999; Hillman 1999). These new analyses often disproved or greatly nuanced another archaeological assumption: that all flint tools with gloss were agricultural harvesting tools (Juel Jensen 1988, 1994; Van Gijn 1999). Historical and ethnographic studies (*e.g.* Sigaut 1978) came to similar conclusions, that 'sickles' were sometimes used to obtain other materials such as reeds and straw or were not even sickles at all, but rather inserts in threshing tools (Anderson and Inizan 1994). In addition, specialised threshing tools, threshing floors and storage facilities were increasingly being found in the archaeological record in the Near East and southern Europe (Avner 1998; Skakun 1999; Gast and Sigaut 1979–85; Kuijt and Finlayson 2009; Miret i Mestre 2006; Cunningham 2011). These were identified by means of experiments (Anderson *et al.* 2004), ancient descriptions and ethnography (Kardulias and Yerkes 1996; Grégoire in Anderson and Inizan 2004; Ataman 1999). Experiments were also part of the research carried out to investigate the effects of different grinding and pounding tools on the grain and to seek to find a full range of wear and archaeobotanical criteria to identify the function of ancient tools (Procopiou *et al.* 2002).

Examination of historical documents such as the earliest cuneiform tablets, combined with these experiments, gave further insight into grain processing and social organisation during the third millennium in the Near East (Grégoire 1999). The same applies to agricultural techniques and tools from Medieval Europe, thanks to the study of historical texts (Comet 1992) and ethnographic and archaeological investigation (Mingote Calderón 1996).

Diet and the **social context of food** are other rich topics of research, involving historical documents (Carpinschi 2002), ethnoarchaeobotanical studies (Sarpaki 2000), and archaeobotanical analysis including funerary offerings (Marinval 1993; Rottoli and Castiglioni 2011) and of luxury foods (Bakels and Jacomet 2003; Van der Veen 2003; Palmer and Van der Veen 2002; Van der Veen 2008). New methods for identifying microscopic food remains (*e.g.* Winton and Winton 1932, 1935; Gassner *et al.* 1989) used the analysis of plant tissue fragments, phytoliths and starch (Dickson 1987; Hansson and Isaksson 1994; Cummings 1992; Perry *et al.* 2007; Gong *et al.* 2011; Valamoti *et al.* 2008; Henry *et al.* 2009; Revedin *et al.* 2010) or the chemical analyses of pottery residues (Evershed *et al.* 2008), all of which opened up new avenues of research.

The complexity of agricultural processes and their organisation within particular communities and societies requires a broad analytical scale on which to investigate them. The archaeology, anthropology and history of the landscape has been popular for some thirty years, linked by a common concern with going 'beyond the site' and situating human activity in its broadest context (Kardulias 1994). During the last twenty years there has been a considerable output of theory and methodology for investigating social and cultural as well as physical landscapes (*e.g.* Anschuetz *et al.* 2001; Ashmore and Knapp 1999; Francovich and Patterson 2000; Behre and Jacomet 1991). This has been complemented by a series of highly intensive, interdisciplinary surveys carrying out empirical analyses of particular landscapes within these theoretical frameworks (*e.g.* Astill and Davies 1999; Barker 1996; Cherry *et al.* 1991; Given and Knapp 2003; Kohler-Schneider 2001; Fischer and Rösch 2010; Stika *et al.* 2008).

This rapid overview shows some of the important scientific results of the above research approaches, and the many new insights they have provided into the human experience of agriculture. These approaches, however, have far greater potential to grow further if allowed to work closely together. Archaeological data and historical records on their own are often not specific enough for the identification of particular agricultural practices. Scientists analysing pollen, phytoliths, seeds and wear traces on tools can now identify particular species and techniques and develop sophisticated taxonomies, but are often forced by practical limitations to work in relative isolation from other interpretative sources. In both cases, disciplines have usually become highly specialised, and paradoxically isolated by their very development and institutionalisation. This has often created a barrier to the investigation of the broader agricultural system with all its actors, strategies and landscapes.

The Origin of the Programme

During the 1980s and early 1990s, a European working group met which was concerned with 'agro-(for agrarian) archaeology', combining archaeologists, archaeobotanists, ethnographers, historians and agronomists, around experiments in archaeology. The principal impetus was given by Jutta Meurers-Balke in Cologne, the late Sytze Bottema in Groningen, Holland, the late Peter Reynolds of Butser farm in England, Grith Lerche and the late Axel Steensberg in Copenhagen, and the late François Sigaut in Paris, France.

Following on this, in 1998 a new group concerned with preserving the knowledge and cultural heritage of agricultural processes began to meet, its members combining several different strands of research from the earlier group. In addition, its organisers felt it was urgent to record the knowledge and skills of agriculturalists that still use or remember 'traditional', nonindustrial techniques, and to find a means for this knowledge to enter into the mainstream of various disciplines in a dynamic way. At the time, such an ambition was largely outside the mainstream tendencies of the different disciplines.

This group, called 'Early Agricultural Remnants and Technical Heritage' (EARTH) was recognised by the Sub-Committee on Cultural Heritage of the Parliamentary Assembly of the Council of Europe, who formally admitted EARTH as a member of the PACT alliance of networks, and to the Fédération Européenne des Réseaux Européens de Coopération Scientifique et Technique de Coordination. EARTH was also formally accredited to the Parliamentary Assembly of the Council of Europe. Unfortunately, however, these affiliations did not include funding for a project.

Preparation of this ESF (European Science Foundation) programme took several years: The first EARTH committee meetings were held in St. Vallier de Thiey, Southern France in 1998, funded by the French CNRS (Centre National de la Recherche Scientifique), then in Copenhagen in November 1999 and at Butser Ancient Farm in England in January 2001, culminating in an ESF-funded preparatory workshop at St. Martin de Vésubie, Southern France, in November 2001.

In parallel, from 2002 to the present, the CNRS funded a group of twenty French EARTH members working in France in different disciplines and field areas as a 'Groupement de Recherche' (GDR 2517, directed by P. Anderson). Its aims, similar to those of the European-based ESF programme, have been to use various media to record preindustrial agricultural activities from ancient times and the present day, prepare a database for films and images, and to feed this information into the EARTH network as a whole. The funding of new field research for its members has contributed to some of the articles in this book.

The Network: Working Methods Leading to this Book

The EARTH Monograph Series, entitled 'The dynamics of non-industrial agriculture: 8,000 years of resilience and innovation', was born from a unique opportunity for interdisciplinary and international (especially inter-European) collaboration between 2004 and 2009 as part of the 'Early Agricultural Remnants and Technical Heritage' (EARTH) Scientific 'à la carte' Programme of the European Science Foundation (ESF, Strasbourg, France), financially supported by 15 European organisations, which were represented by Steering Committee Members of the Programme (p. 371).

This ESF-funded EARTH programme was organised and chaired by Patricia C. Anderson, Nice, France and Michael Given, Glasgow (replaced in 2006 by Leonor Peña-Chocarro, Madrid and Rome).

The originality of this programme lay in the means it provided for unusual networking methods and for innovative forms of output. The aim was to find new common ground for integrating different approaches, viewing agriculture from the standpoint of the human actors involved, and fit this together into a form which could be effectively transmitted for research and teaching, as well as for heritage.

In order to achieve the programme's goals to stimulate the creation of new, integrated interdisciplinary approaches, collaborations and networks, researchers and advanced students were chosen from different disciplines but who worked in an interdisciplinary manner. Some were chosen by the programme organisers, whereas others were suggested to them by the contributing European national organisations (refer to page vi with the ESF Member Organisations), meaning that most of the group did not know one another at the outset. It brought together scientists from a variety of different disciplines of the human and natural sciences: archaeology (including archaeobotany and archaeozoology, microwear analysis of tool function, experimental archaeology), ethnography, history, geography and geology, from about twenty European countries, as well as from North America and the Middle East.

The researchers were organised into three teams, each led by two experts from different countries, while respecting the stipulation of the ESF that as many different disciplinary fields and nationalities as possible be combined in each team, as well as researchers of varied age and at different stages of their career. The European Science Foundation emphasised networking and learning through exchange and communication, and funded annual team meetings and three plenary meetings. Such meetings allowed comparison of different points of view and of agricultural systems in different regions.

Each team approached the theme of agriculture from a slightly different standpoint – that of Crop Choice and Diversity (Book 1 in the series), Skills, Processes and Tools (Book 2 in the series), and Agricultural Landscapes (Book 3 in the series). The networking was achieved largely by workshops held in archaeological and ethnographic field areas or local museums that were directly relevant to the research of certain members (see pp. 374–376). They provided an important opportunity to exchange as well as to talk with local farmers and artisans. Although a few grants were provided for members to meet and work on methodological fine-tuning, no funding was provided by this ESF programme for new field research to authors contributing to this book. Nevertheless, the richness and originality of the scope of the articles presented here was achieved in many cases through networking activities among scientists during the programme. The programme also funded activities for advanced students: to attend hands-on summer schools taught by members and run by Leonor Peña Chocarro in a traditional agricultural area of Spain, and grants to attend the Programme's various meetings. All networking activities were efficiently implemented by Marie Russel, Paris, who served as the programme's coordinator from 2004 to 2009.

Some of the papers in this series address field areas and topics emanating from the workshops, and they were developed gradually over the course of the five years of interdisciplinary exchange. Although the authors come originally from various different fields related to the study of agriculture, they shared an interest in developing a common ground where individual research fields could converge into a broader framework, and provide a new knowledge base. Many contributions integrate field and laboratory methodologies into the case studies, in a manner intended to be accessible to students and researchers in different fields, as well as to an interested general public. Ethnoarchaeological studies figure prominently, and help inject the human perspective into the study of agriculture.

Each article in these volumes has received double-blind peer review from two outside experts. Using numerous illustrations, they provide synthetic, interdisciplinary overviews, or detailed accounts of individual and collaborative research, and in some cases relevant experts outside the programme were invited to participate. However, this book series does not seek to – and indeed cannot – provide complete coverage of all disciplines, research themes, time periods, or geographical regions relevant to the study of agriculture, nor could all the relevant experts in each field be included. Each book should instead be seen as a sampling of exciting ways to explore the subject of preindustrial agriculture and its relevance to life today. This book series is an example of the power of academic networking and of the benefits of approaching the theme of agriculture from different angles and perspectives converging into a common space different from that of the researchers' individual fields of expertise. But it would not have seen the light of day without the devotion of the coordinating editor, Andreas G. Heiss, Vienna, and the language editor, Cozette Griffin-Kremer, Paris. Alexandre Chevalier, Brussels, ensured that abundant maps were made to cover the wide range of areas discussed.

The journey has been arduous, sometimes frustrating, but always fascinating and full of good memories. We hope that the final outcome of our programme, this book series, is of value to other scholars and interested people. It is also our hope that it leads to further work by opening new avenues into the study of agriculture.

Patricia C. Anderson
CNRS, Nice, France

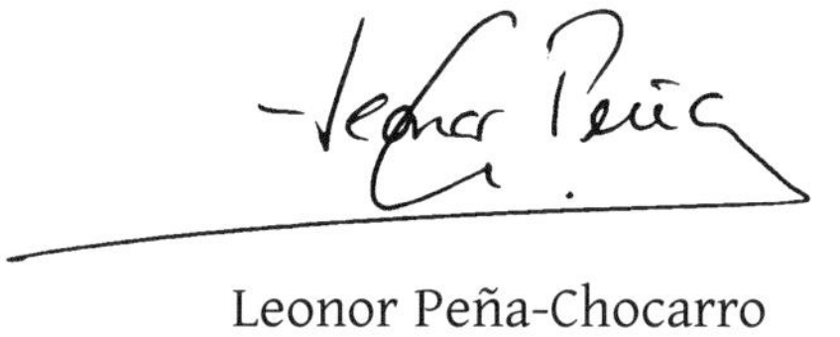

Leonor Peña-Chocarro
CSIC, Madrid, Spain and Rome, Italy

Series Editors of the EARTH Series

Bibliography

Anderson, P. C. (ed.) (1999) *Prehistory of Agriculture. New Experimental and Ethnographic Approaches.* Institute of Archaeology Monograph 40. UCLA, Los Angeles.

Anderson, P. C., Chabot, J. and Van Gijn, A. L. (2004) The Functional Riddle of 'Glossy' Canaanean Blades and the Near Eastern Threshing Sledge. *Journal of Mediterranean Archaeology* 17 (1), 87–130.

Anderson, P. C., Deraprahamian, G. and Willcox, G. (1991) Les premières cultures de céréales sauvages et domestiques au Proche-Orient néolithique: résultats préliminaires d'expériences à Jalès (Ardèche). *Cahiers de l'Euphrate* 5–6, 191–232.

Anderson, P. C. and Inizan, M.-L. (1994) Utilisation du tribulum au début du IIIe millénaire: des lames 'cananéennes' lustrées à Kutan (Nivive V) dans la région de Mossoul, Iraq. *Paléorient* 20 (2), 85–103.

Anschuetz, K., Wilshusen, R. and Scheick, C. (2001) An archaeology of landscapes: perspectives and directions. *Journal of Archaeological Research* 9, 157–211.

Ashmore, W. and Knapp, B. (eds.) (1999) *Archaeologies of Landscape: Contemporary Perspectives.* Blackwell, Oxford.

Astill, G. and Davies, W. (1997) *A Breton Landscape.* UCL Press, London.

Ataman, K. (1999) Threshing sledges and archaeology. In P. Anderson (ed.) *Prehistory of Agriculture.* UCLA, Los Angeles, pp. 211–223.

Avner, U. (1998) Settlement, agriculture, and paleoclimate in Uvda Valley, southern Negev Desert, 6th–3rd millennia BC. In A. Issar and N. Brown (eds.) *Water, Environment and Society in Times of Climate Change. Kluwer,* Dordrecht, pp. 147–202.

Barker, G. (1996) *Farming the Desert: The UNESCO Libyan Valleys Archaeological Survey.* UNESCO, Paris.

Behre, K.-E. and Jacomet, S. (1991) The ecological interpretation of archaeobotanical data. In W. Van Zeist, K. Wasylikowa and K.-E. Behre (eds.) *Progress in Old World Palaeoethnobotany. A retrospective view on the occasion of 20 years of the International Work Group for Palaeoethnobotany* A.A. Balkema, Rotterdam/ Brookfield, pp. 81–108.

Boissinot, P. and Brochier, J.-E. (1997) Pour une archéologie du champ. In G. Chouquer (ed.) *Les Formes du Paysage, Tome II.* Errance, Paris, pp. 35–56.

Bottema, S. (1999) Cereal-type pollen in the Near East as indicators of wild or domestic crops. In P. C. Anderson (ed.) *Prehistory of Agriculture. New Experimental and Ethnographic Approaches.* Institute of Archaeology Monograph 40. UCLA, Los Angeles, pp. 56–66.

Carpinschi, C. (2002) Harvest meals XIXth – XXth centuries. In P. Lysaght (ed.) *Food and Celebration: Proceedings of the 13th International Ethnological Food Research Conference.* Ljubljana, pp. 317–324.

Carrión, J. S., Fuentes, N., González Sampériz, P., Sánchez Quirante, L., Finlayson, J. C., Fernández, S. and Andrade, A. (2007) Holocene environmental change in a montane region of southern Europe with a long history of human settlement. *Quaternary Science Reviews* 26, 1455–1475.

Charles, M., Jones, G. and Hodgson, J. (1997) FIBS in archaeobotany: functional interpretation of weed floras in relation to husbandry practices. *Journal of Archaeological Science* 24, 1151–1161.

Cherry, J., Davis, J., Mantzourani, E. (1991) *Landscape Archaeology as Long-Term History: Northern Keos in the Cycladic Islands from Earliest Settlement Until Modern Times.* UCLA, Los Angeles.

Comet, G. (1992) *Le paysan et son outil: essai d'histoire technique des céréales (France, VIIIe-XVe siècles).* Collection de l'École française de Rome.

Cunningham, P. (2011) Caching your savings: The use of small-scale storage in European prehistory. *Journal of Anthropological Archaeology* 30 (2), 135–144.

Dickson, C. (1987) The identification of cereals from ancient bran fragments. *Circaea* 4 (2), 95–102.

Dickson, J. H. (2011) Why ancient glacier mummies are so special: the ingesta are encoded diaries and maps. In H. Gill-Frerking, W. Rosendahl, A. Zink and D. Piombino-Mascali (eds.) *The Yearbook of Mummy Studies 1.* Verlag Dr. Friedrich Pfeil, München, pp. 45–50.

Evershed, R. P., Payne, S., Sherratt, A. G., Copley, M. S., Coolidge, J., Urem-Kotsu, D., Kotsakis, K., Özdoğan, M., Özdoğan, A. E., Nieuwenhuyse, O., Akkermans, P. M. M. G., Bailey, D., Andeescu, R.-R., Campbell. S., Farid, S., Hodder, I., Yalman, N., Özbaşaran, M., Bıçakcı, E., Garfinkel, Y., Levy, T. and Burton, M. M. (2008) Earliest date for milk use in the Near East and southeastern Europe linked to cattle herding. *Nature* 455, 528–531.

Fischer, E., Rösch, M., Schatz, K. and Stephan, E. (2010) Rinder, Körner, Schweinespeck — Landnutzung und Ernährung in der Eisenzeit. *Archäologie in Deutschland* 2010 (5), 34–37.

Francovich, R. and Patterson, H. (eds.) (2000) *Extracting Meaning from Ploughsoil Assemblages.* The archaeology of the Mediterranean landscape 5. Oxbow Books, Oxford.

Gassner, G., Hohmann, B. and Deutschmann, F. (1989) *Mikroskopische Untersuchung pflanzlicher Lebensmittel.* 5th edition. Verlag Gustav Fischer, Stuttgart.

Gast, M. and Sigaut, F. (1979–1985) *Les Techniques de Conservation de Grains à Long Terme.* 3 vols in 4 parts. CNRS, Paris.

Van Geel, B., Buurman, J., Brinkkemper, O., Schelvis, J., Aptroot, A. van Reenen, G. and Hakbijl, T. (2003) Environmental reconstruction of a Roman Period settlement site in Uitgeest (The Netherlands), with special reference to coprophilous fungi. *Journal of Archaeological Science* 30 (7), 873–883.

Van Gijn, A. L. (1999) The interpretation of sickles: a cautionary tale. In P. C. Anderson (ed.) *Prehistory of Agriculture. New Experimental and Ethnographic*

Approaches. Institute of Archaeology Monograph 40. UCLA, Los Angeles, pp. 254–260.

Given, M. and Knapp, B. (2003) *The Sydney Cyprus Survey Project: Social Approaches to Regional Archaeological Survey.* UCLA, Los Angeles.

Gong, Y., Yang, Y., Ferguson, D. K., Tao, D., Li, W., Wang, C., Lü, E. and Jiang, H. (2011) Investigation of ancient noodles, cakes, and millet at the Subeixi Site, Xinjiang, China. *Journal of Archaeological Science* 38 (2), 470–479.

Grégoire, J.-P. (1999) Major units for the transformation of grain: the grain-grinding households of southern Mesopotamia at the end of the third millenium BCE. In P. C. Anderson (ed.) *Prehistory of Agriculture. New Experimental and Ethnographic Approaches.* Institute of Archaeology Monograph 40. UCLA, Los Angeles, pp. 223–238.

Harlan, J. (1975) *Crops and Man.* American Society of Agronomy, Madison.

Harlan, J. (1999) Harvesting of wild-grass seed and implications for domestication. In P. C. Anderson (ed.) *Prehistory of Agriculture. New Experimental and Ethnographic Approaches.* Institute of Archaeology Monograph 40. UCLA, Los Angeles, pp. 1–5.

Hansson, A.-M. and Isaksson, S. (1994) Analyses of charred organic remains. *Laborativ Arkeologi* 7, 21–29.

Heer, O. (1865) Die Pflanzen der Pfahlbauten. *Neujahrsblätter der Naturforschenden Gesellschaft in Zürich* 1866, 1–54.

Heiss, A. G. and Oeggl, K. (2009) The plant macro-remains from the Iceman site (Tisenjoch, Italian-Austrian border, eastern Alps): new results on the glacier mummy's environment. *Vegetation History and Archaeobotany* 18 (1), 23–35.

Henry, A. G., Hudson, H. F. and Piperno, D. R. (2009) Changes in starch grain morphologies from cooking. *Journal of Archaeological Science* 36, 915–922.

Herbig, C. (2009) Recent archaeobotanical investigations into the range and abundance of Neolithic crop plants in settlements around Lake Constance and in Upper Swabia (south-west Germany) in relation to cultural influences. *Journal of Archaeological Science* 36 (6) 1277–1285.

Hillman, G. C. (1981) Reconstructing crop husbandry practices from charred remains of crops. In R. Mercer (ed.) *Farming Practice in British Prehistory*, 123–62. Edinburgh University Press, Edinburgh.

Hillman, G. C. (1984) Interpretation of archaeological plant remains: the application of ethnographic models from Turkey. In W. van Zeist and W. Casparie (eds.) *Plants and Ancient Man. Studies in* Palaeoethnobotany. A.A. Balkema, Rotterdam, pp. 1–41.

Hillman, G. C. and Davies, S. (1999) Domestication rates in wild wheats and barley under primitive cultivation: preliminary results and archaeological implications of field measurements of selection coefficient. In P. C. Anderson (ed.) *Prehistory of Agriculture. New Experimental and Ethnographic Approaches.* Institute of Archaeology Monograph 40. UCLA, Los Angeles, pp. 70–103.

Hopf, M. (1954) Anatomische Untersuchungen an Weizenspelzen und -Körnern verschiedener Polyploidiestufen als Vorarbeit für die Bestimmung prähistorischer Funde. *Der Züchter* 24 (6), 174–180.

Hopf, M. (1955) Formveränderungen von Getreidekörnern beim Verkohlen. *Berichte der Deutschen Botanischen Gesellschaft* 68 (4), 191–193.

Jacomet, S. (2009) Plant economy and village life in Neolithic lake dwellings at the time of the Alpine Iceman. *Vegetation History and Archaeoabotany* 18 (1), 47–59.

Jones, G. (1984) Interpretation of archaeological plant remains: ethnographic models from Greece. In W. van Zeist and W. Casparie (eds.) *Plants and Ancient Man. Studies in Palaeoethnobotany.* A.A. Balkema, Rotterdam, pp. 43–61.

Jones, G. (1987) A statistical approach to the archaeological identification of crop processing. *Journal of Archaeological Science* 14, 311–23.

Jones, G., Charles, M., Bogaard, A. and Hodgson, J. (2010) Crops and weeds: the role of weed functional ecology in the identification of crop husbandry methods. *Journal of Archaeological Science* 37 (1), 70–77.

Juel Jensen, H. (1988) Functional analysis of prehistoric flint tools by high-power microscopy: a review of west European research. *Journal of World Prehistory* 1, 53–88.

Juel-Jensen, H. (1994) *Flint Tools and Plant-Working. Hidden Traces of Stone-Age Technology.* Aarhus University Press, Aarhus.

Kalis, A. J., Merkt, J. and Wunderlich, J. (2003) Environmental changes during the Holocene climatic optimum in central Europe- human impact and natural causes. *Quaternary Science Reviews* 22 (1), 33–79.

Kardulias, P. N. (ed.) (1994) *Beyond the Site: Regional Studies in the Aegean Area.* University Press of America, Lanham.

Kardulias, P. N. and Yerkes, R. (1996) Microwear and metric analysis of threshing sledge flints from Greece and Cyprus. *Journal of Archaeological Science* 23, 657–66.

Kohler-Schneider, M. (2001) *Verkohlte Kultur- und Wildpflanzenreste aus Stillfried an der March als Spiegel spätbronzezeitlicher Landwirtschaft im Weinviertel, Niederösterreich.* Mitteilungen der Prähistorischen Kommission 37. Verlag der Österreichischen Akademie der Wissenschaften, Wien.

Korobkova, G. (1981) Ancient reaping tools and their productivity in the light of experimental tracewear analysis. In P. Kohl (ed.) *The Bronze Age Civilisation of Central Asia.* M. E. Sharpe, New York, pp. 325–349.

Kuijt, I. and Finlayson, B. (2009) Evidence for food storage and predomestication granaries 11,000 years ago in the Jordan Valley. *Proceedings of the National Academy of Sciences* 106 (27), 10966–10970.

Lerche, G. (1994) *Ploughing Implements and Tillage Practices in Denmark from the Viking Period to About 1800, Experimentally Substantiated.* Poul Kristensen, Copenhagen.

Lerche, G. and Steensberg, A. (1980) *Agricultural Field Shapes.* National Museum, Copenhagen.

Meurers-Balke, J. and Lüning, J. (1999) Some aspects and experiments concerning the processing of glume wheats. In P. C. Anderson (ed.) *Prehistory of Agriculture. New Experimental and Ethnographic Approaches.* Institute of Archaeology Monograph 40. UCLA, Los Angeles, pp. 238–254.

Mingote Calderón, J. L. (1996) *Tecnología Agrícola Medieval en España.* Ministerio de Agricultura, Pesca y Alimentación, Madrid.

Miret i Mestre, J. (2006) Sobre les sitges i altres estructures excavades al subsòl. *Cypsela* 16, 213–225.

Palmer, C. and Van der Veen, M. (2002) Archaeobotany and the social context of food. *Acta Palaeobotanica* 42 (2), 195–202.

Pearsall, D. M. (1988) Interpreting the Meaning of Macroremain Abundance: The Impact of Source and Context. In C. A. Hastorf and V. S. Popper (eds.) *Current Paleoethnobotany. Analytical Methods and Cultural Interpretations of Archaeological Plant Remains.* Prehistoric Archaeology and Ecology University of Chicago Press, Chicago/London, pp. 97–118.

Peña-Chocarro, L. (1999) *Prehistoric agriculture in Spain: the application of ethnographic models.* BAR International Series 818. Archaeopress, Oxford.

Perry, L., Dickau, R., Zarrillo, S., Holst, I., Pearsall, D. M., Piperno, D. R., Berman, M. J., Cooke, R. G., Rademaker, K., Ranere, A. J., Raymond, J. S., Sandweiss, D. H., Scaramelli, F., Tarble, K. and Zeidler, J. A. (2007) Starch Fossils and the Domestication and Dispersal of Chili Peppers (*Capsicum* spp. L.) in the Americas. *Science* 315, 986–988.

Procopiou, H., Anderson, P. C., Formenti, F. and Juan-Tresseras, J. (2002) Étude des matériaux transformés sur des outils de mouture : identification des résidus par analyse chimique et par observation en microscopie optique et électronique. In H. Procopiou and R. Treuil (eds.) *Moudre et Broyer. L'identification fonctionnelle de l'outillage de mouture dans la Préhistoire et l'Antiquité.* Vol. 1: Editions CTHS, Paris, pp. 111–127.

Revedin, A., Aranguren, B., Becattini, R., Longo, L., Marconi, E., Mariotti Lippi, M., Skakun, N., Sinitsyn, A., Spiridonova, E. and Svoboda, J. (2010) Thirty thousand-year-old evidence of plant food processing. *Proceedings of the National Academy of Sciences* 107 (44), 18815–18819.

Reynolds, P. (1974) Experimental Iron Age Storage Pits. *Proceedings of the Prehistoric Society* 40, 118–131.

Reynolds, P. J. (1979) *Iron Age Farm: the Butser Experiment.* British Museum Publications, London.

Reynolds, P. J. (1981) Deadstock and livestock. In R. Mercer (ed.) *Farming Practice in British Prehistory.* Edinburgh University Press, Edinburgh.

Reynolds, P. J. (1999) Crop yields of the prehistoric cereal types emmer and spelt: the worst option. In P. C. Anderson (ed.) *Prehistory of Agriculture. New Experimental and Ethnographic Approaches.* Institute of Archaeology Monograph 40. UCLA, Los Angeles, pp. 267–275.

Rollo, F., Ubaldi, M., Ermini, L. and Marota, I. (2002) Ötzi's last meals: DNA analysis of the intestinal content of the Neolithic glacier mummy from the Alps. *Proceedings of the National Academy of Sciences of the United States of America* 99 (20), 12594–12599.

Rösch, M., Ehrmann, O., Herrmann, L., Bogenrieder, A., Goldammer, J. P., Hall, M., Page, H. and Schier, W. (2002) An experimental approach to Neolithic shifting cultivation. *Vegetation History and Archaeobotany* 11, 143–54.

Rottoli, M. and Castiglioni, E. (2011) Plant offerings from Roman cremations in northern Italy: a review. *Vegetation History and Archaeobotany* 20 (5), 495–506.

Sarpaki, A. (2000) The study of palaeodiet in the Aegean: food for thought. In S. J. Vaughan and W. D. E. Coulson (eds.) *Palaeodiet in the Aegean. Papers from a colloquium held at the 1993 meeting of the Archaeological Institute of America.* Oxbow Books, Oxford, pp. 115–121.

Semenov, S. (1964) *Prehistoric Technology.* Moonraker Press, Bradford-on-Avon.

Sigaut, F. (1999) Yields, sowing, and fertility: analytical significance of yields. In P. C. Anderson (ed.) *Prehistory of Agriculture. New Experimental and Ethnographic Approaches.* Institute of Archaeology Monograph 40. UCLA, Los Angeles, pp. 275–281.

Sigaut, F. (1978) Identification des techniques de récolte des graines alimentaires. *Journal d' agriculture traditionnelle et de la botanique appliquée* 24 (3), 145–161.

Skakun N. N. (1999) Evolution at agricultural techniques in Eneolithic (Chalcolithic) Bulgaria: data from use-wear analysis. In P. C. Anderson (ed.) *Prehistory of Agriculture. New Experimental and Ethnographic Approaches.* Institute of Archaeology Monograph 40. UCLA, Los Angeles, pp. 199–211.

Steensberg, A. 1943. *Ancient harvesting implements. A study in archaeology and human geography.* Copenhagen.

Stika, H.-P., Heiss, A. G. and Zach, B. (2008) Plant remains from the early Iron Age in western Sicily: differences in subsistence strategies of Greek and Elymian sites. *Vegetation History and Archaeobotany* 17 (Supplement 1), 139–148.

Valamoti, S.-M., Samuel, D., Bayram, M. and Marinova, E. (2008) Prehistoric cereal foods from Greece and Bulgaria: investigation of starch microstructure in experimental and archaeological charred remains. *Vegetation History and Archaeobotany* 17 (Supplement 1), 265–276.

Van der Veen, M. (2003) When is food a luxury? *World Archaeology* 34 (3), 405–427.

Van der Veen, M. (2008) Food as embodied material culture:

diversity and change in plant food consumption in Roman Britain. *Journal of Roman Archaeology* 21, 83–109.
Winton, A. L. and Winton, K. B. (1932) *The Structure and Composition of Foods. Volume I: Cereals, Starch, Oil Seeds, Nuts, Oils, Forage Plants. Vol. 1.* John Wiley & Sons, New York.
Winton, A. L. and Winton, K. B. (1935) *The Structure and Composition of Foods. Volume II: Vegetables, Legumes, Fruits. Vol. 2.* John Wiley & Sons, New York.
Vuorela, I. (1986) Palynological and historical evidence of slash-and-burn cultivation in south Finland. In K.-E. Behre (ed.) *Anthropogenic indicators in pollen diagrams.* A. A. Balkema, Rotterdam/Boston, pp. 53–64.
Zohary, D., Hopf, M. and Weiss, E. (2012) *The domestication of plants in the Old World.* Oxford University Press, Oxford.

SECTION 1

Introduction

1 The Dimension of Tools, Skills and Processes: Exploring Diversity

Patricia C. Anderson, Annelou van Gijn, John C. Whittaker and François Sigaut(†)

Introduction and Objectives

Writing a book on agricultural tools, skills and practices is ambitious because the subject is vast. It has been studied in a variety of ways: archaeologists search for the first sickles, historians examine the documents produced by farmers, scholars, and governments, ethnographers study living agricultural systems in various parts of the world. The uniqueness of this book lies in its interdisciplinary approach: the question of agricultural tools, skills and processes is addressed from a variety of approaches ranging from archaeological and historical, to ethnographic and even agronomical perspectives. Each and every one of these disciplines have addressed the issue of agricultural tools and processes and published extensively on it. However, the EARTH project was innovative in the sense that it brought together people from different parts of the globe, with different research experiences. These people were not bound by common research questions or methods, but had to find a way to communicate their knowledge to each other and find common ground. This has led, we believe, to an original collection of case studies which enlightens and complements the more general textbooks that have appeared on this topic. In one sense the case studies in this book form an illustration of the kind of information that can be obtained by the various disciplines and methods of analysis used. In this way the book can be regarded as a methodological overview. However, the contributions also constitute windows into the vast variety of technological choices people made both in the past and in the present with respect to their agricultural practices. In doing so this book also shows the complexity of such choices, and how they change through time, shaped by the interrelationships between technology and other aspects of the human world, including social organisation and cultural rules, as well as peoples' interactions with the natural environment (climate, topography, plants and animals).

The aim of our book is to explore and where possible explain the diversity of agricultural tools and processes through space and time. Agriculture encompasses both plant and animal husbandry, which are intertwined and interdependent (Bieleman 1992). For this volume, however, animal husbandry will be only marginally touched upon, the focus being on plant husbandry. This more restrictive definition of agriculture will emphasise the growing of cereals, although several contributions address other crops such as fruit and wild grasses. The book will not attempt to duplicate the numerous books that have been written on the agricultural process (*e.g.* Anderson 1999; Anderson *et al.* 2003; Damania *et al.* 1998; Zohary *et al.* 2012; Redman 1978; Cauvin 1994; Harris and Hillman 1989; Harlan 1975), nor to provide an exhaustive overview of agricultural tools and practices in the regions of focus. Instead, it departs from the knowledge and expertise of the various contributors to illustrate the difference in choices made in the agricultural process. In this, the people are central, and the case studies presented have attempted to elucidate these choices either by studying the 'physical' remains of agricultural technology (the tools, the plant remains, the depictions in historical documents), by examining iconography and texts in historical documents, or by interviewing and observing present-day actors. This brings us to another objective of this book, which is to show the range of fields which study agricultural phenomena: case studies come out of history, ethnography, agronomy, archaeology and art history. Consequently, various methods of analysis are represented as well, such as experimental archaeology and functional analysis of artefacts (use-wear and residue studies) or archaeobotany. An auxiliary objective therefore is to illustrate the kinds of information each discipline can contribute to questions of agricultural skills, knowledge and practices.

The geographical span of the book will be largely confined to Europe and the circum-Mediterranean areas. There is, however, the occasional venture to other continents as a comparison or to reinforce points we believe to be important. Our chronological focus starts with the Natufian, dated in the Near East around 12,000 BCE when we have the first evidence of 'manipulation' of the environment. A number of case studies addressing the agricultural process derive from various periods of the Neolithic, the period of the first farmers. The designation 'Neolithic' is, however, not without conceptual problems (Jennbert 1997). It is usually defined as the period during which the first farmers appeared, but clearly in many areas, especially in northern and western Europe, there is a long period of transition during which people probably had knowledge of farming but did not substantially change their hunter-fisher-gatherer way of life for a substantial period of time (Louwe Kooijmans 2007; Zvelebil 2000). However, this book is explicitly not intended to address the Neolithisation process, dealt with in numerous archaeological works (*e.g.* Fokkens *et al.* 2008; Price 2000; Whittle and Cummings 2007). Apart from case studies from prehistoric contexts, a number of studies concern Medieval and sub-recent times, relying on various historical sources such as works of art, old text fragments and museum pieces. Lastly, there is the ethnographic present, largely focused on the circum-Mediterranean area, illustrating agricultural practices in present-day societies. Some examples of the latter are presented in film clips, which can be watched by following the links in the referring datasets of our online database 'Columella' and the dedicated links pointing to Vimeo online service

The book has two main parts. The first part follows the different steps of the agricultural process: preparation of the fields (Chapter 2), tending the crops (Chapter 3), harvesting (Chapter 4), threshing (Chapter 5), conservation and storage (Chapter 6) and, last, processing the crops (Chapter 7). Obviously, there is not space here to exhaustively cover all the various steps, but case studies are used to illustrate diversity and aspects of choice, as well as different ways to approach research into agricultural processes and techniques. The second part of the book has a thematic approach. Chapter 8 addresses the transmission of skills and knowledge and shows the social embedding of agricultural practices. Chapter 9 explores how closely religious and legal aspects are related to agricultural practices. Finally, Chapter 10 discusses the long-term processes, the time depth of agricultural practices and the process by which some of them ended or were transformed, and how new inventions were incorporated into existing technological systems.

Studying the Agricultural Process: some Key Theoretical Concepts

Adopting agriculture changes the rhythm of the yearly cycle of hunter-gatherer-fishers, a lifestyle that lasted for millennia. Hunter-gatherers usually have a more or less mobile life-style determined by the availability of the different resources through time and across space. This mobility pattern allowed to some extent for a pastoral component, as animals can sometimes be moved when people move. However, mobility is more difficult for people growing crops as these require various activities that are widely separated throughout the year. From the preparation of the fields, to sowing the seeds, tending and harvesting the crops and, lastly, processing them, a period of at least several months passes. Throughout this period, time and effort need to be invested in the fields, requiring continued return or presence throughout the year. Orchestrating the agricultural tasks through the crop's growing season is very difficult in a mobile lifestyle because of time/space restrictions (Carlstein 1982). Growing crops can thus severely restrict mobility, limiting the possibilities of obtaining other resources that require movement in space, or which consume time during crucial periods in the agricultural process, such as harvesting. It is thus no surprise that agriculture usually involves a relatively sedentary way of life, although in (sub-) tropical conditions this is not always the case (see for instance Politis 1996).

Each step in the agricultural process requires the investment of knowledge and skills. In pre-industrial agricultural societies, tradition was the backbone of agricultural practices. Knowledge about the land and the weather, the seasons and the life cycle of the crops, and of course about the necessary tools and techniques, was transmitted from generation to generation. From a very young age children became familiar with the agricultural way of life, gradually incorporating the skills until they were 'second nature' and became part of their 'muscle memory' (see Chapter 2 and 8). Agriculture was very much part and parcel of social and ideological life. This was especially so because it is inherently precarious: when sowing the crops there is no guarantee that they will come to fruition a number of months later. Historical accounts of famine and warfare or raiding are numerous. The entire process is therefore surrounded with social habits and regulations, legal restrictions and religious prohibitions and rituals (see Chapter 3 and 9).

Agriculture as a technological system is also a cultural construct, which gradually develops through time. It is often considered to be inherently conservative. However, the contributions to this volume show that the apparent conservatism is a vision enforced by law, and that individual farmers do indeed innovate. Such innovations are incorporated more generally when deemed suitable and fitting with the existing technological system (Lemonnier 1986).The technological choices made are varied and complex and not do not always follow the 'logic' of contemporary western rationale, yet they constitute a coherent system that can be studied. A key concept in analysing complex technological systems such as the agricultural process is that of the *chaîne opératoire* (the 'operational sequence', a term forged by Leroi-Gourhan 1964), a concept that requires further elucidation.

The *Chaîne Opératoire* of the Agricultural Process

In agriculture, as in all economic activities, production is the result of a number of operations that follow each other in ordered sequences. What has not been sown cannot be harvested, and sowing requires that the field has been previously prepared according to specific rules that depend on the soil, the climate, the season, and so on, and these rules are based on the experience people have gathered over generations. That is what the concept of an operational sequence (the *chaîne opératoire*) is about. This implies the sequence of technical operations, the human actions, but also the tools, products, and by-products of each action, and the way the operations are embedded in culture, particularly the rules which guide the sequence.

The problem is that complete sequences are often quite difficult to observe. Successive operations may happen to be quite widely separated in time or space (or both), so that it may be difficult or even impossible for the casual observer to acquire a complete view of the whole process. In that respect, archaeologists are typically casual observers, because they come long after the action has been carried out. But willingly or not, ethnographers are often casual observers too, inasmuch as they cannot possibly always be in the right place at the right time. It is because of all this that the concept of the *chaîne opératoire* is useful. Taken separately, operations cannot be completely understood. A view of the whole is necessary to understand each of its parts – and conversely.

The case of agriculture is especially complex because different kinds of sequences are mixed together in people's daily routines in ways that are not easy to disentangle. In metallurgy, for example, operations are rather simply ordered from ore extraction to making finished metal items, and since they are usually done by craftsmen, the logic of the operational sequence is not especially difficult to find. In agriculture, there are many different sequences that happen to cross each other or to partially coincide in time and which must be made compatible within people's daily or seasonal agendas. The problem was not clearly understood before the 1960s, when new techniques of programming provided engineers with mathematical tools to tackle it.

We do not propose to use such techniques for the understanding of ancient agricultural systems, we simply want to stress the fact that what is called 'agriculture' is usually a complex of many tasks which are combined by people in ways that may be quite difficult to understand by outsiders.

Generally speaking, it may be added that agricultural systems are usually made up of at least three different kinds of operations. The first are seasonal and are done in the fields, from field clearance or first ploughing to harvest. The last are daily or weekly chores and are done in or near the household, such as preparing food, spinning, weaving, etc. And in between there are a lot of tasks that may be done in quite different ways according to time and place. Threshing, for example, may be done in or near the field, within a few weeks immediately following the harvest, or it may be done little by little, as a daily routine which will take place over the whole year. There are many such details. They concur to make agricultures in different societies very different things.

One important consequence is also that the concept of 'agriculture' does not have the same relevance from one place to another. In Europe, threshing clearly belongs to 'agriculture' proper; in many parts of Africa, southern Asia and pre-Colombian America, threshing is (was) included in the sphere of household tasks, which are the responsibility of the housewife, like milling, cooking, etc. The usual limits between what is agricultural and what is not are themselves the result of modern history in Europe. They cannot be extrapolated without any modification to other times and places.

The Social Organisation of Production

A number of agricultural tasks may or must be performed by individuals working alone. But even in those cases, working alone is the exception. The rule is rather that of working in teams, because it makes the work more pleasant, more secure, and more efficient. We expect digging a field with a spade to be an individual task, and in many cases it cannot be done otherwise. But in other cases, several persons, from three up to eight or ten, unite their efforts to cut and turn the same slice of earth at once, which may then be up to two or three metres long. Usually, however, teamwork implies cooperation among people doing different sub-tasks. The man who guides the plough is assisted by a boy who leads the animals, the person who reaps the grain with a sickle is assisted by someone who binds the sheaves, and so on.

Tasks should also be analysed according to their size (the amount of work to be done at the same time), which is often a consequence of seasonality. As has been already said, it is not the same thing to thresh the entire harvest of cereals at once in late summer, as to thresh a few pounds each morning for the meal of the day. In the first case, threshing is emphatically men's work (with some exceptions). If women do cooperate, it is in relatively minor tasks. In the second case, threshing is women's work and a man might be the object of sexual jokes if he happened to take any part in it. Similar differences can be observed within the borders of Europe. In most rural areas, for instance, bread making is a woman's job (as is nearly everything belonging to food preparation). In towns of some size, it became a craft from at least the early Middle Ages on, and of course the craftsmen (the bakers) are men. But in areas where rivers freeze in winter, stopping the mills, bread has to be made in large quantities for the cold months, and it is again a man's job.

Another important point is the degree of skill involved in each task, and the degree of apprenticeship it requires. Daily chores may sometimes require a long apprenticeship, during which time skill is acquired progressively by performing the chores as needed. Such chores may be performed by anyone, but allow women and children with less skill or developing skills to contribute to the farm labour. Whereas when the task is seasonal, intensive, task-specific training becomes more important, and differences in skill more decisive. Broadcast sowing may be one of the best examples. People are not occupied in sowing more than one or two months in the year, but since the stakes are so high, sowing requires the most skilled workers. Because broadcast sowing is technically very difficult, it follows that prior training here is of crucial importance. There are unfortunately few documents on this theme, but it cannot be by chance that broadcast sowing is almost without exception a man's job. When for some reason men are absent, the sowing may be done by women, but their way of doing it will not be the same as men's.

Methodological Approaches towards Studying Agricultural Tools and Practices

This book is based on research from different fields of study, notably archaeology, history, art history, palaeobotany, ethnography and agronomy. All of these disciplines encompass a range of methods and theories, discussed extensively in basic textbooks. Clearly this is not the place to dwell on them and we assume that most readers will be acquainted with these fields. Where necessary, specific information is provided in the individual chapters. However, a few specialist methods of study which are especially prevalent in this particular book will be briefly described here. For the most part, these concern methods utilised in the discipline of archaeology. Although the basic idea of archaeological study of the past is familiar to most people, archaeological approaches to material artefacts and subsistence processes are often useful in understanding the current world, and a number of technical approaches used by archaeologists are not very familiar to those outside the discipline. It is our contention throughout this book that all the methodologies can be used by all the different fields. Just as archaeological analyses are improved when texts produced by or describing an ancient society can be consulted, so too historians will find it informative to examine not only documentary sources, but the material objects, structures, and landscapes that people use in their lives.

Ethnoarchaeology

Ancient cultures are often interpreted in terms of observations made by ethnographers studying pre-industrial cultures in recent times. Ethnoarchaeology uses such observations of current cultures specifically for the purpose of archaeological interpretation, and often focuses on learning what kinds of physical remains we can expect to see as the results of particular human actions and organisational structures. Numerous collections of representative ethnoarchaeological studies have been published (Anderson 1999; David 2001; Donnan and Clewlow 1974; Kramer 1979; Gould and Schiffer 1981; Staski and Sutro 1991; Terradas 2005; Yellen 1977).

Archaeologists have found it useful to perform their own ethnographic studies because until a decade or so ago, ethnography had moved away from detailed observation of all aspects of a culture to more focused and limited research, often with an emphasis on the non-material aspects of culture. While archaeologists are interested in the cultural rules, beliefs, and symbols that shape human behaviour, they cannot observe them directly, and thus emphasise observations of the material results of human behaviour. Artefacts, structures, residues, traces of wear, and the effects of human action on the natural environment can all be directly observed in the archaeological record. In the best cases, these can be interpreted in more human and cultural terms, but only if there is a firm foundation of understanding of how human behaviour is linked to material results in the modern world. Ethnoarchaeological studies are thus uniformitarian in nature. They attempt to produce analogical arguments allowing us to interpret material evidence in terms of human behaviour by comparison to behaviours and their results that we have directly observed, under the assumption that similar behaviours in the past will have produced similar results.

Ethnoarchaeological examples may be quite simple. For example, we can understand how sickles may be used in a variety of ways by working with living harvesters, and use that information to interpret the form of prehistoric sickles and the wear that develops on them as they are used. More complex examples support social and cultural interpretations. We expect from ethnographic observations that a particular kind of social organisation will often favour similar living arrangements, economic organisation, and ritual behaviour wherever it occurs. When we recognise the characteristic physical remains in archaeological contexts, we interpret them in terms of such known social forms. Specific ethnoarchaeological studies have approached issues such as the social organisation of production, the learning of skills and transmission of technological information, and the development of stylistic variation in artefacts and their social uses (Longacre and Skibo 1994; Hodder 1982; Stout 2002; Weisner 1983).

There is some danger in ethnoarchaeology that these modern analogues may confine us or bias us toward recognising only phenomena that have been observed in modern times, while we admit, at least theoretically, that some human behaviour in the past may not have modern analogues. This is especially true when we are considering ancestral human forms, and past societies that have been extinct for millennia.

Experimental Archaeology

Like ethnoarchaeology, experimental archaeology makes modern observations to produce analogies useful for interpreting the archaeological record. There are a number of common variants. Numerous representative collections of the work of experimental archaeology have been published in the decades since it became common in the 1960s (Coles 1973, 1979; Ferguson 2010; Ingersoll *et al.* 1977; Mathieu 2002; Reynolds 1979; Anderson 1999).

The most common form of experimental archaeology, with roots running back into the nineteenth century is the replicative experiment (Coles 1973; Johnson 1978; Schiffer 2009). Many of the artefacts recovered in archaeological contexts are not in use today, and their functions may be quite obscure. Replicative experiments recreate ancient tools, use them in a variety of ways, and attempt to understand their functions, evaluate their efficiency, and see what other material evidence of their use is produced. Some replicative experiments are limited to simple questions: can a flint axe successfully cut down a large tree? In more elaborate cases, a large part of the *chaîne opératoire* of an artefact can be studied by replication, from its manufacture and the resulting traces and by-products, to its use, which also produces damage, wear, and modifications of whatever material the tool is used to work, to its finally being discarded and the way discard treatment affects the tool. A replication is considered successful if the material results of each step of the *chaîne opératoire* match the archaeological evidence. The studies of sickles and threshing sledges discussed in later chapters are good examples of this approach. Archaeological, textual, and ethnoarchaeological evidence is used to reconstruct the form of sickles and sledges where only flint blades survive. These hypothetical complete tools are then used under a variety of conditions, to learn how they might have been used, and how efficient they could have been. We see the effects on the plants that are harvested and processed, and the wear that develops on the tools, and we can look for and interpret similar evidence in archaeological sites, allowing us to, for example, distinguish sickle blades from threshing sledge inserts. Such practical experiments also allow us to propose new hypotheses for which there might be archaeological evidence. For instance, does a threshing sledge require animal traction, or can it be effectively pulled by humans alone?

A related focus of archaeological experimentation has been with the properties of materials. How does a polish form on a flint surface when a tool is used, and how does it vary with different tool materials? What is the effect on strength, porosity, and heat resistance of adding different kinds of temper to clay when making pots, and was it likely that ancient potters could recognise these effects and control them intentionally (Longacre 1991)? When grain is ground in a stone quern, and cooked in a ceramic pot, what are the effects on its nutritive value, and what residues might remain in the pot and survive on the surfaces of sherds?

Other common experimental archaeological projects deal with natural processes in the formation of sites and artefacts. Taphonomy is the study of the processes of burial, but applies more widely to the processes affecting bones and artefacts and the formation of archaeological and geological deposits. In American archaeological tradition, these are often referred to as formation processes (Schiffer 1987). Binford (1983) coined the term 'middle range research' to refer to studies that link physical processes and human behaviour to specific material evidence.

All such concepts recognise that we do not observe past behaviour directly, and that an archaeological site is never an intact or complete record of what people did there. Natural processes and human activities have always interfered in the time between the deposition of artefacts and their recovery and interpretation by archaeologists. Accordingly, we need to understand processes such as how chemical and biotic decay affects different objects differently, how erosion moves artefacts and disturbs their contexts, what animals and plants do to archaeological deposits, and why some artefacts that we know should be common in the past do not appear in a particular archaeological site. These issues, and many more similar ones, have been examined by naturalistic observations in the modern world or by artificial controlled experiments. In the chapters to follow, such things as our understanding of phytoliths (see definitions below), seeds, pollen, and other plant remains in archaeological sites is heavily dependent on such experimental studies, as is our ability to explain the formation of different kinds of wear on flint tools.

Use-Wear Analysis

Use-wear analysis of prehistoric tools developed in the 1970s, following particularly on the work of Semenov (1964) as a result of the wish to have more empirical data on past human behaviour (Keeley 1980; Odell 1977, 2001). Until then the function of tools was usually inferred on the basis of a morphological similarity to a present-day tool, either one from the researcher's own experience or from an ethnographic context. This resulted in classifications of tools into categories with explicit functional connotations like scraper, knife, axe and so forth. As early as the 1930s, people like Semenov, who studied and replicated Palaeolithic technology in Russia, and Curwen who studied flint sickles, found that wear traces develop as a result of use (Curwen 1930; Semenov 1964). These include edge removals (also referred to as use retouch), striations, edge rounding and polish. These traces vary in appearance depending on the activity carried out and the contact material involved and can be studied by means of stereomicroscopes (magnifications of 5–160 times) and incident light or metallographic microscopy (magnifications normally used ranging from 50–400 times). The variation in wear traces permits researchers to differentiate for instance between tools used for carving bone and those for harvesting cereals. The first use-wear studies were on flint tools (Keeley 1980; Anderson-Gerfaud 1981; Juel Jensen 1994; Van Gijn 1990; Vaughan 1985), but implements made of a variety of other raw materials, such as obsidian (Hurcombe 1993), bone (*e.g.* Maigrot 2001; Van Gijn 2007; Stordeur and Anderson-Gerfaud 1985), shell (Cuenca Solana *et al.* 2011; Lammers 2007) or even coral (Kelly and Van Gijn 2008) or broken sherds (Van Gijn and Lammers-Keijsers 2010), also develop characteristic wear traces after use. By carrying out such microscopic examinations it was possible, for instance, to detect harvesting tools and threshing sledge inserts made of flint or chert (Chapters 4 and 5; Anderson *et al.* 2004; Anderson and Inizan 1994; Ataman 1999; Skakun 1999). Also pertinent to this book, use-wear studies of querns have provided many insights into grain processing and treatment of grinding tools (see Chapter 7; Delgado 2008; Hamon 2004; Procopiou 1998; Procopiou *et al.* 2002; Verbaas and Van Gijn 2007). Although use-wear studies have greatly contributed to our knowledge of the place of different tools in the technological system and the activities they were involved in, it is not often possible to arrive at detailed inferences as to which type of plants were worked, using optical microscopy alone. Other methods such as confocal microscopy are being tested to better distinguish materials worked by flint tools (*e.g.* Evans and Donahue 2008), but study of residues is the principal means of refining the interpretation of tool function.

The scanning electron microscope (SEM) was first used to study agricultural tools, particularly to look for characteristic residues of the material processed by the tool (*e.g.* Stordeur and Anderson-Gerfaud 1985; Anderson-Gerfaud 1986; Kamminga 1979). This analysis, like analysis using optical microscopy, involves looking at points on the used area of the edge and surface of the tool for residues of the worked material in comparison with other areas of the tool unaffected by the use. The virtually infinite depth of field of the SEM, combined with its ability to reach thousands of times magnification, means that residues may be identified according to their distinctive morphology, or by using elemental analysis to find their characteristic inorganic chemical components (*e.g.* Anderson 1980, 2013; Stordeur and Anderson-Gerfaud 1985; Langejans 2011).

Phytolith and Starch Grain Analysis

In addition to organic remains, principally lipids (see Condamin *et al.* 1976; Procopiou *et al.* 2002; Evershed *et al.* 2008; Regert *et al.* 1999), and plant microfossils such as pollen, starch and phytoliths can be found as residue on prehistoric tools used for plant processing. In addition, although macroremains of grain and other plant parts can be precious indicators of plant processing technology in an archaeological context, these will not be discussed here (see EARTH Series Volume 1, and references in Chapter 4 and Chapter 7, this volume). We will limit our discussion in these pages to phytoliths and starch, which can be powerful proxies (*i.e.* indicators) for plant processing (see Chapters 4, 5, and 6).

Phytoliths are plant silica bodies that are formed when silica is taken up into the plant from the ground water and precipitates in or between cells, with these deposits forming casts or imprints of the cellular structure of the plant (Pearsall and Piperno 1993). They are especially abundant in the Poaceae/Gramineae (grass family), which includes wild grasses and domesticated cereals. As they are inorganic, phytoliths can be preserved for thousands of years, long after the plant has decomposed. Phytoliths can have a variety of three-dimensional shapes, and can for example be described as dendriform, ornamented rods, dumbbell, conical, cylindrical, etc. In addition, phytoliths are sometimes preserved joined together in anatomical position in spodograms[1] (*e.g.* Miller Rosen 1999; Anderson 2003). The morphology of phytoliths may indicate the processing of specific plant families. Although silica incrusts the epidermal (skin) tissues in roots, stems and leaves, it is particularly the glumes of cereals which produce phytoliths which are most characteristic of a particular plant type. They are identifiable to the family and genus levels, and even in some cases to species level. Phytoliths in glumes are easily distinguished from those in the leaves and stems, which can help in determining particular plant processing techniques, identify storages of grain (using glume phytoliths as a proxy), or material used for roofing, flooring, mats, and so on.

Interpretation of phytoliths in the archaeological realm is carried out by comparison with phytoliths extracted from living plants. The phytoliths can be extracted from the material resulting from certain kinds of plant processing and use, either from carrying out experiments (Procopiou *et al.* 2002; Vargiolu *et al.* 2003; Avner *et al.* 2003), or from ethnology today (cf. D'Andrea 2003; Peña-Chocarro and Zapata 2003; see also Chapters 4, 5, 6 and 7, this volume). This fresh plant material undergoes chemical extraction processes and, usually, separation processes using centrifugation and heavy liquids, in order to isolate the phytoliths and eliminate the organic material, so that this reference material is comparable to archaeological samples (Pearsall and Piperno 1993). Phytoliths have been extracted from soil samples in order to obtain information, for instance on past climatic conditions, in conjunction with pollen, but our application here concerns these remains extracted from soil samples from structures such as silos and other storage containers or areas, from threshing floors, or even ancient fields (Anderson 2003; Haffner *et al.* 2003; Cummings 2003; Procopiou *et al.* 2002). Phytoliths have also been extracted from the residue adhering to ground stone tools, such as querns (Nieuwenhuis and Van Gijn 2008; Procopiou *et al.* 2002). Examining phytoliths produced under experimental conditions allows us to recognise similar phytoliths in archaeological contexts in order to interpret residues of specific stages of cereal processing such as threshing and winnowing, or dehusking and grinding (Anderson 2003; Procopiou *et al.* 2002; Chapters 5 and 6, this volume).

Extracting phytoliths from soil samples or from the residue present on grinding stones requires using various acids which destroy any organic material and other minerals beside silica, then further separation using centrifugation and heavy liquid separates out quartz grains (Pearsall and Piperno 1993) in order to obtain 'clean' samples where phytoliths are easy to see on a glass slide or in the SEM.

Another procedure to extract phytoliths refrains from using acids so as to preserve starch grains in the residue (Torrence *et al.* 2010; Fullagar and Field 1997). This is especially relevant for the study of querns (Nieuwenhuis and Van Gijn 2008; Langejans 2006). Starch grains have been observed, for example, on querns using iodine and the polarising microscope, which can show the special optical properties (the birefringence, or double light refraction) of starch. Experiments have also successfully characterised starch and its stages of degradation during grinding or cooking (Procopiou *et al.* 2002; Valamoti *et al.* 2008; see also Chapter 6, this volume).

The interpretation of prehistoric tool function gains in precision when wear traces are correlated with residues (Akerman *et al.* 2002; Anderson *et al.* 2004; Fullagar 1993, 1998; Kealhofer *et al.* 1999), particularly when possible contamination is controlled for, such as in recent research into taphonomy (*e.g.* Langejans 2011).

Conclusion

Each chapter in this volume presents a series of examples illustrating the agricultural process or the social significance of agricultural tools and practices. The first chapters follow a *chaîne opératoire* (operational sequence) common to most agriculture, from preparation of the ground through processing of the harvest. We do not attempt to cover the sequence exhaustively, or confine our view to a single place or crop. This book instead is concerned with the choices made by people, the adaptive and cultural reasons why a particular agricultural system proceeds as it does. Variation in our expertise and interests and in the availability of information from other scholars also affects coverage. Throughout, we are interested in the culture of agriculture, especially the technology and human organisation of crop growing. The first chapters are more technological in focus, but the last highlight more theoretical, non-material issues such as skill and traditional knowledge, learning and cultural change, and the ritual, symbolic, and legal elements of crop growing.

From a seed grows a plant, processed to nourish a family. From the knowledge of specific examples, general understandings may be developed, and that has been our guiding principle in the EARTH project.

Chapter Note

1 Phytolith skeletons comprising several or more silicified epidermal cells linked together in anatomical position

Chapter Bibliography

Akerman, K., Fullagar, R. and Van Gijn, A. L. (2002) Weapons and *wunan*. Production, function and exchange of Kimberley points. *Australian Aboriginal Studies 1*, 13–42.

Anderson, P. C. (1980) A testimony of prehistoric tasks: diagnostic residues on stone tool working edges. *World Archaeology* 12 (2), 181–194.

Anderson P. C., Beyries, S., Otte, M. and Plisson, H. (eds.) (1993) *Traces et fonction : les gestes retrouvés.* 2 volumes. ERAUL 50, Liège.

Anderson, P. C. (ed.) (1999) *Prehistory of Agriculture. New Experimental and ethnographic approaches.* Institute of Archaeology Monograph 40. University of California, Los Angeles.

Anderson P. C. (1999) Experimental cultivation, harvest and threshing of wild cereals. Their relevance for interpreting the use of Epipalaeolithic and Neolithic artifacts. In P. C. Anderson (ed.) *Prehistory of Agriculture. New Experimental and ethnographic approaches.* Institute of Archaeology Monograph 40. University of California, Los Angeles, pp. 118–139.

Anderson, P. C. (2003) Observations on the threshing sledge and its products in ancient and present-day Mesopotamia. In P. C. Anderson, L. S. Cummings, T. K. Schippers and B. Simonel (eds.) *Le traitement des récoltes : un regard sur la diversité du Néolithique au présent.* ADPCA, Antibes, pp. 417–438.

Anderson P. C. (2013) Neolithic tools used for stripping ears from hulled cereals: an update. In P. C. Anderson, C. Cheval and A. Durand (eds.) *Regards croisés sur les outils liés au travail des végétaux / An Interdisciplinary Focus on Plant-Working Tools.* XXXIII rencontres internationales d'archéologie et d'histoire d'Antibes. Éditions APDCA, Antibes, pp. 89–103.

Anderson, P. C., Chabot, J. and Van Gijn, A. L. (2004) The functional riddle of 'glossy' Canaanean blades and the Near Eastern threshing sledge. *Journal of Mediterranean Archaeology 17* (1), 13–87.

Anderson, P. C., Cummings, L. S., Schippers, T. K. and Simonel, B. (eds.) (2003). *Le traitement des récoltes : un regard sur la diversité du Néolithique au présent.* ADPCA, Antibes.

Anderson-Gerfaud, P. C. (1981) *Contribution méthodologique à l'analyse des microtraces d'utilisation sur les outils préhistoriques.* Ph.D. thesis, Université de Bordeaux I, Talence. 2 volumes.

Anderson-Gerfaud, P. C. (1986) A few comments concerning residue analysis of stone plant-processing tools. In L. Owen and G. Unrath (eds.) *Technical aspects of microwear studies on stone tools.* Early Man News 9/10/11, 69–87.

Avner, U., Anderson, P. C., Bui-Thi, M., Chabot, J. and Cummings, L. S. (2003) Ancient threshing floors, threshing tools and plant remains in Uvda Valley, southern Negev desert, Israel. A preliminary report. In P. C. Anderson, L. S. Cummings, T. K. Schippers and B. Simonel (eds.) *Le traitement des récoltes : un regard sur la diversité du Néolithique au présent.* ADPCA, Antibes, pp. 455–476.

Bieleman, J. (1992) *Geschiedenis van de landbouw in Nederland 1500-1950. Veranderingen en verscheidenheid.* Boom, Meppel.

Binford, L. R. (1983) *In Pursuit of the Past: Decoding the Archaeological Record.* Thames and Hudson, New York.

Carlstein, T. (1982) *Time resources, society and ecology. On the capacity for human interaction in space and time. Volume 1: Preindustrial societies.* George Allen & Unwin, London.

Cauvin, J. (1994) *Naissance des divinités, naissance de l'agriculture.* CNRS, Paris.

Christiansen, M., Walter, P. H. and Menu, M. (1992) Usewear characterisation of prehistoric flints with IBA. *Nuclear instruments and methods in physics research B* 64, 488–493.

Coles, J. (1973) *Archaeology by Experiment.* Charles Scribner's Sons, New York.

Coles, J. (1979) *Experimental Archaeology.* Academic Press, New York.

Condamin, J., Formenti, F., Metais, M. O., Michel, M. and Blond, P. (1976) The application of gas chromatography to the tracing of oil in ancient amphorae. *Archaeometry* 18, 195–201.

Cuenca Solana, D., Zugasti, I. G. and Clemente Conte, I. (2011) The use of mollusc shells as tools by coastal human groups. *Journal of Anthropological Research* 67, 77–102.

Curwen, E. C. (1930) Prehistoric flint sickles. *Antiquity* 9, 62–66.

D'Andrea, A. C. (2003) Social and technological aspects of nonmechanised emmer processing. In P. C. Anderson, L. S. Cummings, T. K. Schippers and B. Simonel (eds.) *Le traitement des récoltes: un regard sur la diversité du Néolithique au présent.* ADPCA, Antibes, 47–61.

Damania, A., Valkoun, J., Willcox, G. and Qualset, C. (1999) *The origins of agriculture and crop domestication.* ICARDA, Aleppo.

David, N. (2001) *Ethnoarchaeology in action.* Cambridge University Press, Cambridge/New York.

Delgado Raack, S. (2008) *Practicas económicas y gestión social de recursos (macro)líticos en la prehistoria reciente (III-I milenios AC) del Mediterráneo Occidental.* Ph.D. thesis, Universitat Autònoma Barcelona.

Donnan, C. B. and Clewlow, C. W. (1974) *Ethnoarchaeology.* UCLA, Los Angeles.

Evans, A. A. and Donahue, R. E. (2008) Laser scanning confocal microscopy: a potential technique for the study of lithic microwear. *Journal of Archaeological Science* 35, 2223–2230.

Evershed, R. P., Payne, S., Sherratt, A. G., Copley, M. S., Coolidge, J., Urem-Kotsu, D., Kotsakis, K., Özdoğan, M., Özdoğan, A. E., Nieuwenhuyse, O., Akkermans, P. M. M. G., Bailey, D., Andeescu, R.-R., Campbell. S., Farid, S., Hodder, I., Yalman, N., Özbaşaran, M., Bıçakcı, E., Garfinkel, Y., Levy, T. and Burton, M. M. (2008) Earliest date for milk use in the Near East and southeastern Europe linked to cattle herding. *Nature* 455, 528–531.

Ferguson, J. (ed.) (2010) *Designing Experimental Research in Archaeology: Examining Technology Through Production and Use.* University Press of Colorado, Boulder.

Fokkens, H., Coles, B., Van Gijn, A. L., Kleijne, J., Ponjee, H. and Slappendel, C. (eds.) (2008) *Between foraging and farming. An extended broad spectrum of papers presented to Leendert Louwe Kooijmans.* Leiden Universiteit.

Fullagar, R. (1993) Taphonomy and tool-use: A role for phytoliths in use-wear and residues analysis. In B. Fankhauser and R. Bird (eds.) *Archaeometry: Current Australian Research.* 22nd ed. Research School of Pacific Studies, Canberra, pp. 21–27.

Fullagar, R. (ed.) (1998) *A closer look. Recent studies of Australian stone tools.* University of Sydney, Sydney.

Fullagar, R. and Field, J. (1997) Pleistocene seed-grinding implements from the Australian arid zone. *Antiquity* 71, 300–307.

Van Gijn, A. L. (1990) *The wear and tear of flint. Principles of functional analysis applied to Dutch Neolithic assemblages.* Analecta Praehistorica Leidensia 22, Leiden.

Van Gijn, A. L. (2007) The use of bone and antler tools. Two examples from the Late Mesolithic in the Dutch coastal zone. In C. Gates St.-Pierre and R. Walker (eds.) *Bones as tools. Current methods and interpretations in worked bone studies.* BAR International Series 1622. Archaeopress, Oxford, pp. 81–92.

Van Gijn, A. L. and Lammers-Keijsers, Y. (2010) Toolkits for ceramic production: informal tools and the importance of high power use-wear analysis. *Bulletin de la Société préhistorique française* 107 (4), 755–762.

Gould, R. and Schiffer, M. (eds.) (1981) *Modern Material Culture: The Archaeology of Us.* Academic Press, New York.

Hamon, C. (2004) Le statut des outils de broyage et d'abrasion dans l'espace domestique au Néolithique ancien en Bassin Parisien. *Notae Praehistoricae* 24, 117–128.

Harris, D. R. and Hillman, G. C. (1989) *Foraging and Farming: The evolution of plant exploitation.* Unwin and Hyman, London.

Harlan, J. R. (1995) *The living fields. Our agricultural heritage.* Cambridge University Press, Cambridge.

Hodder, I. (1982) *Symbols in Action: Ethnoarchaeological Studies of Material Culture.* Cambridge University Press, Cambridge.

Hurcombe, L. (1993) The restricted function of Neolithic obsidian tools at grotta Filiestru, Sardinia. *Traces et Fonction: les gestes retrouvé.* ERAUL 50, Liège, pp. 87–95.

Ingersoll D. W., Yellen, J. E. and Macdonald, W. (1977) *Experimental Archaeology.* Columbian University Press, New York.

Jennbert, K. (1997) Mentality and the social word. The Mesolithic/Neolithic transition in Southern Scandinavia. In A. L. van Gijn and M. Zvelebil (eds.) *Ideology and social structure of stone age communities in Europe.* Analecta Praehistorica Leidensia 29. Leiden, pp. 51–56.

Johnson, L. L. (1978) A History of Flint-Knapping Experimentation, 1838–1976. *Current Anthropology* 19 (2), 337–372.

Juel Jensen, H. (1994) *Flint tools and plant working. Hidden traces of stone age technology.* Aarhus.

Kamminga, J. (1979) The nature of use-polish and abrasive smoothing on stone tools. In B. Hayden (ed.) *Lithic use-wear analysis.* Academic Press, New York, pp. 143–158.

Kealhofer, L., Torrence, R. and Fullagar, R. (1999) Integrating phytoliths within use-wear/residue studies of stone tools. *Journal of Archaeological Science* 26, 527–546.

Keeley, L. H. (1980) *Experimental determination of stone tool uses. A microwear analysis.* University of Chicago Press, Chicago.

Kelly, H. and Van Gijn, A. L. (2008) Understanding the function of coral tools from Anse a la Gourde: an experimental approach. In C. L. Hofman, M. Hoogland and A. L. van Gijn (eds.) *Crossing disicplinary boundaries and national borders. New methods and techniques in the study of archaeological materials from the Caribbean.* Alabama University Press, Tuscaloosa, pp. 115–124.

Kramer, C. (1979) *Ethnoarchaeology: Implications of Ethnography for Archaeology.* Columbia University Press, New York.

Lammers, Y. (2007) *Tracing traces from present to past: a functional analysis of pre-Columbian shell and stone artefacts from Anse à la gourde and Morel, Guadeloupe.* University Leiden, Leiden.

Langejans, G. H. J. (2006) Starch grain analysis on Late Iron Age grindstones from South Africa. *Southern African Humanities* 18 (2), 71–91.

Langejans, G. H. J. (2011) Discerning use-related micro-residues on tools: testing the multi-stranded approach for archaeological studies. *Journal of Archaeological Science* 38 (5), 985–1000.

Lemonnier, P. (1986) The study of material culture today. Toward an anthropology of technological systems. *Journal of Anthropological Archaeology* 5, 147–186.

Leroi-Gourhan, A. (1964) *La geste et la parole I. Techniques et language.* A. Michel, Paris.

Longacre, W. A. (ed.) (1991) *Ceramic ethnoarchaeology.* University of Arizona Press, Tucson.

Longacre, W. A. and Skibo, J. M. (eds.) (1994) *Kalinga ethnoarchaeology: expanding archaeological method and theory.* Smithsonian Institution Press, Washington.

Louwe Kooijmans, L. P. (2007) The gradual transition to farming in the Lower Rhine Basin. In A. Whittle and V. Cummings (eds.) *Going Over. The Mesolithic-Neolithic transition in north west Europe.* Oxford University Press, Oxford, pp. 287–309.

Maigrot, Y. (2001) Technical and functional study of ethnographic (Irian Jaya, Indonesia) and archaeological (Chalain and Clairvaux, Jura, France, 30th century BC) tools made from boars' tusks. In B. Beyries and P. Pétrequin (eds.) *Ethno-archaeology and its transfers.* BAR International Series 983. Archaeopress, Oxford, pp. 37–65.

Mathieu, J. R. (2002) *Experimental archaeology: replicating past objects, behaviors, and processes.* BAR International Series 1035. Archaeopress, Oxford.

Miller Rosen, A. (1999) Phytoliths as indicators of prehistoric irrigation farming. In P. C. Anderson (ed.) *Prehistory of Agriculture. New Experimental and ethnographic approaches.* Institute of Archaeology Monograph 40. University of California, Los Angeles, pp. 193–199.

Nieuwenhuis, C. J. and Van Gijn, A. L. (2008). Cereal harvesting and processing at the Middle Neolithic site of Schipluiden, a coastal site in the Lower Rhine basin. In L. Longo and N. Skakun (eds.) *'Prehistoric Technology' 40 years later. Functional studies and the Russian legacy. Proceedings of the International Congress Verona (Italy), 20-23 April 2005* (pp. 453–455). BAR International Series 1783. Archaeopress, Oxford.

Odell, G. H. (1977) *The application of microwear analysis to the lithic component of an entire prehistoric settlement: methods, problems and functional reconstructions.* Ph.D. thesis, Harvard University, Cambridge.

Odell, G. H. (2001) Stone tool research at the end of the millennium. Classification, function, and behavior. *Journal of Archaeological Research*, 9 (1), 45–100.

Peña-Chocarro, L. and Zapata, L. (2003) Post-processing of hulled wheats. An ethnoarchaeological approach. In P. C. Anderson, L. S. Cummings, T. K. Schippers and B. Simonel (eds.) *Le traitement des récoltes : un regard sur la diversité du Néolithique au présent.* ADPCA, Antibes, pp. 99–115.

Pearsall, D. M. and Piperno, D. R. (1993) The nature and status of phytolith analysis. In D. M. Pearsall and D. R. Piperno (eds.) *Current research in phytolith analysis: application in archaeology and paleoecology.* Vol. 10. University of Pennsylvania, Philadelphia, pp. 9–18.

Politis, G. G. (1996) Moving to produce: Nukak mobility and settlement patterns in Amazonia. *World Archaeology* 27 (3), 492–511.

Price, T. D. (ed.) (2000) *Europe's first farmers.* Cambridge University Press, Cambridge.

Procopiou, H. (1998) *L'outillage de mouture et de broyage en Crète minoenne.* Ph.D. thesis, Université de Paris.

Procopiou, H., Anderson, P. C., Formenti, F. and Juan-Tresseras, J. (2002) Etude des matières transformées sur les outils de mouture : Identification des résidus et des traces d'usure par analyse chimique et par observations en microscopie optique et électronique. In H. Procopiou and R. Treuil (eds.) *Moudre et Broyer. Moudre et broyer. L'interprétation fonctionnelle de l'outillage de mouture et de broyage dans la préhistoire et l'Antiquité.* CTHS, Paris, pp. 111–128.

Redman, C. I. (1978) *The rise of civilization. Early farmers to urban society in the Near East.* Freeman, San Francisco.

Regert, M., Dudd, S. N., Pétrequin, P. and Evershed, R. P. (1999) Fonction des céramiques et alimentation au Néolithique final sur les sites de Chalain. De nouvelles voies d'étude fondées sur l'analyse chimique des résidus organiques conservés dans les poteries. Revue d'Archéométrie 23, 91–99.

Reynolds, P. J. (1979) *Iron-age farm: the Butser experiment.* British Museum Publications, London.

Schiffer, M. B. (1987) *Formation Processes of the Archaeological Record.* University of New Mexico Press, Albuquerque.

Schiffer, M. B. (2009) Ethnoarchaeology, Experimental Archaeology, and the 'American School.' *Ethnoarchaeology* 1 (1), 7–26.

Semenov, S. A. (1964) *Prehistoric technology.* Moonraker Press, London.

Staski, E. and Sutro, L. D. (eds.) (1991) *The Ethnoarchaeology of Refuse Disposal.* Arizona State University Anthropological Research Papers 42. Arizona State University, Tempe.

Stordeur, D. and Anderson-Gerfaud, P. C. (1985) Les

omoplates encochées néolithiques de Ganj Dareh (Iran). Etude morphologique et fonctionnelle. *Cahiers de l'Euphrate* 4, 289–313.

Stout, D. (2002) Skill and Cognition in Stone Tool Production: An Ethnographic Case Study from Irian Jaya. *Current Anthropology* 43 (5), 693–722.

Terradas, X. (2005) *Lithic Toolkits in Ethnoarchaeological Contexts*. BAR International Series 1370. Archaeopress, Oxford.

Torrence, R., Kononenko, N. and White, P. (2010) Archaeological Research in West New Britain and the National Museum, November 2010. Report submitted to official organizations in Papua New Guinea.

Vargiolu R., Zahouani, H. and Anderson, P. C. (2003) Etude tribologique du processus d'usure des lames de silex et fonctionnement du tribulum. In P. C. Anderson, L. S. Cummings, T. K. Schippers and B. Simonel (eds.) *Le traitement des récoltes : un regard sur la diversité du Néolithique au présent*. ADPCA, Antibes, pp. 439–454.

Vaughan, P. (1985). *Use-wear analysis of flaked stone tools*. University of Arizona Press, Tucson.

Verbaas, A. and Van Gijn, A. L. (2007) Querns and other hard stone tools from Geleen-Janskamperveld. In P. Van de Velde (ed.) *Geleen-Janskamperveld*. Universiteit Leiden, pp. 191–204.

Weissner, P. (1983) Style and Social Information in Kalahari San Projectile Points. *American Antiquity* 48 (2), 253–276.

Whittle, A. and Cummings, V. (eds.) (2007) *Going Over. The Mesolithic-Neolithic transition in Northwest Europe*. Oxford University Press, Oxford.

Yellen, J. (1977) *Archaeological Approaches to the Present: Models for Reconstructing the Past*. Academic Press, New York.

Zohary, D., Hopf, M. and Weiss, E. (2012) *The domestication of plants in the Old World*. Oxford University Press, Oxford.

Zvelebil, M. (2000) The social context of the agricultural transition in Europe. In C. Renfrew and K. Boyle (eds.) *Archaeogenetics: DNA and the population prehistory of Europe*.

SECTION 2

The Agricultural Process: Tools and Techniques in Cultural Context

2 From Land Clearance and Preparation to Sowing

2.1. INTRODUCTION

Inja Smerdel and Grith Lerche

The basic resources of pre-industrial societies were water and land, with their fauna and flora, and the sun's caress. Besides gathering, fishing, hunting, and numerous crafts, as Fernand Braudel said in his *Structures of Everyday Life* (1988, 136, 182), Western civilisation was based on a 'plant determinism' of wheat and grass, and on farm life supported by 'ploughing and pasturing.' If we continue slightly poetically, as Alexander Fenton put it in *Scottish Country Life* (1999, 1), 'as far back as history and archaeology take us, in all countries of Europe, the basic essential was the way in which the arable, producing grain, was integrated with the pasture, producing grass to feed stock'. For centuries, in the eternal cycle of the four seasons and the alternation of everyday and seasonal activities, many tools and implements hardly changed, because they were preserved as they were by their basic function. 'Agriculture evolved to suit the needs of the communities it served in different regions, being a system of overlapping systems that varied in sophistication according to local resources and terrain, administrative organisation, access to labour supply and markets, and earlier historical circumstances' (Fenton 1991, 1).

All agricultural systems probably experienced a virginal phase in which settlers in a new environment looked around, sifted the soil between their fingers, and made their first decisions on how to prepare the land for tilling, and for what crops. They were challenged to choose, invent, or adapt tools for various working processes, including those animals which would provide draught power. These processes of choice and adaptation continued, strongly influenced by local environmental conditions, and as agricultural systems stabilised or evolved, so too did their cultural contexts, the social, legal, religious, and personal settings of farm life.

Agriculture and the issues of its development and processes have been studied for generations now by the various humanities, social sciences and agronomy, applying different approaches and methods. Numerous ethnological, folk, and applied development studies all over the world, on a national or local level, begin by describing and explaining the preparatory steps of cultivation: removing stones, building enclosure walls, felling and burning woods, cutting sod, digging and ploughing, applying manure to foster fertility, harrowing to prepare the soil for sowing or to cover the seeds. These operations usually form a 'first chapter' of material culture in a study of any agricultural system.

In the past century, ards and ploughs were the central focus of research on the theme of traditional agriculture. Ards and ploughs indeed are material witnesses to life's cycles, implements from ancient times which were instrumental to the development of civilisation. It is hardly surprising that they have become the subject of many studies from at least 1773, when Andreas Berg's work on ploughing implements appeared (*Methodus investigandi origines gentium ope instrumentorum ruralium,* Bratanić 1955, 269), to recent times (Leser 1931; Haudricourt and Delamarre 1955; Lerche 1994; Bourrigaud and Sigaut 2007).

Ploughs and ards were considered important enough to be called the *Herzstück* (the centrepiece) of Europe's farming culture (Bratanić 1952, 208, referring to Koren 1952). Nevertheless, and in spite of the fact that the studies concentrate on such a basic cultural element with rich meanings, many of them are characterised by 'narrow views, characteristic (over-)precision, and repetition' (Sigaut 2005, 7–8): they focus on forms, dimensions, names of parts, the geographical spread of individual ploughing implements, and the search for their origin. They particularly attracted representatives of the cultural-historical trend in European ethnology. Such studies of ploughing implements proliferated in the decades around the mid-20th century, but later largely dried up as theoretical changes in ethnology brought to the forefront different research themes and approaches. But given our interest in 'man as the bearer of culture' in relation to 'his cultural and natural environment as shown by everyday routines' (Slavec-Gradišnik 1995, 130), we can hardly afford to neglect such research on pre-industrial ploughing implements, embedded as they are in the totality of the daily life of farming populations, and through their social and spiritual elements affecting human connections with the natural environment.

This is also the case with all the other agricultural operations in the process of land clearance and preparation. Manual tilling techniques – spade and hoe cultivation – are less researched than ploughs, with some exceptions (Steensberg and Lerche 1977; Fenton 1977; Sigaut 1977). Fire-clearance husbandry (in which mostly manual tilling techniques prevail), has received more attention. Harold Conklin, in *The Study of Shifting Cultivation* (1963), listed about 1300 books, papers and reviews as well as some unpublished but publicly available works on 'swidden-agriculture'. To his list much was added during the following decades; including some landmark treatises based on a comparative approach, such as *L'agriculture et le feu* by François Sigaut (1975), and *Fire-Clearance Husbandry* by Axel Steensberg (1993).

Dealing with such an abundance of previous work on land clearance and preparation, we wonder what we can add to this wealth of information, deliberation, hypothesis, findings, and conclusions. The contributions to this chapter aim at understanding tools, working processes, and other cultural elements of land clearance and preparation in their articulation with the natural environment and social structures. We explore the functioning of different parts of individual tools and their development, how people work with them, developing special skills and habitual gestures, and the social relations of people involved in processes using the tools, as well as the role of individuals in developing or improving various tools. As we see it, there is no doubt that many diverse factors influence agricultural processes, including the historical, social, legal, and religious milieu, but the role of place, the impact of the natural environment, the soil, the geomorphology, the climatic conditions, must be considered basic to all, the most influential setting for human action, inventiveness and survival strategies.

In the contributions included in this chapter, we do not tackle the entire process from land clearance and preparation to sowing. We set forth some examples of our recent archaeological, historical and ethnological research, differing in scientific approach and methodology, that illuminate what we consider some of the important patterns. In rather different ways, we all strive to understand and to explain the role of place, the impact of various 'faces' of the natural environment on the working processes we do research on.

Sabine Beckmann thus addresses the problem of *perivoloi*, which have been interpreted as Minoan defensive systems, using recent land use to explain on the basis of topography in a dry climate that they were in fact enclosure walls, protecting the scarce soil behind them against erosion and supporting animal husbandry. Inja Smerdel examines fire-clearance husbandry as a response to specific environmental and social circumstances which persevered until almost 1970 in a remote mountainous area. Patricia Anderson takes us to present-day Tunisia, where non-mechanised land preparation operations work superficial soils with the ard. She explains the ard as a multifunctional implement which is made to suit the requirements of the farmer, his field and his animals, and is especially adapted to the land. In her other contribution, Inja Smerdel presents five examples of different ploughing tools – among them two ards that basically correspond to the established definition of this implement,[1] and yet reveal some particularities resulting from the impact of the dynamics of long term use in mountainous environments and from the inventiveness of the makers. She introduces these five implements as eloquent primary material sources, interpreted through oral testimonies of ploughmen, trying to understand what objects can tell and people explain about past ploughing techniques, after they have gone out of use. Grith Lerche then discusses the reconstruction of a medieval cultivation system of ploughing with a wheel plough in ridges and furrows. She uses an experimental approach to link the knowledge of the ancient farmers about their natural environment and their abilities to cope with its diversity to new data and experience, initiating a discussion on what made the old open-field system sustainable for hundreds of years until the dawn of industrialisation.

As already explained, we do not tackle the whole chain of operations in the process from land clearance to sowing in the contributions included. There are only some words on sowing operations, scattered in the texts on fire-clearance husbandry in Slovenia and on ploughing with the ard in Tunisia. And yet the only example from the African continent (Tunisia), viewed in the context of Mediterranean agricultural practices, represents a strong reminder of the diversity of pre-industrial, non-mechanised agricultures. Irrigated and tropical systems have other options, which should not be neglected in general deliberations on agricultural diversity.

The intense everyday dialogue between man and his natural environment, resulting in continuity, resilience, and cultural choices, is still basic for survival in many countries. Awareness of this dialogue is becoming more and more crucial in our mechanised, digitalised world. Plants simply cannot grow and animals cannot be bred inside computers, however much the latter may help 'improve' their genetic structure. Traditional knowledge on land preparation operations might here and there even become essential for survival again; or might at least prove that there are good reasons for the endurance of traditional systems.

2.2. MINOAN ENCLOSURE WALLS

Sabine Beckmann

In Crete, field walls are usually taken to be of relatively recent date, and studies on Minoan material are thus non-existent. However, it is known that the building of terrace walls goes back at least to Bronze Age times (Frederick and Krahtopolou 2000; Chryssoulaki 1999; Moody and Grove 1990).

In the Agios Nikolaos mountain area of Crete (600–1000 m; Fig. 2.1), my studies (Beckmann 2012a, b) show that a series of enclosure walls are Minoan in date. The massiveness of kilometres of these enclosure-walls (*perivoloi*) is one piece of evidence. They also typically define the perimeters of plots of land which each include a Minoan dwelling site datable to the first half of the 2nd millennium BCE, in an area otherwise nearly uninhabited for ages. *Perivoloi* enclose plots of land (1 to 7 hectares each) of different quality, including rocky areas, some small colluvial valleys, and slopes that are often carefully terraced to optimise agricultural use. Obviously they were placed systematically, often just leaving free the space of a passage or 'road' (2–3 m), connecting or leading to dwelling sites characterised mostly by simple coarse cooking pottery. The area shown in Fig. 2.2 alone, (about 1.5 square kilometres out of over 30 square km containing Minoan *perivoloi*) contains evidence of *ca.* 30 km of Minoan enclosure walls. Pieces of the *perivoloi* can still be seen in many places, as there was seldom a reason to remove them in later times when the area was used for mixed agriculture and shepherding since the middle of the 19th century. Parts of these Minoan structures are even visible on maps (noted as fieldwalls or paths) and on aerial photographs as light or dark lines. The 'walls' mentioned in archaeological literature for the area, and sometimes explicitly interpreted as part of a defensive system, were probably parts of *perivoloi* the authors saw (Evans 1895; Taramelli 1899; Brown 2001).

Fig. 2.1. Location of Agios Nikolaos on the island of Crete, Greece. Map: R. Lugon, J.-C. Loubier and A. Chevalier.

Function

The Minoan structures (walls, roads, and buildings) were systematically planned around Protopalatial

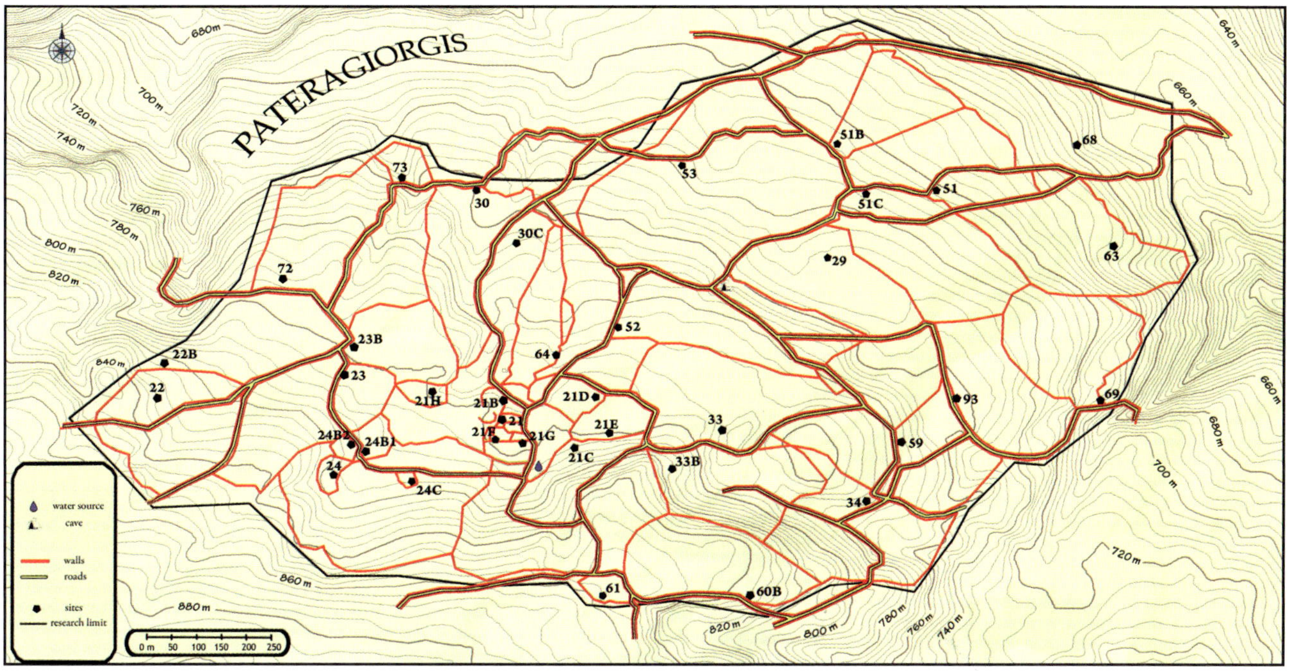

Fig. 2.2. Map showing Minoan structures and land-use.

times (2000–1800 BCE), and used while the area was inhabited for at least 200 years, and sometimes longer. As the climate is known to have been similar to recent times in the Middle Bronze Age (Moody 2000) and characteristics of the landscape do not allow much variety in animal husbandry, an ethnoarchaeological study of pre-industrial Cretan mixed mountain agriculture elucidates the enclosures' probable function (Beckmann 2012a), especially when taking into account that Cretan agriculture until WWII was still mainly on a Bronze Age technical level (Allbaugh 1953, 284). In the 19th and 20th centuries, mixed husbandry prevailed in the area. Farmer/shepherds would have up to fifty sheep and/or goats, while cereals and sometimes vines grew on small fields, usually identical with the Minoan ones. Old enclosure-walls were often re-used in pre-industrial times, built high enough to manage animals. A wall only 1.2 m high is enough to keep sheep out or in. Walls used in mixed husbandry (and especially for goats) would often have an extra layer of thorny twigs on top, increasing the height without needing extra breadth. Dry stone walls can only be built to a certain height related to their width. A wall of 0.8 m width cannot be built higher than 1.2 m (Hylton 1984, 487). This corresponds to the probable original height of the Minoan enclosures, judging from the fallen stones. Most of the Bronze Age enclosure bases still visible in the area are *ca.* 0.8–1.0 m broad, implying an original maximum wall-height of 1.55 m (Figs. 2.3A,B).

Often, enclosed fields were ploughed and sown (usually with barley) in winter while the area's animals could browse outside (winter and spring being the green and fertile time of the year in Crete). After the harvest of the cereals in June, when the uncultivated mountainside's browsing capacity had been exhausted, enclosures were opened for a few hours per day to let the animals browse on the straw or stubble (depending on how much of it was left for the animals during harvest).

Minoan dwelling sites were often placed next to some well-terraced plots of arable land (Fig. 2.3B). Some Minoan roads and *perivoloi* run through the alluvial terraced winter run-off valleys, making it clear that these have never had enough water to destroy the built structures during the last 3500 years and thus presenting a new argument for a similar climate (in terms of precipitation) in this area since their erection.

Be it for grazing or agriculture, enclosure walls protect the soil behind them against erosion by wind or rain (Grove and Rackham 2003). The many

A

surviving Bronze Age fields prove the success of this strategy. Enclosure walls could also protect soil from drying out by sun and strong winds. Ten percent less wind is equivalent to a 15–20% larger amount of precipitation (Krusche *et al.* 1982, 52–53). Walls or hedges result in 50% less wind within a distance of 8 times their height (*i.e.* over 10 m behind a wall of 1.3 m height), whereas the amount of dew is three times larger (Krusche *et al.* 1982, 52–53). These are important gains for a windy mountain area in a dry climate. Again, we can still see until today the increased fertility promoted by walls in the lusher plant growth near even the ruins of the *perivoloi,* often visible as dark lines marking the walls on aerial photographs.

Although the size and extent of *perivoloi* in the very East of Crete (Chryssoulaki 1999) has suggested a defensive function to some scholars, they are better interpreted as evidence of a successful strategy of mixed agriculture and herding in the Bronze Age, analogous to ethnographically documented practices in recent times.

B

Fig. 2.3. A) Typical Minoan enclosure wall/*perivolos*. B) 'Minoan field' still in use.

2.3. FIRE-CLEARANCE HUSBANDRY IN SLOVENIA: THE MARIJA REKA CASE STUDY

Inja Smerdel

Introductory Words

The Marija Reka case study on fire-clearance husbandry began in 2007 as a by-product of my research on the working relationship between man and oxen in Slovenia. Examining the metaphorical meanings of yokes, and their secondary functions, I traced an old photograph from 1955 depicting two guests at a wedding in Maria Reka in the Zasavje region, yoked in a bow yoke, joking and running around a threshing device. Finding the bride and the bridegroom from that long-ago wedding still on their lonely farm pinned to the steep slopes of the Reka mountain, while recording their subsistence farming practices, I became aware that they and their neighbours practised fire-clearance husbandry almost until the last quarter of the past century. This early agricultural technique had not previously been known in this area, although it commonly occurs in the northern mountainous zones of Slovenia.

Man's ability to produce and manipulate fire seems to be one of the most conspicuous of his cultural traits. Fire has always been a powerful tool for a variety of tasks, among them techniques of cutting down and burning trees, scrub and other kinds of vegetation. Preparing the land for cultivation by using fire releases nutrients in the top soil, and the remaining ashes act as fertiliser. This was commonly understood in many cultures based on generations of experience all over the world. In addition, burning acted as the best natural pesticide, producing crops that were practically free of weeds after the first burn. Fire-clearance can be found 'from North and South America, through Africa, South Asia, the Philippines, Melanesia, China and Korea to Siberia and Europe' (Steensberg 1993, 211, 11–13). Numerous works that tackle its spatial, temporal and cultural diversity (see Introduction, this chapter) can confirm a rather established opinion, that fire-clearance may mostly be regarded as a cross-cultural component of farming systems. As such, it was specific for Europe from the Middle Ages and onwards. In the Classical world, Virgil (*Georgics*, Book I), poetically wrote that 'it has been useful to fire barren fields, and burn the light stubble in crackling flames'; and Columella, the most famous of the Roman agrarian authors (Book II), mentioned clearing land by two different methods, 'either by tearing out the trees by the roots and removing them or, if they are few, by simply cutting them down, burning them and ploughing them under' (Steensberg 1993, 188, 189).

In *L'agriculture et le feu*, François Sigaut (1975) attempts to clarify, explain and define the basic fire-clearance techniques, introducing the question of the real meaning of the French term *écobuage* (paring and burning), rather often imprecisely explained as clearing the ley-land by fire. A detailed and well-illustrated description of paring and burning appeared in *The Book of the Farm* by Henry Stephens (1855) and Axel Steensberg (1993, 158) describes it as follows: '... paring is a removal of the surface of the ground, with what may be growing upon it at the time; and burning is the reduction by fire to a state of powder, of what has been pared off. The object of the process is to obtain possession of the soil pared for arable purposes, and paring and burning will certainly insure a crop in advance for one season at least'. Sigaut's comparative analyses finally enabled a clear differentiation and definition of two main fire-clearance methods: paring and burning (*écobuage* in Sigaut's terminology) and fire-clearance of woodlands (*essartage*). The latter has been common at times in most European countries and until lately in three large regions: the North (Scandinavia, Russia and the Baltic states), the mountains of Central Europe (the Ardennes, Schwarzwald, Austrian Alps, Czech mountains etc.), and the mountains of southern Europe (especially in France, and Spain). Sigaut further developed the following general phases of the process of *essartage*: choosing the place, felling trees, drying of wood, chopping up the wood, spreading it out on the ground (using or selling what was good for other purposes), burning, cleaning, piling up and re-burning the remaining wood, and constructing plots (Sigaut 1975, 19–21). In 1980, in a fundamental work, *The Agriculture of the Ancient Slavs (Zemědělství starých Slovanů)*, Magdalena Beranová (1980, 295) defined five principal forms of fire-clearance husbandry: 1 – the burning of climax forest; 2 – the burning of thickets, scrub and undergrowth; 3 – the burning of turf (sod) and of raw humus by using piled up wood, straw, dry grass or other fuel; 4 – the burning of peat and of the upper layer of drained moors; 5 – the burning of grassland and of grassland undergrowth, a method including the burning of grass on grazing areas and of stubble. Axel Steensberg's comprehensive work *Fire-Clearance Husbandry* (1993) deals with all the elements of fire-clearance mentioned above from a world-wide comparative perspective. Steensberg kept to the basic distinction between fire-clearance of woodlands and paring and burning, noting that 'albeit there are transitions between the two techniques' (Steensberg 1993, 13). He stressed the diversity and contextual complexity of fire-clearance, and the importance of individual case studies 'which might elucidate how varying man's decisions had to be from day to day and often from one hour to another because of unpredictable weather conditions, unexpected events and other obstacles in his environment' (Steensberg 1993, 12). The example of Slovenian Marija Reka fire-clearance husbandry further shows how geomorphology, climate, vegetation and access to cheap labour might vary much from place to place even inside the same small province.

The Historical, Regional and Social Framework of Fire-Clearance Husbandry in Slovenia

Cultivating various crops on *terra nuova* or virgin land (meaning reclaimed woodland), using *buoi rossi* or 'red oxen' (meaning fire, Emilio Sereni 1981, 3), was an early agricultural practice of the Slovenes. Lacking material evidence or written sources, it has not yet been ascertained whether fire-clearance husbandry was their basic agricultural practice or only an additional technique, when the Slovenes occupied the eastern Alps between the last decades of the 6th and the 8th century. In archaeological and historical circles, fire-clearance is assumed to be a secondary technique, based on finds (for example symmetrical shares and coulters from the 1st century BCE, from Idrija by Bača, Reka by Cerkno, Vrhovlje by Kojsko, Črniče and Šmihel; Fig. 2.4), and evidence of well developed Celtic and Roman agriculture in the area (Baš 1954/1984, 176–184, 186–192; cf. Grafenauer 1970, 212–214; Smerdel 2008, 94–95).

Slovene fire-clearance history has been studied by Baš (1954), Predan (1956), and Grafenauer (1971), among others. Baš (1954/1984, 190) asserted that the Slovenes 'inhabited the plains, where they continued with well developed Roman agricultural practices and with their own ones. Later, spreading out from their primary settlements ..., they colonised the mountainous areas by applying fire-clearance', and cultivated the land mostly by manual tilling techniques. This colonisation between the 13th and 15th centuries seems to be well supported with evidence in medieval urbaria.[2] For example the Gornji Grad (Oberburg; Fig. 2.4) urbarium from 1426, lists farmers with names like *Laznik* (Swiddener), and *Požarnik* (Fire) (Orožen 1876, 222, 224, 226, 233, 261, 267). Testimony to fire-clearance husbandry in Slovenia also includes 41 toponyms such as *Laze, Lazi* and the like, scattered all over Slovenia except in the Soča valley and Karst, which recognisably place this agricultural practice in the mountainous areas (Baš 1954/1984, 184).

In some Slovenian regions, fire-clearance of woodlands was practised into the decade after the Second World War, until in 1953 the new and more restrictive Law on Forests appeared. On Ratitovec, along the upper Meža, Mislinja and Kokra rivers, and

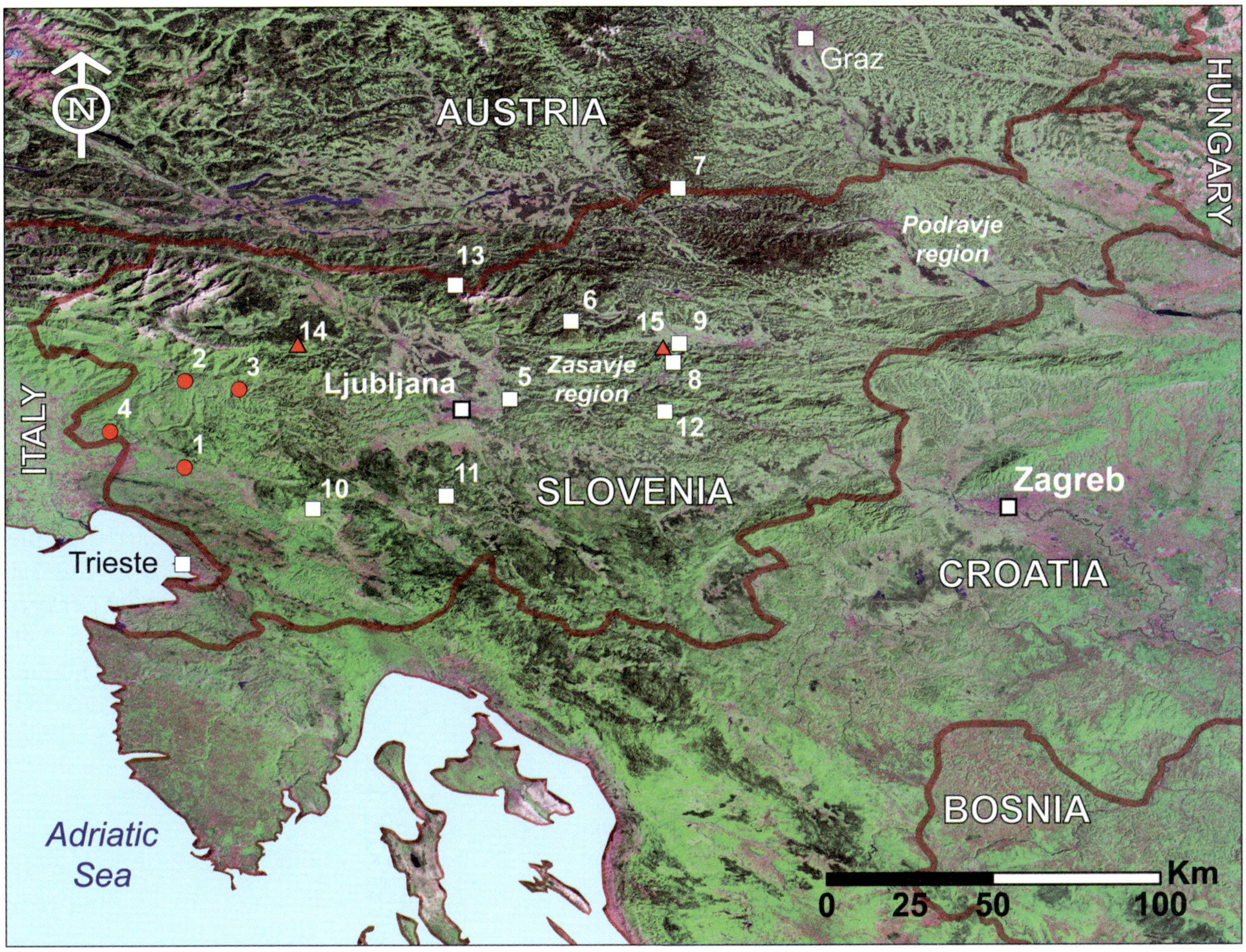

Fig. 2.4. Map of Slovenia with the towns of 6) Gornji Grad; 9) Prebold; 12) Trbovlje; the places of 1) Črniče; 2) Idrija by Bača; 3) Reka by Cerkno; 4) Vrhovlje by Kojsko; 10) Šmihel pod Nanosom; 8) the village of Marija Reka with the homestead by Pečnik; the regions of Zasavje and alpine Podravje, and the mountains of 14) Ratitovec and 15) Reška planina mentioned in Chapter 2.3; and the places of 5) Dol near Ljubljana; 13) Zgornje Jezersko; 7) Mlake; 11) Štrukljeva vas; 8) Marija Reka mentioned in Chapter 2.5. Map: R. Lugon, J.-C. Loubier and A. Chevalier.

in the upper Savinja, Paka and Boljska river basins (Fig. 2.4), the tradition of burning the *lazi, požari, novine* or *velne* (thickets, scrub, and undergrowth) endured (Figs. 2.5A, B, C), as in the alpine Podravje area (Fig. 2.4), the burning of *frate* – or climax forest (Fig. 2.6), that is, the second and first of Beranova's 'five principle forms' of fire-clearance. The latter was a kind of forest husbandry, applied only in coniferous woodland, where areas with mature forest were fully cleared; after burning, the ash was mixed into the soil by hand-hoeing and usually rye seed was planted mixed with seeds of pine (cf. Steensberg 1993, 135, 128). In the first mentioned swiddens, in which the area was used for grazing after harvesting one or two crops and then slowly transformed into open forest in cycles fifteen to thirty years long, the main crops were millet, buckwheat, rye, oats and barley, sometimes wheat, and here and there potatoes.

Fire-clearance was known as 'pulling the porridge out of hell' (Predan 1956, 26), hard and rather dangerous work in which only skilled men were involved. Its social context was twofold: on the lonely mountain farms it was an additional agricultural practice with which farmers periodically added an outfield (later becoming a grazing area) to permanently tilled fields outside the forest. All the work was done by the family members with mutual aid of their neighbours. In the mountain villages, for example in the Ratitovec area, fire-clearance was practised mainly on commons or on pastures of the wealthy farmers; it was carried out by the poor – smallholders, cottagers and crofters, especially

Fig. 2.5. Fire-clearance by *Mavrič*, in St Florjan above Gornji Grad, in 1950s. Usually they used straw for lighting fires. Some people had to guard the edges to prevent potential fire from spreading into the adjacent forest. While the glowing roll was being pulled downwards with a clearly visible iron hook, the flames were burning upwards. (Photo by J. Božič, 1950; from the Celje Regional Museum archives)

the charcoal-burners, as in other countries (Acosta Naranjo 2001; Steensberg 1993; Frühwald 1987). Working on *lazi* offered them additional means to survive: first by 'cooking' (burning) and selling the charcoal, then by cultivating potatoes for two years followed by a third year of rye. Pastures cleared in this way were exploited for at least ten years as cattle-grazing areas by the farmers (which was their main purpose), and in the following twenty years, they gradually became overgrown with thickets again, ready to be reclaimed and burned (Baš 1954/1984, 178).

Fig. 2.6. A *frata* (an area of climax forest made fully bare) in Robanov kot in the upper Savinja valley. (Photo by Ilovar, 1933; SEM Archives)

The Case of Marija Reka Fire-Clearance Husbandry

Marija Reka is a scattered settlement of remote, lonely mountain farms mostly on the sunny-side slopes of the Reka mountain (910 m). This area in the neighbourhood of the upper Boljska river basin is on the southern edge of the Celje basin, above the Mala- and Velika Reka river banks at an altitude of 300 to 780 metres, with a continental climate. Steep slopes (up to 60 degrees) on limestone and dolomite are mainly covered with deciduous wood in the form of scrub and thickets as well as forest dominated by beech, chestnut, white beech, acacia, poplar, dogwood and hazel. Two principal economic activities of the population were (and on the remaining farms still are) forestry and livestock breeding (Savnik 1976, 507). Agriculture in the area up to the 1950s was entirely self-sufficient, fragmented and polycultural, the farms being rather larger than the Slovenian average, with small farmers owning about 20 ha of land, middle-sized ones about 40 ha, and the biggest owners having up to 90 ha.

Although the area seems to be geographically, culturally, and economically uniform, the population developed some individual economic and survival strategies. While, for example, farmers from the left side of the Velika Reka river did practise fire-clearance husbandry in their woods as well as maintaining permanent fields, the ones from the right side of the river, on the contrary, did not apply swiddening in their forests. In the latter, chestnut wood prevailed, ensuring owners good income from selling it for tannin production. The bride and bridegroom from the 1950s whom I met high on the left bank of the Velika Reka river on their homestead by *Pečník* (Fig. 2.4), as well as their neighbours from all the other adjacent farms, vividly remember how they burned and tilled swiddens or *velne* until the 1960s. Their memories of the working process describe a local situation and technical solutions, which can be partly analysed by considering the general phases of fire-clearance developed by Sigaut (1975). Nevertheless, their response to specific environmental and social circumstances evidently reached beyond technical solutions. Fire-clearance husbandry seemed to be their choice of survival strategy within their place and socio-historical context. These choices answer two questions: Why did they keep applying this early agricultural technique until the last quarter of the past century? And why and when did they quit?

Choosing Place

They searched out previously unburned areas, 'where we thought the soil would be fertile' and judged that 'the millet would bring an abundant crop'. The quality of the soil was decisive, even if the place chosen on their property was situated in a gorge, with no possibility of transporting firewood out. A good *velna*, also locally called *trebež* (reclaimed place), was at least 50 m wide and 80 m long. 'A longer *velna* would give us more crops.'

Felling Trees

In the autumn 'each year we cut down about 20 ares and made a *velna*.' For a *velna* of 80 ares (or 0.8 hectares), eight to ten men had to fell thickets for three days, using axes or hand saws. They tried to cut the trunks as low as possible, 'almost into the ground'. Only the biggest farmers hired workers from elsewhere (for example miners from Trbovlje or some 'homeless ones' that worked 'just for food and a shirt' in the decade after World War II).

Drying Wood

In winter, the cut branches 'needed to freeze a little bit' and afterwards dry – until burning the *velna* at the end of April and beginning of May. But some deciduous trees had to be girdled to kill them so they could dry out for a year before burning. Thus the farmers 'whitened' or ringed acacia and 'mountain poplar', because if felled fresh, the stumps of these trees would immediately bud new branches which would then hinder the growth of crops.

Spreading Wood out on the Ground (using and selling what was good for other purposes)

After cutting off the branches of the felled trees and spreading the brush, they transported home what was good for firewood. Part of this they themselves 'cooked' into charcoal in winter, selling it by the kilogram (for example, to the blacksmith in Prebold; Fig. 2.4). Charcoal brought them better income and was easier to transport than wood.

Burning

In spring, they 'waited until the forest became green' (the trees around the reclaimed area had to have fresh leaves which helped to prevent potential fire spread) and also for 'a calm, windless day' before beginning the dangerous work of burning. On the upper part of the *velna* slope, they put together 3 to 4 m long branches and thinner trunks into a bunch, a roll or *rolna*. The men working the *velna* had to be not more than 3 to 4 m apart. The burning began on the upper part of the *velna* by lighting the *rolna*, usually with straw (Fig. 2.5A). The burning *rolna* was pulled or rolled steadily down the slope, using special iron hooks (Figs. 2.5C, 2.7A, B, C.). Burning time depended on the length of the *velna*. A *velna* of 60 to 70 m long needed about one hour, while longer ones took from 2 to 3 hours. When they had burned to the bottom, the men had lunch and about an hour later the *velna* had cooled down. The men then took their hoes and began digging (Fig. 2.7D). These toothed hoes were well adapted to hand-hoeing in the swiddens where the soil was full of tree-roots (cf. Forni 1982; Steensberg 1986, 137). The work was finished after 5 or 6 hours, at the end of the afternoon. All the work was done by men from the homestead that burned a *velna* and by those from the neighbouring farms. Helping each other was natural and indispensable in those mountains and partly explains why fire-clearance was sustainable in Marija Reka, contrary to some authors' argument that fire-clearance husbandry defies economic logic because the costs of labour are much higher than the possible profit (*e.g.* Hlubek 1846; Baš 1954/1984).

Fig. 2.7. A, B) Drawings of the burning technique (after fieldwork sketches by I. Smerdel, 2007): A) shows how on the upper part of the *velna* slope they put together branches and thinner trunks into a bunch, a roll or *rolna*, and how the men working the *velna* had to be no more than 3–4 m apart. B) the side view shows the angle of the *velna* slope on which the burning *rolna* was pulled or rolled steadily down with a special iron hook. C) The principal tool – the iron hook or *akel*, *hakel*, *krivelj* (from Ger. *Haken* = hook, Slo. *kavelj*); from the homestead by Čérjet, presented by Kondi Kos. Hooks were mainly made by the blacksmith from Prebold; their price was something like the price of a pair of shoes. Each man that came to burn a *velna* brought his own hook, and the owner of the swidden had to have some spare ones aside. The iron part, the hook, is approximately 70 cm long; the handle, a hop pole made of strong young-dried spruce, was usually up to 5 m long; the longer it was, the better – definitely safer, but requiring rather skilled men. D) The toothed hand-hoe applied for manual tilling of a *velna* after burning, presented by Kondi Kos (Photos by I. Smerdel, 2007).

Cleaning, Piling Up and Re-Burning

The Marija Reka farmers did not need to execute this phase of work. One burning by their techniques sufficed.

Constructing Plots

They also did not construct fences to protect the swiddened areas from animals, as was done in other parts of Slovenia (Ložar 1944, 120; Baš 1954/1984, 181) and Europe (Steensberg 1993, 139). Deer did not damage millet and wheat, preferring instead to graze alfalfa or beans, wild boars did not appear in the Marija Reka area until the 1970s, and the mice which cause only minor damage were left to feast.

Marija Reka Velna Cycle and Labour Division

The burning and digging was done only by men. Youngsters had to be at least sixteen years old before they were allowed to join men in carrying out this hard and dangerous work. In sowing and covering the seeds with rakes, men and women took turns. When they harvested millet or wheat, reaping with sickles was only woman's work.

The farmers applied the following cycle in cultivating the *velnas*. In the first year, millet was planted. After digging the burned area, they waited just a day or two before sowing, 'so that the seeds wouldn't glow.' The millet was reaped at the end of August and into September. In October, winter wheat, which they preferred to barley, was sown. For the second year, wheat or barley was planted. If the soil was favourable, two crops were taken from the new area in two years. In the third year and onward, after the wheat was harvested, they used the *velna* for grazing or even for fodder. 'The goal for the future was to make that land a meadow for some time,' mowing for hay and later for litter, as fern began to grow in. After ten to fifteen years a *velna* or a *trebež* needed to be slowly transformed again into open forest, sometimes by planting firs to produce a coniferous forest. And then the swiddening cycle started anew.

In Conclusion

The Marija Reka case generally supports Kaland's (1979) and Iversen's (1941) assumption that land clearing by fire in much of Europe was intended first and foremost to produce open areas for cattle grazing (in Steensberg 1986, 83). It also clearly represents a local situation which must be 'viewed as one of a series of tactical solutions to a particular problem – not as a remnant from some once-universal stage of agriculture' (Rowly-Conwy 1981 in Steensberg 1986, 134). The Marija Reka farmers remember practising fire-clearance husbandry 'forever' to until the last *velna* was burnt in the 1960s on one of the biggest farms by *Narat*. Fire-clearance husbandry was a response to their particular problems, a permanent shortage of food and their complex social organisation. In the remote mountain settlement with no proper roads until the 1960s, they continued to cultivate swiddens to produce additional food for themselves and their cattle; cultivating their permanent fields, far less fertile than those in the valley, simply did not suffice. The typical small Marija Reka farms of about 20 ha had up to 3 ha of arable land in permanent fields, about 3 ha of meadows and pastures, and all the rest was woodland. Thus they kept on utilising their scrubland and thickets, good almost for nothing else; the nutrients released by burning gave them up to one third more produce on swiddens than they could harvest on other fields. Neither selling cattle (worn out oxen) nor wood was enough of a supplement. On each farm, there were ten to eighteen people, the elderly, parents, children and unmarried uncles and aunts. 'We had to produce everything, but even everything was not enough. If we were six, we were already a little bit hungry...', is how they remember their everyday life. Even the new Law on Forests in 1953 did not stop them burning the *velnas*. Although it strictly forbade 'to reclaim forests or to change them into another culture' they did not have any problems with the authorities when they burned scrub and thickets. 'But if you would only touch birch for firewood – an inspector would immediately appear...'

Changing ways of life and land management slowly brought an end to fire-clearance husbandry. From the 1950s onwards, women and men of Marija Reka became farmer-labourers searching for regular employment in two adjacent towns, Prebold with a textile industry, and Trbovlje's coal mine. On the permanent fields, they began to apply a special 'mountain plough' adapted to the steep slopes (Smerdel, this chapter). This invention of Kondi Kos almost doubled the crops. As cultivation diminished, changes in the landscape occurred: once arable land turned into grazing areas, former alpine meadows became woodland.

The intrinsic significance of *velne* was also pertinent to other spheres of life. Besides providing Marija Reka people and their cattle with needed additional food, the *velne* also had a strong impact on the community's social life. The necessity of cooperating in this hard work formed community bonds. Helping burn a *velna* for the first time was a kind of *rite de passage* for young men. And the women still remember how they went to pick the most beautiful flowers in those areas. *Velne* were indeed an important element of their environment and their daily life.

2.4. WORKING WITH THE ARD IN PRESENT-DAY TUNISIA

Patricia C. Anderson

In the course of research on non-mechanised agriculture between 2005 and the present in the Atlas (Northwestern Tell) in Tunisia, I became interested in the use of the ard (Anderson 2006). The region studied lies approximately 100 km to the west of Tunis and measures 30 × 20 km. The ards here are made by a local carpenter in the city of Beja.[3] The study area has 800 mm of annual rainfall on the average, and is karstic, with limestone bedrock. The soil is principally rich in clay, acidic and fairly shallow. The agricultural system is biennial. Most farms using ard ploughing today have fields on a slope (up to 25%), and are situated at an altitude of 350 to 700 m (Ghozzi 2006). Today the ard is used primarily on sloped and mountain fields because mechanisation is difficult in such areas.

Animals are used for traction but also for most transport of goods and persons, and work is a social occasion. Ards, like draught animals, are loaned between friends and family, and mutual help during the agricultural season plays an important role. The ard is used only by men, often young, who transmit this skill. The young men begin to walk with the older men alongside the ard as children, until when they are 14 or 15 years old and strong enough to hold the ard handle behind the animal team, they try using it little by little, under adult supervision. The gendered restriction of ard use to males is a social convention based upon religious rules. In fact, women are capable of using the ard, and do so on occasion if necessary, for example because of lack of a male family member to work their fields.

Crops include 'macaroni' wheat (*Triticum durum*), barley, oat, pulses, and vegetables (Anderson 2006). The ard is multifunctional, used to prepare the fields before sowing (clearing plants, working in dung and household refuse as fertiliser, then tilling) and for covering the grain of cereals and pulses after they are broadcast sown. The ard is also used, with the addition of two lateral sticks, to harvest potatoes.

The 'chambige' type ard (Fig. 2.8) is assembled of three wooden parts, with the wood (wild olive, Aleppo pine, and eucalyptus) carefully selected by the carpenter for properties of strength and flexibility as needed. He fashions the different parts and assembles this light ard (12–15 kg in weight) in just several hours, and ards are easily assembled and disassembled for transport in the mountain farms. The iron ardshare, with a long tang inserted into the sole and with an arrow-shaped point, is made by the local blacksmith. The share is hafted into a groove in the sole by the carpenter, and protrudes perpendicularly as a direct extension of it. Its beam and sole are joined in a fixed position, at a low angle. The ard has one handle, hafted on the end of the stilt. Its structure and the iron share are particularly

Fig. 2.8. Carpenter in Beja, specialised in agricultural instruments, assembling an ard made from three kinds of wood. The ard shown here has a short extension of the stilt, and is made for one draft animal, as opposed to the longer stilts shown in the other figures, for pairs of animals.

Fig. 2.9. A) Ard pulled by mules in a sloped field in Met Algua. Note the straight yoke, and collars made from blue jean legs stuffed with chopped straw (Photo by P. C. Anderson). B) Side view of the ard in Fig. 2.9A, working on a slope to clean plant material from the field. Note the man directing is leaning into the slope, and is holding a *rieta* (Photo by P. C. Anderson). C) Ard opening a fallow field, in the mountains, pulled by oxen, in Al Asaîdiya (Photo by F. and S. Ghozzi). D) Ard pulled by oxen on mountain farm in Mergaga. Note the form of the yoke, shaped to set on the two animals' necks (Photo by F. and S. Ghozzi).

adapted to the soil type and slope in the region, and because of the fairly superficial soil cover, reflect the need for shallow tilling.

The ard is usually drawn by two animals, although there is a model made with a shorter beam, for traction using only one animal. Mules (Figs. 2.9A, B), or oxen (Figs. 2.9C, D), working in pairs, are used to pull the ard today. In general, each ard is made to suit the requirements of the individual farmer, his field and his animals. A different form of yoke is made by the carpenter for mules than for oxen (compare yokes in Figs. 2.15A and B), and to suit the particular team of animals. Mules work faster than oxen but can only be used if the soil is not too heavy and the slope not too great. Oxen, although slower, are used, and are in fact the only option where the soil is heavy and clayey and there is a marked slope. In these cases, no machinery can be used.

If the field is on a slope, the user needs to lean towards it (Figs. 2.9A, B). No one leads the team from in front; the procedure is entirely controlled by one person from behind. He carries a kind of double tool (*rieta*), an iron spatula mounted in a long, flexible handle (Fig. 2.9B). One end is used to scrape soil from the share, and the other, to encourage the animals. The person using the ard directs mules with one set of commands, and oxen with another (Anderson 2007).[4]

Concluding Comments

In the plains area, today a small metal plough may be pulled by animals in place of the ard. However, even when tractors till with ploughs in the plains, in these same fields the animal-powered ard is used for sowing (for covering over the grain) so that the furrows are as close together as possible. Recently ards are again being used in low-lying fields in the plains for tilling as well, because the rising cost of fuel for tractors is prohibitive, and people still have draught animals (Anderson 2007). An ancient and traditional tool continues in use alongside modern machinery because it is versatile and effective in marginal conditions where tractors are not, and efficient and less expensive for parts of the tilling process even when tractors are used.

The persistence of the ard in northwestern Tunisia needs also to be seen in the context of the region and the operating sequence of manual agricultural processes with which it is linked. First, the Tunisian ard is adapted to the somewhat fragile and shallow soil cover in this region, and to an ongoing manual agricultural cycle. This was driven home when we attempted to conduct experiments in France with the ard from Tunisia, made by the carpenter in Beja, to work fields in southern France that, although formerly tilled with an ard, had not been worked in over 50 years. It was clear to us that the Tunisian ard was adapted exactly to its home environment, and not ours: it was unable to break hard, new ground, but was adapted to working fields that had been tilled over the years. Second, it seems significant that sowing with the ard is followed by manual harvesting techniques, using a sickle for cereals and pulling up pulses by hand, for example, and threshing with either an animal-drawn threshing sledge or by using animal feet alone (see Chapter 5). Local farmers emphasise again and again the importance of the cycle of production of the grain itself, beginning with the animal-drawn ard, as being the absolute determinant of its quality, in terms of robustness and nutritional value. They are also dismayed at the decline in the quality of the soil in places where it is worked mechanically, and chemical fertiliser and weed killers are used. They consider that the taste of the grain is directly affected by abusing the traditional techniques, and that this, along with the disappearance of many of the traditional grain varieties in favour of those that work well with mechanisation, chemical fertilisers and weed killers, has directly affected the taste of food, and made it impossible to make many of the traditional dishes. The ard therefore seems to play a key role in maintaining a certain agricultural cycle and traditional grain varieties.

2.5. PLOUGHING TECHNIQUES IN SLOVENIA: WHAT TOOLS CAN TELL AND PEOPLE EXPLAIN

Inja Smerdel

Research on arable farming – land preparation using various ploughing implements – at least research on the use and social setting of tools, rather than their form alone (see Lerche 1994 and Sigaut 2005 for this distinction), is severely restricted by the place and time of our research. In a Mediterranean country like Tunisia where ploughing with the ard is still common (Anderson, this chapter), an ethnologist can work by participant observation – observing, asking, and even experiencing the work, to understand an agricultural process in its social and environmental context. Such research is not possible in present-day Slovenia or most other European countries. From the 1960s onwards, agriculture became mechanised, replacing draught animals and older ploughing implements with modern ploughs pulled by tractors.

The Slovenian territory saw the spread of asymmetrical ploughs from the 10th and 11th centuries onwards, almost contemporary with the stabilisation of fields and the introduction of three-field rotation with fallowing. Urbaria prove that from the 13th century onwards, ploughing implements were the principal tilling tools all over Slovenia, but heavier wooden ploughs only gradually replaced the lighter, symmetrical ard which required fewer draught animals (Baš 1968, 1–2; Grafenauer 1970, 214). Further changes to ploughing implements came later, in the agrarian-technical revolution of the second half of the 18th and first half of the 19th centuries. Nevertheless, the use of the ard and plough then as well as now depended, among other things, on the nature of the land and the social conditions. For instance, 18th-century records show that in certain areas of slopes with heavy soil, up to half the farms had neither an ard nor a plough, and mattocks and hoes were used instead (Smerdel 1991, 35; after Grafenauer 1970).

A number of historical trends can be seen. First, in mountain areas the ard was gradually replaced by the double plough from the early 19th century onwards (Orel 1955; 1961). This plough with two plough bodies attached to the beam approximately 90° apart, two shares, two coulters, two mouldboards but (usually) only three handles was used for reversible ploughing (Figs. 2.10A–C) and must have been considered an important advance. Secondly, at about the same time, the mouldboard plough prevailed in low-lying areas. It was generally used for conventional ploughing until the Second World War, when reversible ploughing spread to the plains with the introduction of reversible ploughs. Third, in the mid-19th century, industrial production of iron ploughs started in the Austrian realms, making the implement, which could plough deeper and yet required fewer draught animals, cheaper and more affordable to the rural population (Smerdel 1991, 35; after Grafenauer 1970). However, a record from 1940 states that 'unfortunately our farmers have very poor ploughs today, not much better than the ones their distant ancestors had' (Simonič 1940, 10). And reports from some areas in Slovenia set the introduction of iron ploughs in the 1920s and 30s and even then largely limited to prosperous landowners (Smerdel 2008, 98). Lastly, the spread of root and other crops gradually led also to the introduction of new ploughing implements from the 19th century onwards, cultivators and ridging machines for root crops, trench ploughs for hop fields, vineyards and tree crops, and the symmetrical subsoil plough for deep soil loosening (Smerdel 2008). In some cases, older ploughing implements such as ards were retained for use in some stages of a crop, for example for weeding the root crops or digging them up (Smerdel 1991, 33, 35; Orel 1961, 37).

Previous historical studies of ploughing in Slovenia are scarce, and those that exist often describe the forms of ards and ploughs and the geographical spread of individual implements, but only rarely explain ploughing techniques. In Slovenia, it was the ethnologist Boris Orel (1955, 1961) whom we must thank for some descriptions of ploughing, and for organising Slovenian plough scholarship up to that time.

In his Slovenian work, Orel (1955, 1961) recorded data for three levels of analysis suggested by François Sigaut (2005, 7–8):

1. he described existing ards and their parts in detail, noting their dimensions and dialect names (Sigaut's first level, *forme ou structure*), identifying three main types of ards in Slovenia, the Carinthian-Styrian rectangular wheeled ard with a coulter, the rectangular ard with a long yoke beam and the ard without a sole;
2. he collected data on the function of ards and their individual parts (Sigaut's second level, *fonctionnement*);
3. and, most importantly, Orel also studied the use of the ard, the work process (Sigaut's third level, *fonction*), and did not overlook the importance of its manufacture, the social context (who ploughed, who drove the oxen, who dug up the unploughed parts of a field) or its spiritual elements.

When I investigated ard research in 2006–2007 and catalogued the large collection of ards and ploughs in the Slovenian national ethnographic museum (Smerdel 2008), the records of Orel's field research from fifty years ago did not offer much possibility of cross-examining any relevant informant on the subject. Even then his findings on ards and on ploughing were based on 'autopsies and oral information' (Orel 1955, 31; 1961, 15), and not much has been added to them since. Some further data on ploughing implements and techniques in various Slovenian regions, recorded during systematic surveys by so-called 'Orel's Teams' between 1948 and 1961, were found in the general work *Peasant Economy in Slovenia* by Marija Makarovič (1978); and some other relevant information was introduced by Angelos Baš in the exhibition catalogue *Ard and Plough* (1968) and in his book *Slovenian Folk Tradition* (1980).

On the Two Fundamental Ploughing Techniques in Slovenia

In Slovenia, two ploughing methods were commonly applied: *conventional* and *reversible ploughing*. The first method was particularly common on flat land and carried out with a plough with a single plough body; the second method was common on steep fields and done with an ard or a double plough.

In conventional ploughing, one year the field was 'cast', that is the first furrow was ploughed along the right edge of the field, and the mouldboard then turned the furrow slice 'out' to the right side; then the plough was turned to the left at the end of the furrow and the second furrow was ploughed down along the opposite left edge of the field; then followed the third furrow ploughed along the inner or left side of the first furrow, and the fourth furrow ploughed down along the side of the second furrow, and so on until the last two furrows met along the centre line of the field and here left an open furrow (Fig. 2.10Aa, B). In the second year, however, the field was 'gathered', that is, the ploughman started with the first furrow along the centre line of the field, and since the mouldboard turns the furrow slice to the right, hereafter when the plough was turned to the right and the second furrow ploughed down close to the first one, the result was thus a 'gathering' of the two furrow slices. This continued with the third furrow lying to the left of furrow one, the fourth furrow running down along furrow two and so on until the edges of the field were reached and left with open furrows (Fig. 2.10Ab, C). (Smerdel 2008, 61; after Makarovič 1978 and Baš 1980). In this way the field surface stayed flat and even.

In reversible ploughing, the traditional way of using the ard, but also done with a double plough, the ploughman started on one side of the lower edge of the field, proceeded all through the field (turning the furrow slice down slope) and on the other side turned the implement to plough the second furrow in the reverse direction (again turning the

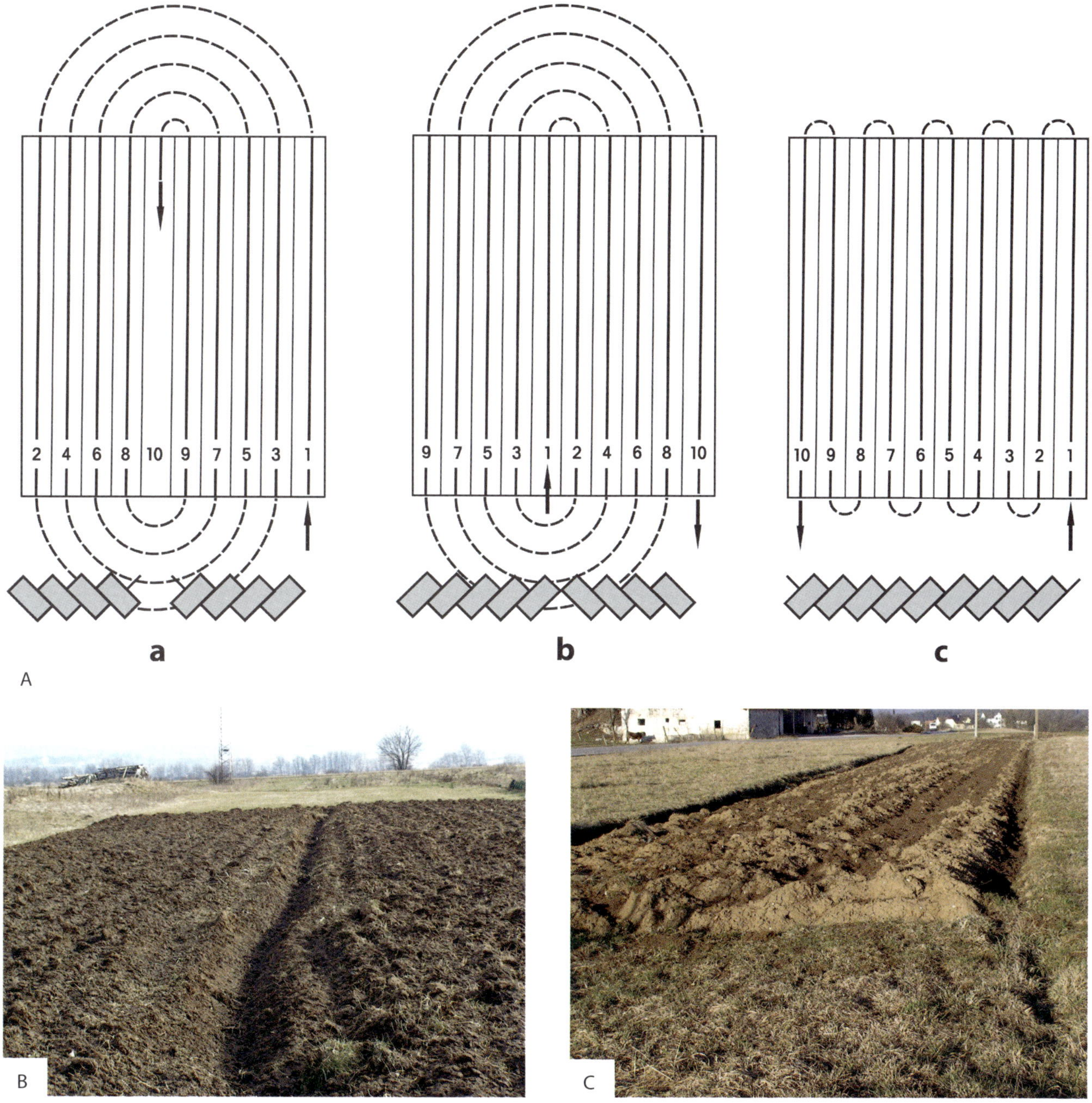

Fig. 2.10. A) Drawings of the two basic ploughing methods, of conventional and reversible ploughing (Smerdel 2008, Fig. 57 and Errata). B, C) Casting and gathering methods of the conventional way of ploughing, still visible on the fields in the Pivka region for example, even when tractor ploughs are used. One year the field is ploughed so that the soil is 'cast', and the next year it is 'gathered' (Smerdel 2008, fig. 58a and b).

furrow slice down slope) and so on until he reached the upper edge of the field. In one direction the right plough body was in action and in the reverse direction the left one. Namely, at the end of each furrow the ploughman had to turn (by gripping the central and another side handle) the whole plough body 90º (Figs. 2.10Ac and 2.13D).

On Ploughing with Five Selected Specimens of Ards and Ploughs

The first four ards and ploughs from the Slovenian museum collection were chosen to illustrate the description above, as they belong to the common types of pre-industrial Slovenian ploughing implements, and because their acquisition records preserve some oral commentary on ploughing with them. Only a few comments are added. The fifth, the 'mountain plough' still in use, presents an example of invention and reveals an individual ploughing technique.

The Symmetrical Rectangular Wheel Ard with a Ristle (Fig. 2.11)

The so-called *špíčmoh* (after *Spitzmaus*, in German a shrew – because of its share's conical shape) from Zgornje Jezersko (Carinthia region; Fig. 2.4), is constructed of four principal parts: sole, beam, stilt and sheath (that form a rectangular frame), all made of deciduous hardwood (beam and sole of white beech, sheath and stilt of beech and mould strokers of elm). The sole, the bottom horizontal part, slid across the soil and penetrated into it. The stilt has a hole in which the plough beam is mounted. A special type of wedge is set into the hole and was used to lower or raise the plough beam to adjust the ploughing depth (for which also the sheath, the link between plough beam and sole was used). A symmetrical ploughing iron, the share is mounted both into the sole and the mould strokers (or ears), which are made of a single piece of wood – reinforced with pieces of iron or sheet metal to prevent excessive wear (Fig. 2.11B). The ard's beam is attached to the forecarriage by the draught chain. The ristle (Fig. 2.11C) is this ard's separate component, its accompanying tool that functioned as a separate coulter, and was used to cut up the soil in ploughing. Its principal parts are the yoke beam (reinforced with an iron ring joint), stilt and ristle (70 cm long). The old inhabitants of Jezersko mostly used this plough on flat or gently sloping terrain, and this explains its heaviness and large size. It was drawn by an ox team. The ard always trailed the ristle (Fig. 2.11D) opening up the furrow, before the ard's share afterwards loosened the furrow soil in the bottom. Such an ard was used for ploughing in Jezersko until the final decades of the 19th century. It was partly used later, but its function had changed into a potato ridging and ploughing implement. (Smerdel 2008, 130–134).

More on ploughing:

> 'The ploughman ploughed with the ard, and the 'cutter' drove the ristle. The ploughman usually was the farmer himself or the main farmhand, the 'cutter' the 'second farmhand' or sometimes a woman. The oxen at the ard and ristle were led by herdsmen. [...] beside the ploughman, the cutter and two herdsmen, there was a fifth man with a hoe. [...] unploughed strips of soil, which the ard had merely scratched, were turned over with a hoe to complete the ploughman's work. [...] The first to go on the field was the cutter, who cut up the soil (furrow) with the ristle. (Hard soil, ley and fallow land was first cut before using the ard.) He was followed immediately (at a distance of 4 to 7 m) by the ploughman with the ard' (Orel 1955, 51–55).

The Symmetrical Rectangular Wheel Ard with a Coulter (Fig. 2.12) from Mlake near Muta (Carinthia region; Fig. 2.4)

It is also constructed of the four principal parts made of deciduous hardwood (the ard beam, sole and stilt are beech, the sheath ash) that form a rectangular frame. Its ard beam passes through a wedged gap in the stilt at the back (for ploughing depth regulation). Through the beam passes the coulter (attached instead of a separate ristle as a forerunner of the ard). The sole is rounded and thicker at the centre; a symmetrical share is set into it. The heel of the sole is metal-reinforced, as well as the long mould strokers (bent upwards and set into the thicker part of the sole) (Fig. 2.12B). The stilt splits into two natural horns, the handles that probably made tilting it easier (Fig. 2.12A and C). The ard beam is attached to the forecarriage through a hole in the axle tree by a draught peg. This ard was still used in 1942, later they ploughed with a double iron plough (Smerdel 2008, 157–159).

On ploughing:

> 'They (the ploughman, the herdsman who led the oxen, and another man or woman with a hoe) started to plough at the lower end, [...] especially on uphill fields. All the furrows were turned to one side (one way or reversible ploughing; Fig. 2.10Ac), the furrow slices turned down when ploughing uphill and fell one upon the other into the trough [...]. Wider furrows were ploughed on flat land, narrow ones uphill. In general, narrow furrows were held to be better, because it was much easier to pass over such furrows with a harrow. [...] Deep or shallow ploughing mostly depended on the quality of the soil. A field with heavy soil was usually ploughed deeper, sandy soil shallower. [...] In turning over and crushing the furrow slices the ploughman added pressure on the mould-strokers with the stilt. [...] By tilting the stilt [...] always to the side to which the furrows turned over and fell, the ard tilted and together with it the rear ends of the mould-strokers and their ears to have a better grip on the slices and turn them over.' (Orel 1955, 51–55, 62).

The ploughman thus had to know how to steer the ard and be skillful at adjusting it to shallow and

A

B

C

Fig. 2.11. A) Rectangular wheel ard, *špičmoh* from Zgornje Jezersko. B) Detail of its symmetrical share and mould strokers, reinforced with pieces of iron or sheet metal to prevent excessive wear. C) Ristle, the *špičmoh's* separate component. D) Demonstration of ploughing with the Jezersko ard and its ristle as a separate ploughing implement (Smerdel 2008, cat. nos. 8 and 9, fig. 37).

D

Fig. 2.12. A) Rectangular wheel ard, *orálo* from Mlake. B) closer view of its sheath and mould strokers, bent upwards at the back and reinforced on the outside to prevent excessive wear (Smerdel 2008, cat. no. 23). C) Demonstration of ploughing with the ard from Mlake; the ploughman walked alongside the stilt, tilting it to the side to which the furrows turned over and fell. (Photo by F. Šarf, 1957)

deep ploughing. When ploughing uphill, he had to move alongside the stilt two furrows below the ard, watch out and keep it in the furrow, and also make sure that the bottom mould-stroker caught the furrow slices accurately with its ear and turned them over (only part of the furrow-slice, only half of it and simply crushed the other half). Such ploughing obviously was an asymmetrical use of an otherwise symmetrical implement (cf. Steensberg 1986, 138–139).

The Asymmetrical Double Wheel Plough (Fig. 2.13) from Dolina (Dol near Ljubljana; Fig. 2.4)

This plough with the popular name *baba* (woman) or *kuzva* (bitch), is made of hardwood (all wooden parts are beech) and has two plough bodies attached to the beam approx. 90° apart (Fig. 2.13B). In the centre of the plough beam, two coulters are mounted, then two sheaths, that are set into the sole. Each sole has a mounted one-sided share. The plough has three handles, all set into the plough beam. The central one ends right under the plough beam, the ends of the side handles are connected with the mouldboards (in some cases side handles were even part of mouldboards; see Fig. 2.13C). The plough was last used around 1907; to plough steep fields 'so that all furrows are thrown down'. In the 1950s, iron double ploughs were used instead (Smerdel 2008, 139–140). Ploughs of this type were applied for reversible ploughing (Fig. 2.10Ac).

Fig. 2.13. A) Double plough, *bába*, *kúzva* from Dolina. B) Closer view on its two plough bodies. C) Model of a double plough from Loka by Mengeš, common in the first half of 19th century (Smerdel 2008, cat. nos. 13 and 2). D) *Tonkovi* from Bukovo ploughing with their double plough for buckwheat. (Photo by B. Orel, 1954) When the left plough body was in action (turning furrows down slope to the left), the ploughman steered the implement with the central and the right side handle; in the reverse direction, when the right plough body was in action, he had to steer the implement with the central and the left side handle.

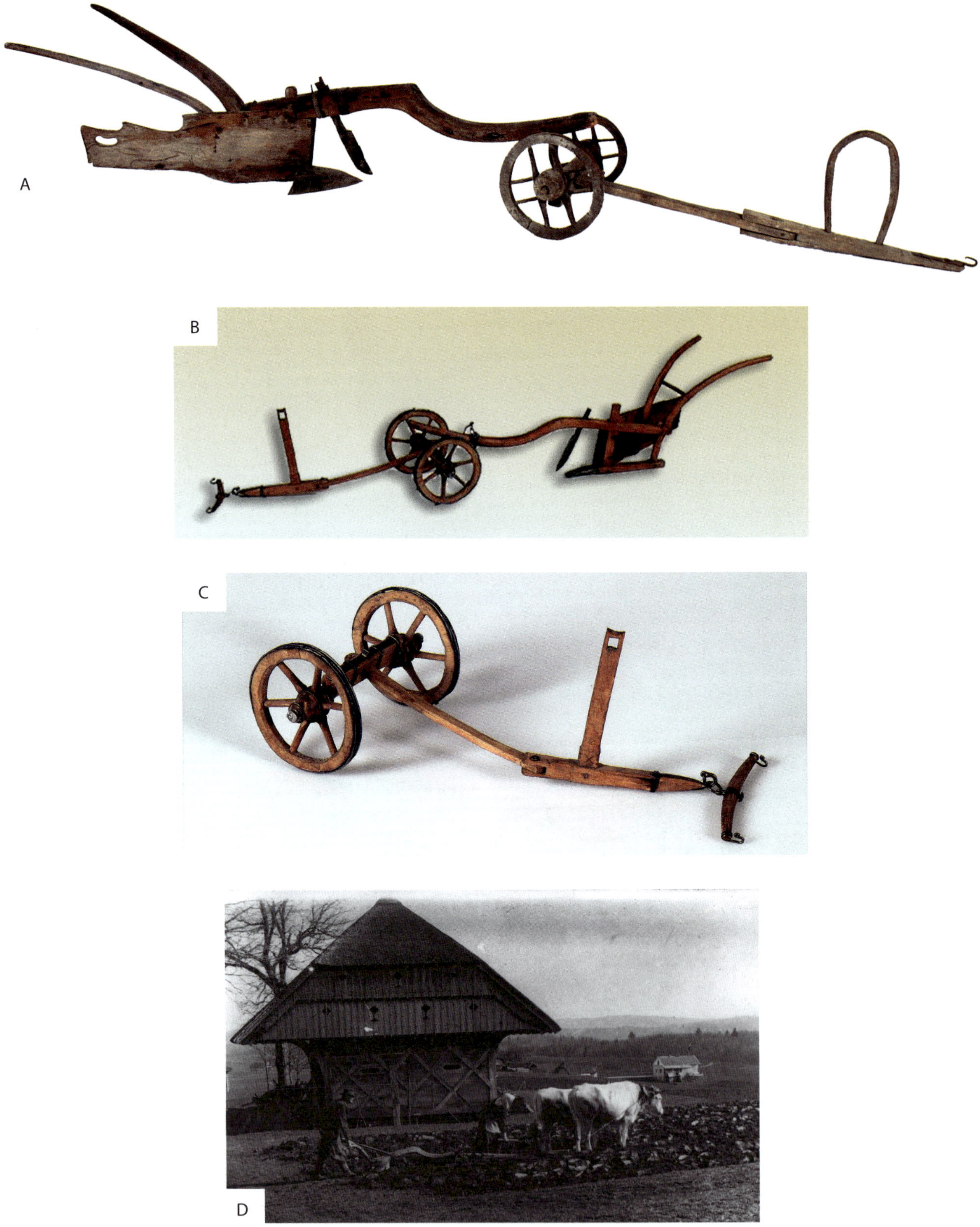

Fig. 2.14. A) Single wheel plough, *drevo* from Štrukljeva vas. B) Model of a single wheel plough from Zadvor near Sostro, acquired by the former Carniolan Provincial Museum in the 1830s (among models of ploughing implements, common in the first half of the 19th century). C) Forecarriage of the same plough model, with a draught beam consisting of two parts – tongue and tongs (common only with single ploughs from the arch stretching from Bela krajina to the Ljubljana environs and on to Notranjsko). A handle of various forms is fixed to the tongues and the ploughman's helper tightens or loosens the tongues to adjust the furrow (Smerdel 2008, 116; cat. nos. 13 and 2). D) Conventional ploughing, keeping the furrow straight with tongues. Dolenjska, around 1916 (Smerdel 2008, fig. 55).

Fig. 2.15. A) Innovative 'mountain plough for steep fields', made by Kondi Kos from Marija Reka, below the large mountain farm by *Završnik*, where his neighbours still use it (Smerdel 2008, fig. 63). B) Kondi's mountain plough in action on a steep field, steered by master Završnik, the plough's 'pilot' (Photo by I. Smerdel, 2008). C) Drawing of the ploughing method with Kondi's plough (after field sketch by I. Smerdel, 2007).

The Asymmetrical Single Wheel Plough (Fig. 2.14) from Štrukljeva Vas (Notranjska region; Fig. 2.4)

It is made of wood (plough beam, sole and left handle of white beech, right handle of ash, mouldboard of spruce) and consists of an articulated plough beam, into which the coulter is set. The sheath passes vertically through the beam; its lower end is set into the sole, which has a share mounted at its front end. The mouldboard (with a cut-out handle) is fixed to the outside of the right handle and to the sheath; it was originally metal-reinforced. The plough beam is metal-reinforced at the back end and on both sides of the inserted coulter. The heel of the sole and the left handle are also metal-reinforced. The plough's forecarriage has two cross-spoke wheels, of which the right (running in the opened furrow) is bigger than the left, which keeps the axle horizontal. The plough beam is fixed to the forecarriage with a twisted hazel shoot (a collar) as a security release. An extension (called 'tongue') is set into the axle tree at a right angle. The front extension (called 'tongs') is movable to about 240°. A wooden, curved pole is set into it vertically to let the ploughman's aid (who tightens or loosens the tongs) adjust the furrow (Fig. 2.14A–D). The implement was made around 1895 in Štrukljeva vas and was used until 1940 (Smerdel 2008, 176–178). Ploughs of this type were used for conventional ploughing (Fig. 2.10Aa, Ab, B, C).

The Asymmetrical Two-Furrow Iron Plough (Fig. 2.15)

This is an individual invention, the so-called 'mountain plough', made by Kondi Kos (born in 1934) on an isolated farm in Marija Reka above Prebold (Styria region). The inventor is a farmer and was a fitter at the Trbovlje Mine (Fig. 2.4) from the 1960s onwards. He started thinking about making his own special plough to ease his part-time farm work on the steep mountain fields of Marija Reka. (Fig. 2.15A). He built it all by himself without a construction plan, improving the prototype by trial and error. It had two shares and two mouldboards (two left ones and two right ones) and could plough two furrows at a time, each 40 cm wide. His invention made it possible to plough 35 cm deep, which was necessary for potatoes and maize, but could not be achieved with the old double wheel ploughs pulled by teams of oxen. After he perfected his plough in the late 1960s, he and his wife lent it to their neighbours and other local farmers. This 'mountain plough' made ploughing much easier and increased yields by fifty percent (Smerdel 2008, 99).

In ploughing with Kondi's invention, a tractor was standing on the uphill right edge of a field, equipped with a winch, and pulled the plough together with its 'pilot' (the sitting ploughman) up the field with a rope-pulley. Strong wooden poles, inserted 80 cm into the ground, were set every 3 m at the upper edge of the field. (Fig. 2.15C) Three times two furrows were ploughed 'on one pole' (the 'pilot' alone was able to move the plough for 80 cm with a steering-handle). Then the rope-pulley had to be moved to the next pole. Going downhill, the ploughman lifted the plough (both shares and mouldboards) with a handle, and while the winch let the plough go down, he walked alongside (or in front of it) on the unploughed ground. (Fig. 2.15B) As Kondi's plough had two left shares and mouldboards and two right shares and mouldboards, they ploughed three years with the left ones and then the following three years with the right ones – 'to come back'. After roads were constructed on the upper edges of some fields the farmers do not need the poles and the winch any more. All they have to do is to move the tractor.

Can Tools and People Tell Us Enough?

The descriptions of the selected ploughing implements and the associated oral information demonstrate several facts. The details of the environment (slopes and mountain fields, flatland fields) affect the choice of plough type and the ploughing technique. Ards and ploughs were locally made of hardwood and their individual parts had to be reinforced according to their function, for example iron fittings on the exposed parts to reduce wear. The choice of traction animals related also to ploughs and techniques, as did the social context of ploughing (who ploughed, who drove the oxen, and who dug up the unploughed parts of a field).

The selected implements could also initiate a provocative discussion concerning introduction and acceptance or rejection of innovation such as the popular double wheel plough, which was already recorded in the north of Slovene Styria in the 18th century (Bratanić 1952, 210; after Koren 1950), which has indeed been considered an important advance in some Slovenian areas (Smerdel 2008, 98), and which – in its iron variant – in some places is still applied on sloped fields. And then, there is such an innovation as the two-furrow iron plough from Marija Reka, an individual invention, which remained in use only in its narrow local frame. In the first case, Orel also recognised during his research, for example, that more advanced tools are not always the most adequate choice. 'It is interesting to note that, even in places where they already had a double plough, people still liked to use the ard, especially when the oxen were not really strong enough for the plough'. In the opinion of older people 'the advantage of the ard was that the ploughed furrows were softer,' better for the harrow, and fields ploughed with an ard favoured the growth of the grain (Orel 1955, 52–58; see Anderson, this chapter). In the second case, a mixture of unfavourable circumstances prevented the spread of Kondi's two-furrow 'mountain plough for steep fields', for which he was even awarded the title 'Innovator of the Year' in 1996 by the mayor of his local municipality. In those days, he had over one hundred potential buyers but no available and proper producer at hand. Nowadays, the buyers are gone, as cultivation on the steep slopes diminished. Inventions always needed to fit a particular situation...

Nevertheless, finally an answer to the concluding question should follow: Can tools and people tell us enough? – When a researcher is unable to observe a working process such as the ploughing operation and its technique, and must rely only on seeing individual implements and on reading records with more or less reliable explanations from people who once worked with them, the quality of the information depends largely on the questions asked by previous researchers. Although recorded memories may be extremely rich in contextual information, the modern researcher's findings depend partly on his or her capability to interpret them. Although analysing and interpreting all the relevant documentation and material evidence can help to understand and explain a ploughing technique, it is still far from the best way to document the process. Experiments, such as described in the next contribution, could be used to improve our understanding of surviving ploughing implements.

2.6. AN EXPERIMENTAL APPROACH TO MEDIEVAL CULTIVATION: THE DANISH WHEEL PLOUGH AND TILLAGE PRACTICE

Grith Lerche

Tools and implements are made with a functional purpose. They extend our limbs and strength and by the skill in their handling enable us to execute desired work to attain our goals. These may include operations in sequences which need logistic planning. Cultivation and ploughing are such work processes. Different agricultural systems relied on the experience, knowledge and skill developed by farmers to cope with environmental challenges. What is left to us are the remnants of their ancient agricultural tools and traces of the cultivation systems as fossil fields. These remnants are evidence of previous agrarian technology, mute witnesses to the work processes of farmers that are seldom mentioned in written sources. Experiments are one way to test conclusions and interpretations of the appearance and functions of a tool. In particular, traces of use and wear on ancient tools are a focus of many imitative experiments. By executing *controlled experiments,* the researchers also get operational data and measurements on subjects they hitherto had mere guesses and hypotheses about, for example, how a working process could be performed, the duration of an action and the time it took to develop certain traces of use or wear.

Why Reconstruct Medieval Wheel Plough Cultivation in Ridges and Furrows?

In the 1970s, for some years my research was concentrated on the history of agricultural implements and field structures. I had studied ploughing with the ard during field work in the Middle East, filmed ploughing with swing ploughs in Denmark and examined Danish archaeological finds of plough irons and important parts of wheel ploughs dated by ^{14}C to the medieval period (Lerche and Steensberg 1983; Lerche 1970; 1994; 1995). These pebble-studded plough parts urged me to study their traces of use and hard wear. So far they are unique as the only finds of wooden soles of a medieval plough in Europe (Fig. 2.16). Fossil fields are preserved in undisturbed land as long, narrow, parallel, more or less ridged strips. I had investigated some, but ridged fields were mostly unrecorded and seldom dated, usually interpreted as medieval and early modern fields. Their wavy profile has been interpreted as the result of the specific ploughing pattern when using a wheel plough for a long time. I decided to test whether a trustworthy replica of a medieval plough could be made with wheels and irons, and whether it would plough satisfactorily. Fig. 2.17 shows our experimental plough. Could it turn the furrow slice well, and would there remain traces of wear comparable to those on the archaeological finds? How long would it take to get measurable wear? And when executed in accordance with sources and earlier interpretation, could the ploughing pattern alone create ridged strips, and how soon would the wavy contours known from the fossil fields be achieved (Lerche 1986)? I had the opportunity to try to answer these questions in a long-term ploughing experiment carried out from 1979 to 1982. The main results of the experiment are summarised

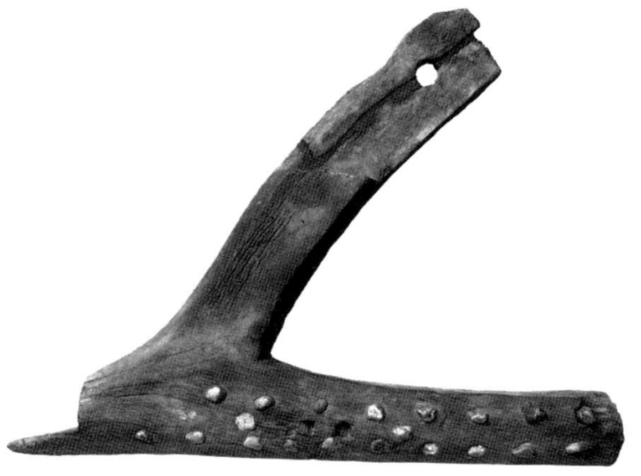

Fig. 2.16. The bog find of the Andbjerg plough. ^{14}C date calibrated to 1450 CE. It is a pebble-studded bifurcate of beech used as the sheath/sole of a plough. It was the model for the replica sole/sheath. Photo. National Museum of Denmark. (Lerche 1994, fig. 3.13).

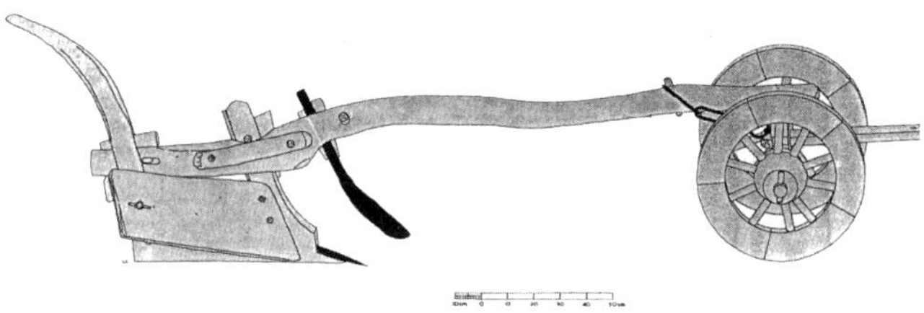

Fig. 2.17. The replica of a medieval wheel plough made based on finds. Here in its unused stage ready for the ploughing experiment. Measured by Mogens Brahde and Grith Lerche.

below (details in Lerche 1994; 1986; 1991; 2001) in tables along with my remarks. I focus on the most interesting observations for considering new interpretations of how original ploughs might have been used. This elaborate documentation of simple, ancient agrarian activities is the scientific way to obtain factual data and to test hypotheses attempting to understand the creative ways humans have made decisions to survive. We can thereby rediscover some lost knowledge, to understand some of the conditions and reasons for choices that were logical and rational to people in their time.

The Experiment, Main Results and Data

The ploughing was carried out during five seasons on a 2.5 ha field which was ploughed as 15 narrow strips, each 12 m wide. Every ploughed furrow was counted. Due to the width of the furrows, the numbers per strip differed and thus the metres ploughed, but every metre ploughed had an impact on the plough parts as traces of wear. Fig. 2.18 shows how much was ploughed in each season and in each strip. The third season was shortened, because the crucial part of the plough – the sheath/sole made of a bifurcated branch of oak – broke when it hit a boulder and the horses pulled too strongly. Oxen might have stopped. The same happened in the fifth season, which put an end to the test ploughings (Fig. 2.19A). But these accidents provided the opportunity to compare development of wear on two different plough soles. In Fig. 2.20 the result is summarised in square metres, in metres and number of furrows per season. These data were the basis for analysis of the development of the traces of use and measured wear on the exposed plough parts (Figs. 2.19B–F) and for the effect on the profile of the field (Figs. 2.21A–D).

A wheel plough is a much more complex implement than an ard (Fig. 2.17). The plough proper is a square framed wooden construction with irons and wheels,

Strip no.	1	2	3	4	5	6	7	8	9	10	11	12	13	14	15	
Strip length [m]	111	112	114	115	117	120	123	126	129	132	134	136	137	138	138	
Seasons:																*Length ploughed:*
0, 1979	3774	672														4446 m
1, 1979	4329	5208	6270	4945	6201	4140	6396	6048	5934	6072	6298	6256	6028	/521	5451	87097 m
2, 1980	4218	4256	5472	4485	4680	5040	4920	4284	4644	5280	5360	5440	4658	4968	5520	73225 m
3, 1980	4107	5208	6270	5175	5265	4800	-	-	-	-	-	-	-	-	-	30825 m
4 ,1981	4218	3808	4560	5175	4095	5040	6150	5544	5934	-	-	-	-	-	-	44524 m
5, 1981	4329	4032	4332	4140	4914	4320	4920	5544	5676	3168	1608	-	-	-	-	46983 m
Total																287100 m

Fig. 2.18. Metres ploughed in the strips in spring and autumn seasons 1979–1981. Each strip measured 12 metres in width.

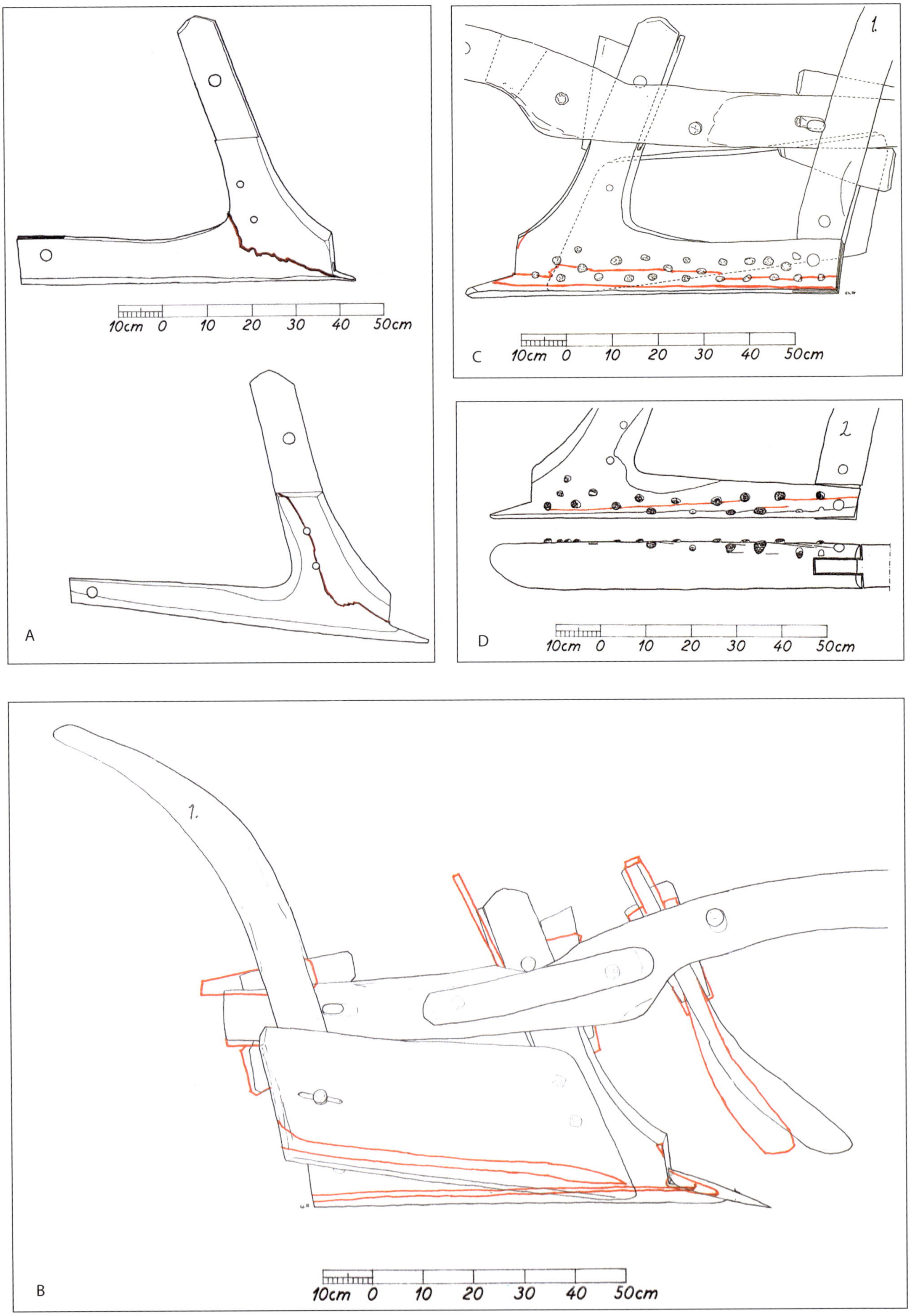
10cm 0 10 20 30 40 50cm
A
1.
C
10cm 0 10 20 30 40 50cm
2
D
10cm 0 10 20 30 40 50cm
1.
B
10cm 0 10 20 30 40 50cm

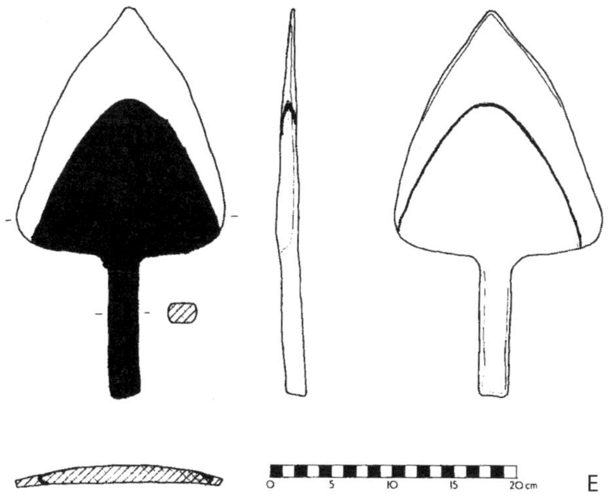

E

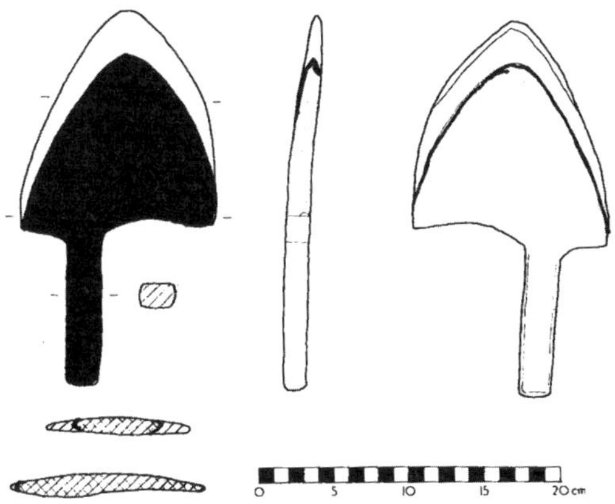

F

Fig. 2.19. A) The two worn and broken sheath/soles of the test wheel plough. Breaks are along the red lines. Sole 1 is at the top, sole 2 beneath. (Lerche 1994, fig. 8.18, 8.30). B) The drawing shown in black is of the main parts of the plough in position before any use and in red the recorded changes of the shape and size of the wooden parts and irons after use and wear for 3 seasons. The reduced size of sole 1 and of the under edge of the mouldboard is evident as well as the diminished length of the share and the coulter. (Lerche 1994, fig. 8.14). C) Drawing C, shows in black the pebble-studded left side of the unused plough sheath/sole **1**. In red is the final stage of the sole after 3 seasons of ploughing. In spite of the inserted pebbles the wear is considerable and 4 pebbles have fallen out. The two red lines indicate the new position of the underside of the sole and of the left edge which now has reached the lower row of pebbles. The very point, which supports the share, is now nearly not exiting and this is crucial to the lifetime of the sole. D) Drawing D is of sheath/sole **2**, but shown in black in its already worn shape after the first season of ploughing, season 4. The left lower edge line has already crept up and touches the lower row of pebbles and two have fallen out and others are now not positioned at the side but more at the bottom of the sole. The red broken lines indicate the 'rounded' edge line after the second season of use, season 5. The critical spot for the lifetime of this sole is the joint between stilt and sole and when wear makes this unstable, and not so much the diminishing of the point. (Lerche 1994, figs. 4.12, 8.35 and 8.45). E, F) The unworn size of the share copies 1 and 3 before use in the experiments is seen in the outer black contour lines. The reduced size after use is seen as the black inner figures to the left, and in the drawings to the right the inner contours indicate the rear sides as worn. Share 1, was forged as a symmetrical share, share 3, as a true copy of a share find from Odense (Fig. 2.19 A–D, also see Lerche 1994, figs. 8.51 and 8.56).

asymmetrical due to its mouldboard. Oxen or horses provide traction power. Plough parts are exposed to different degrees of wear during the ploughing process and this gives the parts different life spans (Fig. 2.19A–F, Fig. 2.22A, B). On the same plough, a new sole could be combined with a 100-years-older beam and a 10-years-older stilt (Lerche 1994, 268–269). The experiment also showed how important the regulation of the ploughing depth and the shape and size of the sole and share were for the impact of wear on these parts (Figs. 2.19A–F). Depth regulation is performed by changing the position of the wheels underneath the beam as well as by changing the wedges in the stilt beam joint (Lerche 1994, 16–17). For the life-time of a sole, the size of its point is crucial. If it was worn down or damaged, a share could not be held in position to undercut the furrow and the farmer had to renew his plough sole. And if a share became worn down to a point where

Seasons for experiments	Furrows ploughed	Ploughed length in metres	Ploughed area in square metres
0, 1979	40	4446	1344
1, 1979	694	87097	22548
2, 1980	585	73225	22548
3, 1980	268	30825	8268
4 ,1981	374	44524	12804
5, 1981	391	46983	15996
Total	2352 furrows	287100 m	83508 m^2

Fig. 2.20. Season summaries of the ploughing experiment.

it did not protrude enough beyond the very point of the sole, the share had to be changed (Fig. 2.22A, B). This was the reason for discarding many of the medieval plough parts we now have as finds (Lerche

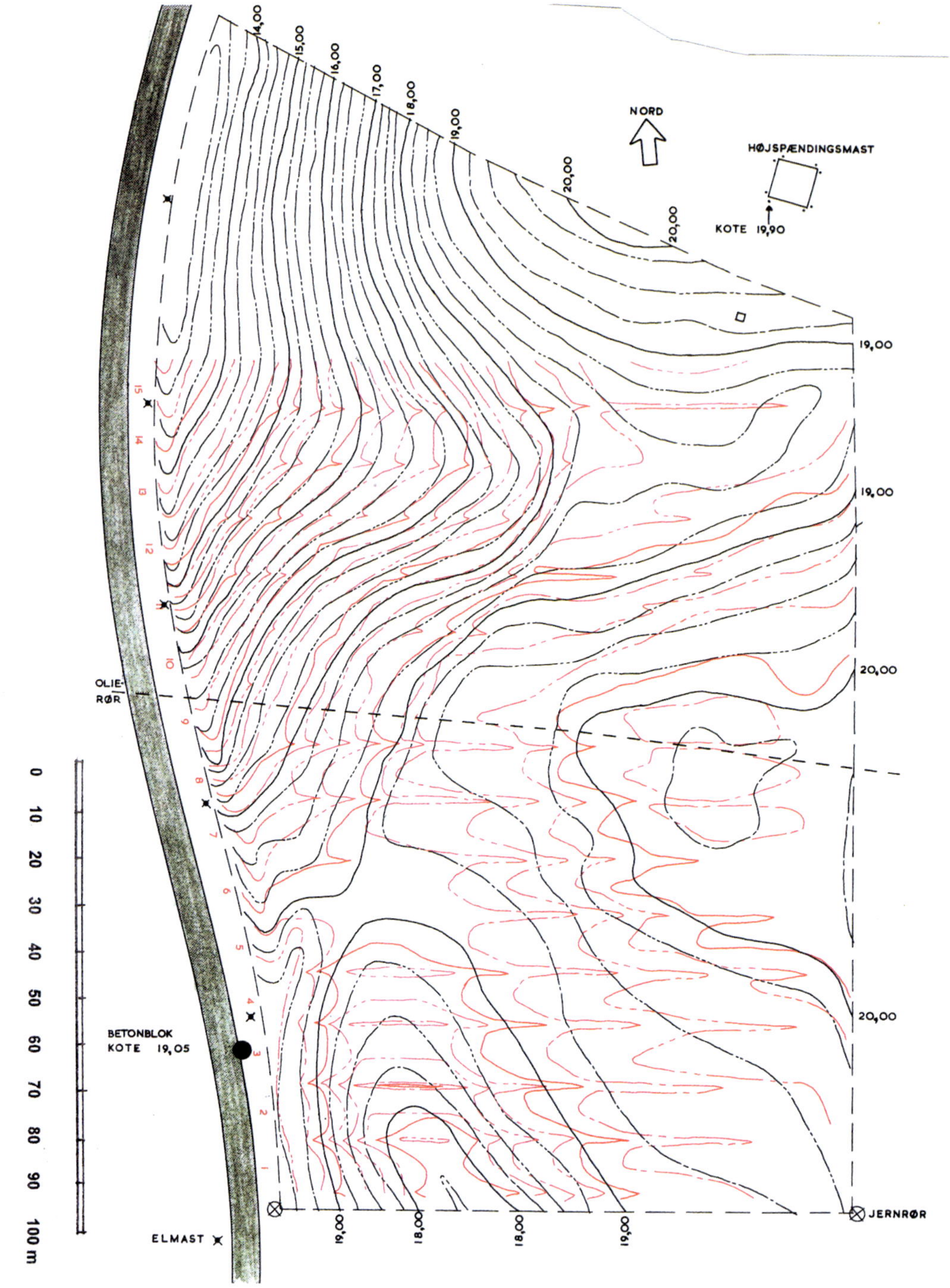

A

Fig. 2.21. A) The mapped field for the ploughing experiment. The contours in black indicate the situation before the tests began and in red how the position of these contours was changed by the impact of ploughing in ridges and furrows in 15 strips. The positions of strip ridges in No.1–9 ploughed 4–5 times are most obvious, whereas the rest are mostly indicated through the position of the open furrows (Lerche 1994, fig.5.25). B) The positions of the strip divisions were still visible at the surface of the old test field 27 years later, and the test field now resembles the appearance of some of the ancient fossil fields found in woods. (Photo G. Lerche 2008). C)The experimental field as measured by GPS 27 years after the last test ploughing stopped. The position of the open furrows is indicated by yellow lines. The position of the ridges of strips 1 to 11 was specially measured and is marked here with crosses in lines. Strip 1 is at the bottom, and the blue figure here following contour 19 m is the low area often flooded during winter time. Measured by Bo Jensen and digitalised by Lea Meistrup Larsen in the spring of 2008. D) In this 3D model of the surveyed test field after 27 years of lying fallow after the ploughing in strips had stopped, the ridges and furrows pattern of the 6 strips which were ploughed 5 times are most visible. The ridges turn up as orange areas. Made by Lea Meistrup Larsen based on Bo Jensen's measurements in 2008.

B

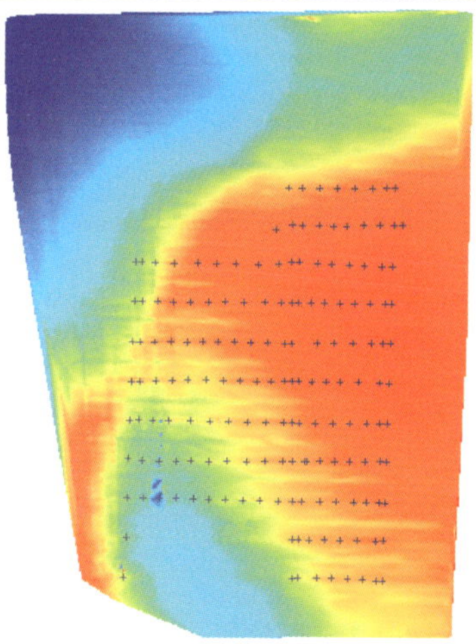

C

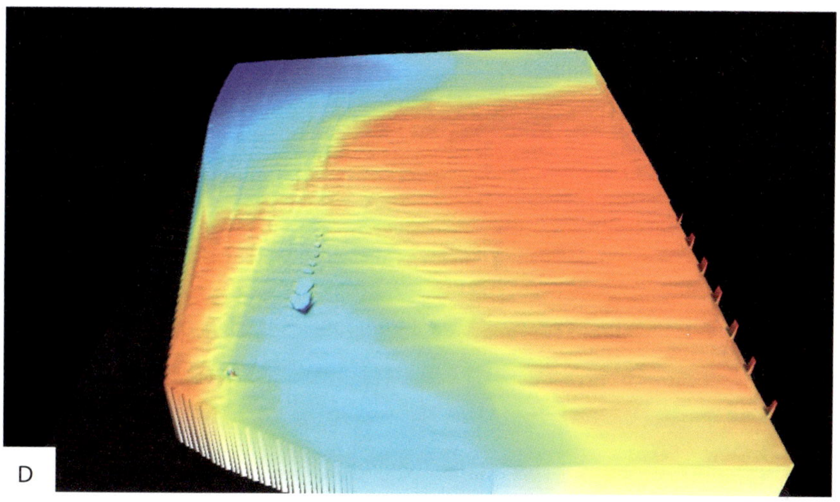

D

1994, 24–40, 208–233). But how long did it take to get the heavy wear which determined the use-life of a plough and its exchangeable parts, and what size was the area that could be ploughed with it?

The many kilometres of experimental ploughing affected the important parts of the plough in loss of material by wear (Figs. 2.23 and 2.24). The two soles were used for different lengths of ploughing, but the one less used was the one most worn (Fig. 2.19C and D). This is a hint to researchers not to generalise too much about where and how quickly wear occurs. Sole 2's endurance was explained by the fact that fresh oak was used, and the plough was regulated to go steadily and more on the heel. Five iron shares were used. No. 6 was of better iron (Lerche 1994, 54, 187) and did not get worn out. It was the only one used on plough sole 2, whereas 4 shares had to be put on plough sole 1 to last the 3 seasons. Shares 1–3 were changed when they were too worn to bite the soil. Share 1 was forged as a symmetrical share shaped to a length which the model, the Odense share, might have had before it got its worn shape (Fig. 2.19E, F). We documented how much iron was actually lost, worn away from the original length. One other thing affecting share

Fig. 2.22. A) The impact of wear at sole 1, at the coulter and the share is evident after having ploughed the first 525 furrows in the first season. The share will soon have to be exchanged, since it is not protruding enough ahead of the point of the sole. Fig. 2.19B shows a smooth worn underside of the sole. Soil adheres to the sheath front and between the pebbles. This slows down wear speed and has an impact on the position of wear marks. It is in spring 1980, during season 2 with share 2, a copy of the Nyborg share, attached. Fig. 17C shows the rear end after season 2. (Lerche 1994, figs. 6.27, 8.39a, 8.39b).

life time was the moment when the share no longer fitted into the worn sole (Fig. 2.22A and B). Shares 1 and 2 started with quite different lengths, but ended up at the same length when they were worn out, in spite of differences in the length they ploughed, as indicated in Fig. 2.23. Shares 3 and 4, copies of the same Odense share, had not much protruding length to wear before they reached the critical point. Their weight also ended closer to the others. The loss of iron from the shares and coulter was impressive, but there would still be enough iron left for more years of ploughing. The many tests showed that a wheel plough was constructed for shallow ploughing, to go steadily in stony and weed-infested soil, and it functioned best at furrow depths between 8 and 15 cm and widths between 24 and 29.5 cm (Lerche 1994, 106–112). Ploughing deeper increased the need for traction power too much. Average traction power varied from 135–170 kg in season 2 to 200–225 kg in season 3. At maximum, it reached 450 kg. This would have needed up to 6 strong oxen for traction (Lerche 1994, 93–96, 305).

Life Time and Costs in Equipment

Fig. 2.25 summarises the different use-life times of plough soles and shares calculated by length, area and numbers of furrows ploughed. The total area ploughed during the tests was 8.3 ha. Plough 1 had ploughed approximately 5.4 ha when it broke. Based on the wear pattern, I would estimate that its sole could have ploughed another 1.5 ha before the tip of the sole was worn out. Plough sole 2 could have ploughed another 1.5 ha before the wear had reached the nail in the joint between the sole and the stilt, a critical point for stability (see Fig. 2.19D). Thus, a total of 11–12 ha could have been ploughed with the given plough parts. However, five shares

Share	Length of use	Length of items, before/after	Reduced in length	Original width	Reduced in width	Weight before use	Weight after use	Loss of iron
No. 1	82 km 6 m	32 cm/24.2 cm	7.8 cm	17 cm	5.8 cm	2380 gr	1422 gr	-958 gr
No. 2	57 km 360 m	26.7 cm/24.3 cm	2.4 cm	14.5 cm	4.3 cm	1612 gr	1364 gr	-248 gr
No. 3	41 km 537 m	25.2 cm/22.2 cm	3 cm	13 cm	3.5 cm	1244 gr	1068 gr	-176 gr
No. 4	19 km 344 m	24 cm/23.2 cm	0.8 cm	12.6 cm	1 cm	1442 gr	1354 gr	- 88 gr
No. 6	91 km 507 m	31.5 cm/28.1 cm	3.4 cm	18 cm	5.3 cm	2850 gr	2386 gr	-435 gr#
Coulter	271 km 876 m	70.3 cm/63.5 cm	6.8 cm	-	-	6400 gr	5832 gr	-543 gr¤

Plough	Length of use	Length of items, before/after	Reduced in length	Original width	Reduced in width	Weight before use	Weight after use	Loss of wood
Sole 1	195 km 593 m	82 cm/76 cm	6 cm	-	-	-	-	1622 cm^3
Sole 2	91 km507 m	83 cm/81 cm	2 cm	-	-	-	-	1744 cm^3

Total loss was 464 gr. but 29 gr. iron was not lost due to wear but because a bit was cut off the shaft.
¤ Total loss of weight was 668 gr. but 125 gr. was not lost due to wear but because a bit of the heel had to be cut off.

Fig. 2.23. Impact of use during the ploughing experiment on the size and weight of the plough irons and the soles.

Shares coulter	Ploughed ha	Loss in gram	Gram per ha	Gram per km
Share 1	2.1	958	457	11.68
Share 2	1.8	248	138	4.32
Share 3	1.2	176	147	4.25
Share 4	≥ 0.4	88	≤ 220	4.55
Share 6	≥ 0.9	435	≤ 150	4.75
Coulter	7.9	543	69	1.99

Fig. 2.24. Loss of iron when ploughing, total in gram per hectare and kilometre.

Ploughed with	Furrows total	Metres total	Square metres total	Estimates in hectares
Plough 1	1587 ½	195593 m	54708 m^2	5.4 ha
Plough 2	765	91507 m	28800 m^2	2.9 ha
Both 1+2	2352 ½	287100 m	83508 m^2	8.3 ha
Share 1	676	82006 m	-	2.1 ha
Share 2	466	57360 m	-	1.8 ha
Share 3	300	41537 m	-	1.2 ha
Share 4	165	19344 m	-	0.4 ha + more
Share 6	765	91507 m	-	2.9 ha + more
Coulter	2229	271876 m	-	94.7 % of all

Fig. 2.25. Life time of plough parts in operation.

had already been in use for the 8.3 ha. For 11–12 ha, the estimate is that eight shares would have been involved, whereas the same coulter could have been used for at least another 10 ha. The economic costs in just the iron shares would thus have meant considerable expenses for a medieval farmer. The loss in iron by ploughing is thought-provoking (Fig. 2.24). But iron from two worn shares was raw material enough to get one new share forged at the blacksmith (Lerche 1994, 209, 213). For a year's cultivation, a medieval farmer might need an extra sole or sheath if one broke in the middle of a season. But breakage might not have been so likely if the farmer was using oxen which stopped for obstacles, whereas horses drew through.

The Ard in the Tool Box of the Medieval Farmer

Analysis of share finds make it evident that the arrow-shaped ones with very long tangs were not fit for the type of plough we have so far found. Another plough type or an ard must have also existed (Lerche 1994, 210–213). Preserved square-framed ards called 'crooks' were used in Zealand and Scania (Fig. 2.26) well into the 19th century alongside the plough proper, as a kind of one-toothed harrow, but until recently there have only been illustrations and no finds from the medieval period. However, the missing link, the medieval ard, did turn up in an excavation in Tårnby in 1994. This find was a bifurcated piece of beech made into a sheath/sole and dated to 1420 cal. CE. It had an oblique point with a penetrating hole for a symmetrical share with a long tang. It had no landside pebbles, no traces of a mouldboard, no asymmetrical wear, but the analysis showed it had been held in a slanting position to try to push the soil to the side, which caused it to break (Fig. 2.27A, B) (Lerche 2005, 311–322; Svart Kristiansen 2005).

Farm Size and the Yearly Cultivation

The new data from the experiment on the cost in worn tools to plough a certain area of land are a good basis for evaluating what was practical, possible and realistic to manage yearly for a farmer in 14th to 17th-century Denmark. Detailed information about arable land, numbers, width and length of the strips of the farms in Danish villages is found in the model books for Christian V's land register of 1688 (Frandsen 1983; Pedersen 1975). Used cautiously, its information can also indicate situations in older periods. The farmer of an average-sized farm in Zealand villages in 1682 would probably have had to plough in one season (autumn or spring) what the experiment achieved in ploughed hectares (8.3 ha) in five seasons (effective 120 hours). In the village of Hejninge (Fig. 2.26), an average farm size in 1682 was 24.1–28.8 ha. With an average of 90 strips distributed into the common three-field rotation system used for spring and winter crop and fallow, the farmer had 30 arable strips or 8–9.6 ha per village field. The length of strips was often 250–300 m or about twice as long as the test strips and the field size was often between 3150 and 4852 square metres (Rise and Steensberg 1951, 153–163; Lerche 1994, 63–64). In the village nearby, Store Valby (Fig. 2.26), the average size of a farm was 18–24 ha and the average arable area per farm in its three village fields was 5.4 ha, 5.8 ha and 7.4 ha, respectively (Steensberg 1974, 41–46).

The Ploughing Pattern and the Ridged Field Profile

Ploughing a narrow strip with a mouldboard plough started down the middle of the strip and by always turning to the right at the ends the two furrow

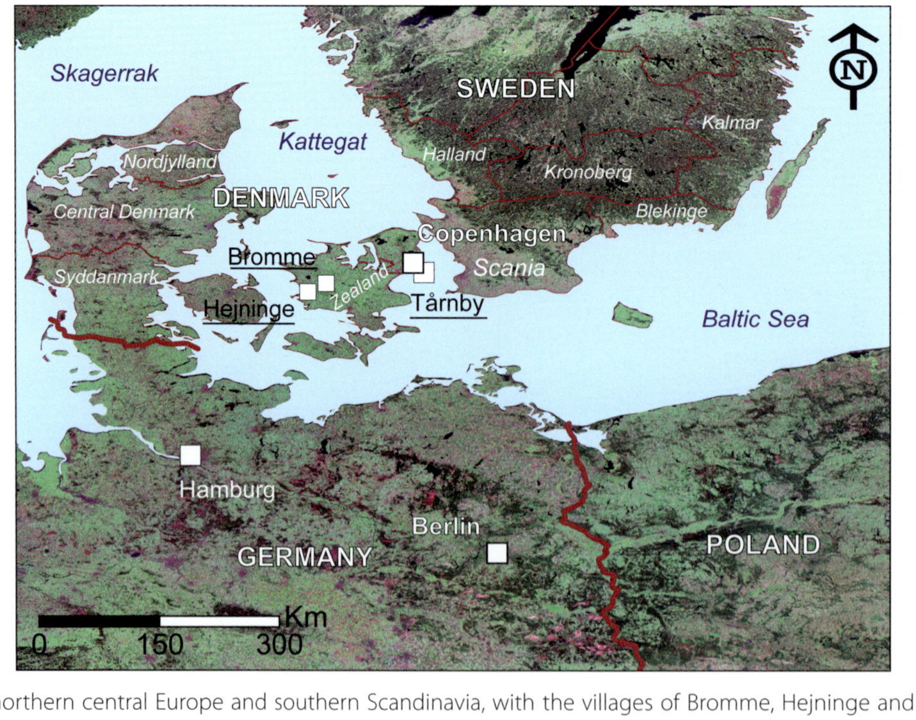

Fig. 2.26. Map of northern central Europe and southern Scandinavia, with the villages of Bromme, Hejninge and Tårnby, as well as the island of Zealand (Sjælland) and Scania (Skåne). Map: R. Lugon, J.-C. Loubier and A. Chevalier.

slices turned towards each other and little by little accumulated a ridge. This procedure went on to the border of the next strip which was ploughed in the same way and both strips ended with open furrows (Fig. 2.28A–D). In earlier periods the fertile mould was rather thin and shallow ploughing in this pattern was a method to accumulate mould in the middle of the strip, to use the open furrows as surface drains and to avoid destroying all plant roots and ploughing up raw soil. Underground pipe drains were not introduced until much later (Lerche 1994, 19–20). This ploughing pattern was used on all 15 test strips and, when repeated, slight ridged profiles divided by open furrows became visible to the naked eye at the surface of the field. After the test ploughings had stopped in 1982, the position of the strips appeared on the surveyed plan as notches and noses in the contours indicated in red (Fig. 2.21A). The contours in black show the situation of the field surface when surveyed before the experiment. Thus, the ploughing method could create a beginning wavy surface like those we find in the fossil fields.

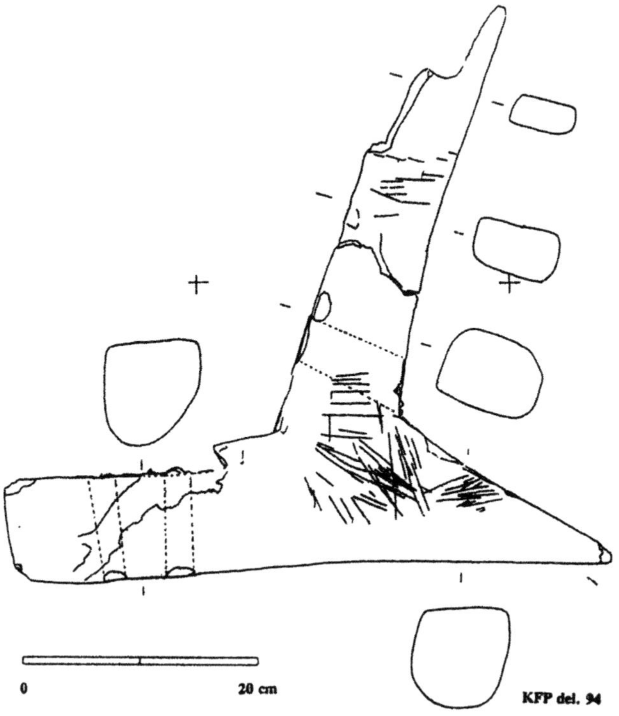

A

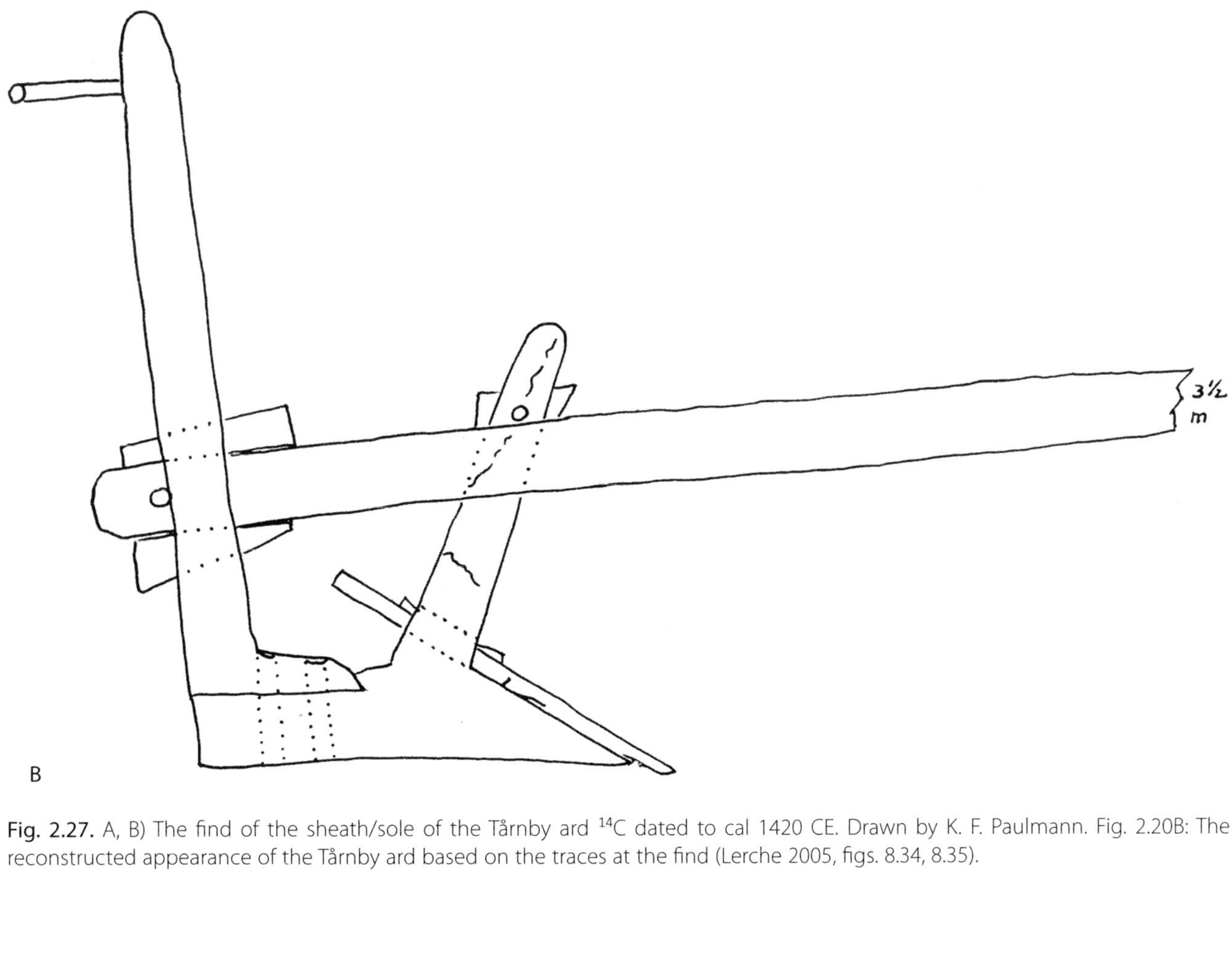

B

Fig. 2.27. A, B) The find of the sheath/sole of the Tårnby ard ^{14}C dated to cal 1420 CE. Drawn by K. F. Paulmann. Fig. 2.20B: The reconstructed appearance of the Tårnby ard based on the traces at the find (Lerche 2005, figs. 8.34, 8.35).

It appears from the test results that one could already, after only 4 to 5 ploughings, get both visible and measurable ridges and furrows, whereas when only ploughed twice the position of open furrows was the only thing to be seen by the naked eye. But still being visible decades or centuries later is another thing. However, the test field has never been cultivated, only grazed after the tests. In 2008, after 27 years as fallow the position of the strips and the ridged contours remain visible to the naked eye (Fig. 2.21 b) and the field has been surveyed again this time with modern GPS methods (Figs. 2.21C and D). The position of the open furrows is clearly visible in yellow in the 3-D model of the GPS survey. The documentation after this long interval is valuable to understand the problem of age in relation to height of fossil fields. The test field today could resemble an ancient situation where strips and furlongs in remote areas of the village had been left for years to regenerate fertility, to be cultivated again for only two years with buckwheat and then left to rest again, as the land register of 1682 indicates was true for many furlongs of the Zealand village Bromme (Fig. 2.26). Today, many fossil strips and furlongs exist in Bromme plantation, and the first large scale investigation and excavation of this kind of field system is now underway. The results from the experiment are being used in Bromme

Fig. 2.28. The functionality of the wheel plough was tested in different soil stages and also the need of draught power. A team of from two to four strong horses of the Jutland breed was tried. A, B) in a loose spring soil (spring 1980). When ploughing in the ridge field pattern one starts along the middle of a strip, and by turning to the right all the time until the border to the open furrow of the adjacent strip the wavy profile and accumulation of a ridge will appear. C) The advantage over the ard to have both a vertical cutting coulter and a mouldboard to turn over the furrow even in a grass and weed covered field is demonstrated in season 5 in 1981. See also Chapter 2.5 by I. Smerdel. D) Ploughing is close to ending up with open furrows between two strips. Spring 1980. (Photos by G. Lerche).

as a reference, and they also play a role in a Swedish experiment testing medieval ard shares at Järvafället.

Conclusion

The results obtained by the experimental approach to medieval plough cultivation have shed new light on the technical capacity, limits and advantages of the proper plough compared to the ard in the ability to break up virgin land and till land after long fallow periods. The plough could enable peasants as well as lords to plan when to increase the area of cultivation for more food production, the basis for the rapid development of urban cultures in Europe at the dawn of the first millennium. The old peasants had knowledge of the environment and worked with its diversity. The new data and experience allow us to discuss what made the old open-field system sustainable for hundreds of years.

The extensive open-field village system with the wheel plough as a well-adapted but costly implement was replaced by an intensive market-oriented farming system on individual farms at the dawn of industrialisation in Europe. Some of the more important interrelated elements that changed thinking and farming practice included attempts to optimise the fertility of soils on the newly re-allotted fields in Denmark by using the manure from animals, the majority of which were now stalled for the summer. This also minimised the need for grazing areas, and instead crops could be grown on these fields which were integrated into a new cereal and root crop rotation system. To minimise the costs in draught-animals and manpower, the introduction of new ploughs run by only one driver and two horses was one important parameter. These more efficient ploughs were swing ploughs with mathematically designed mouldboards, able to cope with the new demands to plough deeper and to turn over the tough weed and grass-infested furrows in autumn.

The data and experiences from this experiment have consequences for future considerations and analysis of how medieval and post-medieval farming could have been carried out, and the impact ploughing practices had on the resources and the landscape (Jensen 2010, 19, 209, 258). In periods of change, knowledge of the past is a powerful pool of information for understanding interactions and making decisions for the future. And in the future, it is not wise to be ignorant or act irresponsibly with our environment, when we know the consequences.

2.7. CONCLUDING REMARKS

Inja Smerdel

To conclude this chapter on the agricultural process from land clearance and preparation to sowing – in which, as already mentioned in the introduction, we do not tackle the whole chain of these operations but set forth only some examples discussing the preparatory steps of cultivation, ploughings, and various related tools – only a few ending remarks can be articulated, pithily summarising some of our findings. They may be written with different words, they are based on differing scientific approaches and methodologies, but these findings nevertheless – among other things – rather harmoniously show that the role of place (the impact of the natural environment, the soil, the geomorphology, the climatic conditions) can indeed be considered the most influential setting for human action, inventiveness and survival strategies.

Sabine Beckmann for example succeeds in interpreting the stone walls of Crete, the *perivoloi*, as enclosure walls (rather than defensive structures) – evidence of a successful strategy of mixed agriculture and herding in the Bronze Age, analogous to ethnographically documented practices in recent times, and (both today and in ancient times) strongly related to a windy mountain area in a dry climate. With her Marija Reka case study, Inja Smerdel adds a crumb to the assumption that land clearing by fire in much of Europe was intended first and foremost to produce open areas for cattle grazing, and explains how fire-clearance husbandry was a response to particular problems of the Marija Reka farmers, inhabitants of a remote mountain settlement, dependent on cultivation of extremely steep sloped fields. She also reveals that the intrinsic significance of swiddens or *velne* was multiple to the Marija Reka people: besides providing them and their cattle with needed additional food, the *velne* also had a strong impact on the community's social life and were indeed an important element of their environment and their daily life. Patricia Anderson concludes her explanation on working with the ard in present-day Tunisia by describing how an ancient and traditional tool continues in use alongside modern machinery because it is versatile and effective in marginal conditions on sloped and mountain fields (where tractors are not); how it is efficient and less expensive for parts of the tilling process even when tractors are used, and also, how it plays a key role in maintaining a certain agricultural cycle and traditional grain varieties. In her other contribution, on the ploughing techniques in Slovenia, Inja Smerdel shows how the descriptions of the five selected ploughing implements and the associated oral information can demonstrate several facts: for example, that the details of the environment affect the choice of plough type and the ploughing technique; and also, that some implements – for example, the ards described – with their particular structural parts, provide helpful testimony to the development of the ard in certain European regions, due to mountainous terrain, soil, and the inventiveness of its makers (thus charmingly showing that farmers did not make their tools according to our definitions). She further stresses that although analysing and interpreting all the relevant documentation and material evidence can help to understand and explain a ploughing technique, it is still far from the best way to document the process. Thus she introduces the debate on the experimental approach, which is then meticulously described and interpreted by Grith Lerche, concerning cultivation with a medieval Danish wheel plough. The results obtained by this method have shed new light on the technical capacity, limitations and advantages of the proper plough compared to the ard in the ability to break up virgin land and till land after long fallow periods. These results also clearly show that the old peasants had knowledge of the environment and worked with its diversity.

If we further follow the links on the chain of agricultural operations we are aware that some of them are missing in our chapter (applying manure to foster fertility, harrowing, sowing...). Thus – after summarising what we learned from all these case studies – let us briefly point out some necessities and possibilities of future research. One of the necessities should definitely be to initiate a discussion on the so far relatively neglected manual tilling operations and techniques, on hand tools (what made them sustainable – persistent and yet resilient for hundreds of years) and through them, on the concept of endogenous development. Another necessity seems to be the study of various ways of fostering fertility; not only concerning knowledge about all the variety of manures, their interaction with diverse soils, and their impact on different yields – but also on the importance and the diversity of fallowing (as an example see Morlon and Sigaut, 2008). As some other options, more research on agricultural diversity could be pointed out; or, for example, the study of fossil fields and other anthropogenic processes in archaeological landscapes, by applying modern research methods such as LiDar (Light Detection and Ranging) or ALA (Airborne Laser Altimetry). Those may in some cases enable us to reach so far unknown or unexpected historical findings about individual agricultural operations (for example, the recent discovery of clear traces of ridges and furrows in Slovenia).

But now, here comes the next link in the chain of agricultural operations – tending the crops. And if it is man's response to the natural environment (the role of place) that seems to be central in the contributions presented above, it is man himself that now becomes crucial... With a variety of physical, material actions to prevent damage to the crops; with all his knowledge about a specific natural environment – its animals and weather conditions; with his superstitions, religious beliefs, and necessary legal deeds...

Chapter Notes

1 An ard, or scratch-plough, differs from the commonly known turnplough (or mouldboard plough) by the way of its operation: it merely traces a furrow into the ground instead of inverting the soil. Its construction is also much simpler. It is a symmetrical ploughing implement, while a plough is an asymmetrical tool.

2 Catalogues of seignorial rights and possessions.

3 See http://www.youtube.com/watch?v=B3z8HVw09pw for a film clip by the author, showing the assemblage of this ard in Beja, Tunisia.

4 See https://vimeo.com/88640257 for a short film showing the use of the 2-animal ard, on a sloped field, here pulled by mules in Met Algua, Tunisia.

Chapter Bibliography

Acosta Naranjo, R. (ed.) (2001) *Memoria de la tierra, campos de memoria, los agroecosistemas tradicionales de Tentudía.* Centro de Desarrollo Comarcal de Tentudía, Tentudía.

Allbaugh, L. G. (1953). *Crete, a case study of an underdeveloped area.* Princeton, Princeton University Press.

Anderson, P. C. (2006) Non-mechanised processing and storage of cereals, grasses and pulses used for fodder, fuel, food and crafts: Examples from N. Tunisia, Atlas Region, Northwestern Tell. In Z. F. Ertuğ (ed.) *Proceedings of the 4th International Congress of Ethnobotany (ICEB 2005),* ISBN: 975-807-153-X , Ege Yaynlar, pp. 223–231.

Anderson, P. C. (2007) Le travail à l'araire aujourd'hui en Tunisie, le point de vue d'une ethnoarchéologue, In R. Bourrigaud and F. Sigaut (eds.) *Nous Labourons. Actes du Colloque, Techniques de travail de la terre, hier et aujourd'hui, ici et là-bas,* Nantes, Nozay, Chateaubriant, 25-29 octobre 2006. Centre d'Histoire du travail, Nantes, pp. 247–258.

Archaeological Visibility. In P. Halstead and C. Frederick (eds.) *Landscape and Land Use in Postglacial Greece.* Sheffield, pp. 79–94.

Baš, A. (1968) *Ralo in plug. Vodnik po razstavi.* Slovenski etnografski muzej, Ljubljana.

Baš, A. (1980) Poljedelstvo. In A. Baš (ed.) *Slovensko ljudsko izročilo.* Cankarjeva založba, Ljubljana, pp. 22–30.

Baš, F. (1954/1984) Pripombe k požigalništvu. *Slovenski etnograf* 6–7(1953–1954): 83–102. Reprinted in A. Baš (ed.) (1984) *Stavbe in gospodarstvo na slovenskem podeželju. Izbrani etnološki spisi.* Slovenska matica, Ljubljana, pp. 175–194.

Beckmann, S. (2012a) *Domesticating mountains in middle bronze Age Crete. Minoan agricultural landscaping in the Agios Nikolaos region.* PhD Thesis, University of Crete. Stable URL: http://phdtheses.ekt.gr/eadd/handle/10442/29129 (Greek National Archive of PhD Theses).

Beckmann, S. (2012b) Bronze Age Landscape and Resilience: 4.000 Years of Tradition? In J. Kneisel, W. Kirleis, M. Dal Corso, N. Taylor and V. Tiedtke (eds.) *Collapse or Continuity? Environment and Development of Bronze Age Human Landscapes.* Proceedings of the International Workshop 'Socio-Environmental Dynamics over the Last 12,000 Years: The Creation of Landscapes II (14th–18th March 2011)' in Kiel, Volume 1. Universitätsforschungen zur prähistorischen Archäologie 205. Habelt, Bonn, pp. 35–54.

Beranová, M. (1980) *Zemědělství starých Slovanů.* Academia, Praha.

Bourrigaud, R. and F. Sigaut (eds.) (2007) *Nous Labourons. Actes du Colloque Techniques de travail de la terre, hier et aujourd'hui, ici et là-bas,* Nantes, Nozay, Chateaubriant, 25–29 octobre 2006. Centre d'Histoire du travail, Nantes.

Bratanić, B. (1952) Plug i ralo. *Slovenski etnograf* 5, 208–236.

Bratanić, B. (1955) Istraživanje oraćih sprava na međunarodnoj osnovi. *Slovenski etnograf* 8, 269–271.

Braudel, F. (1988) *Strukture vsakdanjega življenja: mogoče in nemogoče. Materialna civilizacija, ekonomija in kapitalizem, 15.-18. stoletje,* 1. in 2. knjiga. ŠKUC, Filozofska fakulteta, Ljubljana.

Brown, A. (2001) *Arthur Evans's travels in Crete 1894-1899.* (BAR international series 1000), Archaeopress, Oxford.

Bull, I. D., Betancourt, P. P. and Evershed, R. P. (2001) An Organic Geochemical Investigation of the Practice of Manuring at a Minoan Site on Pseira Island, Crete. *Geoarchaeology*: An International Journal, Vol. 16, No. 2, 223–242.

Conklin, H. C. (1963) *The Study of Shifting Cultivation.* Technical Publications, Washington.

Chryssoulaki, S. (1999) Minoan Roads and Guard Houses – War Regained. In R. Laffineur (ed.) *Polemos. Le contexte guerrier en Égée à l'âge du bronze. Actes de la 7e Recontre égéenne internationale Université de Liège, 14-17 avril 1998 I, Aegaeum 19, (Annales d'archéologie égéenne de L'Université de Liège),* ed., Liège-Austin, Texas, pp. 75–85, pl. 6–9.

Evans, A., Myres, J. L. (1895) A Mycenaean Military Road in Crete. *The Academy*, No. 1204 (1st of June), 469–470. (The whole text given also in: Brown, 2001)

Fenton, A. (1977) Team Cultivation with the Spade in Scotland. *JATBA* (Journal d'Agriculture Traditionelle et de Botanique Appliquée) publication with CNRS, Vol. 24, No. 2–3, April–September. Paris, pp. 77–86.

Fenton, A. (1999) *Scottish Country Life.* Tuckwell Press, East Linton.

Forni, G. (1982) Il trapasso da 'Caccia/Racolta' a 'Coltivazione/Allevamento'nell Ambito della 'Burning Economy' dal Prossimo Oriente alla regione Alpina. In *Studi in onore di Ferrante Rittatore Vonwiller*, Porta Prima Vol. 1. Lito-tipografica G. Malinverno, Como.

Frandsen, K.-E. (1983) *Vang og Tægt. Studier over dyrkningssystemer og agrarstrukturer i Danmarks landsbyer 1682-83.* Bygd: Esbjerg. Summary: Vang og Tægt. Studies in Field Systems and Agrarian Structures in the Villages of Denmark in 1682/83.

Frederick, C. and Krahtopoulou, A. (2000) Deconstructing Archaeological Terraces: Examining the influence of Construction Method on Stratigraphy, Dating and Archaeological Visibility. In P. Halstead and C. Frederick (eds.) *Landscape and Land Use in Postglacial Greece.* Sheffield Academic Press, Sheffield, pp. 79–94.

Frühwald, H. (1987) *Brandwirtschaft in der Steiermark. Eine volkskundliche Bestandsaufname.* PhD thesis, Universität Graz.

Ghozzi, F. (2006) *La feuille 1:50000 de Zaouiet Medien (Tunisie NO): Étude archéologique à travers un Système d'Information Géographique (S.I.G.) de la protohistoire jusqu'à la fin du Moyen Âge,* Unpublished doctoral thesis, département d'Histoire, Université de Nice/ Sophia Antipolis.

Grafenauer, B. (1970) Poljedelsko orodje. In P. Blaznik, B. Grafenauer and S. Vilfan (eds.) *Gospodarska in družbena zgodovina Slovencev,*1. DZS, Ljubljana, pp. 201–218.

Grove, A. T. and Rackham, O. (2003) *The Nature of Mediterranean Europe. An Ecological History*, 2nd ed. New Haven & London.

Haudricourt, A-G. and Delamarre, M. J-B. (1955) *L'homme et la charrue á travers le monde.* Gallimard, Paris.

Hlubek, F. X. (1846) *Die Landwirtschaft des Herzogthums Steiermark.* Graz.

Hylton, W. H. (1984) *Alles zum Selbermachen.* Frankfurt.

Iversen, J. (1941) Land occupation in Denmark's Stone Age. Danm. Geol. Unders. 2 Rk nr. 66. C. A. Reitzel, Copenhagen.

Jensen, J. E. (2010) *Gensidig afhængighed - en arv fra fortiden. Danmarks middelalderbyer - et vidnesbyrd om spredning af vestlig civilisation*, English summary, Odense universitetsforlag, Odense.

Kaland, P. E. (1979) Landskapsutvikling og bosetningshistorie i Nordhordalands lyngheiområde. In R. Fladby and J. Sandnes (eds.) *På leiting etter den eldste garden.* Universitetsforlaget, Oslo.

Koren, H. (1950) *Pflug und Arl. Ein Beitrag zur Volkskunde der Ackergeräte.* Otto Müller Verlag, Salzburg. (Veröffentlichungen des Institutes für Volkskunde; Band 3)

Krusche, P. and M., Althaus, D. and Gabriel, I. (1982) *Ökologisches Bauen.* Wiesbaden/Berlin.

Lerche, G. (1970) The Ploughs of Medieval Denmark. *Tools & Tillage* 1(3), 131–150.

Lerche, G. (1986) Ridged Fields and Profiles of Plough-Furrows. Ploughing Practices in Medieval and Post-Medieval Times. A Study in Experimental Archaeology. *Tools & Tillage* 5(3), 131–156.

Lerche, G. (1991) Wear-Marks Produced in Plough-Furrows and on Front of Mouldboards. Conclusions based on Experiments with Replicas of a Medieval Plough. *Archeologie Experimentale* 2, 218–222. (Corrections to p. 221 fig. 4 weight 375 gr; fig. 6 240 gr).

Lerche, G. (1994) *Ploughing Implements and Tillage Practices in Denmark from the Viking Period to About 1800, Experimentally Substantiated.* The Royal Danish Academy of Sciences and Letters Commission for Research on the History of Agricultural Implements and Field Structures 8., Poul Kristensen, Herning.

Lerche, G. (1995) Radiocarbon Datings of Agricultural Implements in Tools & Tillage 1968–1995. Revised Calibrations and Recent Additions. *Tools & Tillage* 7(4), 172–205.

Lerche, G. (2001) L'expérimentation, une source vivante pour l'histoire de l'outillage. In *L'outillage agricole médiéval et moderne et son histoire*. Presses Universitaires du Mirail, pp. 69–93.

Lerche, G. (2005) *Ard.* In M. Svart Kristiansen (ed.) *Tårnby. Gård og landsby gennem 1000 år.* Jysk Arkæologisk Selskab, Århus, pp. 311–322.

Lerche, G. and A. Steensberg (1983) Tools and Tillage in Iran. Observations made in 1965 in the province of Kerman. *Tools&Tillage* 4 (4), 217–249.

Leser, P. (1931) *Entstehung und Verbreitung des Pfluges.* Anthropos ethnologische Bibliothek, Band III, Heft 3. Münster.

Ložar, R. (1944) Pridobivanje hrane in gospodarstvo. In R. Ložar (ed.) *Narodopisje Slovencev,*1. Založba Klas, Ljubljana, pp. 98–191.

Makarovič, M. (1978) *Kmečko gospodarstvo na Slovenskem.* Mladinska knjiga, Ljubljana.

Martín Galindo, J.-L. (195?) Los cultivos sobre cenizas en la provincial de León. In *Artículos geográficos sobre la provincial de León*, Cuaderno 1. Valladolid, pp. 5–16.

Moody, J. and Grove, A. T. (1990) Terraces and Enclosure Walls in the Cretan Landscape. In S. Bottema, G. Entjes-Nieborg and W. van Zeist (eds.) *Man's Role in the Shaping of the Mediterranean Landscape*. Rotterdam, pp.183–191.

Moody, J. (2000). Holocene Climate Change in Crete. In P. Halstead and C. Frederick, (eds.) *Landscape and Land Use in Postglacial Greece* (Sheffield Studies in Aegean Archaeology 3), Sheffield, Sheffield Academic Press: 52–61.

Morlon, P. and Sigaut, F. (2008) *La troublante histoire de la jachère. Pratiques des cultivateurs, concepts de lettrés et enjeux sociaux*. Educagri éditions-Éditions Quæ, Dijon-Versailles.

Orel, B. (1955) Ralo na Slovenskem. *Slovenski etnograf* 8, 31–68.

Orel, B. (1961) Ralo na Slovenskem. *Slovenski etnograf* 14, 15–40.

Orožen, I. (1876) *Das Benediktiner Stift Oberburg*. Im Selbstverlage, Marburg.

Pedersen, H. (1975) *De Danske Landbrug. Fremstillet på Grundlag af Forarbejderne til Christian V's Matrikel*. Landbohistorisk Selskab, Copenhagen. (repro. first publ. 1928)

Predan, D. (1956) Požarjenje. In *Kotnikov zbornik, Narodopisni zapiski z Gornjegrajskega in Kozjanskega*. Mestni muzej, Celje, pp. 25–28.

Rise Hansen, C. and Steensberg, A. (1951) *Jordfordeling og udskiftning. Undersøgelser i tre sjællandske landsbyer*. The Royal Danish Academy of Sciences and Letters. Hist. philolog. serie 2(1) E. Munksgaard, Copenhagen.

Rowly-Conwy, P. (1981) Slash and Burn in the Temperate European Neolithic. In R. Mercer (ed.) *Farming Practice in British Prehistory*. Edinburgh University Press, Edinburgh, pp. 85–96.

Savnik, R. (ed.) (1976) *Krajevni leksikon Slovenije*. DZS, Ljubljana.

Sereni, E. (1981) *Terra nuova e buoi rossi, e altri saggi per una storia dell'agricoltura europea*. Giulio Einardi editore, Torino.

Sigaut, F. (1975) *L'agriculture et le feu. Role et place du feu dans les techniques de preparation du champ de l'ancienne agriculture européenne*. Mouton & Co, Paris.

Sigaut, F. (1977) Introduction. *JATBA* (Journal d'Agriculture Traditionelle et de Botanique Appliquée) publication with CNRS, Vol. 24, No. 2–3, April–September. Paris, pp. 69–74.

Sigaut, F. (2005) « Labourer, pour quoi faire ? Les fonctions des instruments aratoires dans les agricultures d'autrefois » (DVD, le dossier préparatoire au colloque « Techniques de travail de la terre, hier et aujourd'hui, ici et là-bas », Nantes-Nozay-Châteaubriant, 25–28 octobre 2006)

Simonič, P. (1940) *O obdelovanju njivske zemlje*. Ljubljana. (Knjižnica kmečke mladine; 3)

Slavec Gradišnik, I. (1995) Med narodopisjem in antropologijo: o razdaljah in bližinah. In R. Muršič and M. Ramšak (eds.) *Razvoj slovenske etnologije od Štreklja in Murka do sodobnih etnoloških prizadevanj*. Slovensko etnološko društvo, Ljubljana, pp. 125–140.

Smerdel, I. (1991) Prelomna in druga bistvena gospodarska dogajanja v zgodovini agrarnih panog v 19. stoletju na Slovenskem. *Slovenski etnograf* 33–34 (1988–1990), 25–60.

Smerdel, I. (2008) *Orala. Zbirka Slovenskega etnografskega muzeja. / Ploughing Implements. The Collection of the Slovene Ethnographic Museum*. SEM, Ljubljana (Knjižnica Slovenskega etnografskega muzeja / Slovene Ethnographic Museum Library; 10)

Steensberg, A. (1986) *Man the Manipulator. An Ethno-Archaeological Basis for Reconstructing the Past*. The National Museum of Denmark, Copenhagen.

Steensberg, A. (1993) *Fire-Clearance Husbandry. Traditional Techniques Throughout the World*. Poul Kristensen, Herning.

Steensberg, A. and Lerche, G. (1977) The Use of Single and Double Paddle-Spades in Papua-New Guinea and Europe. *JATBA* (Journal d'Agriculture Traditionelle et de Botanique Appliquée) publication with CNRS, Vol. 24, No. 2–3, April-September. Paris, pp. 131–137.

Steensberg, A. and Østergaard Christensen, J. L. (1974) *Store Valby. Historisk-arkæologisk undersøgelse af en nedlagt landsby på Sjælland*. The Royal Danish Academy of Sciences and Letters. Hist. philolog. serie, 8(1) E. Munksgaard, Copenhagen.

Svart Kristriansen, M. (ed.) (2005) *Tårnby. Gård og landsby gennem 1000 år*. Jysk arkæologisk selskab, Århus. (Captions and summaries in English: Tårnby. Farm and village through 1000 years.)

Taramelli, L. (1899) 'Ricerche archeologiche Cretesi'. *Monumenti Antichi* 9, Milano, pp. 285–446.

3 Tending the Crops

3.1. INTRODUCTION

José Luis Mingote Calderón

In traditional agriculture, the time between crop sowing and harvesting is considered a risky period during which the success of the crop is at stake. Adverse weather conditions, animal depredation, pests, warfare, and raids are common causes of crop loss or complete failure during this span. Studies in this chapter explore some of the technical solutions and physical activities undertaken by traditional farmers to tend and protect their crops, and some of the non-physical, social and ritual means by which agricultural success was ensured.

In the past, technical solutions to protect crops were limited to building fences and physical boundaries or placing scarecrows to keep birds away. There were also tasks which could be considered essential steps in the chain of activities involved in caring for a crop. These include activities such as weeding and irrigation to promote healthy crop growth. In addition, a large number of social regulations and legal measures prohibited incursions and invasions of fields either by animals or people. There were also rituals and practices carried out

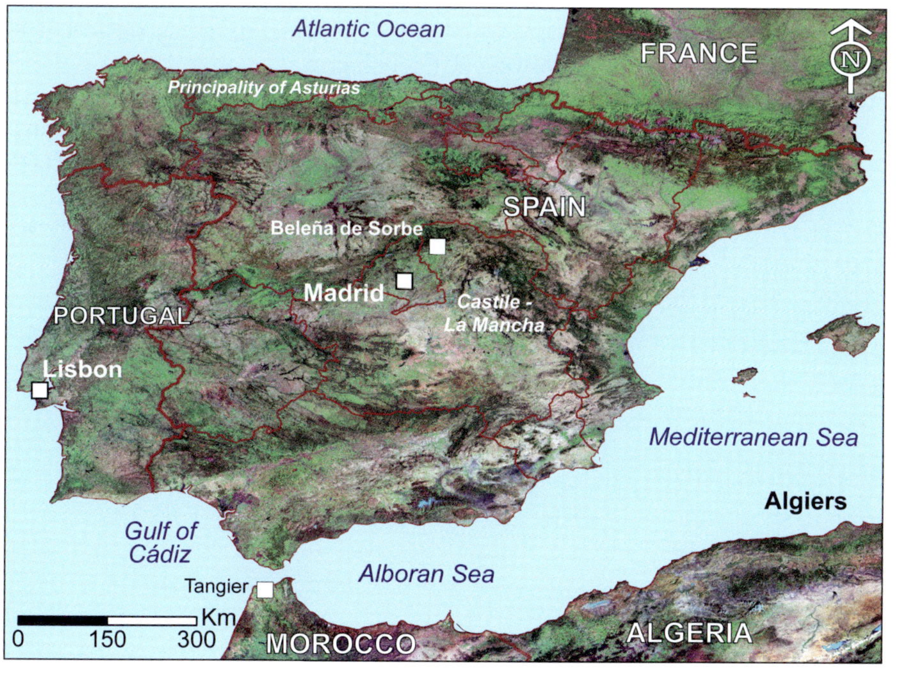

Fig. 3.1. Map of the Iberian Peninsula with the mentioned town of Beleña de Sorbe as well as the Principality of Asturias. Map: R. Lugon, J.-C. Loubier and A. Chevalier.

when technology or social regulations were clearly insufficient. It is striking, however, that many of the agricultural tasks involved in crop care have not been considered of interest for research, and therefore there is little literature focusing on the different attitudes towards crop tending. Traditional weeding, for instance, provides a good example of the scant research interest aroused by this type of activity. Weeding has seen little study in modern research, perhaps as a result of various factors related to the simplicity of the tools involved (hoes, worn out sickles, unspecified tools or even the use of hands) which have not attracted much interest. A further reason has been the involvement of women in this activity and its consequent low social value and meagre rewards for the workers. An exception has been the use of weed seeds among other palaeobotanical remains in reconstructing past agricultural systems from archaeological evidence (Jones *et al.* 1999; Charles *et al.* 2002; 2003; Charles and Jones 1997). Nevertheless and despite this lack of interest, weeding is often important for the success of crops, and frequently requires much time and labour. Its importance can sometimes be recognised in iconography, for instance, in the Romanesque calendar of the Spanish church of Beleña de Sorbe (Guadalajara province; Fig. 3.1) where the month of June is represented by the task of weeding (Fig. 3.2).

Fig. 3.2. Illustration of weeding in the Beleña de Sorbe church (Guadalajara, Spain). Image: J. L. Mingote Calderón.

Another point of interest lies in the relationship between the various treatments given to crops and the nature of the landscape matrix in which the crops were embedded, something very apparent in the case of irrigated spaces. Irrigation was, in fact, another way of caring for the crops or for other productive areas such as meadows, and could be considered of key importance for pre-industrial agriculture in dry areas. Water extraction systems were many and varied. Wells, waterwheels, canals, and *qanats* have been studied from historical and archaeological points of view. Irrigation and its physical structures have transformed entire landscapes and social systems, and the literature of irrigation is too vast to cite here.

Traditionally, irrigation has been linked to specific crops, in Spain the so-called *cultivos de regadío* (irrigated crops) as opposed to *cultivos de secano* (dryland crops) (Watson 1983). In the Mediterranean region, irrigation played an important role under Islamic rule. In fact, Islamic influence is still visible in the spatial organisation of many villages where irrigation systems were established. Terraced landscapes with hydraulic systems create an unmistakable image in areas of Islamic presence. After the Christian conquest, many Islamic farmers remained in the territory and many of the systems remained unaltered while others changed, giving rise to a mixture of traditions. In addition, the importance of the Islamic influence is particularly striking in areas where the invaders never settled (Barcelò *et al.* 1996; Glick 1988 and 1992; Dias 1953; Caro Baroja 1983a, b, c, d, e). Arizaga Castro (this chapter), however, shows that interest in irrigation goes beyond dry lands and that it also played an important role in Atlantic climates.

Besides irrigation there were other ways of tending and protecting crops. Fences or barriers for closing fields were widely used in the past. In many cases the decision to use fences depended upon the type of crop sown in the particular area. In general, winter crops such as winter cereals were sown in open fields without fences, whereas irrigated crops have been commonly protected by fences or barriers. Fencing was made using a vast range of products depending on the availability of resources

Fig. 3.3. Enclosing field wall made with stones (Prádena del Rincón, Madrid, Spain). Image: J. L. Mingote Calderón.

Fig. 3.4. Fence made with thorn bushes (Berzosa de Lozoya, Madrid, Spain). Image: J. L. Mingote Calderón.

in the region. Stones (Fig. 3.3) and plant material such as thorny bushes (Fig. 3.4) were used to limit field access and defend the agrarian space. The fence represented not only a physical boundary but also an ideological limit, as it was forbidden to steal another's gains. What is more, the sacred character given to the crop sown or stored or even to the field boundaries is a common element in many cultures. In Asturias in Spain (Fig. 3.1) it is possible to see *hórreos* (granaries) protected by crosses on the doors. In Morocco, there are areas of concentration of granaries that are protected by religious beliefs (González Urquijo *et al.* 2005). The union between the physical and the ideological spheres reflects the marriage between the material and the immaterial in any culture's attempts to protect crops in field and in storage. Scarecrows (Fig. 3.5) are another good example.

Fig. 3.5. Scarecrows (Vilanova, Asturias, Spain). Image: J. L. Mingote Calderón.

Nevertheless, together with the rich variety of technical actions aiming at controlling and protecting crops, farming groups practice all sorts of 'tactics' that are more in the realm of ritual. Le Bonniec's (1958) study of the cult of Ceres provides many instances of relations between agrarian activities and rituals in the Roman world. The throwing of handfuls of grain across the field accompanied by verbal offerings, prior to sowing, is a good example of the sort of conduct we are dealing with. It could be argued that such actions have a pagan or superstitious character. As such they are thus considered far from current 'modern' and 'urban' thinking, and scholars tend to separate these activities from their technical, practical content. The clear-cut distinction between both spheres (the ritual and the technical) has not always been as categorical nor has it been obvious in all cultures. In fact, the dichotomy modern scholars observe derives from the 'triumph of reason' (Western reason, of course) in the Age of Enlightenment. It shows, in addition, the gap between the new industrialised society and the pre-industrial mentality. The modern view is that technological progress is not linked to immaterial aspects or supernatural issues. Additionally, there is a sense of ethnocentrism which places human beings at the centre of progress: they are seen as fully responsible for their actions and they choose and materialise the options for achieving economic development. From this perspective adopted by modern research, information related to protection and control over crops and harvests in works examining religiosity, mythology or folk customs and beliefs is separated from such information in studies of agricultural techniques. There are, nonetheless, some exceptions. As Malinowski (1925) demonstrated long ago, human cultures always mix the practical and the ritual, the technical and the spiritual in the pursuits of daily life, even if we moderns consider them separate spheres. In A.-G. Haudricourt and M. J.-B. Delamarre's (1955, 371–384) work on ploughing, a section devoted to beliefs, rites and ceremonies is included, demonstrating that this type of information needs to be documented in the studies of techniques. The authors discuss the mythical origins of the plough and add data on various aspects related to its categorisation as a magic implement, the bans related to it, the ceremonial character of the first ploughing and its many religious or magic aspects.

In Isager and Skydsgaard's (1995) work on Greek agriculture, there is an important section focusing on the relationship between gods and agriculture. However, whenever ritual practices are recorded in documentary sources describing 'technical' aspects of agriculture, they are usually discredited and disdained. They tend to be left aside with absolute indifference. In fact, in Roman or Medieval agricultural treatises, the authors that overuse this type of information are considered second class authors. Good illustrations are given by opinion on the *Geopónika* (for a current translation see Dalby 2011) or on Palladius' *Treatise on Agriculture* (translation by Owen 1807), which contains a most interesting chapter on remedies used in horticulture and agriculture (I, XXXV). In both works, techniques and rituals coexist, and this is not unusual. Authors such as Columella, who has a great reputation for his precise technical information, reflect on the need for rituals before starting a task in order to guarantee a good harvest (Mingote Calderón 2007). It is worth noting that the same authors who describe how to perform the work also give instructions on the rituals – some are compulsory – so that abundant harvests can be achieved. It is clear, therefore, that in many cultures and places it is possible to find mixtures between know-how and beliefs applied to technical processes. There are differences, similarities and a great deal of continuity. In any case, all cultures share a common element: the belief in supernatural forces that control the life of human beings. The idea of personal freedom from social or religious rules is a modern concept distant from the world analysed here. Society and political and religious powers (again this division is a modern concept), exert a strong pressure on farmers, leading to the continuity of cultural patterns and behaviours which only disappeared when the pressure decreased or ceased to exist.

The papers to follow all reflect the intertwining of technical practices, social organisation, and ritual belief. Arizaga Castro discusses Spanish irrigation systems, considering both the physical structures, and the way social practices and legal formulae control their use and distribute water. Mingote Calderón aims to highlight the importance ideology had in pre-industrial societies and its use in the shaping and timing of agrarian practices. He considers religion and the ritual practices associated with protecting crops; these also form systems of social control. As Schippers shows, calendrical systems not only reflect necessary agricultural know-how about the seasonal cycle, but moral and religious beliefs as well.

3.2. WATER AND LAND PREPARATION: IRRIGATION IN THE NORTHWESTERN IBERIAN PENINSULA

Álvaro R. Arizaga Castro

In this paper I will approach land preparation from a particular perspective: traditional irrigation techniques. One of the main farm tasks has always been the management of water. In traditional agriculture, everything was designed for a correct combination of sun, air and water. Among these three factors, the use of water tends to leave a more concrete archaeological record. Water has been an object of interest for a range of studies concerning agriculture, archaeology and history in general, as well as for traditional agriculture in particular. Some of these studies are specifically relevant for this contribution (Ciriacono 1998; García 2007; Glick 1972; González and Malpica 1995; Michelsen 1986; 1987; 1995; Rivera 2009; Veiga *et al.* 1995; Wang and Wang 1991). Probably the most famous reference to irrigation is Wittfogel's *Oriental despotism: a comparative study of total power* (1957), which argues that irrigation is the starting point for the development of state systems. However, we shall not attempt to deal with such an influential book and its associated controversies here, but consider irrigation from a much simpler, more modest point of view, with a micro-historical approach mainly related to anthropology in a local context.

Some quite old and well-documented institutions related to water in Spain are still alive and active. One of the most famous is Valencia's Meadow Court for Watering (Tribunal de las Aguas de la Vega de Valencia), well researched decades ago by Thomas F. Glick (1970; 1996). In 2009, UNESCO added this institution to the Representative List of the Intangible Cultural Heritage of Humanity, under the name 'Irrigators' tribunals of the Spanish Mediterranean coast: the Council of Wise Men of the plain of Murcia and the Water Tribunal of the plain of Valencia (Fig. 3.6). This constitutes official recognition of the importance of traditional knowledge about the management of water, even for the present-day economy. To explain why traditional water systems are still important, we will look at traditional irrigation techniques in Galicia, located in the northwestern corner of the Iberian Peninsula. Specifically, this contribution will examine a small area of southern Galicia and northern Portugal, the Minho region (Fig. 3.6), where irrigation had a huge importance in the context of very intensive agriculture, at least from the 17th century onwards (Wateau 2000).

The Minho region is characterised, from a general agrarian point of view, by 'a conjunction of many different kinds of crop- and cattle-raising in a context of historical predominance of small-scale household farming under the conditions of an Atlantic agriculture, in which water is not a scarce resource, but is subjected to a growing demand from the increasing pressure on Galician agriculture from the 18th century onwards... Abundant rain does not exclude a strict regulation of water management' (Sánchez and Fernández 1999, 339). As mentioned before, nowadays the typical irrigated Spanish region is the Mediterranean coast (Valencia, Murcia, Almería; Fig. 3.6), a much dryer area. Apart from the amount of available water, one of the first differences between our study area and the Mediterranean system is that Portuguese and Galician traditional irrigation has had little connection to, or recognition in, the contemporary world. Innovation here was apparently separate from the urban world, and not regulated or administered

Fig. 3.6. Map of the Iberian Peninsula with the mentioned cities of Arbo, Corzos and Vide, the region of Minho as well as the political units of Almería, Galicia, the Region of Murcia and the Valencian Community, as mentioned in the text. Map: R. Lugon, J.-C. Loubier and A. Chevalier.

by official or state irrigation authorities (this kind of state irrigation from the 17th and 18th centuries was described by Eiras 1990, 31–84). As a consequence, irrigation in this area constitutes an endangered heritage. 'Many hydraulic structures present in the Galician landscape are disappearing or being abandoned – from the irrigation systems to the mills and many other machines, and with them many agrarian exploitations and pastures are disappearing as well. In parallel, social and cultural relationships related to the employment of water, very important to understand the rural community, are also vanishing' (Sánchez Regueiro 2001, 757). Apart from the loss of intangible heritage, the abandonment of techniques, channels, pits and the entire irrigation system, along with changes in the use of woodland, causes some important problems (for instance, those related to erosion after forest fires). The traditional drainage system of the land, as a whole, is being abandoned (Fig 3.7).

The origins of the Galician irrigation structures date at least to the Middle Ages (9–10th century), as has been proven for the cultivation terraces (Sánchez and Fernández 1999, 350). Paula Ballesteros Arias (2010) has researched the structure and dating of these Galician terraces, which were sometimes built in close relationship with irrigation structures. Irrigation has usually been related to Arabic or monastic origins, but in fact the introduction of maize constitutes the starting point for the massive development of irrigation (Fig. 3.8). When maize was brought in and adopted from overseas, it had a formidable impact in 17th- and 18th-century Galicia. Maize explains the spread of raised granaries (named *hórreos* or *canastros*), but also the beginning or, at least, the improvement of a huge structure for the management of water. Irrigation seems to have been absolutely necessary for the survival of a multiple crop cultivation system dominated by maize.

The Galician traditional landscape was studied by Abel Bouhier (1979) in a superb book. Following his work, we know that the main irrigation areas were and still are located in southwestern Galicia (Bouhier 1979, 670). The core of this area, the lower Minho valley, is now mainly known for wine cultivation, but in the past there was in fact a very intensive agriculture combining multiple crops. The 18th-century context is quite well known from the Catastro de Ensenada (1752), a key documentary source for understanding crops, fields, agriculture and taxes in the whole of Spain for this period. Bouhier showed there were almost no changes, in many ways, until the 1960s. Irrigated lands started to slowly appear in other regions as well, but, in the main, the model developed since the introduction of maize survived until the mid-20th century (Sánchez and Fernández 1999, 346–347).

In the context of an increasingly intensive agriculture, irrigation was mainly used for maize fields and pastures. Pérez García (2004) studied

Fig. 3.7. An old and decayed double 'mine' for conducting water in O Eido de Ribarteme, As Neves (Pontevedra, Galiza, Spain). Image: Á. R. Arizaga Castro.

Fig. 3.8. Irrigation structure in Candás, Rairiz da Veiga (Ourense, Galiza, Spain). Image: Á. R. Arizaga Castro.

Fig. 3.9. Watering and draining fields for pasture in Corzos, A Veiga (Ourense, Galiza, Spain). Image: Á. R. Arizaga Castro.

the extreme differences in production between irrigated and non-irrigated areas: 'to have irrigation available in the region of the Lower Minho River implies a yield not easily reachable in the context of Ancient Regime agriculture ... The difference between watering and not watering, while not as spectacular as in the Mediterranean world, was very significant, and we know it because it was attested in many legal disputes' (Pérez García 2004, 561–562). In folk life, water was then, and still is, considered 'the blood of the earth' (Sánchez Regueiro 2001, 759), 'blood of the earth' in a double sense indeed, because blood means 'life' but also 'war'. There were multiple types of conflicts involving watering (Candal 1993): arguments about water allocation and distribution among neighbours, concerning the construction of new channels, about hydraulic mills and their water needs, and so on. Some people thought that only plots of land planted in cereals were important enough to employ water, but other people argued that water was a privilege strictly associated with the land: when a plot of land had water rights, they prevailed independently of the crops or the uses of that land. Some people thought distribution of water must be reviewed as time passed by, but others thought tradition was immovable. Many of the conflicts arose between 1650 and 1720, when the model for irrigation was being created, coinciding with a new agrarian model that privileged spring-sown cereals and maize in contrast with the old model that privileged winter cereals (Pérez García 2004, 564) (Fig. 3.9).

There were two main objectives of water control systems: controlling the excessive amount of water in the wet periods of the year (mainly in winter) and managing the resource during dry months (mainly in summer). Water needed to be managed not only when there was a deficit of it, but also to deal with any surplus (Sánchez and Fernández 1999, 348). Water management was an activity with different 'rhythms' depending on the time of the year. Water was completely limited, divided and controlled during the dry season. From March to September, some of the religious festivals defined the timing of water control activities. In some places this 'roll' (*rolda*) starts on the 1st of July, but in many places water 'flows' from the 24th of June to the 8th of September, from the feast of Saint John to the feast of Our Lady's Birth, two of the most important religious dates for a Catholic peasant society (Rivas 2000, 89). Water tended to be used and divided strictly for irrigating all day and all night, but in some places the use of water at night was also allowed to the millers. Of course, not everything was always divided: at times some sources were considered 'free water'. Sometimes people could wait for hours during the night in order to turn water to their land, trying to mislead neighbours and getting access before or just after them. This water was called 'removable water' (*auga de quita-requita*; Fraguas 1996, 16), 'turn-return water' (*auga de tornatorna*) or 'catching water' (*auga da pillota*). In a sense, it was a competition that was not always fair or honest.

In general, water tended to be strictly divided in the summer season. We can imagine the importance for these societies of measuring time with precision. There were many traditional ways to calculate time: when the sun strikes a particular stone, when the face of the king is visible in a coin, or when a person's shadow reaches eight feet long (Rivas 2000, 64). Another interesting system was to throw a leaf or piece of wood on the channel and when the leaf arrived at the plot of land it was time to change or turn the water in the source or pit. Time relating to watering was usually measured not only in imprecise periods like 'half a day' or 'one hour' but in exact minutes: from the beginnings of this model it was increasingly necessary to use modern clocks. In many parishes, we may study watering and irrigation through landscape archaeology by looking at the location of 18th-century sun dials (Fig. 3.10). Sun dials were located at central places, like the church, the spring, or important raised granaries, places where people customarily meet and discuss questions related to watering. New solutions were developed later: in a few cases, some parishes constructed mechanical clocks with a bell tower located in houses at high places. The family living in these strange houses resembling chapels took care of the clock (Rivas 2000, 62) (Fig. 3.11). As we can imagine, discussions and arguments about timing irrigation must have been incessant.

Although summer was the main irrigation period, when conflicts were most likely to occur, it is very important to stress that water management was not only addressed to the summer drought. Winter irrigation was very important too, and this has not always been well understood by scientific researchers. As has been shown for many different countries (Michelsen 1986; 1987), 'the obvious assumption that inside Europe irrigation belongs only to the semi-arid zones of the Mediterranean countries does not hold good' (Michelsen 1987, 243). In some interior and wetter regions of Galicia, irrigation was linked to the increase in cattle-raising and thus pastures from the 18th century onwards. In pastures, humidification helps against frost: there was a proverb that said 'he who does not water at Christmas, in August will not reap' ('*Quen en Nadal non regha, en Agosto non segha*'; Rivas 2000, 20). 'The pastures are muddy areas cultivated for

Fig. 3.10. Sun clock above a raised granary, located in the middle of the main irrigated area (O Regho) in the parish of Tortoreos, As Neves (Pontevedra, Galiza, Spain).Image: Á. R. Arizaga Castro.

Fig. 3.11. A clock tower built as part of a private house, probably at the end of the 19th century, specifically for timing irrigation and quite far away from the parish church (Tortoreos, As Neves, Pontevedra, Galiza, Spain). Image: Á. R. Arizaga Castro.

cattle fodder for the long winter. They are carefully looked after and they receive the majority of water during the year. In some villages irrigation is not applied to them during summer. In other places they receive half the water of the rest of the year. People take care of them especially in autumn, to make them sprout and green' (Cerrato 2007, 27–28). Some good examples of this may be seen in a recent photographic essay by Luis Higinio Flores Rivas, *Paisaxes humanizadas, terras vividas*. (Flores 2008, 54–55, 126–127, 136–143). Irrigation and drainage techniques were quite precise and complex, having many other uses apart from simply watering. For instance, on terraced fields in China, irrigation has been described as a simple but sophisticated way of regulating temperature (Wang and Wang 1991, 196–198). Coming back to Galician agriculture, water coming from calcareous sources helps to prevent soil acidification, one of the most important problems for this region. There were indeed different kinds of waters: those which came through villages or roads, carrying with them fertiliser from dung, were called 'gracious waters' (Sánchez and Fernández 1999, 360). Apart from that, irrigation pits and structures were sometimes used to ret flax. In some places, it was very important to throw straw, or sometimes fern, on the ground to avoid erosion prior to the first watering or 'opening water' (*rega de abrir*) (Sánchez and Fernández 1999, 359; Rivas 2000, 83). In some places, there were special pits called 'air pits' (*pozas de aire*), with a clever vacuum system to control the filling and emptying of the pit almost without human intervention (Cerrato 2007, 37–38).

The techniques for redirecting water were very diverse, but we may observe two main types. One relied on the use of irrigation channels or ditches (*acequias*) and one tended to use pits (*pozas*) located near the sources of water or near the plots of land. These types were not exclusive, but depending on the parish or region, one of them tended to be preferred. As we have been suggesting, it is clear that there was no common system for the whole irrigated area in northwestern Iberia. In fact, each parish had its own system, its own techniques and its own division of water. Some aspects were common, but diversity was the rule. Rivas (2000) records these multiple solutions, taking into account the spatial distribution of linguistic peculiarities. There were usually fully different systems even for neighbouring parishes.

A micro-historical approach to the situation of three parishes in Arbo, Vide and Tortoreos, in As Neves, southern Galicia provides a good example (Fig. 3.1). The complex watering system there is quite old, well attested, and well preserved, enabling us to understand how water was distributed, when, why and for whom. First, the Levadas or 'bringings' of Arbo were a very complex system, in some ways representing the 'perfection' of the irrigation system. They were studied from a scientific point of view a long time ago (Vázquez 1946), when the system was still fully operational. Vasquez explains how an irrigation system developed in such a complex manner that any single change was not only difficult but painful and even dangerous, sometimes causing violent reactions. The research was developed in three parishes in the municipality of Arbo. The most interesting part of the system affected the parish also called 'Arbo' but not the whole of the parish, only a long tract from the northwest to the southwest, where the main channel was situated. It was a system already considered 'immemorial' at the time the research was done. The irrigation area was divided in several groups, then subdivided into more and more parts, depending on proximity to the main channel.

The irrigation system was completely operational for not more than two months, from July 11th to September 11th. On the Sunday before July 11th, there was a big gathering, usually around a church or religious place, and one individual was chosen to be responsible for conducting water distribution during that period. He was named the *levador* ('carrier', 'leader' or 'driver'). The position was allocated by a reverse auction, fixing a maximum cost and giving the title to the cheapest offer. It was allocated to one person but, in fact, the job used to be done by at least two people, given the fact that it implied day and night work with no rest for the couple of months involved. The *levador* was also given special political power in the community, reviewing and clarifying every problem or argument and being the most important witness when the ordinary legal system had to be involved.

Some days before the auction, all the peasants were required to work together to clean the biggest common channel. When it was decided to pay for it, the whole cost was divided among the members depending on their benefit, that is, depending on the amount of water they enjoyed. The same division was applied to pay the *levador*, who used to be paid in corn, but, in the middle of the 20th century, was starting to be paid in cash.

The auction was held on Sunday before July 11th or, in the past, always on June 29th, Saint Peter's day. It was in July when the *levador* started his job and leadership, but the rights of watering started even before, on June 26th. This starting day was completely immovable, because it affected some water used by three mills and changing it would not have been acceptable to the heirs holding these milling rights. The difference between the starting day for watering and the starting day for the 'leader' is explained because his job was especially important for the driest months: those first fifteen days were still, in general, not a moment for arguments.

The regular period of watering lasted for seven full days, starting again from the beginning on the eighth day. Some argued to extend the system to eight watering days, shared proportionally, but, again, the milling rights impeded such a change. Every farmer knew exactly to the minute the amount of time allotted to watering his plots of land. Each one was in charge of opening the water for his plot after the *levador* gave the order. Sometimes this order was given in person and sometimes using the sound of a horn.

There were many particular but very important additional rules, some worth reviewing. For instance, time was measured exactly when the flow of water in the main channel was cut: bringing the water to the plot was a matter for each owner, obviously more laborious and difficult as the distance was longer. Apart from the usual water allocated, some plots had the right to additional leftover water coming from secondary channels after their time passed by and the main channel was opened again. There were some plots with special privileges, for example, having the right to start watering at an exact hour: this meant that some other plots could be left without water in times of shortage. At such times, owners or farmers who had many plots might decide to shift time between some of their plots, increasing or decreasing the watering depending on their needs. But the whole amount of time for each owner never changed. In rainy summers, the owners might decide not to use the water allocated to them, in full or in part: sometimes the *levador* divided the flow of water, trying not to exceed the ideal quantity while benefiting several owners. But in dry summers, the flow of water was slower and some of it was lost in the channel: it was then when incidents might occur involving incompetent *levadores* or farmers 'stealing' water from the main channel. Farmers maintained a constant and strict surveillance of the channels. Water and its rights could be sold, included with or apart from a plot of land: in these last cases, it was 'time' that was sold, affecting a great deal the value of the land. Rights to the land and rights to the water were completely attached but, in a mainly illiterate society, that was not always properly written down. This difficult transition between oral laws and traditions and modern legal contracts was a huge source of conflict, surely one of the causes of the decline of the whole irrigation system.

The oldest preserved irrigation document for the parish of Arbo dates from 1843, and gives the names of every plot and the exact time allocated. There were seven 'houses' or sections: the first one started on June 26th at dawn, involving 21 plots for 24 hours and 20 minutes. After it, came the second 'house', with 11 plots and 23 hours and 42 minutes. The third was 16 plots in 23 hours and 58 minutes. Fourth, 20 plots in 24 hours and 22 minutes. Fifth, 16 plots in 24 hours and 4 minutes. The sixth section, 24 plots in 23 hours and 32 minutes. Last, the seventh section included 22 plots in 26 hours and 17 minutes. Then, the system started again from the beginning. This whole operation comprised 170 hours and 22 minutes, but seven days only have 168 hours. The difference was covered by some leftover water affecting some plots and flowing for those 2 hours and 22 minutes. That is a good evidence of the precision and 'perfection' of this system. It is also indeed a system resulting from strictly divided property, a land shaped by many smallholders. The owner who enjoyed the maximum amount of water only had two hours and twenty minutes of irrigation each week, including all his plots. And the most amazing fact of this irrigation system was that there were four owners who had only half a minute of water. Half a minute, thirty seconds, not less, not more...

In the neighbouring parish of Vide, the old priest's house has now been turned into a museum, 'The House of the Water'. In 1904, Vide constituted one of the first recognised irrigation communities in

Spain at that time, writing a letter 'to the sons of the parish of Vide, living in Lisbon, Río Janeiro and Argentina', published in a small leaflet. The letter explains that they needed some help from an economic point of view, and they needed the approval of the people living abroad, as they were also owners of the land and water rights. This shows how migration caused a huge impact on traditional agriculture. A precise and carefully regulated system started to die as people migrated, because their rights survived in the homeland, and their lands were abandoned as a consequence. In this case, the people at home were able to communicate with the emigrants, and restored the agricultural irrigation system, but this was not always possible.

In Vide, there were even some folktales linked to irrigation: 'the irrigation channel of Vide is very ancient; it comes from our ancestors... People from the parish of Setados and people from the parish of Vide were competing for it. They were working together, but from opposite sides, digging faster and faster, trying to be the first to arrive... and then the people from Setados went for lunch, and Vide won. This is the reason why the channel now belongs to Vide'. The irrigation structure in Vide is a system without pits, mainly based on this big channel, commonly said to be of Roman origin, but probably dating to the Middle Ages.

In the nearby parish of Tortoreos, where pits rather than channels prevailed, we recovered this statement: 'watering was like a puzzle: all night thinking about water. I was in bed, remembering and repeating again and again 'tomorrow at six in the morning that pit, at seven thirty open another pit, not too far away, closing it at nine.... the day after tomorrow it will be another couple of pits... and every Monday it is that pit that is quite far away... and so on'. The calendar was completely full of annotations...'. To show this striking complexity, we transcribe an example: a document describing the irrigation system that affected a small tract of land (600 m2) in the parish of Tortoreos, As Neves municipality, in southwestern Galicia. The land was named Turreiro ('tower'), because it was located beside the church. This family had 21 very small plots of land, each one with a system similar to that described below. We follow a literal transcription, as dictated by the last person to irrigate this land, who wanted her descendants to write it down, remember it and keep their rights safe. It was indeed the first time this common law was written down on a piece of paper, at the end of the 20th century.

'Tower. Water from the Lords' Pit. It starts on Saturday. Each 15 days. It has water during half an hour, once by day and once by night. If it is by night, we are the first irrigating; if it is by day, we are the last ones. This water serves every plot of land in this place, called O Regho ('The Watering'). Water from the Rozadas' Mine. To know who starts, it is needed to draw lots for it. Each 18 days, 24 hours of water. It is only for the inheritors of O Regho. Close it at 9 in the morning until 9 in the morning the next day. Water from the Oak's Pit and others. It is for every land in O Regho. Draw to start. Each 8 days, on Sunday, once 'the big roll' and once 'by pits'. When it is 'the big roll' it starts by the Upper Lands or the Lower Lands. When it is by pits, it is a pit for each land (Upper Land, Middle Land and Lower Land). The big roll is made with the Oak's Pit, the Mangarela Pit, another small one nearby and the Pit of O Regho. When we have the big roll, half of it is for us, together with Mr. Campos, but the other half is for the inheritors of Manuel Francisco and Digna, the Blacksmith's daughter'.

Of course, this system is not alive any more. It was so precise and exact that it was unable to adapt itself to the new conditions of modern times. When a small piece of land was abandoned or destroyed, the full system should have been rearranged to make a new watering division. But destructions and abandonments happened too fast. The plot discussed above was one of the best for this family; it was impossible to think of it as a car park beside the church, which it is now. But the old woman

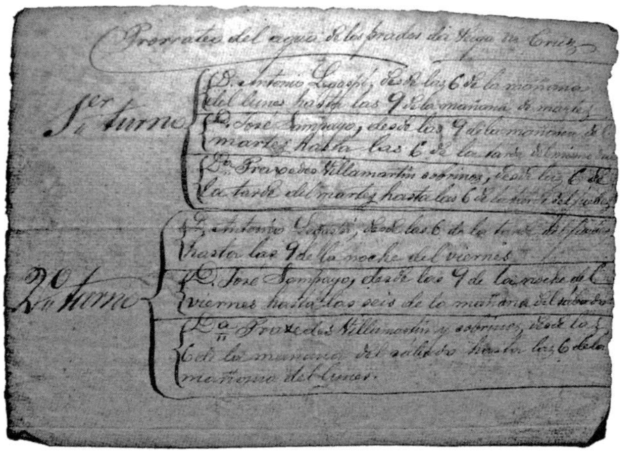

Fig. 3.12. Informally written division of irrigation water in Veiga da Cruz, Meira (Lugo, Galiza, Spain) Image: M. Á. Abraira Pérez.

still wanted to dictate these words and rights, to leave them as a testimonial, as something really important for her (Fig. 3.12).

One of the reasons for watering being relatively neglected in ethnographic research is that water was mainly in the feminine sphere. Women were responsible for washing clothes, and also for opening the sources of water and the pits. Sometimes this job was done by women and children. In fact, in a context of massive migration, as in Galicia in the 19th and 20th centuries, the predominance of women implied that they did almost every job involved in agriculture. An important reason for the disappearance of the irrigation system was the end of the communal system. It was necessary for all people of the parish to work together to clean and prepare the channels and pits. In a context of capitalism and individualism, that became impossible. In many other places there was an official person assigned to organise the watering, taking care of the structures and writing contracts and agreements down, but that was not the case for this region. Another reason for the decline of the Galician communal system was probably the excessive division of the land. Such small plots of land were, in the end, not profitable, above all in a context where the young labour force was migrating massively. Nowadays, this situation is subtly changing, although traditional Galician agriculture is still declining. But, as a significant recognition of the importance of water, many hydraulic engineering structures are being restored these days (mills, fulling mills, forges). Traditional agriculture is increasingly being studied and recorded, somehow still surviving. Educational resources concerning water and watering are becoming available (*e.g.* http://centros.edu.xunta.es/iesasangrina/auga/).

Finally, we want to stress the importance of looking at the use of water in traditional agriculture to understand how peasants tended to be respectful of their ecological resources. It is easy to see through ethnographic research that they did not understand water only from an economic point of view, related to production, but as a communal, social, natural and economic heritage (Sánchez Regueiro 2001, 758). We cannot look in detail at the cultural consequences of these facts, but we must say at least that water was considered almost sacred in folk life. It is easy to find interesting cultural relationships between saints and water or between water and sex (for Galician examples, see Fraguas 1996). There were very common oral folktales involving beautiful and terrible magic women, in some ways similar to the Greek and Roman nymphs or sirens (a folk theme that is widespread throughout Europe). In Galicia, they were usually named *mouras*, meaning 'black women' and also 'Moorish women', living under fountains and sometimes becoming snakes. From a spiritual point of view, water meant life and death, but also sex (fertility) and magical power (sacred spirits). The main sources of water were sometimes called 'the mothers' and 'blood of the earth' was really an appropriate name for the water that shaped agriculture and society (Cerrato 2007, 29).

3.3. RITUALS FOR HARVEST PROTECTION IN PRE-INDUSTRIAL CULTURES OF THE IBERIAN PENINSULA

José Luis Mingote Calderón

Introduction

In preindustrial societies, belief in the interference of supernatural powers in people's lives is still strong. Data from historical sources and ethnographic information show that rituals are strongly linked with technical processes. In fact, the intrusion of supernatural powers affects all situations, from crises like droughts, storms or plagues to mundane technical processes, and it could be said that rituals are as inseparable from the 'practical' work of farming as the other side of a coin. Well-documented rituals in many cultures sometimes show enormous similarities across time and space. However, on some occasions, differences are visible, especially when dealing with official religious practices. The coexistence of official and non-official rituals is particularly striking, and at times opens the door to different perceptions of reality. In fact, many of the non-official rituals are unconventional, and do not relate to the dominant religion. In any case, the behaviour of pre-industrial farmers clearly reflects their wish to benefit from the supernatural forces that could influence the harvest.

Rituals to Protect Field Boundaries and Crops

A general but fundamental aspect of agrarian ritual is the protection of cultivated field boundaries and crops. This is a theme which is associated with an enormous variety of rituals. At the end of the 19th century, E. López Morán (J. Costa 1981, 266–267) described how the respect shown towards stone boundaries in the León mountains (Spain; Fig. 3.13) was related to moral and religious attitudes. In his text, López Morán referred to the Roman world, but he also mentioned tales that of old were told to the children in the region. These were accounts of events in which a member of the community re-appeared to his neighbours after his death, asking them to put a boundary mark that had been moved away in the past back into its original position. The tales reflect the idea of sin and, at the same time, have the role of inspiring children to behave properly and to follow the established moral rules. In rural pre-industrial European societies, using concepts of sin to instill respect for field boundaries is a common practice. In fact, folklorists from the 19th and 20th centuries collected many tales, not only in Hispanic environments, referring to this idea (Fig. 3.14). The existence of Purgatory allows the transgressor to come back to the earth to fix a problem, to atone for a sin. Moreover, re-establishing the proper order frees the soul, so the way to heaven is open. It is clear that we are dealing with an ideological construction of the church that, in turn, makes use of ancestral beliefs such as the one referring to the return of souls to the earth. It must be stressed that cultivated field boundaries are loaded with values that go beyond their technical aspects. Therefore, their protection is closely related not only to legal attributes but also, and explicitly, to moral aspects. The concept of respect towards field boundaries is not only associated with the Medieval Christian world. A. M. di Nola (2006, 213–221) shows the role of sacred protection also in the Greek and Roman world. In the Roman period, many aspects related to the protection of properties, including fields, people and future harvests, were associated with the rite of *suovetaurilia*, the sacrifice

Fig. 3.13. Map of the Iberian Peninsula with the mentioned cities of Cinfães, Linyola, Madrid, Pla de Barcelona and San Cugat del Vallès, the Urgell area, as well as the political units of Asturias, the Basque Country and Castile and León as mentioned in the text. Map: R. Lugon, J.-C. Loubier and A. Chevalier.

of a pig (sus), a sheep (ovis) and a bull (taurus). These were rituals in which limits were blessed and fields purified by the sacrifice of different animals for protection, to insure a good harvest and to read the omens. J. Bayet (1984, 96) indicates that the festivity of the *Ambarvalia*, during which the *suovetaurilia* was carried out, goes back to Indo-European ritual practices and other *lustratio* (purification ceremonies) from Roman times. Cato gave details about the ritual (1979, 120 and 122, in CXLI) and indicated that such ritual purification assured the protection of the earth and at the same time the production of a safe harvest. In addition to requesting protection against natural phenomena, people appealed for favours related to harvests and workers. According to Cato, there was also the possibility that the wish would not be granted, but there were various alternatives which helped to overcome any lack of success of the first attempts.

In the Classical world, the very existence of deities associated with boundaries reflects the importance that limits had at the time. In the same way, there are many medieval legends and later tales which describe the punishment received by those who dared to move boundary stones. These guilty parties are prohibited from entering Paradise. In Catholic countries, the cultivated field boundaries have an almost sacred character related to the marks used to indicate and fix their limits. This symbolic aspect of boundaries appears also in other types of rituals related to field protection. One of the rituals, that of the blessing of fields which is performed in May, shows how areas under divine protection are clearly delimited within the boundaries of the municipality. Only areas within those boundaries are blessed. This is an orthodox ritual which includes texts written by the Catholic Church together with a particular set of official symbols and elements. Thus, crosses in wood or wax are understood as reference marks that watch over and signify the blessed fields (Fig. 3.15). The Catholic paraphernalia associated with these rites can be observed in pictures from the 19th century.

Among the different rituals for blessing fields for their protection against animals and storms, J. M. Satrústegui (1980, 102–106 and 117–118) described the ritual associated with the Cruz de Mayo, a religious festivity whose origin goes back to the moment Saint Helen found the True Cross. This festival protected crops from ‘plagues, clouds and scabies’. Satrústegui’s testimony is crucial because, as a priest, he participated in some of these festivities in the French Basque area. Sometimes farmers placed blessed olive branches in their fields, although the common practice was to place them at home as a protection against lightning bolts and fires. Satrústegui also stresses a significant belief which appears repeatedly in Christian environments: the devil is the hidden hand behind summer storms. However, his power was restrained if crosses were erected in the field. The presence of the devil as origin of any harm for the farmer or any other

Fig. 3.14. Field boundary in Zureda (Asturias, Spain). Image: J. L. Mingote Calderón.

Fig. 3.15. Cross used to protect fields in Albendiego (Guadalajara, Spain). Image: J. L. Mingote Calderón.

person explains why the solution always had a religious character.

The occurrence of crosses in fields as symbols of protection has a long history. O. Giordano (1995, 49 and 44) indicates that crosses had been used to deflect natural catastrophes since before the Middle Ages. This is clearly reflected in Caesarius of Arles (470/471–543) and Gregory of Tours (538–594). J. M. de Barandiarán (1984, 44) also collected other examples which mentioned the placing of a dog or cat skull as amulet against any kind of evil eye which could affect a cultivated field.

Protection Against Birds During Sowing and Growing

During the sowing period and the growing season, the activity of many animals could be harmful to the crop. Therefore, farmers developed a wide range of rituals which allowed them to fight against animal intrusions. Some years ago, I studied a sowing ritual related to the miracles of a saint who lived in Madrid (Fig. 3.13) during the 11th and 12th centuries and who was canonised as the patron saint of farmers in 1662 (Mingote Calderón 1993). In a late compilation of his miracles by Juan Diácono, Saint Isidore (San Isidro), characterised by his Franciscan spirit, fed hungry birds with grain from a sack. The miracle occurred when afterwards the sack was still full and the grain had not diminished. A common practice to protect the grain during sowing was the throwing of handfuls of grain in several directions. These practices generally went together with specific formulae and gestures. The attitude shown by the saint described above seems to be related to the rituals intended to protect the grain during the sowing season. This rite is particularly striking for several reasons: first, because it was performed before a technical activity (sowing) was carried out; second, because it was aimed at the animals and other forces that could eat the unburied grain; and third, because the sentences that went together with specific gestures included references not only to God but also to thieves, poor people, and the natural elements. All of them should be satisfied with the symbolic portion of the sowing that was given to them. Finally, the rite is also remarkable because, as documented in a previous work, this attitude is commonly found in other geographical areas (Mingote Calderón 1993). A good example appears in a song from San Cugat del Vallés (Barcelona; Fig. 3.13) collected by J. Amades in which birds, worms, fog and farmers are considered consignees of the grain sharing: *Un pels ocells / un per les cuques / un per la neula / i un pel pages* (one part for the birds, one for the worms, one for the fog and one for the farmer). The idea behind this conciliatory ritual was that animals would be satisfied with the sharing and refrain from harming the remaining harvest. This practice is found in many regions of Europe and Africa and many examples also come from earlier times. Claudius Elianus (1989, 643–635), a Greco-Roman author of the 2nd–3rd century, recalled in his *History of Animals*, a type of behaviour very similar to that just described. He obtained the information

Fig. 3.16. Location of the Périgord in south-western France (Dordogne). Map: R. Lugon, J.-C. Loubier and A. Chevalier.

from Theopompus and Lico. According to Elianus, in the Venetian area, rooks were fed with barley pies whose flour had been kneaded with honey and oil. The idea was to avoid birds eating the sown grain. In addition, the field was fenced in with red ribbons to prevent the birds from venturing into it. The same author explains that two or three rooks had the role of informing the others of the non-aggression pact established.

J. Amades (1989, 291) described a further protection ritual in Linyola and villages from the Urgell area (Lérida; Fig. 3.13). In this area, a pitcher with a live toad was buried to prevent birds from eating the grain, and to keep mice away. In Pla de Barcelona (Fig. 3.13), on the last day of the harvest, farmers buried a pot which had previously been used for cooking. During ploughing, the pot would get broken and this was considered a sign of a good harvest (J. Amades 1989, 284). P. Sébillot (2005, 240–241) also referred to similar rites in the Périgord (France; Fig. 3.16), which involved burying frogs in the field. He recalled similar practices from Portugal documented by J. Leite de Vasconcellos, related to frightening off birds. The first consisted of burying a pitcher containing ox bile for a full night. After the burial, a person turned around the field saying: *Passarinhos, dexai o meupainço, / que tem fel! / Ide para o monte / Que tem mel!* (Birds keep away from my millet. It has bile. Go to the mountains where there is honey!) (Leite de Vasconcellos 1986, 200–201). Then, before the harvest, the bile should be removed as it could make the millet bitter. The second took place in Cinfâes (Fig. 3.13), and the system used to keep birds away consisted of carrying a basket (*panela*) with the mouth oriented towards the farmer's breast and turning round the field three times saying: *Passarinhos de arzel (sic), ao monte / Que o monte tem mel / E a minha sementeira / Tem fel* (Birds of *arcel* (unknown word), go to the mountain. There is honey in the mountain and my seeds have bile) (Leite de Vasconcellos 1986, 200–201). C. Pedroso (1988, 206 and 209) also documented similar rituals during the 19th century. There is a long tradition of burying objects with the aim of protecting the crops. It is well documented in Pliny the Elder (1972, 110 and 154–155, in XVIII, XLV and LXX) who recalled different ways of protecting crops.

The attitude towards plagues of locusts has also led to a series of practices with a long tradition in Europe including trials by the ecclesiastic authority. The killing of locusts is well documented. Locusts were ritually destroyed by burning them in ovens, by beating them or by using other animals. These are practices that lasted well into the 20th century (Mingote Calderón 2005). There were also many other rituals associated with fire and water which during the 19th and 20th centuries lost part of their meaning in relation to crop cultivation. However, some authors have stressed the connections between what nowadays is a festive celebration and what used to be part of the agrarian rituals of the past (Taboada Chivite 1980; Caro Baroja 1974; 1980).

From the preceding it is clear that the examples of ritual protection of crops are endless (for an overview of further Spanish examples see Mingote Calderón 1995, 153–162).

3.4. SEASONAL VARIATIONS IN CROP TENDING AND FOLK KNOWLEDGE IN SOUTHERN FRANCE

Thomas K. Schippers

It is common knowledge that the agricultural year-cycle is composed of a series of more or less dense 'temporal islands' (Bourdieu 1976, 43–86) in regard to plant and animal biological cycles as well as to climatic seasonality. Local agro-systems often make complex adjustments, choices and compromises among these various natural cycles that have been socialised in more or less precise 'local knowledge' (Geertz 1983), composed of (oral) prescriptions, (written) regulations, calendars and proverbs. Hesiod's *Works and Days* is a classic example (West 1978). A local agricultural year appears in this regard as a succession of intense 'high times' and periods of daily chores. While sowing and harvesting of plants and the birthing of animals are generally periods of intense human activity and stress, the periods of tending growing crops and animals are marked by everyday routines and boredom.

As an example of these seasonal variations in 'local knowledge' concerning bio-climatic changes and risks, my study of Provençal popular weather forecasting has proved quite enlightening. Out of a corpus of about sixty popular proverbs and sayings concerning the agricultural calendar and/or the weather, gathered from folkloristic texts and by ethnographic fieldwork, an overwhelming majority referred to periods of sowing and harvesting (Schippers 1992). Many of these refer to a day or period in the Catholic calendar of Saints followed by an agricultural prescription or a meteorological observation. Beyond their immediate and explicit message, these proverbs and sayings allow a sense of control over the hazards of nature. They reflect a particular (mental) form of domestication of bio-climatic variability by suggesting the existence of a perfect, ideal-type agricultural year-cycle. Like concrete actions undertaken during the various phases of the plant or animal cycle, these 'fossilised pieces of folk-knowledge' contained in the proverbs and sayings, reflect the permanent, though variable, vernacular attention paid to plants and animals. Here precise, daily bio-climatic observations combine (when possible) with precise actions within a framework of local perceptions of 'concrete and operational times' (Leroi-Gourhan 1965, 144; Schippers 1986, 207–220). However, by distributing the elements of local knowledge over a calendar year, it also becomes evident that some periods totally lack these forms of local knowledge. The analysis of this apparent 'silence' in the local repertoires of folk agro-meteorological wisdom suggests two types of explanations: firstly, it points to periods when no crops are growing or being tended, as in late autumn or winter; secondly, it concerns periods during which little or no action can be undertaken to protect or improve the growth of crops or flock. This seems to suggest a circular causality between local folk knowledge and the possibilities of performing concrete action on crops or animals. This hypothesis seems also to be confirmed by a noticeable difference in the local Provençal society between agriculturists on the one hand and transhumant pastoralists on the other. While the former mainly master the calendar-based proverbs concerning sowing and harvesting plants as well as some long-term folk sayings about 'good' or 'bad' years or harvests, the latter often master a significantly richer oral repertoire of folk meteorological knowledge. This difference points to the difference in the modalities of actions these two groups can undertake concretely to prevent accidents or 'worries' with respect to their crops or their sheep (Schippers 1992). One may conclude from this that tending crops not only consists of concrete actions and skills, but also of the mastering of local knowledge repertoires dealing with the more unpredictable ecological 'hazards' occurring during a yearly cycle. Here folk knowledge not only expresses the mental and social 'domestication' of the various agricultural times but also of the unforeseen and the unpredictable events that characterise every new yearly cycle.

3.5. CONCLUSION

José Luis Mingote Calderón

One of the first conclusions of this chapter is the need to get a deeper and more scientific knowledge about many of the numerous issues included under the heading of 'tending crops'. With the exception of irrigation, the remaining tasks related to the care of crops have been little studied. The reasons are many, including the poverty of the material culture associated with the different steps involved in the process of tending a crop between sowing and harvesting and complexity of the spatial implications related to the study of crop distribution and the physical boundaries (fences) of crops. This kind of spatial approach has not been common amongst historians and anthropologists, although archaeologists have published on this topic quite extensively (*e.g.* Pryor 1998). In the case of scarecrows, the scenario is somehow different. There are many books dedicated to them but the approach has often been more related to their artistic value than to their anthropological implications.

Different perspectives are needed if the aim is enlarging our knowledge of agrarian preindustrial crop-tending. This change of perspective should clearly include the ritual practices encountered within the technical process, because in many cultures, including western ones, the idea of a divinity as the giver of the agrarian products is strongly present, as is belief in malevolent spiritual powers that threaten the crops. This explains why human beings and their work also need rituals to assure a good harvest. Rituals associated with the protection of crops during the growing period in the Roman world are good examples of countering the uncertainty created during the early phases of the growing season, and it should not be forgotten that ritual protection also justifies the tithe that peasants had to pay to the church.

This chapter also shows how, in the future, the sole emphasis on the study of technical aspects needs to be extended to also include ritual aspects of crop protection. Not only did people water and weed their fields, they created many objects that embody references to the idea of protection. The so-called 'ray stones' fixed to ploughs from France, Denmark, Scotland or Ireland cannot be considered senseless elements; they were there to protect the ard both physically and symbolically (Mingote Calderón 1996, 90). The same applies to the badger skins hung from the yoked oxens' manes in the Basque Country (Fig. 3.1) or to the designs of crosses and other religious symbols engraved on yokes in Portugal.

The choices of symbols and materials are not casual. Their meaning refers to the human need to obtain protection for material objects and technical actions from the supernatural forces. The diversity of attitudes towards specific implements or tools seems to be related to technical behaviours followed during the process of cultivation. A good example comes from the work of P. Sébillot (2005, 247–248), on the respectful attitudes towards ploughs, or from that of X. Lorenzo Fernández (1974, 63), who mentions the deference towards the farm cart boxes, fallen into disuse. These are not only peasant superstitions, since the high religious hierarchies had an active involvement in setting up the idea that religion and agricultural techniques were closely intertwined. In fact, in the history of mankind, the subjugation of technique to religion has been very common.

Chapter Bibliography

Alonso Romero, F. (1993–94) El espíritu del grano: tradiciones agrícolas propiciatorias en Galicia y en otras comunidades europeas. *Cuadernos de Estudios Gallegos* 16 (106), 367–389.

Amades, J. (1989) *Costumari català. El curs de l'any. Vol. V. Tardor*. Salvat, Barcelona.

de Aranzadi, T. (1928) Espantajos de ingenio y Monigotes de superstición. In S. de Mugica and T. de Aranzadi (eds.) *Homenaje a D. Carmelo de Echegaray (Miscelánea de estudios referentes al País Vasco)*. Diputación de Guipúzcoa, San Sebastián, pp. 31–40.

Ballesteros Arias, P. (2002) *A paisaxe agraria de Elviña: os elementos e as formas*. Laboratorio de Patrimonio – CSIC, Paleoambiente e Paisaxe (IIT – USC), Santiago de Compostela (available online at http://digital.csic.es/handle/10261/5586).

Ballesteros Arias, P. (2003) *La arqueología en la gasificación de Galicia 17: el paisaje agrario*. Laboratorio de Patrimonio – CSIC, Paleoambiente e Paisaxe (IIT – USC), Santiago de Compostela (available online at http://digital.csic.es/handle/10261/5519).

Ballesteros Arias, P. (2010a) *A paisaxe cultural de Brañas de Sar (Santiago de Compostela): arquitectura da auga*. Laboratorio de Patrimonio – CSIC, Paleoambiente e Paisaxe (IIT – USC), Santiago de Compostela (available online at http://digital.csic.es/handle/10261/27652).

Ballesteros Arias, P. (2010b) La arqueología rural y la construcción de un paisaje agrario medieval: el caso de Galicia. In H. Kirchner (ed.) *Por una arqueología agraria. Perspectivas de investigación sobre espacios de cultivo en las sociedades medievales hispánicas*. BAR International Series 2062. Archaeopress, Oxford, pp. 185–202 (available online at http://digital.csic.es/handle/10261/21663).

de Barandiarán, J. M. (1984) *Diccionario de mitología vasca*. Txertoa, San Sebastián.

Bayet, J. (1984) *La religión romana. Historia política y psicológica*. Cristiandad, Madrid.

Bouhier, A. (1979) *La Galice. Essai geographique d'analyse et d'interpretation d'un vieux complexe agraire*. Imprimerie Yonnaise, La Roche-sur-Yon.

Bourdieu, P. (1976) Le sens pratique, *Actes de la Recherche en Sciences sociales* 1, 43–86.

Bouza Brey, F. (1982) Ritos agrarios propiciatorios del espíritu de la tierra en Galicia. In F. Bouza Brey (ed.) *Etnografía y folklore de Galicia. Vol. 1. Edición preparada por José Luis Bouza Álvarez*. Edicións Xerais de Galicia, Vigo, pp. 93–116.

Candal González, X. M. (1993) Pleitos de aguas en la audiencia coruñesa durante el siglo XVIII, *Obradoiro de historia moderna* 2, 85–103.

Casas Gaspar, Enrique (1950) *Ritos agrarios. Folklore campesino español*. Escelicer, Madrid.

Caro Baroja, J. (1974): *Ritos y mitos equívocos*. Itsmo, Madrid.

Catón, M. P. (1979) [Marcus Porcius Cato] *On agriculture* [Marcus Terentius Varro] *On agriculture*. With an english translation by William Davis Hooper... revised by Harrison Boyd Ash. Harvard University Press / William Heinemann Ltd., Cambridge/London.

Cerrato Álvarez, A. (2007) *Vellos traballos comunitarios do mundo labrego*. Asociación Galega para a Cultura e a Ecoloxía (AGCE), Vigo.

Charles, M., Hoppé, C., Jones, G., Bogaard, A. and Hodgson, J. G. (2003). Using weed functional attibutes for the identification of irrigation regimes in Jordan. *Journal of Archaeological Science* 30, 1429–1441.

Charles, M., Bogaard, A., Jones, G., Hodgson, J. and Halstead, P. (2002) Towards the archaeobotanical identification of intensive cereal cultivation: present-day ecological investigation in the mountains of Asturias, northwest Spain. *Vegetation History and Archaeobotany* 11, 133–142.

Charles, M. and Jones, G. (1997) FIBS in archaeobotany: Functional interpretation of weed floras in relation to husbandry practices. *Journal of Archaeological Science* 24, 1151–1161.

Ciriacono, S. (ed.) (1998) *Land Drainage and Irrigation*. Studies in the History of Civil Engineering, 3. Ashgate, Brookfield.

Costa, J. (1981) *Derecho consuetudinario y economía popular. Tomo II, Zaragoza, Guara*. Elías López Morán contribution about León province, 237–332.

Dalby, A. (transl.) (2011) *Geoponika: Farm Work*. Prospect Books, Totnes.

Eiras Roel, A. (1990) *Estudios sobre agricultura y población en la España Moderna*. Tórculo Edicións, Santiago de Compostela.

Eliano, Claudio (1989) *Historia de los animales*. Edición de José Vara Maldonado. Akal, Madrid.

Espantallos (2002) *Espantallos da provincia de Lugo. Museo Provincial de Lugo*. Deputación Provincial de Lugo, Lugo. [Coord.: Encarnación Lago González].

Farri, S. (1984) *Spanventapasseri*. Testo de Massimo Mussini. Bizzocchi Editore, Reggio Emilia.

Flores, H. L. (2008) Paisaxes humanizadas, terras vividas. Fotografías de Luis Higinio Flores Rivas. Museo do Pobo Galego, Santiago de Compostela.

Fraguas Fraguas, A. (1996) A mitoloxía da auga, in

Díaz-Fierros Viqueira, F. (coord.). *As augas de Galicia.* Santiago, Consello da Cultura Galega, 9–40.

Geertz, C. (1983) *Local knowledge.* Basic Books, New York.

Glick, T. F. (1970) *Irrigation and Society in Medieval Valencia.* Harvard University Press, Cambridge, Mass. (available online at http://libro.uca.edu/irrigation/irrigation.htm).

Glick, T. F. (1972) *The Old World Background of the Irrigation System of San Antonio, Texas.* Texas Western Press, El Paso.

Glick, T. F. (1996) *Irrigation and Hydraulic Technology: Medieval Spain and its Legacy.* Variorum, Aldershot.

González Alcantud, J. A. and Malpica Cuello, A. M. (coords.) (1995) *El agua. Mitos, ritos y realidades. Coloquio internacional. Granada, 23-26 de noviembre de 1992.* Diputación Provincial de Granada - Editorial Anthropos, Barcelona.

González Urquijo, J. E., Ibáñez Estévez, J. J., Peña-Chocarro, L., Moreno García, M. and Zapata, L. (2005) An ethnoarchaeological project in the Western Rif (Northern Morocco): First results. *Quaderni del Laboratorio di Archeobiologia* 1, 21–32.

Haudricourt, A.-G. and Jean-Brunhes Delamarre, M. (1955) *L'homme et la charrue à travers le monde.* Galimard, Paris, Gallimard 1955, re-edited by La Manufacture, 1986.

Isager, Signe and Skydsgaard, Jens Erik (1995) *Ancient Greek Agriculture. An Introduction.* Routledge, London/ New York.

Jones, G., Bogaard, A., Halstead, P., Charles, M. & Smith, H. (1999) *Identifying the intensity of crop husbandry practices on the basis of weed floras.* The Annual of the British School at Athens 94, pp. 167–189.

Le Bonniec, H. (1958) *Le culte de Cérès à Rome. Des origines á la fin de la République.* Librairie C. Klincksieck, Paris.

Le Bonniec, H. and Le Boeuffle, A. (transl.) (1972) *Histoire naturelle. Livre XVII.* Société d'Édition « Les Belles Lettres », Paris.

Leite De Vasconcellos, J. (1986) *Tradições populares de Portugal.* Imprensa Nacional Casa da Moeda (INCM). First edition in 1882, Livraria portuense de Clavel, Porto.

Lema Suárez, X. M. (1996) Os derradeiros batáns, en vías de desaparición. In A. Fraguas Fraguas and X. A. Fidalgo Santamariña (coords.) *Tecnoloxía tradicional. Dimensión patrimonial. Valoración arqueolóxica. Ourense 13 ó 15 de outubro de 1994.* Consello da Cultura Galega, Santiago de Compostela, pp. 143–178.

Leroi-Gourhan, André (1965) *Le geste et la parole: la mémoire et les rythmes.* Albin Michel, Paris.

López Fernández, J. A. (2006) La transformación del paisaje en el regadío tradicional de la huerta de Mula. *Papeles de Geografía* 44, 59–72.

Lorenzo Fernández, X. (1974) El carro en el folklore de Galicia. *Revista de Dialectología y Tradiciones Populares* 30 (1–2), 43–76. Galician version editd by Museo do Pobo Galego, 2003.

Lorenzo Fernández, X. (1982) *A Terra. Editorial.* Galaxia, Vigo.

Malinowski, B. (1925) Magic, Science, and Religion. In J. Needham (ed.) *Science, Religion, and Reality.* The Macmillan Company, New York, pp. 19–84.

Michelsen, P. (1986) Irrigation in the Alps. *Tools and Tillage* 5 (3): 160–173.

Michelsen, P. (1987) Irrigation in Norway and Elsewhere in Northern Europe. *Tools and Tillage* 5 (4), 243–260.

Michelsen, P. (1995) *Vand til tørstig jord. Kunstvandingens historie, hvor, hvornår, hvordan. The where, when and how of irrigation.* Christian Ejlers, Copenhagen.

Mingote Calderón, J. L. (1993) Un milagro de San Isidro relacionado con ritos de protección del grano durante la siembra. *Revista de Dialectología y Tradiciones Populares* 48 (2), 135–153.

Mingote Calderón, J. L. (1995) *No todo es trabajo. Técnicas agrícolas tradicionales.* Diputación de Salamanca, Salamanca.

Mingote Calderón, J. L. (1996) *Tecnología agrícola medieval en España. Una relación entre la etnología y la arqueología a través de los aperos agrícolas.* Ministerio de Agricultura, Pesca y Alimentación, Dirección General de Planificación y Desarrollo Rural, Madrid.

Mingote Calderón, J. L. (2005) Ritos contra los animales. Notas históricas sobre la protección de las cosechas. In J. L. Mingote Calderón (ed.) *Animalario. Visiones humanas sobre mundos animales.* Ministerio de Cultura, Madrid, pp. 113–128.

Mingote Calderón, J. L. (2007) Una aproximación etnológica a aspectos inmateriales asociados a los aperos agrícolas romanos. In C. Fernández Ibáñez (ed.) *Sautuola* XIII, *Metalistería de la Hispania Romana.* Santander, pp. 589- 603.

di Nola, A. M. (2006) *La negra señora: antropología de la muerte y el luto.* Belacqua, Barcelona.

Owen, T. (transl.) (1807) *The Fourteen Books of Palladius Rutilius Taurus Aemilianus, on Agriculture.* J. White, London.

Pedroso, C. (1988) *Contribuções para una mitologia popular portuguesa e Outros Escritos etnográficos.* Publicações Dom Quixote, Lisboa.

Pérez García, J. M. (2004) Entre regar y no regar: la intensa disputa por unos recursos hídricos colectivos escasos en la Galicia Meridional (1600–1850). In F. J. Aranda Pérez (ed.) *El mundo rural en la España moderna. Actas de la VIIª Reunión Científica de la Fundación Española de Historia Moderna.* Universidad de Castilla-La Mancha, Cuenca, pp. 555–572.

Propp, V. Y. A. (1987) *Les fêtes agraires russes.* Maisoneuve et Larose, Paris.

Pryor, F. (1998) *Farmers in Prehistoric Britain.* Tempus, Stroud.

Rivas Quintas, E. (2000) *Rega e outros servicios.* Grafodos, Ourense.

Rivera, J. A. (2009) *La cultura de la acequia. Agua, tierra y comunidad en el Suroeste de los Estados Unidos de América.* Universitat de València, Valencia.

Sánchez Regueiro, A. (2001) O sangue da terra. A auga como elemento artellador da cultura e das relacións sociais no agro galego. In X. Balboa López and H. Pernas Oroza (eds.) *Entre Nós. Estudios de Arte, Xeografía e Historia en homenaxe ó profesor Xosé Manuel Pose Antelo.* Universidade de Santiago de Compostela, Santiago de Compostela, pp. 757–766.

Sánchez Regueiro, A. and Fernández Prieto, L. (1999) El uso de las aguas en el sistema agrario gallego (siglos XIX y XX). In R. Garabou Segura and J. M. Naredo Pérez (eds.) *El agua en los sistemas agrarios. Una perspectiva histórica.* Fundación Argentaria, Madrid, pp. 339–364.

Satrústegui, J. M. (1980) *Sobre etnografía vasca 1. Mitos y creencias.* Txertoa, San Sebastián.

Schippers, Thomas K. (1986) *Temps vécus, temps perçus.* CNRS, Paris.

Schippers, Thomas K. (1992) Le proverbe comme forme de transmission de savoirs, quelques réflexions à propos d'un corpus de proverbes météorologiques. In J.-N. Pelen and C. Martel (eds.) *Les voies de la parole.* Presses de l'Université de Provence, Aix-en-Provence, pp. 99–105.

Sébillot, P. (1985) *Le folklore de la France. La Flore.* Imago, Paris.

Sébillot, P. (2005) *Le paganisme contemporain chez les peuples celto-latins.* Elibron Classics.

Sparnacci, G. and Tricca, G. (1999) *Toscani un po' speciali: gli spaventapasseri.* 2nd edition. Alinea, Firenze.

Taboada Chivite, X. (1980) Ceremoniales ígnicos y folklore del fuego en Galicia. In X. Taboada Chivite (ed.) *Ritos y creencias gallegas.* Salvora, La Coruña, pp. 237–252.

Van Gennep, A. (1951) *Manuel de Folklore Français contemporaine. Tome Premier 5. Les cérémonies périodiques et saisonieners. Cérémonies agricoles de l'été.* Picard, Paris.

Van Gennep, A. (1981) *Manuel de Folklore Français contemporaine. Tome Premier VI. Les cérémonies périodiques et saisonieners 4. Cérémonies agricoles et pastorales de l'automne.* Picard, Paris.

Vázquez Martínez, A. (1946) Regadíos tradicionales. Las 'levadas' de Arbo', *El Museo de Pontevedra* 4, 172–185.

Veiga de Oliveira, E., Galhano, F. and Pereira, B. (1995) *Alfaia Agrícola Portuguesa.* Publicações Dom Quixote, Lisboa.

Violant i Simorra, R. (1981) El demonio vegetativo del trigo: Ritos agrarios de la cosecha en Cataluña. In R. Violanti i Simorra (ed.) *Etnología pallaresa. Homenatge a Ramon Violant i Simorra.* Escudella-2, La Pobla de Segur, pp. 27–90.

Wang, X.-G. and Wang, L. (1991) On the ancient terraced fields in China. *Tools and Tillage* 6 (4), 191–201.

Wateau, F. (2000) *Conflitos e Água de Rega. Ensaio sobre a Organização Social no Vale de Melgaço.* Publicações Dom Quixote, Lisboa.

West, M. L. (1978) *Hesiod, Works and Days*; edited with prolegomena and commentary by M. L. West. Clarendon Press, Oxford.

Wittfogel, K. A. (1957) *Oriental despotism: a comparative study of total power.* Yale University Press, New Haven.

4 Diversity in Harvesting Techniques

4.1. INTRODUCTION: REASONS FOR VARIABILITY IN HARVESTING TECHNIQUES AND TOOLS

Patricia C. Anderson and François Sigaut (†)

Quite often in prehistory (and in history too), inventions are solutions to problems we do not know much about. To the best of our knowledge, there has not been a synthetic study of harvesting techniques. There is great diversity among harvesting techniques which are little known because the focus has been on sickles and other such implements. The contributions in this chapter will focus on different aspects of diversity and variability, using case studies from ethnographic, experimental and archaeological research. They unfortunately do not give exhaustive coverage to all harvesting techniques and tools; in particular, fascinating studies of the archaeology and history of techniques in European Antiquity, the Middle Ages and Early Modern periods (for example Comet 1992; 2003; Reignier 2002a) are given only sparse coverage in this chapter. This introduction seeks to briefly set the stage and review some parameters to be taken into account when studying variability in harvesting techniques and tools.

The use of wild cereals and other grasses for food and crafts will also be examined here, beginning with a brief look at wild grasses in particular: some are important players in the origins of agriculture, and agricultural societies in the past and present continue to use them.

Ethnographic Examples of Harvesting Techniques for Wild Grasses

The use of wild grain preceded cereal cultivation, and in some cases, led to it. Some harvesting techniques used to collect wild grains are precursors or pre-adaptations to domestication. There is much diversity in techniques for harvesting wild grasses in the ethnographic record, for example seed beaters from California, where grains are beaten into a basket, and in Africa, the swinging of baskets into wild grass panicles is a grain gathering technique (Harlan 1999). In Australia, the film 'Desert People' (Dunlop and Martin-Jones 1966) shows stripping of wild grass grain with the hands by grasping stems and pulling upwards.[1] In the film 'Cram Cram', Edmund Bernus (1968; 1993) showed the harvest of this grass (*Cenchrus biflorus* Roxb.) by beating with sabre-shaped wooden implements that break the stems. The grains, attached to stem fragments, are then recovered using a wooden fork, and the plant is threshed. Because of the spiky nature of the plant, hand harvesting is impractical, and as the seeds do not fall at maturity, simply beating the plants will not suffice. Therefore, the technique largely depends on the behaviour and morphology of the particular plant, as also can be seen in a contribution to this chapter where Anderson and

M'hamdi describe an unusual harvesting technique using a wooden or metal stick to gather not the grain, but rather the leaves of the wild grass *Stipa tenacissima* L., or *alfa*, in central Tunisia.

On the other hand, metal sickles with a serrated edge (Anderson *et al*., this chapter) are used today in northwestern Tunisia to gather tough, serrated leaves of the wild grass *Ampelodesmos mauritanica* (Poir.) Dur. et Schinz for fodder and to make baskets. Indeed they would be virtually impossible to harvest by pulling up by hand, however the grain on the tough, resistant stalks is not harvested at all (Anderson 2006; 2014). Similarly, in some Sahel areas (south of the Sahara), *Pennisetum glaucum* (L.) R.Br. (pearl millet) and sorghum are never harvested with a sickle, whereas sickles are used by the same people for harvesting wild grasses in the bush to obtain straw for weaving mats, etc.

Harvesting Technique Versus Product Obtained

The question that can be asked is, 'what was or is sought by the harvest?' There is not however, merely a kind of evolution of harvesting techniques through time; we would like to emphasise in this chapter that in all periods, a diversity of methods and choices exist. These can be expressed using a table, such as one published earlier by one of us (Sigaut 1991), then adapted with a few modifications to research presented in this chapter. Fig. 4.1 shows a way of classifying the harvesting action in relation to what is actually harvested. This table, which in general goes from simplest to most complex harvesting techniques, compares pulling, picking up, beating, stripping off, cutting by friction such as with a sickle, and finally cutting using friction employing a swinging motion, as for a scythe. These methods are correlated with material harvested: whole plants, individual grains or spikelets, whole seed heads, leaves, handfuls or branches of stems. We will follow the order of presentation of methods given in the table in this introduction, giving more detailed coverage to methods and questions not represented in the chapter case studies.

Pulling

Pulling up (Fig. 4.1, square 1A) often means uprooting, or collecting the whole plant, as for pulses and flax, but we use the general term because pulling can also collect only part of the plant. Despite its simplicity, pulling is an efficient way to harvest both wild and domestic cereals, according to experimental programs and ethnographic examples (Anderson 1991; 1992; 1999; Hillman and Davies, 1990; 1992; Anderson and Whittaker, this chapter; Anderson and Peña-Chocarro, this chapter), and that it was as rapid a harvesting method as using a sickle, as long as the soil is sufficiently friable. In particular, pulling up is an effective harvesting method in situations in which the stands or fields are sparse (*i.e.*, less than 300 stems per square metre, see Anderson 1991; 1998; 1999). It is also useful when plants grow in loose soil and are not firmly enough fixed to be held and cut in place. Pulling techniques may well have been used at the beginning of agriculture, as well as later, and should be taken into consideration when estimating any kind of yield in an archaeological context.

In this chapter, Anderson and Peña Chocarro discuss instances in which this technique is used for domestic cereals. In both cases as much straw as possible is required: in southern Syria, macaroni wheat and hulled barley stems are desired for animal fodder and fuel, and in Morocco, einkorn (hulled) wheat stems are used for crafts and basketry. Although this method would not leave archaeological traces of a particular tool, it collects basal nodes, the bottom of the stem which may, after threshing, be winnowed out with the grain, and found in silos, for example (McCorriston 1998). Anderson and M'hamdi (this chapter) show that pulling using a wooden stick also may be used to collect leaves alone (Fig. 4.1, square 1D) but without uprooting the plant, in which case the archaeological evidence, unless the stick were preserved and could be identified, might be limited to phytoliths identified as from a grass leaf, or from actual remains of basketry objects.

Picking Up

Picking up the grain (Fig. 4.1, square 2B), seems to be restricted to wild plants and to particular wild grasses. One of us (Sigaut) has seen two examples of this in Niger. In the first case, windborne wild grass grains that had accumulated behind fences and walls and high points of the terrain, were picked up as a regular harvesting method. In the second case,

Action carried out	What is harvested as a result of the action					
	Whole Plants A	Individual grains, spikelets B	Whole ears (whole seed heads) only C	Leaves only D	Handfuls of stems E	Bunches of stems F
1) Pulling	By hand, ex. pulses, cereals etc.			Ex. using a harvesting stick for wild grass		
2) Picking up		From wild plants				
3) Beating		Wild plants with panicles				
4) Stripping off		By hand, ex. for wild grasses with panicles	For several ears at once of hulled wheats, using ex. mesorias, flint blades or harvesting knives	Ex. by using hands		
5) Breaking off or Plucking			By hand, one by one, ex. hulled wheats, maize			
6) Cutting by pressure only (blade is pressed against the stem under the ear, but does not move along the stems)			Using a harvesting knife, ex. for millets, sorghum			
7) Cutting by friction (blade moves across the stems)					Ex. using a sickle	
8) Cutting using friction with a swinging motion, without grasping the stems						Ex. using a scythe

Fig. 4.1. Proposed classification of harvesting motions and methods versus what is harvested, as attested to in the ethnographic, historical and archaeological records. Modified from Sigaut (1999).

ant stores of grain were raided by people in times of famine; this practice was probably widespread. In prehistory, picking up was suggested as a harvesting method for wild barley, and one study claimed this would have been an impetus to the beginnings of agriculture and domestication (Kislev *et al.* 2004). It is unlikely that we can generalise this theory, as other experiments picking up wild cereal, particularly wild wheat, from the ground (Anderson 1999) did not find this to be an effective method, because wild cereal spikelets, when they fall, tend to disappear into cracks in the ground, and because of their awns or barbs with saw-like edges, are virtually impossible to extricate. Picking up probably only works for large seed bodies, like nuts, or large natural accumulations as in the case described for Niger, above. It is unlikely that this method could be identified in the archaeological record.

Beating

The seed beaters from California mentioned above, as well as the swinging baskets in Africa (Harlan 1999), used to harvest wild grass seeds by a percussive movement, would correspond to square 3B in Fig. 4.1. Beating works in these cases because of the particular morphology of the seed head of these grasses, and would not appear to be a viable method for obtaining most of the grain or spikelets (which are grain enclosed in glumes) from wild or domestic wheat and barley, for which grain ripens unevenly on any one seed head, causing unripe or semi-green grains to adhere and not be removed by a beating motion according to field experiments (Anderson 1991, 1999; Willcox 1999). However, the agronomist Jack Harlan (1967 and personal communication, 1987) taking a different view of yield, found that this technique was the best for obtaining the highest quality and most viable wild wheat grain, as only the ripest grain would be taken in any one pass (as opposed to other techniques such as cutting with a sickle, pulling up or plucking or picking, which would obtain more grain but of uneven maturity, Anderson 1991; 1999; Willcox 1999).

This category also includes beating with paddles for the wild grass *cram-cram*, referred to above, which obtains the grain, but only by breaking the stems, after which a subsequent process separates the seeds from the stems, another example of a process adapted to the morphology of a particular plant. It appears unlikely that wooden beating tools or specific archaeobotanical remains of beating would be preserved or recognised as such in an archaeological context.

Stripping

Stripping off (in French, *érussage)* using the hands or a handheld tool using an upward movement, is used to collect seeds (Fig. 4.1, square 4B) as in Australia (Dunlop and Martin- Jones 1966), or leaves alone (square 4D), as in the case of women who harvest leaves of the indigo plant in the Dogon region in Mali (personal communication, Isabel Brouillet).

Stripping off numerous whole seed heads at once (square 4C) is effective only for grasses with a certain kind of plant morphology: a fragile point at the base of the seed head or rachis, typical of the hulled wheats einkorn, emmer and spelt. Often very simple tools are used. Peña-Chocarro (this chapter) describes ethnographic examples of this technique in the Spanish Asturias region, using a simple wooden tool, the *mesorias*. This is a case where speed in obtaining the seed may be important, avoiding the need to thresh out the grain from the stalks in humid Atlantic climate conditions.

Experiments (Fig. 4.2) have shown that flint blades, used with the edge pressed against the thumb, can effectively be used to pinch and then to strip off, using an upward motion, several seed heads of hulled wheats at once, specifically emmer wheat (Anderson 1999) and einkorn wheat (Méry *et al.* 2004; Anderson 2013, and see also Fig. 4.2).[2] This action produced characteristic bright microwear traces on the blades, running perpendicular to the working edge. Moreover, there appear to be flint blades with small wooden hafts in archaeological sites from the Swiss Neolithic (Anderson 1999) as well as blades from the Neolithic at Mureybet and Abu Hureyra in the Euphrates Valley in Syria, (Anderson 2013) that have these same microwear traces and could well have been used to strip off seed heads of hulled wheats at these sites.

Another tool serving a similar purpose, but found effective only after hulled wheat and hulled barley stems have first been harvested, is a multiple-notched notched bone tool, made from sheep and goat scapula. Many examples were found at the Neolithic site of Ganj Dareh in Iran (Fig. 4.3). Microwear traces and phytolith and silica residues

Fig. 4.2. Experimental harvesting of einkorn wheat in Seranon, southern France, using a flint blade pressed against the thumb while pulling upwards to strip off (*érusser*) several of the seed heads at once as the motion is continued by pulling the tool upwards. Image: D. MacDonald.

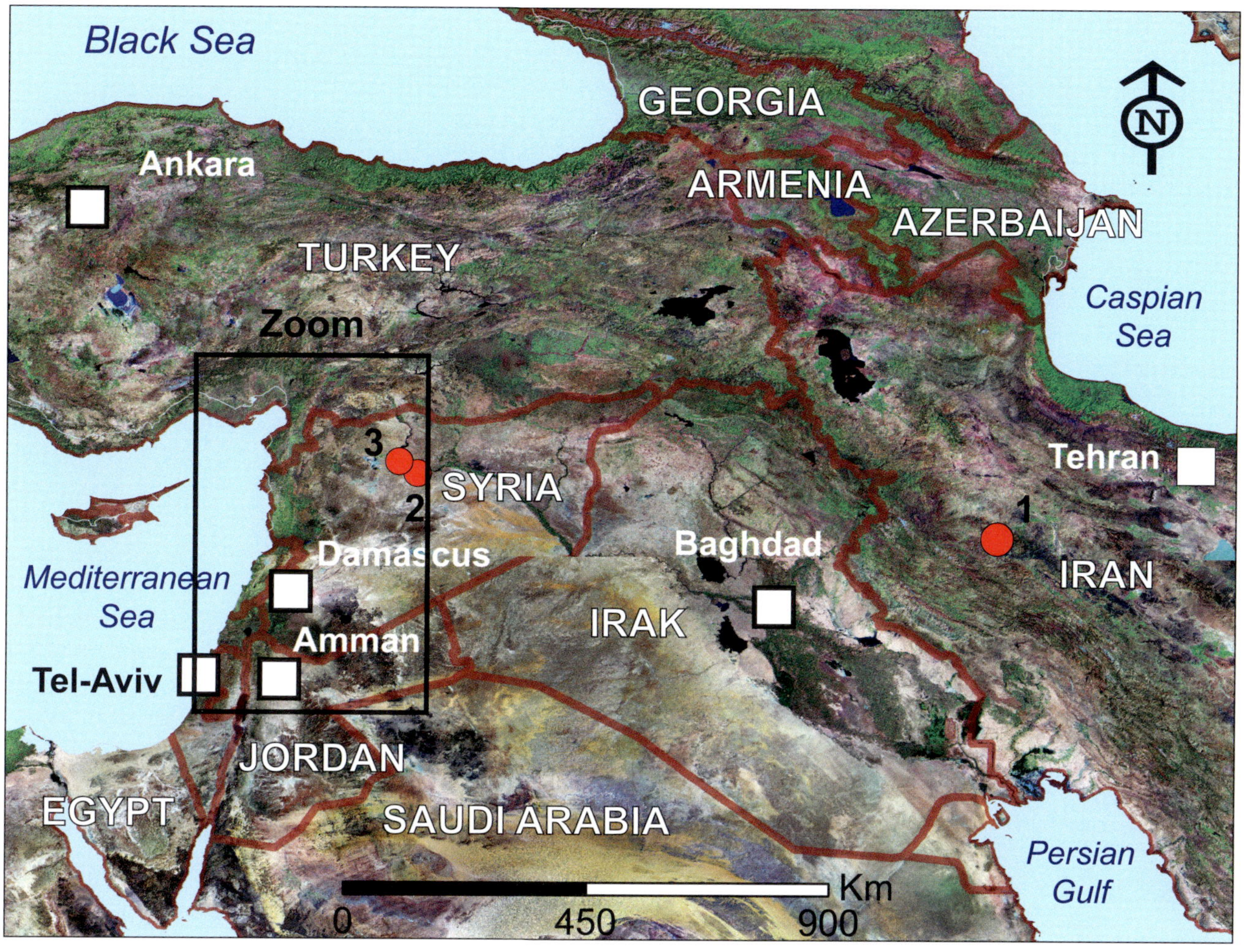

Fig. 4.3. The archaeological sites of 1) Ganj Dareh in Iran; 2) Tell Abu Hureyra in Syria; 3) Tell Mureybet in Syria. Inset box: see Fig. 4.4. Map: R. Lugon, J.-C. Loubier and A. Chevalier.

indicate they were used, pressed against another object such as a long piece of wood or bone, to strip off multiple seed heads at once, from the stems.[3] Experiments found these tools effective and reproduced these traces (Stordeur and Anderson-Gerfaud 1985; Anderson-Gerfaud 1986; Anderson 2013). A similar tool made from scapula but without notches, appears to have been used in Chalcolithic sites in Bulgaria (Skakun 1993, 367–368). Peña-Chocarro (this chapter) discusses the possibility that tools cited by Pliny and Columella, used to strip off grain, could be like the tools above, or less likely, like the *mesorias*. This technique therefore can in some cases leave behind tools with identifiable microwear traces (Stordeur and Anderson-Gerfaud 1985; Anderson-Gerfaud 1986; Anderson op. cit.), and theoretically it could leave behind archaeobotanical remains like those from plucking (breaking off, picking) (see Bakels and Van Gijn, this chapter; Out 2009).

Plucking

Breaking off or plucking or picking one by one of seed heads (square 5C in Fig. 4.1) was found to be an effective method of harvest, albeit slow, for both domestic and wild cereals (Anderson 1999), provided they had a fragile point at the base of the rachis, as do hulled wheats. For wild cereals, plucking the rachis ensured gathering only the ripest seed heads, day by day. In general, this method is effective, as is pulling up the crop, when stands or fields are either dense or sparse (sparse being fewer than 300 stems per square metre, where methods such as sickle cutting or stripping would not work). Sickle harvesting and stripping was found to need a minimum of approximately 600 plants per square metre to function without uprooting the plants (Anderson 1998, 1999). Bakels and Van Gijn (this chapter) describe an archaeological example of stripping or plucking.

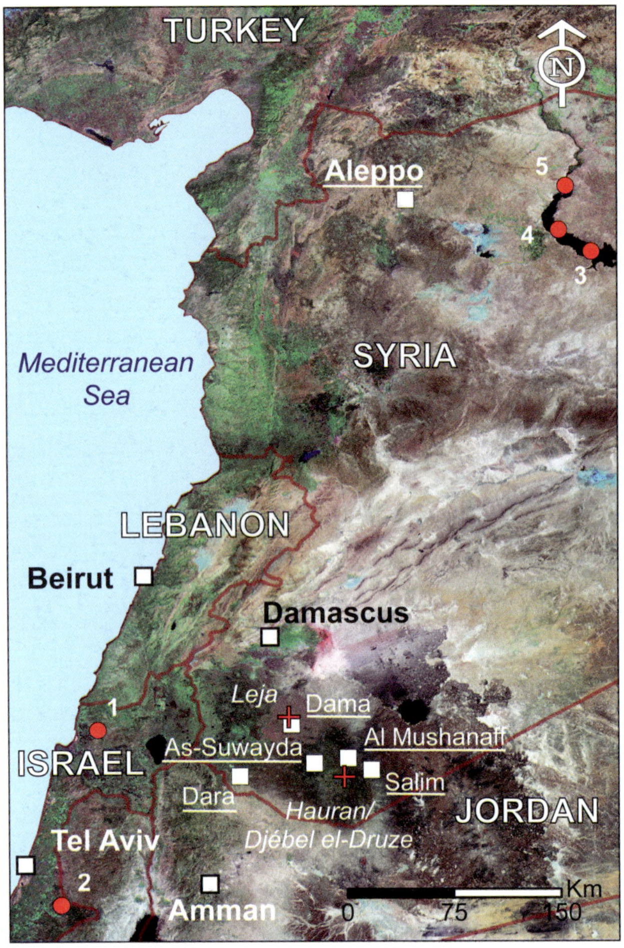

Fig. 4.4. Map of the Near East, with the cities of Aleppo, Al Mushanaff, As-Suwayda, Dama, Dara, and Salim, as well as the Leja and Hauran/Jabal al-Druze (Djébel el-Druze) regions. Archaeological sites are: 1) Hayonim Terrace; 2) Hatula; 3) Tell Abu Hureyra; 4) Tell Mureybet; 5) Tell Halula. Map: R. Lugon, J.-C. Loubier and A. Chevalier.

Cutting by Pressure

Cutting by pressure (6C) involves cutting the stem below the seed head, stem by stem. To the best of our knowledge, this function has not been identified archaeologically to date. Harvesting knives of various shapes are used in this way to gather seed heads that grow in the form of panicles, carrying many grains per seed head, such as for millet and sorghum, in Africa (Mali, Brouillet 2013)[4] and in the Far East (Sigaut 1978). In the countryside near Aleppo, in the north of Syria (Fig. 4.4), one of us (Anderson) saw women using a small kitchen knife, pressing the knife blade against the thumb to cut just below the seed head of unripe bread wheat (*Triticum aestivum* L.) (Fig. 4.5). Some of the 'green' seed heads were being collected to make *frike*, a snack of fried or roasted grain, and the rest of the wheat would be cut later, when it was ripe, near the bottom of the stems, using a sickle.

Cutting by Friction

Cutting by friction (Fig. 4.1, square 7E) is most often represented by the sickle, common around the world in various forms. What was the problem to which the sickle, corresponding to cutting by friction by moving the tool across the stems while holding them in the hand, was the solution? The specific problem that the use of the sickle was designed to solve was possibly the one of *harvesting the straw along with harvesting the grains.* This implies that the straw was valuable enough to warrant the extra work in threshing and transport. Sickles have been well studied ethnographically and experimentally, and represent a solution to several problems of the harvest (Anderson and Peña-Chocarro, this chapter; Anderson and Whittaker, this chapter). Because sickles are relatively distinctive and durable tools,

Fig. 4.5. Near Aleppo, Syria, 1994. Harvesting of ears (grain heads) of bread wheat one by one, at the 'green' stage, using pressure between a kitchen knife and the thumb to cut the stems just below the ear. The green seeds will be fried to make *frike*. Image: P. C. Anderson.

they figure largely in archaeological studies of early agriculture (Gibaja *et al.*, this chapter; Anderson 2000). Archaeological and ethnographic information also shows that sickles and similar tools are not always used to cut only the grain as we expect, even in agricultural societies (Van Gijn, Chapter 9; Anderson *et al.* 2004; Contributions in Van Gijn 1994/1997).

Although many other methods were probably used in prehistoric times to harvest cereals, examples abound of small flakes and blades used in hafts to harvest wild cereals by cutting stems near the ground, in Middle Eastern archaeological sites prior to cereal domestication, such as Hayonim Terrace (Anderson and Valla 1996) and Hatula (Anderson, 1994) in present-day Israel (Fig. 4.4), and at Abu Hureyra and Mureybet in present-day Syria (Fig. 4.4; Anderson 1992; 1999). Many of these individual tools were used for long periods of time, based upon our experiments harvesting wild cereals (Anderson 1991; 1992; 1999), and show great expertise in hafting of stone tool inserts (Anderson and Valla 1996). Experiments have shown that such sickle harvesting of wild cereals may have produced unconscious selection for future morphologically domestic cereals, according to certain scenarios (Hillman and Davies 1992; 1999; Anderson 1999; Willcox 1999). Such harvesting tools persist of course in Middle Eastern Neolithic sites having remains of morphologically domesticated cereal (*e.g.* Anderson 2000). Sickles with particular hafts related to specific harvesting motions are localised in different geographic areas that appear to correspond to distinct pathways of the arrival of the Neolithic economy in southern Europe, specifically Spain and France (Gibaja *et al.*, this chapter).

In general, metal harvesting tools, be they sickles or scythes, are poorly preserved or difficult to identify in archaeological sites. In a case study presented in this chapter (Anderson, Rodet-Belarbi and Moreno-García), metal sickles with teeth or serrations are identified in sites in Southern and Eastern Europe principally due to a proxy tool, a bone anvil used to make the teeth of the blade.

Cutting with a Swinging Motion

The scythe is related to the sickle, as it also represents cutting with friction, but we have distinguished it because it uses a swinging motion and the grain stems are not held by the harvester (Fig. 4.1, square 8F). In fact, there are intermediate tools. A special large sickle, or reaping hook (*volant* in French) was used from the 15th or 16th century by harvesters in the plains of Brittany, Flanders, Auvergne, and the Savoy region of France. The *volant* is between a sickle and a scythe, alternating cutting by friction and a swinging motion (Fig. 4.6).

Fig. 4.6. A farmer in Seranon in the Alpes-Maritimes province of southern France, using the *volant*, a large sickle (or reaping hook) which cuts by alternating friction pulling towards the user, and friction by striking using a swinging motion. Image: H. el Gtari.

Scythes only appear in the Late Iron Age (Second Iron Age) in Switzerland (La Tène period, 5th to 1st century BCE) and were apparently not used for harvesting cereals for a long time. The scythe was used for haymaking from the beginning, then from Carolingian times (8th-9th century), for harvesting oats, and not until the 16th century for harvesting wheat, rye and barley. According to Henning (1991), the scythe with the structure we know today, and scythe anvils, were found in the 7th century in Belgium. The scythe was principally used to make hay for winter and spring fodder. Perhaps its use was restricted because of the cost of its manufacture, or the difficulty of its use compared with sickles, or the fact that it only cuts low on the stalk, whereas the sickle can cut at different levels. The scythe tends to be used by men, whereas the sickle is used by both men and women (Sigaut 1978). The scythe was not known before the 20th century in the Mediterranean area, probably because the main systems of animal husbandry were based upon transhumance between winter and summer grazing areas. The principal fodder used in the Mediterranean is straw finely chopped by treading with animals or the threshing sledge (see Chapter 5). Stubble left in the fields after harvest is also important for grazing. In either case, the efficiency of the scythe is offset by the need for different products of the harvest, a good example of the interconnections between technologies and crops.

People did not begin to harvest cereals with the scythe until rather late, in the 18th or the 19th century, except for certain regions like Poland, which was already exporting large quantities of wheat during the Middle Ages. The earliest cereal harvested in Western Europe using a scythe was apparently spring oats, beginning in the High Middle Ages. These oats were harvested to feed horses, whereas other oats, known as winter oats, were used to feed people, and were harvested using a sickle. Therefore, before the 19th century, the scythe was used only to harvest pasture grasses or oats, for animal fodder.

The scythe adapted to harvesting cereals has a cradle or frame made of wood and rope (Fig. 4.7), against which the harvested stems can be caught, then laid on the ground in an orderly fashion. Today the scythe, but without a cradle, is used in Tunisia for spring (green) harvest of fodder plants such as alfalfa (*Medicago sativa* L.), and green barley and oats, then barley and oats are harvested in the summer after ripening, using a sickle. The sickle is used in preference to the scythe because these crops are harvested ripe for occasional human consumption and animal consumption, as well as for seed, and using the scythe would cause grain spillage (Anderson 2006).

Fig. 4.7. Using a scythe with a cradle, a device for catching the stems as they are cut, used for harvesting of cereals, here bread wheat (*Triticum aestivum*), during a 1999 Festival in La Fuliola, Catalonia, Spain. Image: P. C. Anderson.

As we have seen, different harvesting methods correspond to using the bare hands or using a tool, and are chosen according to what part of the plant is desired; some work only for certain plant morphologies. In cases where there is a choice, the density of the field, the type of soil, and the use to which the stems will be put can be determining factors. Only the scythe is unlikely to have been used during prehistoric times, although only some of the harvesting methods can be identified archaeologically.

The contributions in this chapter will follow the order in which the methods are presented in Fig. 4.1, and concern pulling up, stripping off, picking, and cutting using friction (with sickles).

4.2. HARVESTING BY PULLING UP THE CROP BY HAND: AN 'INVISIBLE' METHOD?

Patricia C. Anderson and Leonor Peña-Chocarro

Introduction

The most commonly observed technique for harvesting cereals in non-mechanised agricultural systems is harvesting with a sickle (this chapter), or in some more recent contexts, with a scythe (introduction, this chapter; chapter 8). However, crops were probably harvested by pulling up, or uprooting, more commonly than is usually assumed. Ethnographic examples, such as those described below, show it can be the technique chosen for the whole crop, rather than simply an occasional harvesting variant. Recent experiments show its feasibility and rapidity for both wild and domestic wheat (Anderson and Whittaker, this chapter; Anderson 1999; Hillman and Davies 1990; 1999).

Pulling up the crop, usually in handfuls, can be done using both hands or one hand. A sickle or other tool may be held in one hand to use as a lever to assist in pulling up the crop, which is grasped in the other hand. A sickle used in this way would probably not have traces recognisable to the microwear analyst, as it does not cut through any plant material. Pulling up the crop is done with the harvester bending at the waist or squatting down near the ground, with these working positions sometimes alternated to avoid fatigue. In some cases, the roots are pulled up along with the stems and seed heads; in others, the harvesting motion of the hands is such that the stems are cracked off at the base, near the soil, without removing the roots. This choice may depend upon the nature of the soil (*i.e.* if it is friable or hard), on climate, on whether the plants have long or short stems and on what follows the harvest, for example how the plants will be threshed and to what use their stems will ultimately be put. This article will provide an ethnoarchaeological approach to understanding why this technique may be chosen, using case studies from Syria, Spain and Morocco.

Examples from Syria

Crops commonly harvested everywhere by pulling up are the pulses such as chickpea (*Cicer arietinum* L.), horse bean (*Vicia faba* L.), and lentil (*Lens culinaris* Medik.). This contribution focuses on pulling up of cereals. Our first example comes from the Hauran, including the Jabal al-Druze region in mountainous southern Syria (Fig. 4.4), where this is the technique of choice for both hard (durum) wheat (*Triticum durum* Desf.), and for hulled barley (*Hordeum vulgare* L. var. *vulgare*). A French military officer states:

> « *dans le Hauran et dans certains villages du Djebel-sud, on ne se sert même pas de faucilles. Les moissonneurs, groupés sur un rang, arrachent à la main les tiges tout entières, pendant que derrière eux attendent les glaneuses qui sont des pauvresses bédouines ou chrétiennes* » (N. Bouron 1930, 337)
>
> 'In the Hauran and in certain villages in the southern Jebel, people do not even use sickles. Harvesters, in a row, pull up whole stems by hand, as poor Bedouin and Christian women wait behind them to glean what remains.'

Bouron considered the method primitive, but uprooting is in fact efficient and widespread. One of us (Anderson) also saw this method of grain harvest used elsewhere in the Hauran, in the Leja region as well as north of Deraa, and even near Aleppo (Fig. 4.4). The context of each instance is slightly different.

Near Al Mushanaff in the high Jabal al-Druze, hulled barley was uprooted in June 2001. In the mountains, barley does not grow very high and can be sparse, so uprooting is a logical way to harvest quickly. In addition, it ensures the collection of as much stem as possible to be chopped up during threshing for animal fodder. The barley crop is now used for animals, as barley is 'famine food' and wheat bread is considered to taste better. As is often the case where pulling up is used, the soil was friable and the uprooting went very quickly. The male and female workers, a family group, said that they harvested from 4–11 in the morning, after which time the heat was too great, then they continued from 3 in the afternoon until dusk and sometimes even into the night, working by moonlight. About 15 handfuls were pulled up before they were laid down in a stack, then the process repeated. No sheaves were tied in any of the cases of uprooting cereals that we observed. In the past, and occasionally today, a threshing sledge was used, working for 10–15 days, 4–5 hours per day (see Anderson, Chapter 5).

Somewhat taller barley was seen being harvested by uprooting, in the plains in 1995, by men and women, undoubtedly families, working together along the Deraa-Damascus highway near Sweida (Fig. 4.8); the man on the left wore protective covers on his fourth and fifth fingers and the women wore rubber gloves.

A father, who was a Sweida Museum guard, and four of his teenage children, were harvesting the family field of tall stalks of macaroni wheat (*Triticum durum*) in a large, dense field near Salim (Fig. 4.4), which grew to a height of 75–90 cm, in the early evening in late June of 2004. They used a motion of pulling towards the harvester, at an oblique angle. If too much soil was taken, they beat stem bases and roots with a stick to remove it. One young girl carried wheat from each harvester to a make a large stack, and carefully arranged seed heads in the same direction, and when the wind rose, she laid basalt stones on the stacks to weight them down. Here too, the father wore special finger protectors of metal on the fourth and fifth fingers, whereas the children wore black rubber gloves, which we learned were sold especially for this purpose.

Another harvester in Salim, a retired truck driver from Lebanon, who was harvesting a smaller field of macaroni wheat in 2004, lived with his wife in a small house and kept cows for milk. He was harvesting macaroni wheat principally for family consumption, and as fodder for his animals. He was alternately pulling up the wheat by hand, and using a tiny sickle in his right hand as a lever while pulling with his left hand, whenever thistles occurred in the field. He also used the sickle to help carry and stack harvested stems. As he pulled up the plants, he tapped dirt off the roots against the heel of his boot.

Finally, in Dama, in the Hauran (Fig. 4.4) in 2004, a couple were pulling up macaroni wheat (Fig. 4.9).[5] The sheaves were not tied, but were stacked on a net which was tied shut into a cylindrical shaped bundle, and this was transported on donkey back to the house or to the threshing floor alongside. They said that because the crop was abundant, they would call in the mechanical thresher to thresh their crop. They still use the threshing sledge with blunt basalt inserts (see Chapter 5) for small quantities of pulses, however, such as broad bean (*Vicia faba*).

Discussing the technique with those harvesting macaroni wheat (*Triticum durum*) always brought a similar response: it is important to harvest the basal node or hard stem base of macaroni wheat, and as much of the stem as possible. Although the wheat is still threshed using a threshing sledge, albeit an unusual variant using the only stone available in the region (rounded basalt chunks), winnowing is used to separate out the hard stem bases, which are burned as fuel to warm houses or for the bread oven, the softer stem fragments which are fed to the animals, and the grain (Anderson 2003).

The technique is occasionally seen in Syria outside the Hauran and Jabal al-Druze (Fig. 4.4): for example, in 1994 near Aleppo, on friable soil, this technique was used as a harvesting alternative to the combine harvester in a year of poor yield (Fig. 4.10).

In each case, in these dry farming areas, the regular use of uprooting seems to correspond with friable soil and the desire to obtain as much stem as possible for animal fodder, fuel, and to temper plaster. Small field size, slope and altitude, may also play a part. This technique appears to correspond to subsistence agriculture, rather than any kind of commercial system; families and neighbours and friends were harvesting for their own use. In several instances, some harvesters were aware of the sickle,

Clockwise from top left:

Fig. 4.8. Near Sweida, Syria, 1995, harvesting hulled barley by pulling up or uprooting by hand, wearing either protection on the fourth and fifth fingers (man on the left), or rubber gloves. Image: P. C. Anderson.

Fig. 4.9. Harvesting by uprooting bread wheat in the early hours of the morning near Aleppo, Syria, in 1996. Image: P. C. Anderson..

Fig. 4.10. Near Dama, Syria, 2004. A couple harvesting macaroni wheat by pulling up by hand. The hard base of the stem is needed as a source of fuel, and will be isolated from the grain and other parts of the stem by cutting and winnowing with a threshing machine. Image: P. C. Anderson.

but said this tool was more useful only where crops are irrigated or stems grew high, although this is contradicted by some of the above examples. In the Hauran (Jabal al-Druze) area of Syria (Fig. 4.4), this technique, other than being used for practical reasons described above, appears to correspond to cultural habit and individual preference.

Examples from Spain and Morocco

Uprooting was used in Spain in areas of the Canary Islands and in particular in Lanzarote Island (Fig. 4.11) where barley grew very short. The small size of the plants made harvesting with sickles an uncomfortable task. In addition, this particular barley from Lanzarote grows in volcanic soils which allow easy uprooting. After harvesting, the bundles of barley are cut right at the middle of the stem by passing them over the toothed edge of an iron blade inserted into the surface of a work bench. This operation removes the roots which would eventually contaminate the cereal if processed together in the threshing yard. This uprooting method is well documented in the 15th-century chronicles which described the ways of life of the aboriginal communities of the Canary Islands. Abreu Galindo, a chronicler of the Spanish conquest, left a clear description of the method: they sowed barley by turning over the land with the horns of goats, and, when ripe, they uprooted it and cleaned it (Abreu Galindo, 1602?/1977).

In other areas of the Iberian Peninsula, uprooted cereals were processed in a different manner to avoid mixing clay attached to the roots with the clean grain. In the Basque Country, for instance, roots were not cut off, but the bundles were struck against a hard object to separate the grains from their glumes. In Morocco, farmers also uproot barley in areas where crops are thin and short due to environmental constraints (Fig. 4.12).

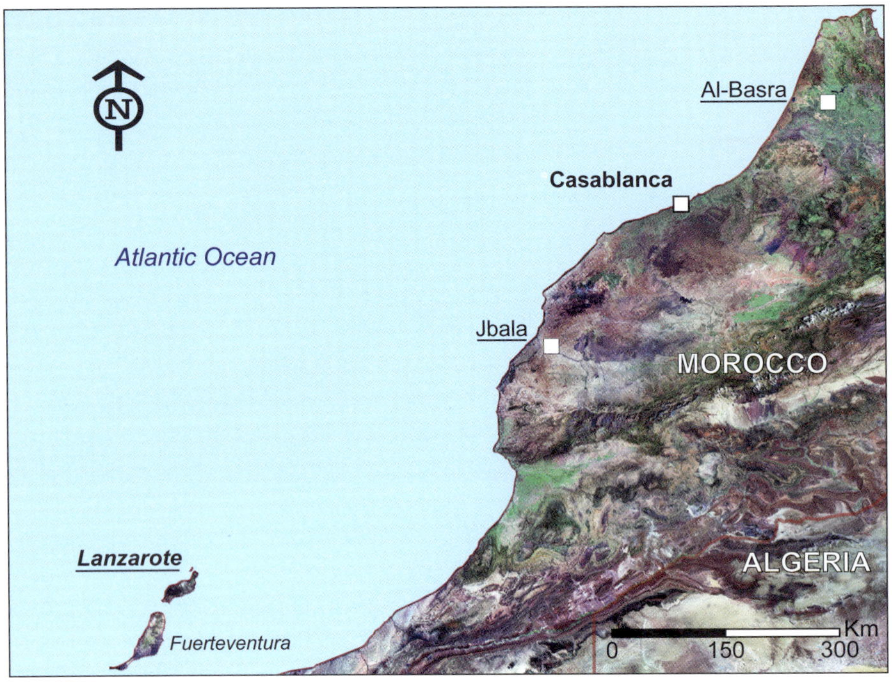

Fig. 4.11. Map of north-western Africa with the Canary Island Lanzarote, and the Moroccan cities of Al-Basra and Jbala. Map: R. Lugon, J.-C. Loubier and A. Chevalier.

Examples from Morocco

Research into traditional agriculture in the western Rif (north Morocco) (Peña-Chocarro *et al.* 2000; 2005; 2009; Peña-Chocarro and Zapata 2003) has emphasised the importance of a marginal crop, einkorn wheat (*Triticum monococcum*), to the local economy. Work carried out in the mountain area of the Jbala (western Rif; Fig. 4.11) where einkorn is still under cultivation and generally uprooted (Fig. 4.13) allowed one of us (Peña-Chocarro) to explore the reasons farmers chose this technique.

Einkorn is a cereal mostly cultivated to obtain raw material for thatching. Its straw is particularly good because of its length and resistance to humidity. Small fields dedicated to einkorn cultivation provide enough material for re-thatching houses every year. In this case, einkorn tends to be harvested by uprooting (or by using sickles and cutting very low on the stem), to obtain long stems for thatching. Uprooting may be done with the hands alone, or also using a type of tool fitted with a bent iron blade (*mugtaf*). The same implement is used also for cutting branches for green fodder. The blade of the *mugtaf* is used as a lever for pulling at the base of the stem. Similarly, when sickles are used during uprooting, these are used by thrusting the pointed end into the ground and pulling out the plant.

Fig. 4.12. Uprooting barley in southern Morocco. Image: L. Peña-Chocarro.

Conclusion

In all these cases, we see that the advantages of uprooting relate to the total length of the stems obtained, but also, to the presence of friable soil and for barley, to its low height. Uprooted stems are longer and more appropriate for thatching or crafts, and provide more material for fuel or fodder in these regions, where trees are rare. Moreover, uprooted straw is protected by the roots attached to

Fig. 4.13. Uprooting einkorn for thatching in the Rif (northwestern Morocco). Image: L. Peña-Chocarro.

the stems, and does not degrade as fast as cut straw. Uprooting is efficient and can be as rapid as using a sickle in loose soils, and, unlike sickles, it works well for sparse crops. As we have seen, pulling up is also used for densely growing, tall crops, and this choice is made for reasons stated above, as well as cultural preference.

However, when plants are harvested by pulling up, this activity may be invisible to the archaeological analyst because it usually will not leave tools that can be identified. If some plant remains, such as of roots or basal (bottom) nodes of the stem are present in the archaeological remains, this may suggest uprooting, since sickle harvesting only rarely collects the base of the plants. Uprooting of barley has been suggested for Bronze Age Syria, based on basal node remains recovered from a silo (McCorriston 1998; Chapter 5, this volume). If harvesting by pulling up the crop is usually an archaeologically 'invisible' activity, this brings into question figures of crop yield or of harvesting intensity that are based solely on remains of sickle blades found in archaeological sites.

4.3. HARVESTING OF THE WILD GRASS ALFA (*STIPA TENACISSSIMA* L.) BY PULLING IN THE HIGH TUNISIAN STEPPE: AN UNUSUAL METHOD

Patricia C. Anderson and Mondher M'hamdi

Introduction

Harvesting of wild grasses is a key to our understanding of human history. The focus is often upon the wild progenitors or ancestors of cereal crops, and on edible plants in general. We sorely lack hypotheses concerning uses of other wild plants, particularly grasses, that turn up in the archaeological record. One example of this is grasses of the genus *Stipa*, whose grains appear in a Mesolithic context in the Middle East, at Abu Hureyra (Hillman 1984; Hillman *et al.* 1987; Willcox 1998, 27), just prior to the appearance of the first cultivated cereals. A species of *Stipa* is discussed in this case study from Tunisia, bearing testimony to the fact that wild grasses continue to be key in agricultural contexts.

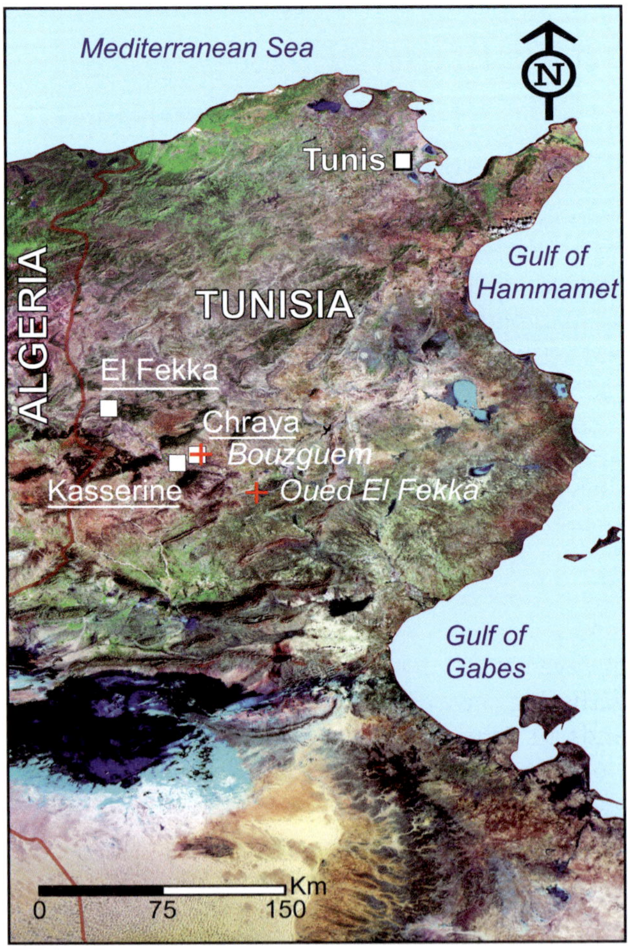

Fig. 4.14. Map of Tunisia with the towns of Chraya, El Fekka, and Kasserine, as well as the Bouzguem plain and the Oued El Fekka. Map: R. Lugon, J.-C. Loubier and A. Chevalier.

Many stone tools found on archaeological sites during the Mesolithic and Neolithic appear to be used for harvesting not only wild cereals, but other wild grasses with tough leaves, judging from their microscopic wear traces (Anderson 1991; 1994; 1996; 1999). In an earlier ethnoarchaeological study, it was found that another grass, *Ampelodesmos mauritanica*, called *diss*, was harvested in northwestern Tunisia using the same sickles, locally made and with a dentate blade (see Anderson *et al.*, this chapter; Anderson 2014) used for harvesting wheat and barley. This kind of harvesting may be represented in archaeological sites. *Ampelodesmos mauritanica* is essentially used for animal fodder and for basketry.

This article presents a case study considering pulling up of just the leaves of a wild grass, *alfa* or *Stipa tenacissima* L., also called esparto grass, a perennial tussock-grass growing in the high altitude (up to 1,544 m) steppic regions of central western Tunisia, in the area of Kasserine (Fig. 4.14). The method of harvesting *alfa* is of particular interest, as it involves using a non-cutting tool that might never be recognised outside the context, and a characteristic harvesting motion.

Alfa grows in the arid and semiarid areas of the Mediterranean basin. In fact the origin of the name of the region, 'the high steppes' comes from the Latin name of *Stipa*. This region has been an area of transhumance and pastoralism, and annual rainfall is only 400 mm. Today *Stipa tenacissima* covers an area of 170,000 ha in the high steppe region of Tunisia. Because it grows on semiarid and arid soil, *Stipa* helps prevent desertification and erosion in this region.

Techniques of Harvest and Uses of *Alfa*

The leaves of *alfa* are cylindrical, resistant, 60 to 70 cm long, and rich in lignin. Despite their rigidity they are sometimes grazed by horses, cows and camels and on rare occasions are gathered for animal fodder. This one-metre high plant grows in tufts (Fig. 4.15). Its grain is not collected or eaten today. At the end of the season, the dry leaves have secondary uses: they can be pulled up and used as tinder for fires (*amadou*), and often are used as a thatch for barns and stables. They are also used to cover chopped straw stores.

Harvesting of *alfa* is the main resource for poor families now, and it is said it provides a source of revenue for over 6000 homes during the main gathering season, which begins in September and ends in February. Groups of harvesters, mainly women and sometimes children and men, are either local families, (Fig. 4.15) or nomadic tribes from areas up to 50 km away.[6] Since 1963, *alfa* has been harvested mainly for factories making paper pulp (cellulose factories). Before this date *alfa* was mostly used by professional craftsman to make basketry objects such as mats, mattresses and baskets. These crafts continue to be practiced, and despite the predominance of use for industrial paper-making, this plant is still harvested by hand with the same techniques and tools.

To harvest *alfa*, the leaves are pulled off by means of a simple but ingenious technique. The harvester grasps a number of leaves (Fig. 4.16), slips the harvesting tool under them and wraps them once or twice around a stick or a bar often made especially for this purpose (Fig. 4.17) which is then pulled up towards the harvester. The leaves separate at a node low on the stems as the upward traction increases, and thus the plant is never uprooted (Fig. 4.18).

Harvesting instruments are variable between areas and people, and according to the availability of different raw materials. Each user makes his own tool, most often of tamarisk (*tarfa*) wood, which is found in wadis and along river banks. Tamarisk wood is often used by farmers to fashion sticks and canes for shepherds. The grooved stick used in El Fekka (Fig. 4.14) for harvesting measures about 30 cm in length, and a notch is made at each extremity (Fig. 4.18). At one end the harvester attaches a loop of string to attach the tool to his wrist, and at the other end, the circular groove serves to prevent the leaves from slipping off the stick while pulling them up. Sometimes harvesters wear some kind of a protection for their hands in the form of a glove (Figs. 4.16, 4.17). There are other variants of this harvesting tool: for example it can be made from other kinds of wood such as willow, or as in Bouzguem (Fig. 4.14), thin metal rods such as those used to reinforce concrete, or reworked screwdrivers were used to harvest in the same manner (Fig. 4.19).

Fig. 4.15. Harvesting of *alfa* (*Stipa tenacissima*), by pulling up, in Bouzguem, central Tunisia in December, 2010 by women in a family, collecting the harvest in large scarves tied around their foreheads or shoulders and back. Image: M. M'hamdi.

Figs. 4.16, 4.17: Harvesting of *alfa* (*Stipa tenacissima*) in El Fekka in 2010, by winding the leaves around a small stick and pulling up, near Kasserine, Tunisia in the high steppe. The leaves are mainly used for making paper pulp and for basketry. Images: S. Ghozzi.

Why resort to this technique, rather than simply harvesting with a sickle? Indeed, sickles are used in this region to harvest other plants such as barley and wheat. In the case of *alfa*, however, a special strategy is required that neatly separates the leaf at a node, rather than actually cutting through it. This is for two reasons:

First, it is considered a simpler method than sickle harvesting, and the harvesters can make their own tools. Second, if the leaves of *alfa* are cut, for example using a sickle, they will have a tendency to split, which will weaken them and render the process of basketry-making difficult or even impossible. The leaves need to be solid, sealed right at the ends for insertion of each leaf in the basketry process. The preservation of cellulose in the lower 15 to 20 cm of the stem is necessary so that the lower part of the stem does not split and remains hard, and this is also an important consideration

Fig. 4.18. The harvesting stick used in the figures above, with a groove at one end for holding the leaves, and at the other end for tying on string to attach the tool to the harvester' s wrist. The image also shows harvested *alfa*, where the point of detachment of the leaves from the stems can be seen to the right. Image: J.- D. Strich.

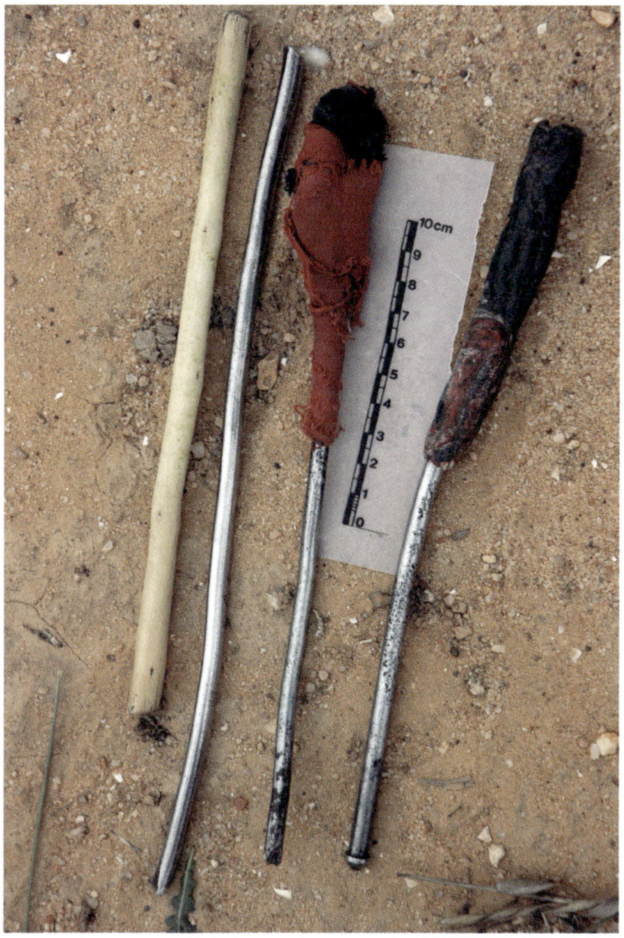

Fig. 4.19. Various homemade harvesting sticks and rods used for harvesting *alfa* by men and women in Bouzguem, Tunisia. Image: M. M'hamdi.

in industrial making of paper pulp from *alfa* eaves.

As the harvest progresses, either the leaves are put into a kind of back sling made by tying a large piece of coloured cloth around the shoulder or the forehead of the harvester (Fig. 4.15), or they are simply laid on the ground. Using several of the leaves woven into a kind of rope, the stems are bound to form a bundle of up to 50 cm in diameter. At the end of the day these bundles are gathered up by the harvester. According to the use to which the leaves will be put, the sheaves are brought home for basketry-making, or more commonly stacked up on a net laid upon the ground (*chebka* or *jaaîba*) which is then rolled up and tied on the back of a pack animal (Fig. 4.20), or occasionally onto an animal-drawn cart. The harvest is then transported each evening to a collection centre and the harvester is immediately paid (1 kg of *alfa* is worth the equivalent of €0,05). Sometimes artisans can come to such centres to buy *alfa* for basketry, but most of it is purchased by factories making paper pulp, which is then exported abroad for paper-making.

Fig. 4.20. The harvested *alfa* is bound into sheaves, and these are tied in a net, which is loaded onto donkey back and carried to the weighing centre, where harvesters will be directly paid by the government, or to the home. Image: M. M'hamdi.

Fig. 4.21. Artisans in Chraya, sitting on mats (*hassir*) made from *alfa* using a loom, where it can be seen that the ends of the *alfa* stems have been left long and untrimmed so the mats are springy and soft. The mats are traditionally used for sleeping and sitting in homes. Small basketry items are also being made from *alfa*. Image: M. M'hamdi.

Local Uses

Alfa has always been used by transhumant Bedouins in this region for making such things as tent ropes. Farmers and nomads make many different objects for daily use: ropes to attach loads to the backs of animals or to hobble animals so they cannot stray far, slings to scare away predators such as birds or larger animals, and basketry for preservation of certain foods or to transport fruit and vegetables during their travels. For farmers, *alfa* baskets are used for provisions or to transport goods on donkey back for sale from village to village. Special double baskets of *alfa* drape over animal backs and are used to transport cereal grain from the underground silos where it is stored (*matmoor*) to where it is to be consumed. A particularly interesting traditional object which is now also sold for the tourist trade is a long mat known as *hassir*, traditionally woven with the long ends of the stems left sticking out and not trimmed off (as documented from Chraya; Fig. 4.14, Fig. 4.21) so that the underside of the mat is soft and springy, good as a mattress for sleeping or sitting. These were used by the Bedouin, spread all over the floor of the house, and lasted about 20 years.

Concluding Comments

Harvesting and use of *alfa* persists today, providing both objects for daily use for people of this region and significant income, particularly because of the paper pulp factories. The local *alfa* craft objects are of varied forms and uses, and most could be imagined to have existed in prehistoric times. Indeed, although the wooden harvesting tools described above might not be detected (but see M'hamdi and Anderson 2013) or preserved, the basketry and particularly floor matting may well be like that sometimes found in archaeological sites as carbonised remains, visible imprints, or remains of phytoliths. The characteristic imprints of plant cell structures represented by phytoliths often allow the identification of at least the general kind of plant, for example as a member of the Poaceae/Gramineae family. The use of *alfa* also reminds us that a great deal of grass harvesting for non-food purposes has undoubtedly occurred throughout time, for making objects essential for daily living, which leave little archaeological or historical trace.

4.4. THE USE OF *MESORIAS* TO HARVEST HULLED WHEAT BY STRIPPING: AN ANCIENT TOOL?

Leonor Peña-Chocarro

Although the study of agricultural techniques is a well-developed field within anthropological studies and to some extent also in archaeology (White 1967), it is also true that, at least in the latter, attention has focused more on issues concerning soil preparation (terracing), irrigation or even crop processing than on the variability of harvest techniques existing in both the archaeological and the ethnographical record. Harvesting has been generally associated with the use of sickles and, with some exceptions (Anderson 1992; Ibañez *et al.* 2001; Peña-Chocarro 1996; 1999), little research has been done into alternative harvesting techniques.

One of the most striking alternative harvest methods in Europe is the use of the *mesorias*, an ancient implement still in use in mountain areas of northern Spain, in Asturias. The *mesorias* consist of two 50–cm long sticks (today often obtained from a broom handle), joined together at the upper part using a piece of leather or a string (Fig. 4.22). They are used to harvest spelt (and in the past also emmer) by stripping the ears off from the stem. The method, as practiced today, consists of several steps. First, farmers gather a bunch of ears between the two sticks and close them around the stems at some distance below the ear (Fig. 4.23). Next, the *mesorias* are pulled up (Fig. 4.24) to the base of the ear where the two sticks are tightly closed. Finally, the farmers keep pulling the *mesorias* up holding them tightly so the cereal ears are snapped off from the straw (Fig. 4.25), falling inside a large basket (*goxa*) that harvesters have next to them (Fig. 4.26), leaving the straw in the field for cutting later for further use as animal bedding or to be burnt to improve soil quality.

An important consideration relates to the type of cereals harvested with this method. Both emmer and spelt are hulled wheats, primitive species which are characterised by glume tenacity and the rachis' semi-fragility. In the wild cereals, rachis fragility allows the spike to shatter into segments when ripe, while in the domesticated species the rachis is tough and at maturity it remains intact. In both emmer and spelt, the rachis is semi-fragile and therefore it breaks easily when some pressure is applied as occurs during harvesting with *mesorias*.

The rarity of this method has drawn the attention of many researchers (Alvargonzález 1908; Bregadze 1982; Buxó i Capdevila 1989; Caro Baroja 1972; 1975; Dantín Cereceda 1941; González Llana 1889; Ibáñez *et al.* 2001; Menéndez Pidal 1993; Ortiz and Sigaut 1980; Peña-Chocarro 1996; 1999; Reigniez 2002b; 2003; Sigaut 1978; Stordeur and Anderson-Gerfaud 1985; Toffin 1983; Vavilov 1926) who have approached their study from various perspectives.

Some of these scholars reported the use of this type of harvesting implement in other parts of the world. Toffin (1983), for instance, described similar tools in Nepal. Made of bamboo or other woods hardened by fire, the *te-shing* or *tep-shing* were used by the Tamang, an indigenous ethnic group settled in central Nepal in the highlands of the Ankhu Khola river. Although no specific references are made to the species involved (wheat and barley are mentioned), this group also uses sickles to collect other cereals, as do many other groups in the region. In any case, the description made of the operation is very much like that in Asturias. Toffin (1983) suggests that the survival of this technique may

Fig. 4.22. *Mesorias*, a pair of wooden sticks joined together by a piece of leather. Image: L. Peña-Chocarro.

Fig. 4.23. and 4.24. Farmers using the *mesorias*. The two sticks are closed around a bunch of stems just below spikes, held tightly with both hands and then pulled upwards. Photos: L. Peña-Chocarro, P. C. Anderson.

be related to past traditions that have continued to exist due to cultural and ethnic factors. He indicates that this tradition has been maintained in isolated areas away from southern influences, while everywhere else the sickle dominates the harvesting techniques used. A possible explanation for the use of this implement may be related to the type of cereal harvested, perhaps a wild species or a semi-brittle domesticate, but no information is available regarding this aspect.

Further west, in Georgia, researchers (Bregazde 1982; Menéndez Pidal 1993; Reigniez 2002; 2003; Sigaut 1978; Steensberg 1943), have reported the use of a similar tool named, according to different authors, šamkvi, šankvi, šnakvi, *shakvi* or *chamkvi* and associated with different wheat species (emmer and combinations of *Triticum timopheevi* Zhuk., *T. palaeocolchicum* Menabde and *T. macha* Dekapr. and Menabde). Very little is known about the origin of this implement.

The analyses of Classical texts and in particular the descriptions of different implements may provide some information about the existence of similar tools in the past. Various authors (Plautus, Servius,

Fig. 4.25. Cereal ears (spikes) are snapped off at the base of the ear while the ear falls into a basket. Image: P. C. Anderson.

Pliny, Columella) mentioned an implement called *mergae* although only Pliny and Columella described their use. In his *Naturalis Historia* (18. 296 after White 1967) Pliny says that 'elsewhere the stalks are cut off at mid-height with the sickle and the ears stripped off between two forks'. Columella instead points to a particular harvesting technique 'Many gather the heads only with forks, and others with combs, an operation which is very easy in a thin crop, but very difficult in a thick one.' (Columella, *De Re Rustica*, 2. 20.3 after White 1967). Attention has been drawn (White 1967) to the difficulties in properly identifying the tool described by Pliny. Two different methods (cutting, and stripping with the so-called *mergae*) seem to be involved, but it is uncertain whether the text refers to an implement like the Asturian *mesorias*. The way the *mergae* seem to operate on cut plants does not match the way modern *mesorias* work. The latter strip cereal ears off while the plant is still attached to the earth, as this facilitates the stripping movement.

Columella's description of the *mergae* as 'forks' and Pliny's account of their use to remove ears from culms which have already been cut bring to mind an implement similar to the one retrieved from Ganj Dareh Tepe in Iran (Fig. 4.3; Stordeur and Anderson-Gerfaud 1985; Anderson 2013) made from ovicaprid scapulae. In addition, Columella's *pectines* (combs) seem to be also present amongst certain communities in Nepal, associated with barley (Toffin 1983), although such implements have not been documented in Spain.

In summary, the available data does not support a match between the *mergae* and the Asturian *mesorias*. In addition, it is clear that already in Classical times alternative harvesting methods were in use, although it remains difficult to correlate the ancient tools with possible modern counterparts. In any case, the *mesorias* remain in use as an alternative to more familiar sickle harvesting of grain, and stripping very likely was practiced in various forms in the past.

Fig. 4.26. Detail of basket (*goxa*). Image: L. Peña-Chocarro.

4.5. EXPERIMENTS WITH HARVESTING TECHNIQUES: NEOLITHIC SICKLES AND UPROOTING

Patricia C. Anderson and John C. Whittaker

Experiments with replicated Neolithic sickles in the summer of 2005 exemplify both the kinds of useful information that such experiments provide, and the problems and ambiguities inherent in archaeological experiments. Whittaker manufactured several flint-bladed sickles modeled after Neolithic specimens from Hacılar Höyük, Turkey (Fig. 4.27, Fig. 4.28B, Fig. 4.29) and similar finds from sites such as Karanovo, Bulgaria (Fig. 4.28B; see Gurova, Chapter 10). The handles were antler (*Cervus elaphus*), with short flint blade segments set in a groove with asphalt, as used in some prehistoric examples. Meanwhile, Anderson, with the help of local farmers in the region of Cailles (Fig. 4.28; Fig. 4.28B), southern France, had grown several plots of einkorn wheat (*Triticum monococcum*) to harvest.

With a team of about a dozen experimenters, we examined several issues:[7]

1. Sickle design. Prehistoric sickles are mostly not as strongly curved as recent metal sickles. The length of cutting edge provided by flint inserts also varies. Some of the Hacılar sickles have all the blades set parallel to the antler to provide a straight cutting edge, while in others, the flints are set in at an angle, giving a toothed effect. We tried both forms (Fig. 4.29). Some flint sickles elsewhere have small teeth worked into the edge of the blades, a variant we did not try. For comparison, we used some modern metal sickles as well, with both toothed and smooth edges.
2. Efficiency. We quantified several trials, with different experimenters using flint sickles, metal sickles, or uprooting the crop in 2 × 3 m plots of *Triticum monococcum* (einkorn wheat), recording time spent and the resulting weight of grain and stems harvested (Fig. 4.30). The uprooting (Fig. 4.31) was based upon Anderson's ethnographic observations of this technique being used in Syria (Anderson and Peña-Chocarro, this chapter; Anderson 2003). Archaeobotanists in our team were sceptical about uprooting or pulling up as a viable option, although as we have seen, it is a common and successful technique ethnographically.

These were small-scale, preliminary experiments, but they allowed several initial evaluations. Much of the variation in sickles through time and geographical space should relate to how they are

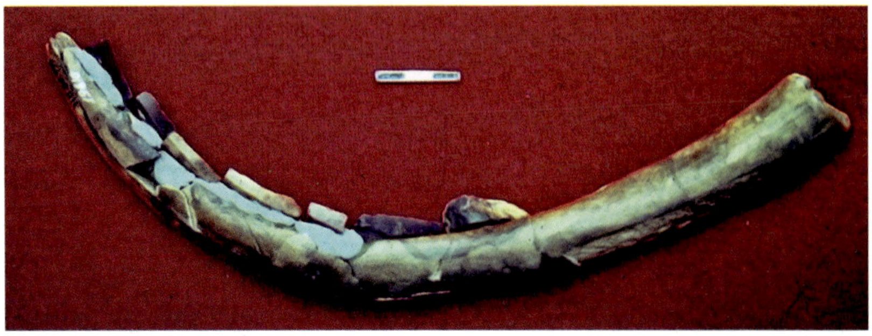

Fig. 4.27. Sickle archived in the Ankara Museum with flint blades and an antler handle, from the late Neolithic levels at Hacılar Höyük, Turkey, dating to about 7000 years before the present. This kind of sickle is like those known later from Karanovo, in the Bulgarian Neolithic. Image: P. C. Anderson.

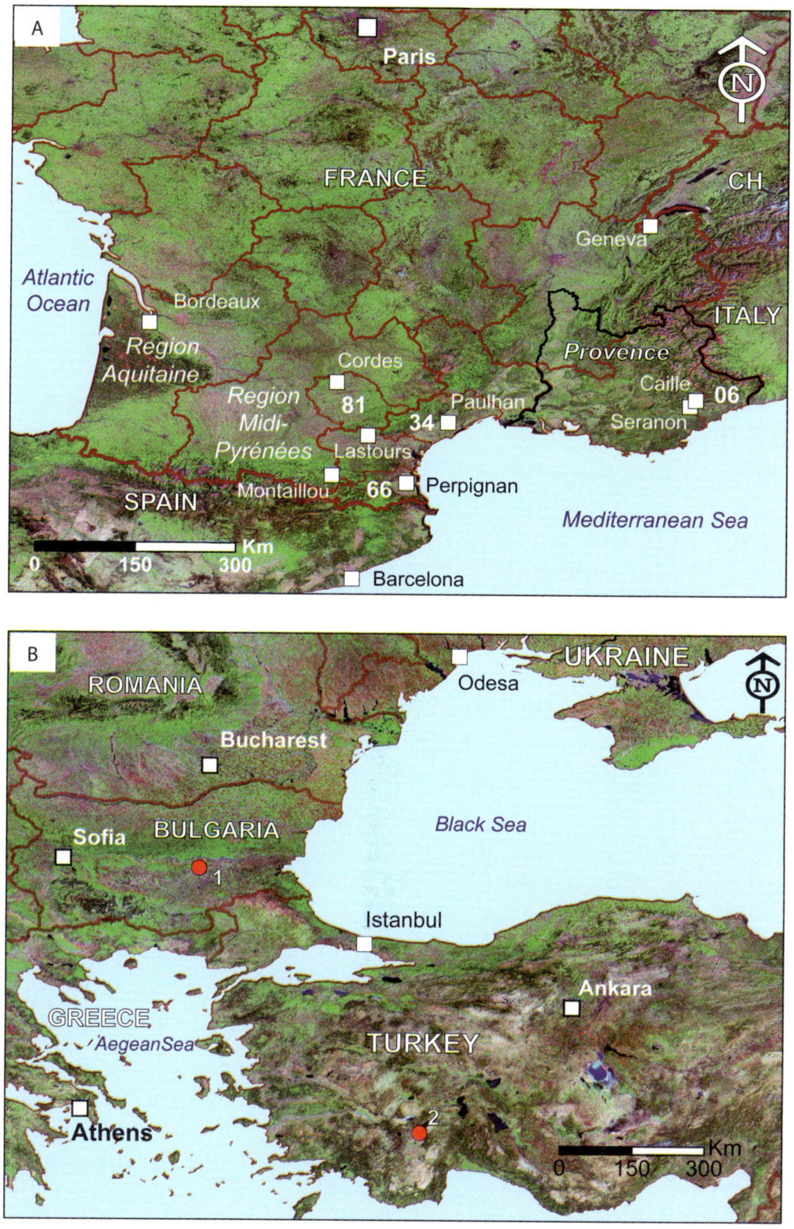

Fig. 4.28. A: Map of France with the Aquitaine and Midi-Pyrénées political regions; the Provence cultural and historical region that includes de Provence-Alpes-Côte d'Azur political region and the Drôme Department; the 06) Alpes-Maritimes; 34) Hérault; 66) Pyrénées-Orientales; and 81) Tarn departments and the towns of Bordeaux, Cailles, Cordes, Lastours, Montaillou, Paulhan, Perpignan, and Seranon. B: The archaeological sites of 1) Tell Karanovo, in Bulgaria, 2) Hacılar Höyük in Turkey. Maps: R. Lugon, J.-C. Loubier and A. Chevalier.

Fig. 4.29. Three sickles made by J. C. Whittaker according to the design of those from Hacılar Höyük and Karanovo, that were used in experiments. Image: P. C. Anderson.

used, on what crops, and under what conditions, but there is plainly a great deal of choice in sickle forms and effective designs, with cultural tradition and individual preference playing a part (see Gibaja *et* al., this chapter). Our team had a diversity of opinions about the sickles we tried. Some preferred the toothed effect of obliquely set blades (Fig. 4.32), while others found the straight cutting edge more effective (Fig. 4.33). The variability in individual productivity meant that we were unable to make any objective comparison of the two sickle styles. In all cases, it was better to have a smooth transition between handle and cutting edge, because a sudden step, or a gap in the line of blades, caught and pulled the wheat stems without cutting and reduced

Experiments comparing individual users and methods	Flint sickle	Metal sickle	Uprooting	Mean
Experiment 1	207.06	96.77	122.03	141.96
Experiment 2	-	-	54.86	-
Experiment 3	170.29	101.33	112.53	128.05
Experiment 4	128.33	112.4	131.43	124.05
Experiment 5	-	-	128.28	-
Experiment 6	127.06	60	64.71	83.92
Mean	158.18	92.63	102.3	-

Fig. 4.30. Results of harvesting: time in seconds to harvest a 2 m x 3 m plot of einkorn wheat. Table: L. Bouby.

Fig. 4.31. Pulling up or uprooting einkorn wheat in experiments. Image: K. Kamp.

Fig. 4.32. Harvesting small handfuls of stems with a single cut per handful, using the sickle with obliquely-set inserts, the middle one in Figure 4.29. Image: K. Kamp.

Fig. 4.33. Harvesting larger handfuls of stems with several cuts per handful, using a sickle with parallel-set inserts, the one to the right of Figure 4.29. Image: K. Kamp.

efficiency. A relatively long cutting edge was more effective than too short an edge. In addition, having an adequate length of handle to grasp in the hand was also a criterion mentioned both here in these experiments, and by harvesters in Tunisia when they were given the sickles to use.

Individual efficiency in using sickles varied enormously, depending on strength, motion, and dexterity, but none of us had enough previous experience with sickle harvesting to feel that we gave the tools a truly fair trial. For instance, one experimenter gripped a large handful of stems and cut through them with several strokes (Fig. 4.33), while others cut smaller numbers with a single stroke (Fig. 4.32). The first individual ended with the most weight harvested per time, and the shortest time per square metre, using either metal or flint sickles or uprooting, but this may be simply because he was a more energetic harvester than others, rather than a more skilful one. The variability in experimenters means that our quantification of efficiency is not applicable to any prehistoric situation, but within our group of experimenters, each individual did better with the metal sickles than with the flint, and uprooting was usually intermediate and closer to the area/time and time/weight achieved with the metal sickles (Fig. 4.30). However, weight/time when uprooting was elevated because roots and some soil were collected (Fig. 4.31) as well as upper stalks.

Experiments like this, with many uncontrolled variables, especially those of human skill and ability, make it difficult to evaluate a tool in a truly scientific and replicable manner. Such experiments are useful for establishing some rough parameters, suggesting testable hypotheses about how tools might be used and resulting wear patterns. For instance, it is clear that the wear on obliquely set blades will differ from that on blades set parallel to the handle (Gibaja *et al.*, this chapter). Even the least controlled experiments can provide invaluable personal experience that leads to more sophisticated understanding of agricultural tools and work, as well as developing the necessary skills for more consistent and realistic experimentation. They can result in a few clear-cut conclusions; at least our experiments served to convince the botanists that uprooting was a viable technique and should be taken into consideration when interpreting archaeological and ethnographic data.

4.6. THE MYSTERY OF THE MISSING SICKLES IN THE NORTHWEST MICHELSBERG CULTURE IN LIMBURG, THE NETHERLANDS

Corrie Bakels and Annelou van Gijn

Neolithisation in the Netherlands was a very gradual process. The first farmers, of the Linearbandkeramik culture (LBK), settled on the fertile loess plateaus in Limburg in the very southern part of the Netherlands (Fig. 4.34), around 5300 cal BCE. They were succeeded by the farming societies of the Rössen and Michelsberg cultures. It was only during the Michelsberg period that we find the first evidence for crop growing further north and west (Bakels and Zeiler 2005; Kubiak-Martens 2006; Out 2009). During these periods there is evidence for several kinds of crops. However, it is not only plants that tell us about agriculture but also sickles. Flint sickle inserts have been found in settlements from the LBK, Rössen and Hazendonk sites (Fig. 4.35; see also Van Gijn 2010). They are, however, mysteriously absent in Michelsberg flint assemblages (Schreurs 1992; Schreurs and Brounen 1998). Also in neighbouring areas of Belgium and the Rhineland, sickles are similarly lacking in the flint assemblages (for instance Spiere 'de Hel'; Vanmontfort 2004, 59–63). What can be the reason for their mysterious absence?

Fig. 4.34. Limburg, The Netherlands. Map: R. Lugon, J.-C. Loubier and A. Chevalier.

During the Michelsberg (MK) period there is ample evidence for the cultivation of cereal crops: wheat (a naked tetraploid species, comparable to *Triticum durum*, as well as emmer and einkorn) and naked barley are the main cereals grown (Bakels 2009). Based on the presence of glumes and rachis remains it can be assumed that these crops were locally grown and not transported from elsewhere. Apart from the tetraploid wheat, all other cultivated species seem to reflect a continuation of the LBK and Rössen agricultural practices. The presumed location of the cereal fields and the condition of their soils are similar too (Bakels 2007). Lastly, the weed species found amongst the crop processing

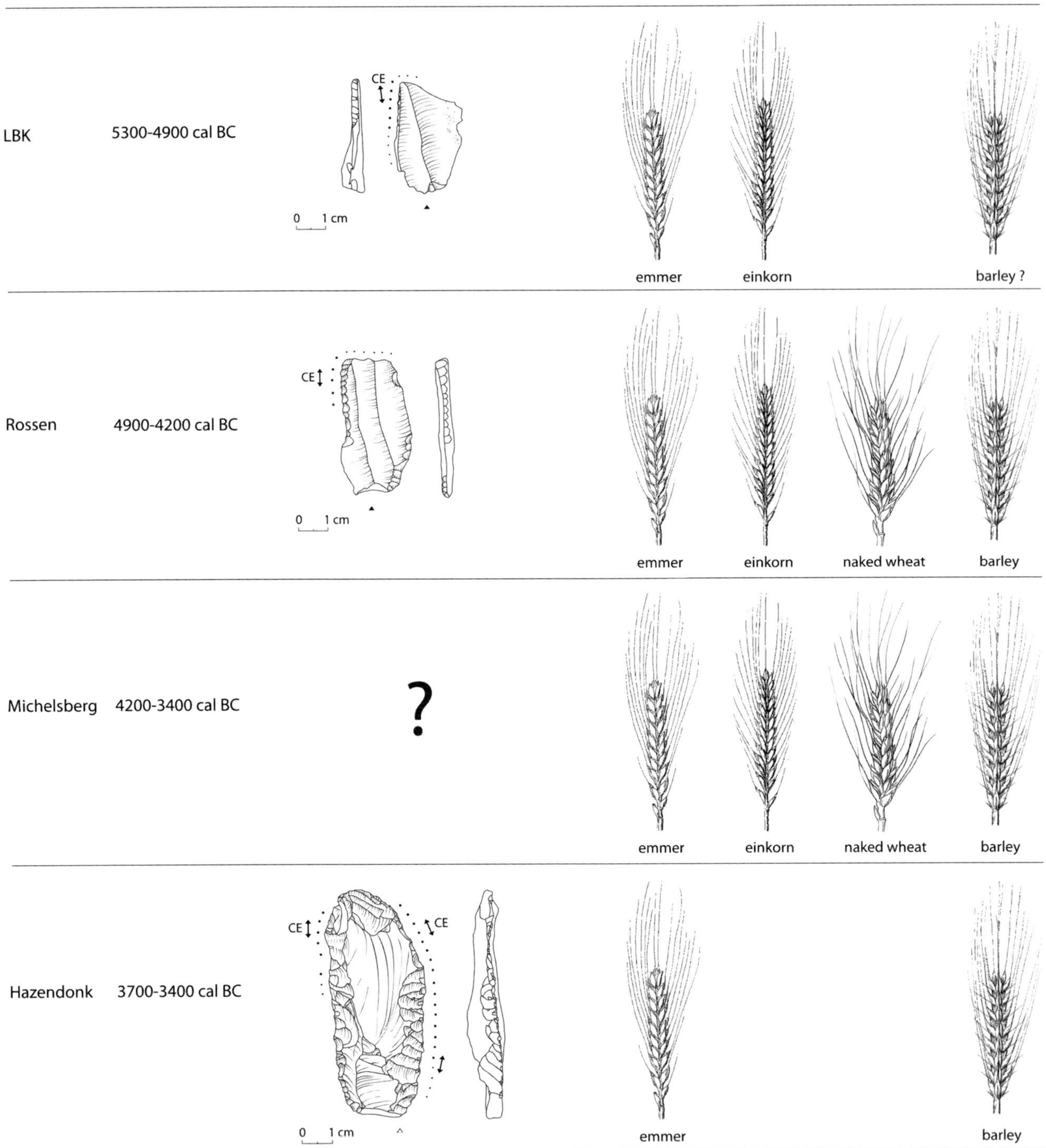

Fig. 4.35. The presence/absence of crop species and sickles during the different cultural periods of the Early and Middle Neolithic of the southern Netherlands (drawing Joanne Porck, Faculty of Archaeology, Leiden University).

waste are identical to what is found for the preceding LBK and Rössen periods. Tall growing or climbing species predominate. The category of species with heights below 50 cm present in the archaeobotanical samples is low and has not increased since the LBK.

As all of these weeds are introduced into the settlements along with the harvested cereals, it is to be expected that we also have continuity in harvesting techniques between LBK/Rössen on the one hand and the MK on the other. However, as we have already mentioned above, sickles are present in the flint assemblages from both LBK and Rössen, but are lacking in those from the MK. This absence of flint sickles is remarkable considering the apparent continuity in other aspects of the agricultural practices, both in the location of the fields, the crops selected and the harvesting height. How can we explain this?

There must have been an alternative harvesting method practiced during the MK period. Uprooting can be rejected because of the absence of low-growing weeds. The answer may lie in an alternative technology like handpicking or the use of harvesting tools such as *mesorias*. Both handpicking and using *mesorias* result in relatively clean harvests. In the few instances in which a concentration of waste derived from crop processing was present, the amount of weeds was indeed relatively small compared to what was found in, for instance, the later phases of the LBK. This may indicate the use of these alternative methods of harvesting. Another option is that they used harvesting knives made of other raw materials than flint. This could include sickles of bone, freshwater mussels, wood or even baked clay. Such sickles have however never been found in MK contexts, despite the great number of sites known so far.

For now, we can only conclude that there must have been another method of harvesting practiced in the Michelsberg period, possibly handpicking. If this is the case, it is strange that the inhabitants of the southern coastal zones, attributed to the contemporaneous and closely related Hazendonk group, did make use of flint sickles for harvesting their crops (Van Gijn *et al.* 2006). In fact, these coastal people imported typical Michelsberg large blade implements which they used unhafted as harvesting implements (Van Gijn 2008), instead of using the LBK-Rössen type of composite sickles. It is also surprising that the Michelsberg people broke with this traditional type of sickle that their LBK and Rössen predecessors used for so long. The only conclusion can thus be: 'different folks, different strokes'.

4.7. NEOLITHIC SICKLES IN THE IBERIAN PENINSULA

Juan Francisco Gibaja, Juan José Ibáñez and Jesús González Urquijor

The First Agriculture in the Iberian Peninsula

The first evidence of agriculture in Iberia is dated to the second half of the 6th millennium BCE. The older dates have been documented on the Levantine coast (Valencia and Alicante), and in the northeast (Catalonia and Aragon). According to the very old chronology of some Neolithic sites in the inner Iberian Peninsula, the spread of agriculture in this area must have taken place early, and quickly (Rojo *et al.* 2008). The staples which were cultivated by the first farmers were quite varied, including cereals (hulled and free-threshing wheats and barley), pulses (pea, lentil, fava bean, vetch, and grass pea), flax and poppy. The variety of staples shows that agriculture was firmly established among the first farming communities (Buxó 1997; Peña-Chocarro 1999; Peña-Chocarro *et al.* 2005; Pérez 2005; Peña-Chocarro and Zapata 2010). Up to the present time, three types of Neolithic sickles have been observed.

Curved Sickles with Oblique Inserts

For this first type of sickle, short flint blade fragments were chosen as sickle elements. These sickle elements are, normally, between 2 and 3 cm long, and around 1 cm wide. Fragments of blades or bladelets intentionally broken were used to obtain this type of cutting element. Typically, the sickle gloss is distributed along two thirds of the cutting edge. Moreover, the gloss appears as a narrow line on the edge and it gets more invasive towards one of the extremities of the element. The other extremity is free of use-wear polish. This distribution of use-gloss indicates that blade fragments were obliquely inserted into the sickle haft. The area free of use traces was inside the mastic which glued the element to the shaft, while the opposite extremity protruded for cutting the cereal stems. In the site of Los Murciélagos de Zuheros, in a context of very good preservation of the organic material because of the dry conditions of the cave, the distribution of mastic remains on several sickle elements confirms that this type of short sickle element was inserted obliquely (González Urquijo *et al.* 2000). On one of the sickle elements where the mastic is especially well preserved, the shape of the base of the mastic remains shows that there was a groove in the shaft to facilitate the hafting of the flint elements (Fig. 4.36). Most probably, these oblique elements were mounted into curved sickle hafts. We have observed experimentally that when using this kind of sickle element in straight sickles, the use-wear traces only show straight striations which are parallel to the edge, whereas when curved sickles are used, besides the more dominant parallel striations, other oblique ones can also be observed. This is due to the different type of motion with which each type of sickle is used, the curving motion made with the bent sickle favouring the creation of some oblique striations (González Urquijo *et al.* 1994).

These oblique striations are present on the archaeological sickle elements. Moreover, one complete sickle was found in the mid-19th century in the dry cave of Los Murciélagos de Albuñol (Granada). The tool is not preserved but we have a drawing made by M. de Góngora (1868/1991) which followed the description of one of the discoverers. The sickle was curved and the cutting edge, made with flint elements, was dentate. The sickle with oblique insertions from the Iberian Neolithic must have looked like the complete sickle found in the lake site of La Marmotta in central Italy (Fig. 4.37; Fugazzola and Pessina, 1999).

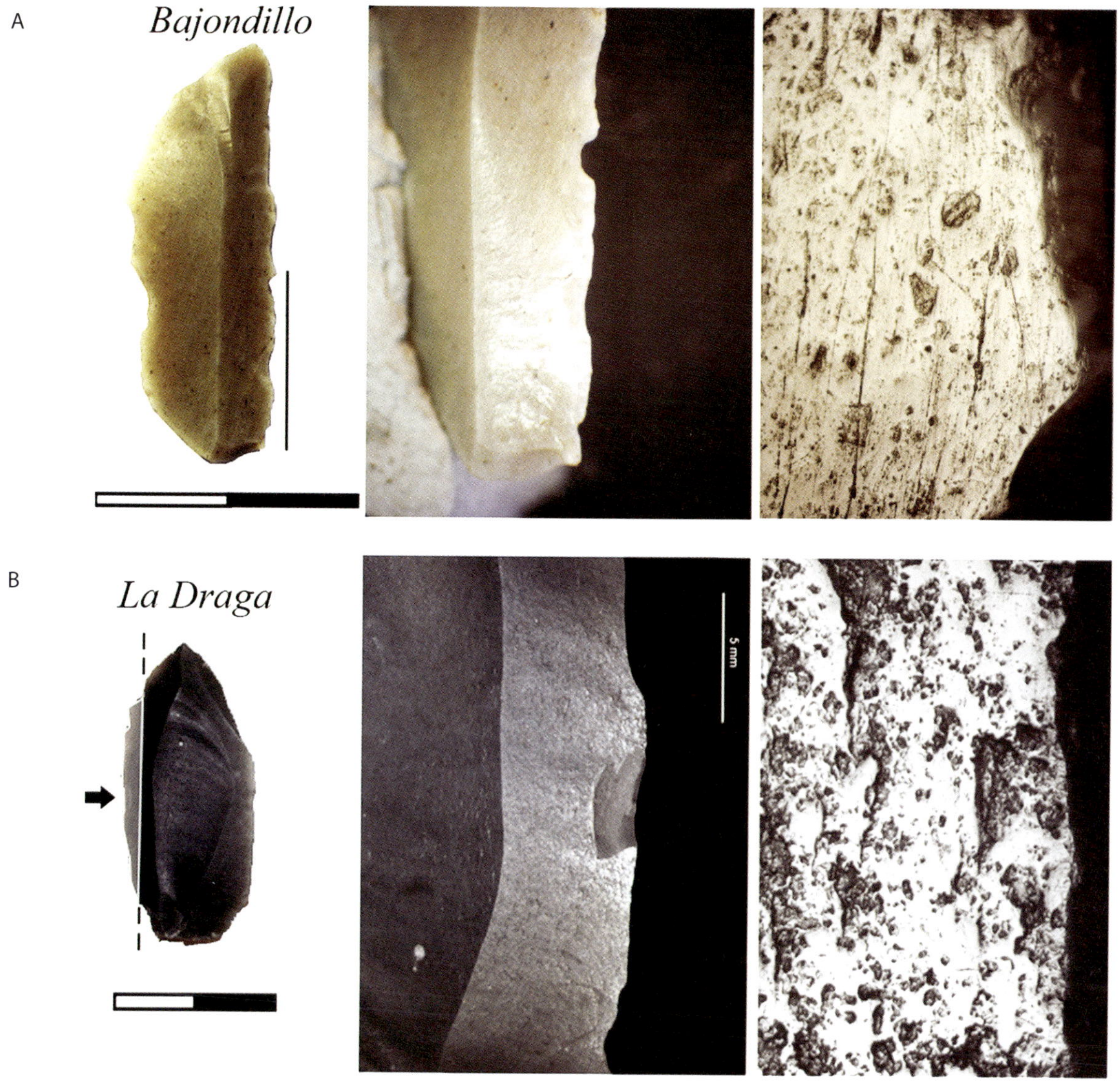

Fig. 4.36. Sickles for harvesting cereals. Use-wear polish enables us to reconstruct the hafting system: A) Sickle from Bajondillo site (Málaga, Spain) with oblique insertion, B) Sickle from La Draga (Girona, Spain) in parallel insertion (Gibaja *et al.* 2010, Palomo *et al.* 2011).

Microscopic analysis of the archaeological elements shows a typical cereal harvesting polish. Certain variability in the quantity of striations can be explained by the diversity in the height at which the cereal stems were cut (cutting near the ground produces more striations) and in the environmental conditions during the period of harvesting (more aridity also leads to more striations). The curved form of the sickle was used to gather the stems, which were held in the bare hand before cutting them with a slightly curving motion. This type of sickle is present in the Early Neolithic sites of Valencia, Andalucía, and the Portuguese Ribatejo (Rodríguez 1999; Rodríguez 2004; Clemente and García 2008; Gibaja *et al.* 2010). It has also been detected in the older Neolithic levels of Los Cascajos, which is located in Navarre, dating to the end of the 6th millennium BCE (Ibáñez *et al.* 2008).

Sickles with One Long Blade in Parallel Insertion

A second category of sickle element in the Neolithic of the Iberian Peninsula is represented by longer blades, between 5 and 10 cm, with sickle gloss distributed parallel to the edge (Fig. 4.36). This distribution shows that the blades were inserted parallel to the haft. The only information on the morphology of whole sickles with this type of flint element comes from the lake site of La Draga (Banyoles, Girona). It consists of an L-shaped

Fig. 4.37. Location of the archaeological site of La Marmotta in northern Italy. Map: R. Lugon, J.-C. Loubier and A. Chevalier.

wooden sickle with the flint blade inserted in the main shaft (Bosch *et al.* 2000) (Fig. 4.38).

The variability of use-wear traces on these long blades is higher than on the short blades inserted obliquely. We observed some long blades having typical harvesting traces. But others show abrasive traces with different degrees of intensity. In some cases these traces are intense and distributed regularly along the edge, in other cases they are mixed with the typical harvesting patterns. When the traces are mixed, the abrasive traces seem to erode the previous harvesting traces. We have reproduced these abrasive traces by cutting stems in contact with sediment on the ground (Clemente and Gibaja 1998; Gibaja 2002; 2003).

Most probably, the long blades in parallel insertion and showing typical harvesting traces were part of a sickle which was similar to the one preserved at La Draga. If this is the case, the sickles were used grasping them by the straight shaft. The cereal stems were gathered with the lateral branch, which is perpendicular to the shaft, and the stems were held in the bare hand. The sickle haft was then turned, so the flint blade faced the cereal stems, and they were cut off. If the blades were just inserted into simple harvesting knives, that is, tools consisting of a straight shaft with a flint blade, the harvesting gesture would be simpler but less efficient as the bare hand would have to gather the stems before they are cut with the tool.

With respect to the blades showing abrasive traces, the reconstruction of their use is more hypothetical. The abrasive wear can be generated while harvesting if the stems are cut very low, in contact with the ground, or in a later processing activity in which the cereal stems are cut while lying on the ground. Cereals are still cut very low in several ethnographic communities in North Africa (Peña-Chocarro *et al.* 2009). A later processing cut would make sense when cereals were harvested with the stems (by low cutting or uprooting) and it was necessary to separate ears and stems for storage or threshing. The storage of complete ears, with a small part of the stem still attached to the basal rachis, is documented in several Neolithic Swiss lake sites (Maier 1999), and stem cutting after harvest is still observed in some ethnographic contexts. The La Draga type sickle cannot be used for these tasks, because of the lateral branch and the central position of the blade in the shaft. Thus, the abraded blades were used in other tool forms. The blades with both harvesting and abrasive traces are present in Neolithic sites of Catalonia and in the southern area of the Basque Country and in the Neolithic levels of the 5th millennium BCE in the site of Los Cascajos (Ibáñez *et al.* 2008; Perales 2010), and could have been the result of the recycling of harvesting blades for the activity which generated abrasive traces. Blades previously used for harvesting were usually recycled for cutting stems on the ground because, in the process of harvesting and processing cereals, the latter was the last technical activity carried out with tools before storing the cereal. Blades were discarded after this activity because it erodes considerably the cutting edge, thus leaving the edge unusable for other functions.

Such abrasive traces have also been explained in other contexts as the result of the use of a threshing sledge (*tribulum*) for cereal processing (Anderson

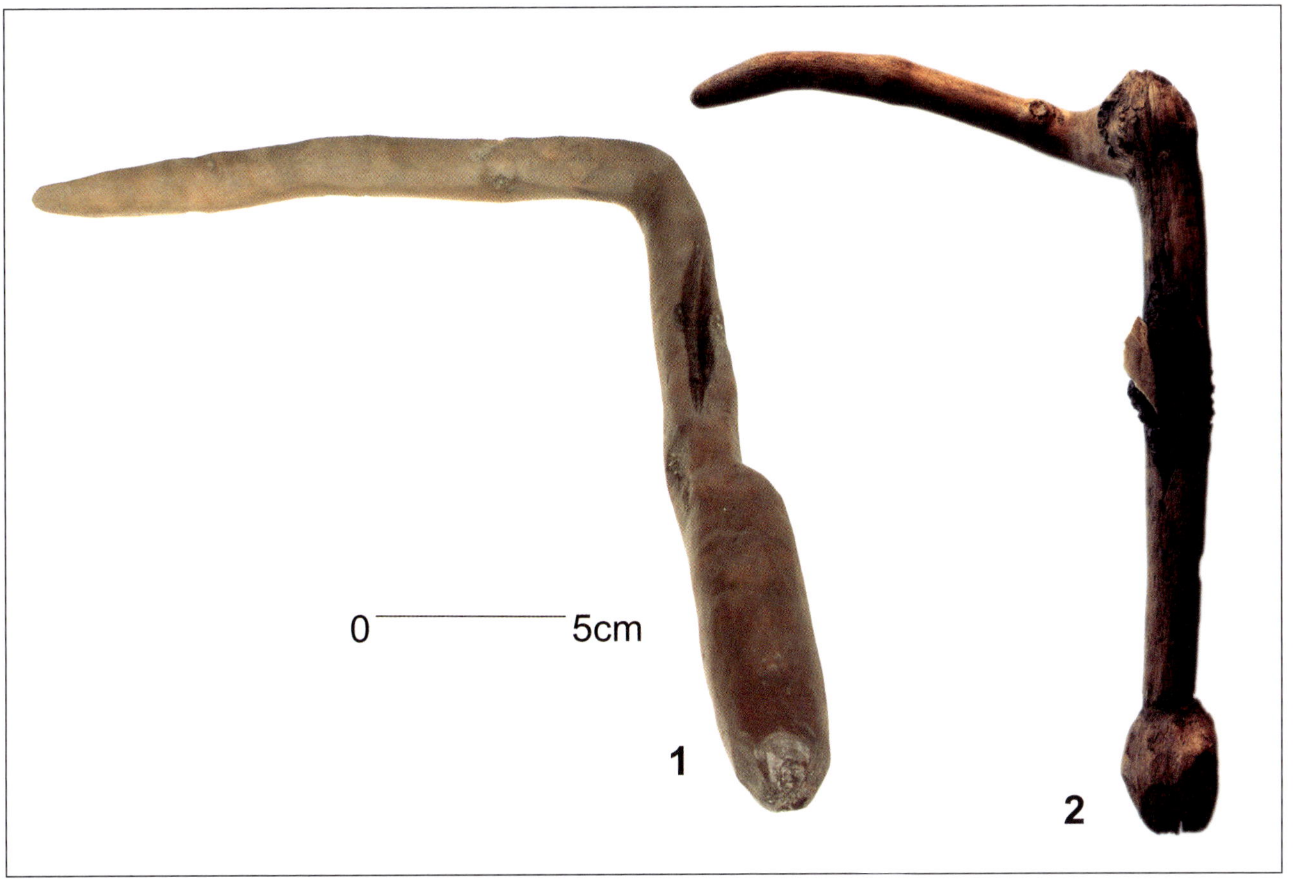

Fig. 4.38. Sickles from La Draga. 1. Sickle haft made of boxwood (*Buxus sempervirens* L.), 2. sickle with flint blade in place (wood from elder, *Sambucus* sp.) Photography by MACB. (Palomo *et al*. 2011).

2000). However, several reasons allow us to propose the use of a simple knife tool for cutting the stems on the ground rather than a threshing sledge for these Neolithic contexts: 1. The intensity of the traces, especially edge rounding, are much more intense in the ethnographic *tribulum* than in our archaeological blades; 2. Ethnographic threshing sledge elements only show abrasive traces, lacking the eroded typical harvesting traces which we have documented for the archaeological blades; 3. Many of our Neolithic blades are quite narrow (less than 1 cm wide), so they could not have been mounted into a threshing sledge plank, as the cutting edge would had been too fragile and too scarcely protruding to be effective; 4. The use of threshing sledges implies that many flint elements were used in a single tool; however, glossed and abraded blades are not common in these Neolithic assemblages.

Sickles with One Long Blade in Oblique Insertion

Sickle blades in this category are long (from 5 to 8 cm) and show a gloss distribution that occupies two thirds or one half of the cutting edge, being more marginal at the proximal end and getting more invasive towards one of the extremities. In this extremity the use-wear polish can be very invasive, up to 1 cm or more. The distribution of the gloss shows that the blade was inserted obliquely to the shaft. Several complete sickles which have been found in La Draga illustrate the morphology of the sickles into which these blades were inserted (Bosch *et al.* 2000, 2006; Palomo *et al.* 2011). They consist of a main straight shaft where the flint blade is inserted obliquely. In the extremity of the shaft opposite to the handle, a lateral branch protrudes perpendicularly to the main shaft, but the oblique blade is inserted on a different plane. Because of this, the gesture of harvesting begins by gathering the stems with the lateral branch, holding them in the bare hand, then turning the sickle 90° to cut the stems with the blade. The long blades which were inserted obliquely always bear typical harvesting traces exclusively, without any abrasive wear.

Besides the wooden sickles of this type found in La Draga, another sickle made of antler, which was found in the Neolithic site of Costamar in Castellón, most probably corresponds to this type. It was made of a fragment of deer horn, which shows a main straight shaft and a lateral branch. In the main shaft is a deep and short incision where the flint blade must have been inserted obliquely (Flors 2010). This type of sickle blade has been documented in the sites of La Draga (Girona), Costamar (Castellón) and some sites in the central Iberian Plateau (the flint mines of Casa Montero, La Vaquera, La Lámpara y La Revilla; see Gibaja 2000; 2008; Gibaja *et al.* 2010; Palomo *et al.* 2011).

Discussion: The meaning of the Variability in Sickle Morphology

The information at our disposal about Neolithic sickles is still quite fragmentary, as there are still many areas of the Iberian Peninsula where data are scarce or non-existent, especially in the northern and western regions (Fig. 4.39). However, we can begin to give a preliminary view of the geographical distribution and suggest hypotheses about the meaning of this distribution.

In some sites, sickles with one blade in parallel insertion are the only harvesting tools. In some other sites where sickles with one blade in oblique insertion were used, sickles with one blade in parallel insertion (with traces of typical harvesting and/or abrasive traces) were also present, as in La Draga and La Vaquera. However, sickles with one long blade, either in parallel or in oblique insertion, and curved sickles with oblique insertions are mutually exclusive. They seem to be part of two different technical traditions of harvesting.

This technical variability does not seem to be linked to the harvesting of different cereal types (Gassin *et al.* 2010). The key for the understanding of this variability is probably in its geographical distribution. Curved sickles with oblique insertions were in use in the south of the Iberian Peninsula. They are part of the Neolithic assemblages in the southern part of Iberia (roughly south of the Tagus-Jucar valley line), plus the older levels of Los Cascajos (Navarre). The sickles with one long blade in parallel insertion have been documented in the northeast of the Iberian Peninsula (Catalonia, Aragón, Alava and Castellón) and in La Vaquera, in the Central Plateau, and in the upper levels of Los Cascajos (Navarre). The blades with one long blade in oblique or parallel insertion have also been documented in several sites of the Languedoc and Provence, in France (Fig. 4.28; Gassin *et al.* 2010).

The spread of agriculture in the Iberian Peninsula must have implied the transport of seeds, which

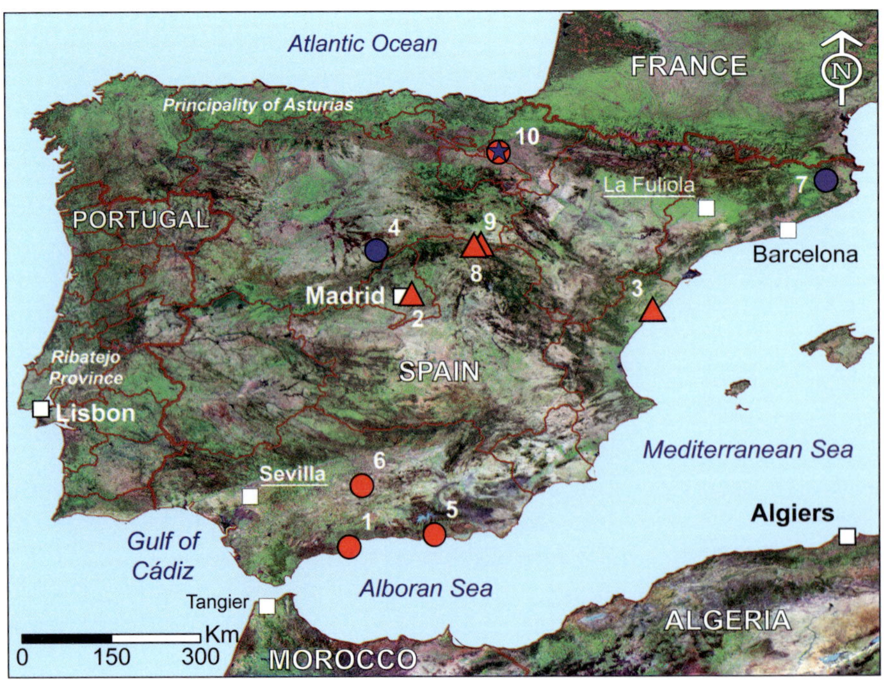

Fig. 4.39. Map showing the distribution of different Neolithic sickle types discovered at various sites on the Iberian Peninsula. Red dots (●): curved sickles with oblique insertions at 1) Bajondillo; 5) Cueva de los Murciélagos de Albuñol; 6) Cueva de los Murciélagos de Zuheros; 10) Los Cascajos. Red triangles (▲): Sickles with one long blade in oblique insertion at 2) Casa Montero; 3) Costamar; 8) La Lámpara; 9) La Revilla. Blue star (★) sickles with one long blade in parallel insertion 10) Los Cascajos. Blue dots (●): sickles with both one long blade in oblique and one in parallel insertion at 4) Cueva de La Vaquera; 7) La Draga. Sites mentioned in Chapters 4.4., 4.7. and 4.8. Map: R. Lugon, J.-C. Loubier and A. Chevalier.

must have been accompanied by the transmission of the knowledge of techniques for cultivating, harvesting, processing, and storage. Because of this, the analysis of the agricultural technology in the first Neolithic groups is a good way to obtain information on the origins and spread of agriculture. The two areas in the south and in the northeast, where the two different types of sickles have been documented, could indicate two pathways of introduction of agriculture into the Iberian Peninsula. This double path of origin could fit well with the older Neolithic sites in the Iberian Peninsula, which are documented almost simultaneously in the northeast and the southeast (Manen *et al.* 2007; Gibaja and Carvalho 2010; Carvalho 2010).

The spread of agriculture in the inner part of the Iberian Peninsula would have been a more complex process. The sickles of Casa Montero, La Vaquera, La Lámpara, La Revilla and the upper levels of Los Cascajos are related to the northeastern types. In the lower level of Los Cascajos, the southern type was in use. It seems as if some specific sites were related to the northeastern focus and others to the southern one. To better understand this question it is necessary to analyse more sites in the inner part of the Iberian Peninsula, taking into account the details of the chronology. As a preliminary hypothesis, we suggest that this distribution of sickle morphology in the inner area of the Iberian Peninsula could correspond to a model of Neolithisation in which pioneering groups coming from the south or the northeast would have colonised the inner Iberian Peninsula, carrying with them their seeds, tools and know-how. This model could explain the patchy distribution of different types of sickles. Moreover the use of the southern type in the earlier Neolithic levels of Los Cascajos and the change to the northeastern type in the upper ones would indicate that the influence of one technical tradition in one area could change through time. The study of Neolithic sickles opens new lines of research into the origins of agriculture, though more sites must be analysed before going further with our interpretations.

4.8. SICKLES WITH TEETH AND BONE ANVILS

Patricia C. Anderson, Isabelle Rodet-Belarbi and Marta Moreno-García

Introduction

Sometimes exploration into harvesting techniques of the past is stymied by poor preservation of tools. For instance, metal sickles with toothed cutting edges are difficult to identify in the archaeological record. Enigmatic bone objects associated with the sickles have been found (Briois *et al.* 1995), but only ethnographic research was able to provide a key to deciphering such objects. This contribution will therefore use an ethnoarchaeological approach, and begin exploration into the time depth and geographic extent of the use of these tools. We describe a present-day example of the use of sickles with teeth in northwestern Tunisia and their manufacture, and another Tunisian example based on iconography, followed by similar examples from France, Spain and Portugal from the last century. Then, the process of teasing out the true function of the enigmatic bone objects will be described. This research is beginning to show a wide distribution of these tools from the Ukraine to the Iberian Peninsula and North Africa and a temporal span beginning in the Hellenistic period.

Ethnoarchaeological Approaches

In northwest Tunisia from 2005–2010 Anderson observed farmers using sickles with a toothed or dentate edge, called a *menjel* (Fig. 4.40). The local manufacture of these instruments by a blacksmith in Beja is of interest, not least of all because of the special bone tools used to form the sickle teeth.

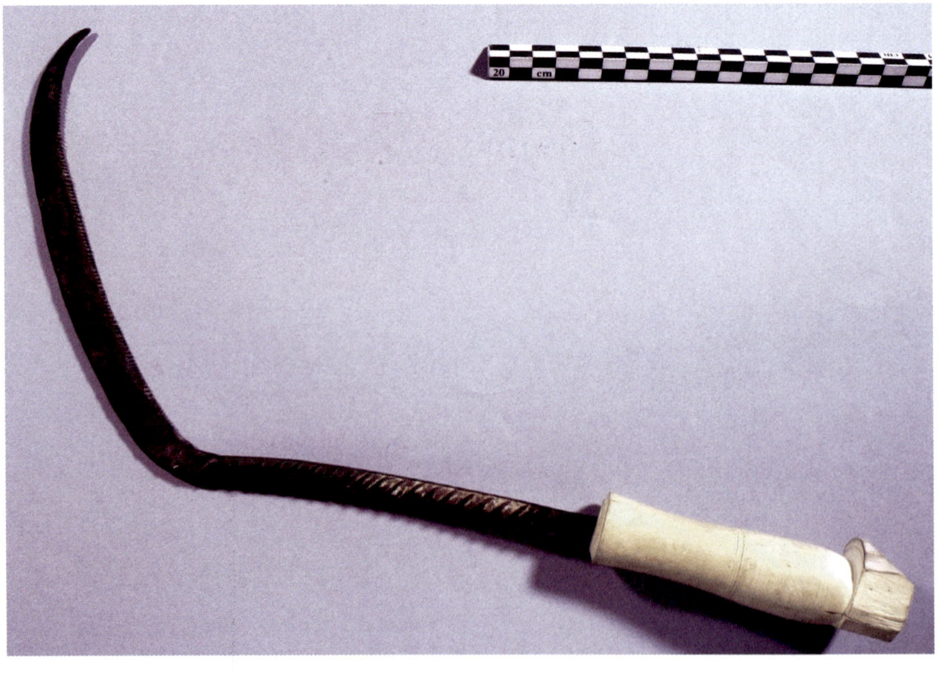

Fig. 4.40. Sickle with dentate edge used in northwestern Tunisia today. Scale divisions = 1 centimetre. Image: J.-D. Strich.

Fig. 4.41. Use of a sickle like that in Fig. 4.40, to harvest hulled barley in Ain Sallem, Tunisia. Image: P. C. Anderson.

These sickles are used to cut a number of plants in this area, on farms from about 200–500 m in altitude. Their principal use is for harvesting cereals, an activity carried out by either men or women, beginning in May for hulled barley. As seen in Fig. 4.41, cereals are grasped in one hand and the sickle cuts by pulling the blade towards the harvester. The stems are cut at mid-height or slightly nearer to the ground, depending upon the preference of the user, but this is done in such a way that stubble is always left for animals to browse. In June and July, the sickle is used in the same way to harvest ripe durum wheat (*Triticum durum*)[8] and oats. Its most common use today is nonetheless for wheat, and it is said that sickles with teeth, as opposed to sickles with a smooth blade, are necessary in this region because the wheat stems are so hard. It is likely that cereals in Tunisia, which has a hot, dry climate, contain more silica in their stems than in countries to the north having cooler, wetter climates, due to greater water transpiration and silica transport (see Anderson, Chapter 5).

The harvested stems of wheat and barley are tied into sheaves using a tie made from the stems, then bound in a net device (*gaiaba* or *tabbaha*) and transported to the threshing floor on mule or donkey back, where wheat is threshed using the threshing sledge, and the other cereals threshed by animal trampling (Anderson, chapter 5). Both the harvesting method of cutting long stems and the threshing technique which cuts the stems into small pieces, are geared to obtaining the best quality dry animal food. The same toothed sickle is also used to harvest sunflower heads, and to clean plants from border areas between sown fields. Men alone use it to harvest leaves of *Ampelodesmos mauritanica*, a wild grass used for animal fodder and for basketry (Anderson 2013).

Sickles having a smooth cutting edge need to be frequently sharpened using a whetstone during work. Sickles with teeth do not need sharpening, but must be re-forged and the teeth redone every several years by a blacksmith. We observed the Beja blacksmith making dentate sickles four times over the past five years.[9] The sickle blade is initially made from rods for reinforcing concrete, shaped by forging (Fig. 4.40 and Fig. 4.42). The same process occurs when the blade needs to be resharpened. After forging, the blade is ground on a wheel, dipped in quick lime, and left to cool further before the teeth are made.

To make the teeth, the blacksmith sits on a chair with a metal frame that supports (Fig. 4.43 and Fig. 4.44) a dromedary (*Camelus dromedarius*) metatarsus. The dry metatarsus has been shaved to create a flat working surface (Fig. 4.45). The smith braces the metatarsus against his chest and the metal support, and lays the sickle blade on it. A leather strap stabilises the arrangement, and the smith controls the tension with his foot (Fig. 4.44). He uses a forged chisel to cut teeth into the edge of the blade while it rests on the camel metatarsal (Fig. 4.46). After cutting into the sickle blade, the chisel strikes the

Fig. 4.42. Shaping of the sickle by hammering after heating in the forge, in Beja, Tunisia. Image: P. C. Anderson.

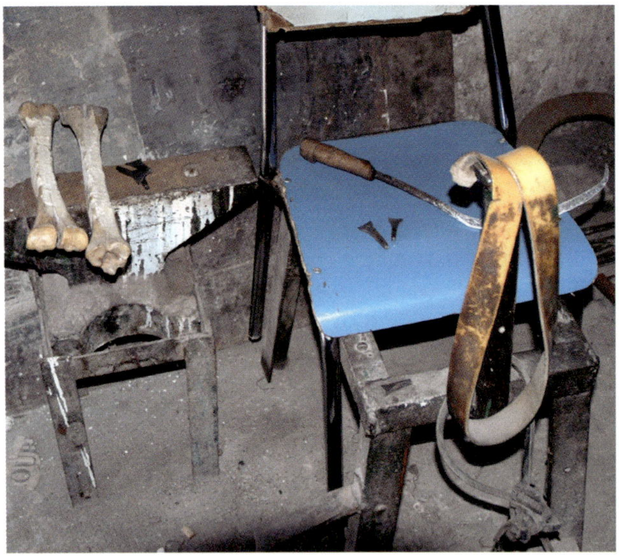

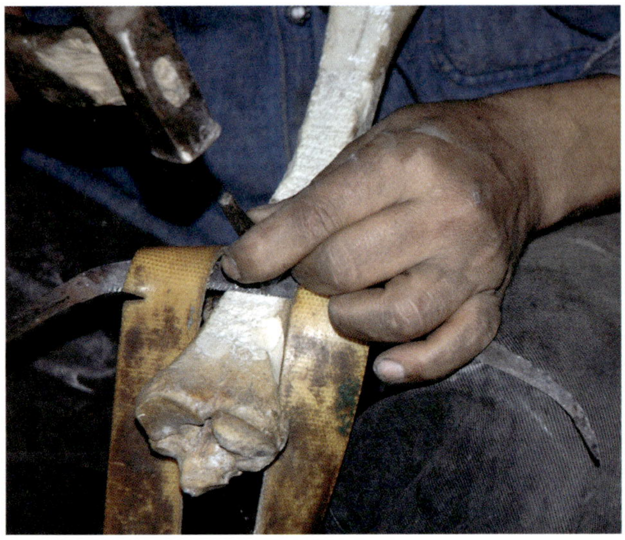

Anti-clockwise from top left:

Fig. 4.43. Tools used by the blacksmith in Beja, Tunisia, to shape sickle teeth: on the anvil at the left, two dromedary metatarsal bones and three chisels. On the seat of the chair, two chisels and the sickle to be sharpened. Note that the chair is placed over an iron structure with four feet, with a vertical rod at the front. On this rod is a leather strap that will be looped around the sickle and through the rubber strap below. Image: P. C. Anderson. **Fig. 4.44**. Making of the teeth on the sickle blade by striking the blade with a hammer and chisel, using the dromedary bone as an anvil. The leather strap is looped around the sickle blade, with the tension regulated by the blacksmith's foot in the lower strap. The blacksmith in Beja, Tunisia is bracing the bone against his chest at one end and the other is set on the vertical piece of the iron frame. Image: P. C. Anderson. **Fig. 4.45**. The dromedary metatarsal bone being used as a sickle anvil. Note the marks from shaping the surface to make it flatter, visible on the articulations. On the surface, note the marks made by the point of the chisel each time it is struck to make a tooth on the sickle edge. Image: J.-D. Strich. **Fig. 4.46**. A close-up view of the work of shaping teeth on the sickle edge, showing the point of the chisel punching out teeth, as it strikes the anvil. Image: P. C. Anderson. **Fig. 4.47**. A close-up view of the indentation marks on the anvil in Figure 4.43, made by the hammer and chisel used to make the dentate edge. This pattern is characteristic of this one area being used more than one time. Image: J.-D. Strich.

bone anvil, making a triangular indentation (Fig. 4.45). As the process is again repeated, moving along the length of the sickle, rows of tiny indentations are created across the bone anvil. When the surface of the bone is covered with marks, the blacksmith goes over the same areas of the bone another time, creating a characteristic surface pattern (Figs. 4.45, 4.46, and 4.47). Eventually the roughened surface will be shaved smooth again, then reused until it is too thin and fragile. The same bone anvil will last several years, depending of course on how many sickles are worked. Bone is perfectly adapted to this function: it is hard enough to furnish a working surface but soft enough to not damage the edge of the chisel. Dromedary metatarsi have always been used, at least within the memory of this blacksmith.

As archaeological finds of metatarsi and other bones with odd indentations accumulated, iconographic data also appeared, although the association between the two was not made until much later (Poplin 2007, 216). Ethnographic observation in the late 1800s of a craftsman in his shop in the Tunis *souk* was published in the form of a painting called *L'affuteur de faucilles* or 'the sickle sharpener' (Lallemand 1890). Today we realise that the craftsman was using the procedure described above to put teeth on the sickle edge, but the picture itself is misleading because it depicts the craftsman tilting up the sickle vertically to show its shape, rather than lying it flat on the wooden support and on the bone anvil in the correct working position. Although the image was probably an accurate depiction, it does not show how the work was done, indeed it is ambiguous about just what work was being done. Therefore, the use or meaning of the anvils found in an archaeological context remained a mystery, awaiting ethnographic observations by persons familiar with the archaeological data (Esteban Nadal 2003; Aguirre *et al.* 2004).

This technique of making teeth on sickles was still in use until the latter half of the 20th century in Spain and Portugal (Moreno-Garcia *et al.* 2007, 195). In Portugal, during the 1950s and 1960s, in the last phase of this practice, craftsman even reformed teeth on mass produced sickles imported from Spain!

In Spain and in Portugal, the process was exactly the same as the one in northwest Tunisia (Veiga de Oliveira *et al.* 1983; Esteban-Nadal and Carbonell-Roure 2004; Aguirre *et al.* 2004; Moreno-García *et al.* 2005): a wooden workbench at the end of which there is a vertical piece of wood was used (not iron, as in the Tunisian example from today in Fig. 4.43), and the work also produced bone anvils with very clear triangular indentations in regular lines.

Interpreting the Archaeological Bone anvils

The first archaeological dromedary bone anvil was recovered from Seville (Spain) and dated to the 11th–12th centuries CE (Moreno-Garcia *et al.* 2007, 196). It was a radius-ulna. The convex surfaces were flattened by shaving, and the regular triangular indentations run transversally across the diaphysis (midsection of a long bone). Because bones used as anvils need to be wide enough to give the craftsman an adequate working surface, and long enough to be held stable, they come from skeletons of large animals, such as camels, equids and oxen. The most valued quality in a bone to be used as an anvil was its toughness (Moreno-Garcia *et al.* 2007, 200).

Bones identified as sickle anvils are mandibles, humeri, radii, femora, tibiae (Fig. 4.48) and

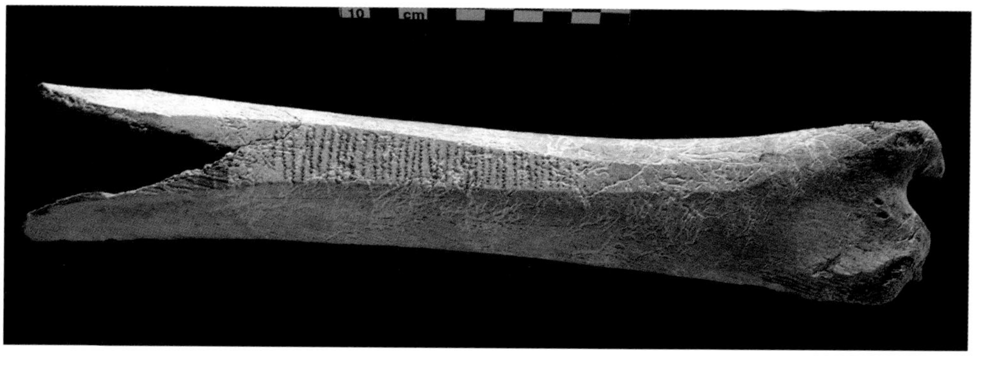

Fig. 4.48. Tibia of an equid, showing that the crest has been removed in order to make a flat working surface for its use as an anvil. From Montaillou, Ariège, France, dating from the end of 13th century to the beginning of the 14th century. Image: J.-D. Strich.

Fig. 4.49. Location of the site of 1) Cegléd-Fertály-földek II in Hungary. Map: R. Lugon, J.-C. Loubier and A. Chevalier.

Fig. 4.50. Close-up view of the indentation marks made by the chisel on the tibia anvil shown in Figure 4.46. From Montaillou, Ariège, France, dating from the end of the 13th century to the beginning of the 14th century. Image: J.-D. Strich.

metapodials used in different proportions in different archaeological sites (*e.g.* Moreno-García *et al.* 2007; Benco *et al.* 2002; Gál *et al.* (2010). French anvil specimens (mandibles), date from the 7th–8th century and the 10th–11th century, and equid hip bones used in this way were discovered in a level dating from the 7th–8th century in Paulhan, France (Rodet-Belarbi *et al.* 2002). The rural Hungarian settlement of Cegléd-Fertály-földek II (Fig. 4.49), dated to 10th–13th-centuries CE, produced ox and horse pelvic bones with punch marks on the surfaces (Gál *et al.* 2010).

The preparation of bones in different times and places varies somewhat. Some bones, like mandibles and pelvic bones, may be used without preparation of a flat surface. The diagnostic mark of all the anvils is the rows of asymmetric triangular indentations resulting from the crushing of the bone material by the chisel (Fig. 4.50), but these marks show great variability. They may be large, or superimposed to the point of no longer forming distinct lines. Sometimes the change in orientation of the point of the triangular indentations shows that once the artisan arrived in the middle of the anvil, he turned it around and continued to work from the other end. The lines are sometimes slightly curved, corresponding to the shape of the sickle blade. These differences do not seem to correspond to any particular chronological or geographic variable, and may reflect differences in skill or individual preference (Moreno-García *et al.* 2006b). Some anvils show traces of the bottom of earlier indentations, indicating that the anvil was shaved to renew its surface. There is no general rule however, as some anvils are barely used. If one surface had only a few lines near one extremity, we could imagine that the anvil was in the course of use. But in the case of several objects, two opposite surfaces are only lightly used.

The Distribution and Antiquity of Dentate Sickles and Bone Anvils

Bone anvils were manufactured in different geographical regions, extending from the Ukraine to the Iberian Peninsula and North Africa, and data show that they have been in use from the Roman period (see below) and the Hellenistic period (Semenov 1964) until today. They serve as a proxy

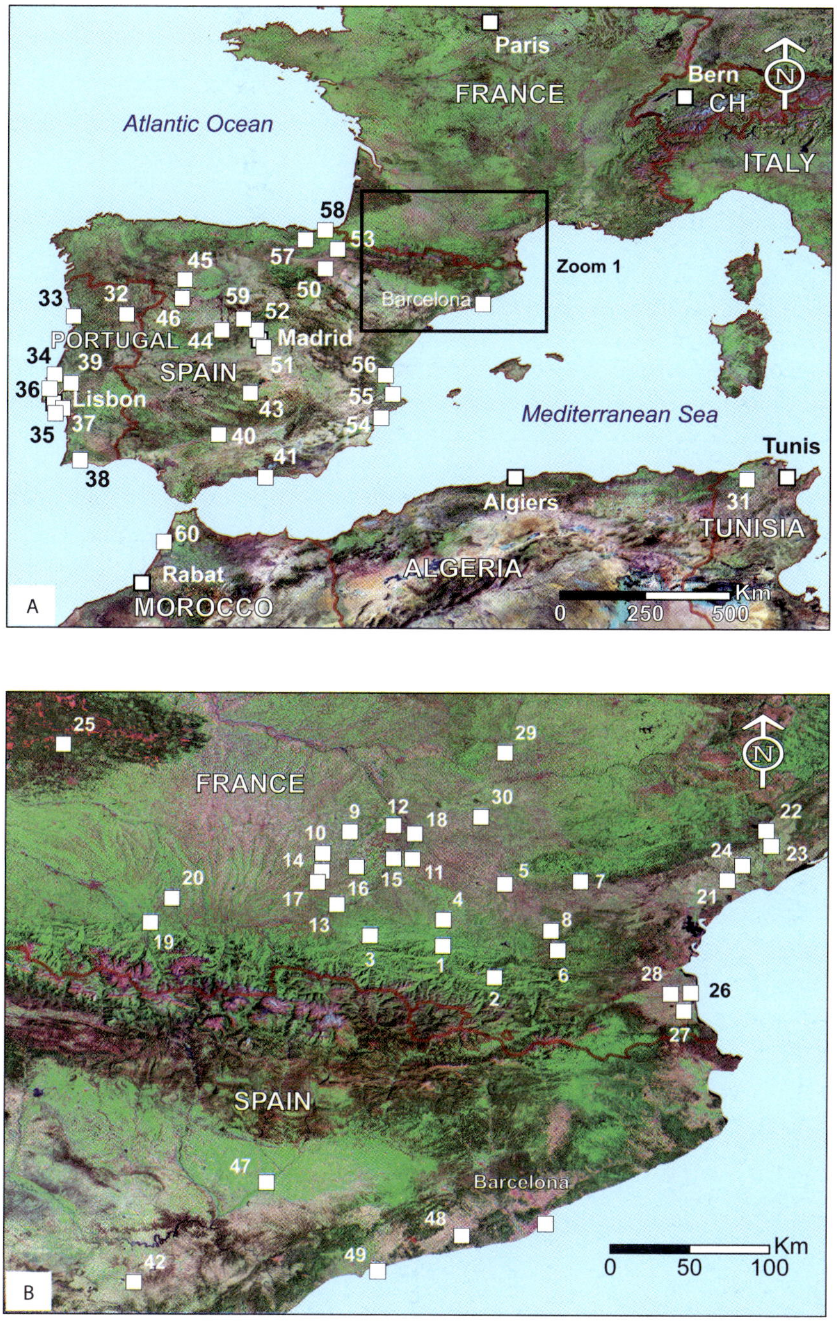

Fig. 4.51. A) Map of archaeological sites in a) Spain, Morocco and Tunisia, and B) in Spain and France, where bone anvils have been found, indicating that sickles with teeth were made there. The numbers correspond to entries in Fig. 4.52.

1) Foix, Ariège (13th–14th c.); 2) Montaillou, Ariège (end 13th–early 14th c.); 3) Montesquieu Avantès, Ariège (undated); 4) Pamiers, Ariège (13th–14th c.); 5) Castelnaudary, Aude (undated); 6) Couiza, Aude (undated); 7) Lastours, Aude (end 12th c.); 8) Limoux, Aude (12th–beginning 13th c.); 9) Isle-Jourdain, Gers (10th–11th c.); 10) Samatan, Gers (11th–12th c.); 11) Clermont-le-Fort, Haute-Garonne (13th c.); 12) Cornebarrieu, Haute-Garonne (Medieval); 13) Martes-Tolosane, Haute-Garonne (undated); 14) Mauvezin-de-l'Isle, Haute-Garonne (12th–14th c.); 15) Muret, Haute-Garonne (14th c.); 16) Rieumes, Haute-Garonne (10th–12th c.); 17) St-Laurent-sur-Save, Haute-Garonne (12th–14th c.); 18) Toulouse, Haute-Garonne (12th–4th c.); 19) Lourdes, Hautes-Pyrénées (undated); 20) Tarbes, Hautes-Pyrénées (12th - 14th c.); 21) Beziers, Herault (end 13th–15th c.); 22) Clermont-l'Herault, Herault (HS, Medieval ?); 23) Paulhan, Herault (7th -8th c.); 24) Servian, Herault (undated); 25) Labrit, Landes (14th–15th c.); 26) Canet, Pyrénées-Orientales (12th–14th c.); 27) Elne, Pyrénées-Orientales (12th–14th c.); 28) Perpignan, Pyrénées-Orientales (low Middle Ages); 29) Cordes, Tarn (14th c.); 30) Lavaur, Tarn (undated); 31) Beja (14th–17th c.); 32) Freixo de Numaõ (undated); 33) Idanha (undated); 34) Mouraria, Lx (14th–15th c.); 35) Palmela (15th c.); 36) Santarèm (18th–19th c.); 37) Sesimbra (undated); 38) Silves (12th–15th c.); 39) Torre Vedras (15th–17th c.); 40) Córdoba, Andalucía (9th–12th c.); 41) Motril, Andalucía (13th–15th c.); 42) Alcañiz, Aragon (16th–18th c.); 43) Ciudad Real, Castilla-La Mancha (12th c.); 44) Avila, Castilla-Leon (13th–15th c.); 45) Benavente, Castilla-Leon (off site); 46) Zamora, Castilla-Leon (16th–18th c.); 47) Lleida, Cataluña (from the 1st/2nd half of 12th c. -1707); 48) Olèrdola, Cataluña (9th–12th c.); 49) Tarragona, Cataluña (Medieval, from 2nd half of 13th c.); 50) Calahorra, La Rioja (off site); 51) Buzanca, S. Martin Vega, Madrid (5th–6th c.); 52) Colmenar Viejo, Madrid (16th–18th c.); 53) Pamplona, Navarra (off site); 54) Alicante, Pais Valenciano (19th - 20th c.); 55) Gandia, Pais Valenciano (13th–15th c., 16th–18th c.); 56) Valencia, Pais Valenciano (11th–12th c., 16th–18th c.); 57) Arrasate-Mondragón, País Vasco (19th–20th c.); 58) San Sebastián, País Vasco (19th–20th c.); 59) Segovia, Castilla y León (Medieval); 60) Al-Basra. Map: R. Lugon, J.-C. Loubier and A. Chevalier.

for dentate or toothed metal sickles. In France, ancient bone anvils occur in a clearly defined zone from Southeast Aquitaine to the Midi- Pyrénées region, with Cordes (Tarn) to the north, Paulhan (Hérault) to the east and Perpignan (Pyrénées-Orientales) to the south (Fig. 4.28; Rodet-Belarbi *et al.* 2007). Several excavation reports from the beginning to the end of the 19th century cited these objects but their function was an enigma. During the last decades of the 20th century, archaeologists working in southeastern France recorded these particular bone objects fashioned primarily from cattle and horse metapodials and interpreted them as polishers, files or sharpeners (Briois *et al.* 1995; Rodet-Belarbi *et al.* 2002). The French objects are often associated with archaeological levels containing iron-working waste. Several objects from a forge site in Lastours (Aude; Fig. 4.28) dating to the 12th–13th centuries, were cleaned metatarsi waiting to be used by the blacksmith.

	France	Date
1	Foix (Ariège)	XIIIth–XIVth c.
2	Montaillou (Ariège)	end XIIIth–early XIVth c.
3	Montesquieu Avantès (Ariège)	undated
4	Pamiers (Ariège)	XIIIth–XIVth c.
5	Castelnaudary (Aude)	undated
6	Couiza (Aude)	undated
7	Lastours (Aude)	end XIIth c.
8	Limoux (Aude)	XIIth–beginning XIIIth c.
9	Isle-Jourdain (Gers)	Xth–XIth c.
10	Samatan (Gers)	XIth–XIIth c.
11	Clermont-le-Fort (Hte-Garonne)	XIIIth c.
12	Cornebarrieu (Haute-Garonne)	Medieval
13	Martes-Tolosane (Hte-Garonne)	undated
14	Mauvezin-de-l'Isle (Hte-Garonne)	XIIth–XIVth c.
15	Muret (Hte-Garonne)	XIVth c
16	Rieumes (Hte-Garonne)	Xth–XIIth c.
17	St-Laurent-sur-Save (Hte-Garonne)	XIIth–XIVth c.
18	Toulouse (Hte-Garonne)	XIIth–XIVth c.
19	Lourdes (Hts-Pyrenees)	undated
20	Tarbes (Htes-Pyrenees)	XIIth–XIVth c.
21	Beziers (Herault)	end XIIIth–XVth c
22	Clermont-l'Herault (Herault)	HS (Medieval ?)
23	Paulhan (Herault)	VIIth–VIIIth c.
24	Servian (Herault)	undated
25	Labrit (Landes)	XIVth–Xvth c.
26	Canet (Pyrenees-Orientales)	XIIth–XIVth c
27	Elne (Pyrenees-Orientales)	XIIth–XIVth c.
28	Perpignan (Pyrenees-Orientales)	Low Middle Ages
29	Cordes (Tarn)	XVth c.
30	Lavaur (Tarn)	undated

	Tunisia	Date
31	Beja	XVth–XVIIth c.

	Portugal	Date
32	Freixo de Numaõ	undated
33	Idanha	undated
34	Mouraria, Lx	XIVth–XVth c.
35	Palmela	XVth c.
36	Santarèm	XVIIIth–XIX c.
37	Sesimbra	undated
38	Silves	XIIth–XVth c
39	Torre Vedras	XVth–XVIIth c.

	Spain	Date
40	Córdoba (Andalucia)	IXth–XIIth c.
40	Córdoba (Andalucia)	Off site
41	Motril (Andalucia)	XIIIth–XVth c.
42	Alcañiz (Aragon)	XVIth–XVIIIth c
43	Ciudad Real (Castilla-La Mancha)	XII c.
44	Avila (Castilla-Leon)	XIIIth–XVth c.
45	Benavente (Castilla-Leon)	Off site
46	Zamora (Castilla-Leon)	XVIth–XVIIIth c.
47	Lleida (Cataluña)	1st–2nd 1/2 XIIth c.
47	Lleida (Cataluña)	2nd 1/2 Xth–XIIth c.
47	Lleida (Cataluña)	2nd 1/2 XIIth–XIIIth c.
47	Lleida (Cataluña)	XIVth–XVth c
47	Lleida (Cataluña)	1580–1707
47	Lleida (Cataluña)	Off site
48	Olerdola (Cataluña)	IXth–XIIth c.
49	Tarragona (Cataluña)	2nd 1/2 XIIIth c.
49	Tarragona (Cataluña)	2nd 1/2 XIVth–1st 1/4 XVth c.
49	Tarragona (Cataluña)	Medieval
49	Tarragona (Cataluña)	Off site
50	Calahorra (La Rioja)	Off site
51	Buzanca, S.Martin Vega (Madrid)	Vth– VIth s.
52	Colmenar Viejo (Madrid)	XVIth–XVIIIth c.
53	Pamplona (Navarra)	Off site
54	Alicante (Pais Valenciano)	XIXth–XXth c.
55	Gandia (Pais Valenciano)	XIIIth–XVth c.
55	Gandia (Pais Valenciano)	XVIth–XVIIIth c.
56	Valencia (Pais Valenciano)	XIth–XIIth c.
56	Valencia (Pais Valenciano)	XVIth–XVIIIth c.
56	Valencia (Pais Valenciano)	Off site
57	Arrasate-Mondragón (Pais Vasco)	XIXth–XXth c.
58	San Sebastián (Pais Vasco)	XIXth–XXth c.
58	San Sebastián (Pais Vasco)	Off site
59	Segovia (Castilla y León)	Medieval

	Morocco	Date
60	Al-Basra	

Fig. 4.52. Sites and dates for finds of bone anvils used to make dentate sickles. See Fig. 4.51 for a map of the sites. Table: I. Rodet-Belarbi.

In Morocco, 186 anvils were found in Al-Basra (also known as Basra Al Hamra; Fig. 4.11) in an Islamic level in association with iron metallurgy (Benco *et al.* 2002). An object of this type was also found in Tunisia in a generally medieval context, in Beja, where the tradition continues to the present day (above).

Over sixty sites in the southwest of France, the Iberian Peninsula, the Maghreb, the Ukraine and Hungary have yielded at least 500 bone anvils (see Fig. 4.51 and Fig. 4.52). The oldest known anvil in the Iberian Peninsula was discovered in Madrid (Buzanca, S. Martín de laVega), and dates from the 5th-6th century, but a particularly large concentration of finds comes from sites dating to the Islamic period, 9th–13th centuries (Moreno-García *et al.* 2006a). In France, although the five anvils dating from the 8th century are the oldest ones known there, most come from levels dating from the 12th-14th century. But the oldest bone anvils documented to date are new discoveries from the Roman period, from Metaponto in southern Italy dated to between 190 BCE–CE 10 or 2070 BP, (Gál and Bartosiewicz 2012) and from Romania from 1st–2nd century CE contexts (Beldiman *et al.* 2011).

The geographic and chronological distribution of these anvils does not always correspond to the distribution of toothed sickles. However, iron tools do not preserve well in the ground, and in many cases may have been destroyed to recover the metal when they wore out. Nevertheless, the bone anvils act as proxies for the presence of toothed sickles, and show that serrated sickles were used over a long period, from the 5th–6th century until around 30 years ago in the Iberian Peninsula, from the 8th century to the 16th century in the South of France and from the 10th century to the 13th century in Hungary.

Even the distribution of bone anvils gives only a partial picture. Their apparent absence in the north of France may not be due to a lack of serrated sickles, but rather because a different material from bone may have been used as an anvil, including woods such as *Cornus mas* (Poplin 2007, 219). The distribution of this wood goes as far south as the Aquitaine region of southern France, near Bordeaux, and this same region is the northern limit of where the bone anvils are found. The preservation of objects made from such woody plants is of course extremely rare in archaeology, so verification of this hypothesis is difficult.

This study has focused on the complex, indirect and interdisciplinary process of identifying an obscure tool, and exploring its use and distribution. The misinterpretations of bone sickle anvils as objects having only symbolic value, or mistaken functional ideas is an important cautionary tale. It shows the immense interpretive value of ethnographic studies when they are rigorously applied to archaeological data. In this case, identifying one tool, the bone sickle anvil, allows us to trace the distribution of the associated toothed sickles, and the agricultural processes that these imply.

4.9. CONCLUSION

Patricia C. Anderson

The case studies in this chapter emphasise variability and choice in harvesting technique, and describe alternate ways to harvest grasses and cereals with simple tools like the *mesorias*, more complex tools such as many forms of sickle, and harvesting with the hands alone. A focus in most of the articles is on the reasons for the choice of a particular harvesting method. This choice often concerns what product or part of the plant is sought, and how it will be used after harvest. The morphology of specific plants may affect the choice of tools and harvesting motions. Local conditions such as the friability of the soil also may determine how a crop can be harvested.

Another thrust of these studies has been to show the importance of multidisciplinary approaches. Ethnography of modern farmers provides a necessary basis for most of our understanding of traditional and ancient agriculture. Here, it also serves more pointedly to examine details of harvesting methods and choices among alternatives, and allows us to interpret obscure artefacts in the archaeological record. The use-wear on tool edges is only interpretable by reference to ethnographic cases and experimental studies that replicate what we see archaeologically. Experiments also allow us to test hypothetical reproductions of ancient tools and understand the potentials and problems of traditional harvesting methods.

The tools themselves imply particular types of agricultural techniques and adaptations. One archaeological case study showed, how differing forms of sickle may indicate two distinct pathways of Neolithisation in the Iberian Peninsula, while another showed, unexpectedly, that one cultural group apparently stopped using sickles for harvest and opted instead for an alternate technique involving harvesting seed heads by hand. Another describes how differing forms of sickle may indicate two distinct pathways of Neolithisation in the Iberian Peninsula.

This chapter presented studies with interdisciplinary approaches being used to solve particular problems, and often uncommon solutions emerged. Many of the case studies emphasised the importance of the harvesting technique in relation to how the produce will be used or the threshing technique that will be applied to it, subjects that will be examined in the following two chapters.

Chapter Acknowledgements

Research presented by L. Peña-Chocarro (Chapter 4.2) was carried out within the projects AGRIWESTMED (ERC-AdG-230561), HAR2008–03976/HIST, HAR2008–01920/HIST and HAR2011–23716 *Nuevos cultivos, nuevos paisajes* and Programa Consolider TCP-CSD2007–00058 from the Spanish Government, as well as within the Research Group at the UPV/EHU IT-288–07 funded by the Basque Government.

The work presented by J. F. Gibaja, J. J. Ibáñez and J. González Urquijo (Chapter 4.7) was carried out within the framework of the following research projects: '*Origins and spread of agriculture in the south-western Mediterranean región*' funded by the European Research Council (ERC-AdG 230561.); '*Las ocupaciones lacustres y la gestion de los recursos entre las primeras sociedades agricolas y ganaderas del NE peninsular: Tecnología de las producciones materiales y usos instrumental. Estrategias agroforestales y ganaderas*' supported by the Ministerio de Ciencia e Innovación (HAR2009–13494–C02–01 and HAR2009–13494–C02–02.); '*The last hunter-gatherers and the first farming communities in the south of the Iberian Peninsula and north of Morocco: a socio-economic approach through the management of production instruments and exploitation of the domestic resources*' Fundaçao para Ciência e a Tecnologia (PTDC/HAH/64548/2006).

P. C. Anderson (Chapters 4.2, 4.3, 4.5, 4.8), and M. M'hamdi (Chapter 4.3) acknowledge the support of the GDR 2517, *Regards interdisciplinaires sur les activités et techniques agricoles anciennes et préindustrielles*, Center for National Scientific Research (CNRS), for ethnoarchaeological fieldwork. P. C. Anderson thanks Catherine Llaty, Nice, France, and Hani Abu-Asaf, Sweida, Syria, for their collaboration in ethnographic fieldwork in the Jebel Druze region, Syria.

J. C. Whittaker (Chapter 4.5) received financial support from the European Science Foundation, EARTH Programme, for a collaborative research visit to carry out joint experiments. Anderson and Whittaker thank Laurent Bouby, Montpellier, France for his help in quantifying the experiments.

Chapter Notes

1 See http://www.abc.net.au/aplacetothink/ #watch/mh_1960/desert/watchVideo.
2 http://vimeo.com/73966655 shows a film clip of an experiment using this harvesting method.
3 http://vimeo.com/73954386 shows a film clip of an experiment using such a bone tool.
4 Please refer to the film clips at http://vimeo.com/73401066 and http://vimeo.com/74133713 on the manufacture and use of this harvesting knife, the kebele.
5 See the film clip at http://vimeo.com/77179882
6 The short film at http://vimeo.com/73974255 shows harvesting and craft use of alfa to make a hassir, described in this text.
7 See the film clip at http://vimeo.com/77179913.
8 See film clip at http://vimeo.com/77179894 showing harvesting of this wheat.
9 See film clip at http://vimeo.com/73954409 of this blacksmith working to put teeth on a sickle edge.

Chapter Bibliography

Abreu Galindo, Fr. J. (1602?/1977) *Historia de la conquista de las siete islas de Canaria.* Reprint: Goya Ediciones, Santa Cruz de Tenerife.

Aguirre, A., Etxeberria, F. and Herrasti, L. (2004) El yunque de hueso para afilar la hoz metálica dentada. *Munibe* 56: 113–121.

Alvargonzález, C. (1908) *La escanda. Su origen. Su cultivo.* El Noroeste, Gijón.

Anderson, P. C. (1991) Harvesting of wild cereals during the Natufian as seen from experimental cultivation and harvest of wild einkorn wheat and micro wear analysis of stone tools. In O. Bar-Yosef and F. Valla (eds.) *The Natufian culture in the Levant.* International Monographs in Prehistory, Archaeological series 1. Ann Arbor, Mi, pp. 521–556.

Anderson, P. C. (1992) Experimental cultivation, harvesting and threshing of wild cereals and their relevance for interpreting the use of Epipaleolithic and Neolithic artefacts. In P. C. Anderson (ed.) *Préhistoire de l'agriculture: nouvelles approches expérimentales et ethnographiques.* Monographies du CRA 6. CNRS, Paris, pp. 179–210.

Anderson, P. C. (1994) Insights into plant harvesting and other activities at Hatoula, as revealed by microscopic functional analysis of selected chipped stone tools. In M. Lechevallier and A. Ronen (eds.) *Le gisement de Hatoula en Judée occidentale.* Israël. Mémoires et travaux du Centre de Recherche Français de Jérusalem 8. Association Paléorient, Paris, pp. 277–293.

Anderson, P. C. (1998) The history of harvesting and threshing techniques for cereals in the prehistoric Near East. In A. B. Damania, J. Valkoun, G. Willcox and C. O. Qualset (eds.) *The Origins of agriculture and crop domestication.* ICARDA/IPIGRI/Grcf/FAO, pp. 141–155.

Anderson, P. C. (1999) Experimental cultivation, harvest and threshing of wild cereals. In P. C. Anderson (ed.) *Prehistory of Agriculture: New Experimental and Ethnographic Approaches.* Institute of Archaeology Monograph 40. University of California, Los Angeles, pp. 104–118.

Anderson, P. C. (2000) Technologie agricole et artisanat a El Kowm 2: L'apport des microtraces d'utilisation sur des outils en silex. In D. Stordeur (ed.) *El Koum 2, Une île dans le désert.* CNRS, Paris, pp. 179–196.

Anderson, P. C. (2006) Non-mechanised processing and storage of cereals, grasses and pulses used for fodder, fuel, food and crafts: Examples from N. Tunisia, Atlas Region, Northwestern Tell. In Ertuğ (ed.) *Proceedings of the IVth International Congress of Ethnobotany (Iceb 2005).* Ege Yaynlar, 223-231.

Anderson, P. C. (2013) Neolithic tools for stripping ears from hulled cereals: an update. In P. C. Anderson, C. Cheval and A. Durand (eds.) *Regards croisés sur les outils liés au travail des végétaux (An Interdisciplinary Focus On Plant Working Tools).* XXXIIIe Rencontres Internationales d'Archéologie et d'Histoire d'Antibes, Editions ADPCA, Antibes, pp. 89–102.

Anderson, P. C. (2014) Uses of the wild grass *Ampelodesmos mauritanica* in northwestern Tunisia today. Chapter 6: A versatile world: examples of diversity in plant use. In A. Chevalier, E. Marinova and L. Peña-Chocarro (eds.) *Plants and People: Choices and Diversity through Time.* EARTH Series Volume 1. Oxbow Books, Oxford, pp. 287–292.

Anderson, P. C. and Valla, F. (1996) 'Glossed tools' from Hayonim Terrace: Blank choice and functional tendencies. In S. K. Kozlowski and H. G. Gebel (eds.) *Neolithic Chipped Stone Industries of the Fertile Crescent and their contemporaries in adjacent regions.* Studies in Near Eastern Production, Subsistence and Environment 3. ex oriente, Berlin, pp. 341–362.

Anderson-Gerfaud, P. (1986) A few comments concerning residue analysis of stone plant-processing tools. In L. Owen and G. Unrath (eds.) *Technical aspects of microwear studies on stone tools.* Early Man News 9/10/11, 69–87.

Bakels, C. C. (2007) Nature or culture? Cereal crops raised by Neolithic farmers on Dutch loess soils. In S. Colledge and J. Connolly (eds.) *The origins and spread of domestic plants in southwest Asia and Europe.* Left Coast Press, Walnut Creek, pp. 343–347.

Bakels, C. C. (2009) *The western European loess belt. Agrarian history, 5300 BC - AD 1000.* Springer, Dordrecht/ Heidelberg/London/New York.

Bakels, C. C. and Zeiler, J. (2005) The fruits of the land: Neolithic subsistence. In L. P. Louwe Kooijmans, P. W. Van den Broeke, H. Fokkens and A. L. van Gijn (eds.) *The prehistory of the Netherlands.* Amsterdam University Press, Amsterdam, pp. 311–336.

Beldiman, C., Sztancs, D.-M., Ruscu-Bolindet V. and Achim, I. A. (2011) Skeletal technologies, metal-working and wheat harvesting: ancient bone and antler anvils for manufacturing saw-toothed iron sickles discovered in Romania. In J. Baron and B. Kufel-Diakowska (eds.) *Written in bones. Studies on technological and social contexts of past faunal skeletal remains.* EICAZ, Uniwersytet Wrocławski, Instytut Archeologii, Wrocłam, pp. 173–186.

Benco, N. L., Ettahiri, A. and Loyet, M. (2002) Worked bone tools: linking metal artisans and animal processors in medieval Islamic Morocco. *Antiquity* 76, 447–457.

Bernus, E. (1968) *Cram-Cram*, CNRS audiovisuel, Paris, film 16 mm, colour, 12 min.

Bernus, E. (1993) *Touaregs nigériens: unité culturelle et diversité régionale d'un peuple pasteur.* Harmattan, Paris.

Bosch, A., Tarrùs, J. and Chinchilla, J. (2000) *El poblat lacustre neolític de La Draga: excavacions de 1990 a 1998. Monografies del CASC 2.* Centre d'arqueologia subaquàtica de Catalunya, Museu d'Arqueologia de Catalunya, Girona.

Bosch, A., Chinchilla, J. and Tarrùs J. (2006) *Els objectes de fusta del poblat neolític de la Draga. Excavacions de 1995-2005.* Monografies del CASC 6. Centre d'arqueologia subaquàtica de Catalunya, Museu d'arqueologia de Catalunya, Girona.

Bouron, N. (1930) *Les Druzes. Histoire du Liban et de la Montagne Haournaise.* Berger-Levrault, Paris.

Bregadze, N. A. (1982) *Sketches on the agroethnography of*

Georgia. Academy of Sciences, Georgian Soviet Socialist Republic, Tbilisi.

Briois, F., Poplin, F. and Rodet-Belarbi, I. (1995) Aiguisoirs, polissoirs médiévaux en os (Xe-XIVe s.) Sud-Ouest de la France. *Archéologie du Midi Médiéval* 13, 197–213.

Brouillet, I. (2013) Un outil de moisson du mil au Mali aujourd'hui : le 'kebele'. In P. C. Anderson, C. Cheval and A. Durand (eds.) Regards croisés sur les outils liés au travail des végétaux (An Interdisciplinary Focus On Plant Working Tools). XXXIIIe Rencontres Internationales d'Archéologie et d'Histoire d'Antibes, Editions ADPCA, Antibes, pp. 61–74.

Bruyn, A. (1984) Een vuurstenen 'sikkel' uit Medemblik. Jaarverslag ROB, 89–94.

Buxò, R. (1997) *Arqueología de las plantas*. Crítica, Barcelona.

Buxó i Capdevila, R. (1989) La présence de l'épeautre (*Triticum spelta* L.) dans l'alimentation en Espagne: les temps anciens et les temps modernes. In J. P. Devroey and J. J. Van Mol (eds.) *L'épeautre*: histoire et ethnologie. Editions Dire, Treignes, pp. 107–122.

Caro Baroja, J. (1972) 'Escanda'. *Gran Enciclopedia Asturiana*. Vol. VI. Gijón, 121–122.

Caro Baroja, J. (1975) *Los pueblos de España*. Itsmo, Madrid.

Carvalho, A. F. (2010) Le passage vers l'Atlantique: le processus de néolithisation en Algarve (sud du Portugal). *L'Anthropologie* 114, 141–178.

Clemente, I. and Gibaja, J. F. (1998) Working Processes on Cereals: an Approach through Microwear Analysis. *Journal of Archaeological Science* 25 (5), 457–464.

Clemente, I. and Garcia, V. (2008) Yacimientos arqueológicos de la Bahía de Cádiz. Aplicación del análisis funcional a los instrumentos de trabajo líticos del Embarcadero del río Palmones, La Mesa y La Esparragosa. In J. Ramos (coord.) *La ocupación prehistórica de la campiña litoral y banda atlántica de Cádiz : aproximación al estudio de las sociedades cazadorasrecolectoras, tribalescomunitarias y clasistas iniciales*. Consejería de Cultura, Sevilla, pp. 185–198.

Comet, G. (1992) *Le paysan et son outil. Essai d'histoire technique des céréales*, Ecole française de Rome, Rome.

Comet, G. (ed.) (2003) *L'outillage médiéval et moderne et son histoire*, Presses universitaires du Mirail-Toulouse, Toulouse.

Dantín Cereceda, J. (1941) Distribución geográfica de la escanda asturiana. *Estudios Geográficos* 5, 739–797.

Dunlop, I. and Martin-Jones, J. (1966) *Desert People: A day in the life of two nomadic families of the Australian Western Desert*. Australian national film Board. Australian Commonwealth Film Unit for the Australian. Institute of Aboriginal Studies. 49 minutes. http://www.abc.net.au/aplacetothink/#watch/mh_1960/desert/watchVideo

Esteban Nadal, M. (2003) *Ossos de ferrer mil.lenaris. Actes del II Congrés d'Història d' Alcanar. 1,2 i 3 de desembre de 2000*. Ajuntament d'Alcanar, Alcanar, pp. 111–120.

Esteban Nadal, M. and Carbonell Roure, E. (2004) Saw-toothed sickles and on bone anvils: a medieval technique from Spain. *Antiquity* 301 (78), 637–646.

Flors, E. (2010) *Torre la Sal (Ribera de Cabanes, Castellón). Evolución del paisaje antrópico desde la prehistoria hasta el medioevo*. Monografies d'arqueologia i prehistòria castellonenques 8. Servicio de Investigaciones Arqueológicas y Prehistóricas – Servicio de Publicaciones, Diputación de Castellón.

Fugazzola, M. A. and Pessina, A. (1999) Le village néolithique submergé de La Marmotta. Le Néolithique du Nord-Ouest méditerranéen. *Actes du XXIV Congrès Préhistorique de France*, pp. 35–38.

Gál, E., Kovács, E., Kováts, I. and Zimborán, G. (2010) Early medieval (10th–13th century) bone anvils from Hungary: another example for the use of animal bones. In J. Gömöri and A. Körösi (eds.) *Bone and Leather: History, Archaeology and Ethnography of crafts utilizing raw materials from animals*. Publication Series of the Working Committe for Industrial Archaeology and Archaeometry and the Working Committe for Handicraft History at the Regional Centre of the Hungarian Academy of Sciences, Veszprém (VEAB), Budapest, pp. 117–126.

Gál, E. and Bartosiewicz, L. (2012) A radiocarbon-dated bone anvil from the *chora* of Metaponte, southern Italy. *Antiquity*, 86 (331). URL: http://antiquity.ac.uk/projgall/gal331/

Gassin, B., Bicho, N., Bouby, L., Buxo, R., Carvalho, A. F., Clemente, I., Gibaja, J. F., Gonzalez Uquijo, J., Ibañez, J. J., Linton, J., Marinval, P. H., Marquez, B., Peña-Chocarro, L., Perez Jorda, G., Philibert, S., Rodriguez, A. and Zapata, L. (2010) Variabilité des techniques de récolte et traitement des céréales dans l'occident méditerranéen au Néolithique ancien et moyen: facteurs environnementaux, économiques et sociaux. In A. Beeching, E. Thirault and J. Vital (dir.) Économie et société à la fin de la Préhistoire. *Actualité de la recherche*. Documents d'Archéologie en Rhône-Alpes et en Auvergne Nº 34. Publications de la Maison de l'Orient et de la Méditerranée, Lyon, pp. 19–37.

Gauquelin, T., Jalut, G., Iglesias, M., Valle, F., Fromard, F. and Dedoubat, J. J. (1996) Phytomass and carbon storage in the *Stipa tenacissima* steppes of the Baza basin, Andalusia. *Journal of arid Environments* 34, 77–286.

Gibaja, J. F. (2000) La función del instrumental lítico tallado de La Draga (Banyoles, Pla de l´Estany). In A. Bosch, J. Tarrús and J. Chinchilla (eds.) *El poblat lacustre neolític de La Draga: excavacions de 1990 a 1998*. Monografies del CASC 2. Centre d'arqueologia subaquàtica de Catalunya, Museu d'Arqueologia de Catalunya, Girona, pp. 206–213.

Gibaja, J. F. (2002) Las hoces neolíticas del noreste de la Península Ibérica. *Préhistoire anthropologie méditerranéenne* 10, 83–96.

Gibaja, J. F. (2003) Comunidades Neolíticas del Noreste de la Península Ibérica: una aproximación socioeconómica a partir del estudio de la función de los útiles líticos. BAR International Series 1140. Archaeopress, Oxford.

Gibaja, J. F. (2008) La función del utillaje lítico documentado en los yacimientos neolíticos de Revilla del Campo y La Lámpara (Ambrona, Soria). In M. Rojo *et al.* (eds.) Paisaje de la memoria: Asentamientos del neolítico antiguo en el Valle de Ambrona (Soria, España). Universidad de Valladolid, Valladolid. *Arte y Arqueología* 23, 451–493.

Gibaja, J. F. and Carvalho, A. F. (2010) The last hunter-gatherers and the first farming communities in the South of the Iberian peninsula and North of Morocco. In J. F. Gibaja and A. F. Carvalho (eds.) *Promontoria* 15.

Gibaja, J. F., Ibañez, J. J., Rodriguez, A., Gonzalez, J. E., Clemente, I., Garcia, V. and Perales, U. (2010) Estado de la cuestión sobre los estudios traceológicos realizados en contextos mesolíticos y neolíticos del sur peninsular y noroeste de África. In J. F. Gibaja and A. F. Carvalho (eds.) The last hunter-gatherers and the first farming communities in the South of the Iberian peninsula and North of Morocco. *Promontoria* 15, 181–189.

Van Gijn, A. L. (1988) The use of Bronze age flint sickles in the Netherlands: a preliminary report. In S. Beyries (ed.) *Industries lithiques; tracéologie et technologie.* BAR International Series 411. Archaeopress, Oxford, pp. 197–218.

Van Gijn, A. L. (1999) The interpretation of 'sickles': a cautionary tale. In P. C. Anderson (ed.) *Prehistory of Agriculture: New Experimental and Ethnographic Approaches.* Institute of Archaeology Monograph 40. University of California, Los Angeles, pp. 363–372.

Van Gijn, A. L. (2008) Exotic flint and the negotiation of a new identity in the 'margins' of the agricultural world. The case of the Rhine-Meuse delta. In H. Fokkens, B. Coles and Van A. L. Gijn (eds.) *Between foraging and farming. An extended broad spectrum of papers presented to Leendert Louwe Kooijmans.* Faculty of Archaeology. Leiden, pp. 193–202.

Van Gijn, A. L. (2010) *Flint in focus. Lithic biographies in the Neolithic and Bronze Age.* Sidestone Press, Leiden.

Van Gijn, A. L., Van Betuw, V., Verbaas, A. and Wentink, K. (2006) Flint. Procurement and use. In L. P. Louwe Kooijmans and P. F. B. Jongste (eds.) *Schipluiden a Neolithic site on the Dutch north sea coast (3800-3500 BC),* Analecta Praehistorica Leidensia 37/38, 129–166.

de Gongora, M. (1991) *Antigüedades prehistóricas de Andalucía.* Facsimile (1st ed.: Madrid, 1868). Archivum 27. Universidad de Granada, Granada.

González Llana, J. (1889) *Manual de agricultura práctica de la provincia de Oviedo.* Madrid.

Gonzalez, J. E., Ibañez, J. J., Peña-Chocarro, L., Gavilan, B. and Vera, J. C. (1994) Harvesting Tasks in the Neolithic Levels of 'Los Murcielagos' Cave: an Archeobotanical and Functional approach. *Helinium* 34 (2), 322–341.

Gonzalez, J. E., Ibáñez, J. J., Peña-Chocarro, L., Gavilan, B. and Vera, J. C. (2000) El aprovechamiento de recursos vegetales en los niveles neolíticos del yacimiento de Los Murciélagos, en Zuheros (Córdoba). Estudio arqueobotánico y de la función del utillaje. *Complutum* 11, 171–189.

Harlan, J. R. (1967) A Wild Wheat Harvest in Turkey. *Archaeology* 20: 197–201.

Harlan, J. R. (1999) Harvesting of Wild-grass Seed and Implications for Domestication. In P. C. Anderson (ed.) *Prehistory of Agriculture: New Experimental and Ethnographic Approaches.* Institute of Archaeology Monograph 40. University of California, Los Angeles, pp. 1–5.

Henning, J. (1991) Fortleben und Weiterentwicklung spätrömischer Agrargerätetraditionen in Nordgalllien – Eine Mähsense der Merowingerzeit aus Kerkhove (Belgien). *Acta Archaeologica Lovaniensa* 30, 49–59.

Hillman, G. C. (1975) The plant remains from Tell Abu Hureyra. In A. M. T. Moore, G. C. Hillman and A. J. Legge (eds.) The excavation of Tell Abu Hureyra in Syria: a preliminary report. *Proceedings of the Prehistoric Society* 41, 70–73.

Hillman, G. C. and Davies, M. S. (1990) Measured domestication rates in wild wheats and barley under primitive cultivation, and their archaelogical implications. *Journal of World Prehistory* 4 (2), 157–219.

Hillman, G. C. and Davies, M. S. (1999) Domestication rate in wild wheats and barley under primitive cultivation: preliminary results and archaeological implications of field measurements of selection coefficient. In P. C. Anderson (ed.) *Prehistory of Agriculture: New Experimental and Ethnographic Approaches.* Institute of Archaeology Monograph 40. University of California, Los Angeles, pp. 70–102.

Ibáñez, J. J., González Urquijo, J., Peña-Chocarro, L., Zapata, L. and Beugnier, V. (2001) Harvesting without sickles. Neolithic examples from humid mountain areas. In S. Beyries and P. Pétrequin (eds.) *EthnoArchaeology and its Transfers.* BAR International Series 983. Archaeopress, Oxford, pp. 23–36.

Ibáñez, J. J., Gonzalez, J. E., Gibaja, J. F., Rodriguez, A., Marquez, B., Gassin, B. and Clemente, I. (2008) *Harvesting in the Neolithic: characteristics and spread of early agriculture in the Iberian peninsula. Prehistoric Technology. 40 Years Later: Functional Analysis and the Russian Legacy.* BAR International Series. Archaeopress., Oxford, pp. 183–195.

Jacob-Friesen, K. H. (1955) Einführung in Niedersachsens Urgeschichte, Teil I, Steinzeit. Hildesheim.

Kislev, M. E., Weiss, U. and Hartmann, A. (2004) Impetus for sowing and the beginning of agriculture: Ground collecting of wild cereals. *PNAS* 101 (9), 2692–2695.

Kubiak-Martens, L. (2006) Botanical remains and plant food subsistence. In L. P. Louwe, Kooijmans and P. F. B. Jongste (eds.) *Schipluiden. A Neolithic settlement on the Dutch North Sea coast c. 3500 cal BC*. Faculty of Archaeology. Leiden, pp. 317–338.

Lallemand, C. (1890) *Tunis et ses environs.*

Maier, U. (1999) Agricultural activities and land use in a Neolithic village around 3900 B.C.: Hornstaad Hörnle I A, Lake Constance, Germany. *Vegetation History and Archaeobotany* 8, 87–94.

Manen, C., Marchand, G. and Carvalho, A. F. (2007) Le Néolithique ancien de la péninsule Ibérique: vers une nouvelle évaluation du mirage africain. In J. Evin (ed.) *Actes du XXVIe congrès préhistorique de France-Avignon 2004*, pp. 133–151.

Menéndez Pidal, G. (1993) *España en sus caminos*. Caja de Madrid, Madrid.

Mery, S., Anderson, P., Inizan, M. L., Lechevallier, M. and Pélegrin, J. (2007) A Pottery Workshop with flint tools on blades knapped with copper at Nausharo (Indus Civilization, *ca.* 2500 BC). *Journal of Archaeological Science* 34, 1098–1116.

M'hamdi, M. and Anderson, P. C. (2013) Approche ethnoarchéologique d'outils et techniques de moisson de l'alfa (*Stipa tenacissima*) dans la région des Hautes steppes en Tunisie: Premières observations et analyses. In P. C. Anderson, C. Cheval and A. Durand (eds.) *Regards croisés sur les outils liés au travail des végétaux (An Interdisciplinary Focus On Plant Working Tools)*. XXXIIIe Rencontres Internationales d'Archéologie et d'Histoire d'Antibes. Editions ADPCA, Antibes, pp. 75–88.

McCorriston, J. (1998) Landscape and Human-environment Interaction in the Middle Khabur Drainage from the Neolithic Period to the Bronze Age. In M. Fortin and O. Aurenche (eds.) *Espace naturel, espace habité en Syrie du nord (10e-2e millénaires av. J.-C.)*, Société canadienne des études mésopotamiennes et Maison de l'Orient Méditerranéen, Québec/Lyon, pp. 43– 53.

Moore, A. M. T., Hillman, G. C. and Legge, A. J. (1975) The Excavation of Tell Abu Hureyra in Syria: a Preliminary Report. *Proceedings of the Prehistoric Society* 41, 50–77.

Moreno-García, M., Pimenta, C. M. and Ruas, J. P. (2005) Safras em osso para picar foicinhas de gume serrilhado ... a sua longa história! *Revista Portuguesa de Arqueologia*, 8, 571–627.

Moreno-Garcia M., Esteban Nadal, M., Pimenta, C., López Gila Ma, D. and Morales, A. (2006a) Los yunques en hueso en la Península Ibérica: estado de la cuestión, In *Animais na Pré-historia e Arqueologia da Peninsula Ibérica.* Actas do IV Congresso de Arqueologia. Faro, pp. 247–262.

Moreno-García, M., Pimenta, C., Gonçalves, M. J., Pires, A. (2006b) Evidência arqueozoológica de um oficio na Silves islâmica: ossos, foices e ferreiros, *Xelb* 6: 281–292.

Moreno-Garcia M., Pimenta C. M., Lopez Aldana P. M. and Pajuelo Pando, A. (2007) The signature of a blacksmith on a dromedary bone from Islamic Seville (Spain). *Archaeofauna* 16, 103–202.

Ortiz, F. and Sigaut, F. (1980) La moisson de l'epeautre avec les 'mesorias' dans deux villages asturiens. *Bulletin de la Societé d'Ethnozoologie et d'Ethnobotanique* 8, 2–4.

Out, W. A. (2009) *Sowing the seed? Human impact and plant subsistence in Dutch wetlands during the Late Mesolithic and Early and Middle Neolithic (5500-3400 cal BC)*. Archaeological Studies, Leiden University, Leiden University Press.

Palomo, A., Gibaja, J. F., Pique, R., Bosch, A., Chinchilla, J. and Tarrus, J. (2011) Harvesting cereals and other plants in Neolithic Iberia: the assemblage from the lake settlement at La Draga. *Antiquity* 85, 759–771.

Peña-Chocarro, L. (1996) In situ conservation of hulled wheats species: the case of Spain. In S. Padulosi, K. Hammer and J. Heller (eds.) *Hulled Wheats: Proceedings of the First International Workshop on Hulled Wheats, Castelvecchio Pascoli, Tuscany, Italy*. IPGRI, Rome, pp. 129–146.

Peña-Chocarro, L. (1999) *Prehistoric Agriculture in Southern Spain during the Neolithic and the Bronze Age: the application of ethnographic models*. BAR International Series 818. Archaeopress, Oxford.

Peña-Chocarro, L. and. Zapata, L. (2003) Post-harvest processing of hulled wheats. An ethnoarchaeological approach, In P. C. Anderson, L. S. Cummings, T. S. Schippers and B. Simonel (eds.) *Le traitement des récoltes: un regard sur la diversité, du Néolithique au présent. Actes des XXIIIe rencontres Internationales d'archéologie et d'histoire d'Antibes, 17-19 octobre 2002*. Éditions APDCA, Antibes, pp. 99–114.

Peña-Chocarro, L., Zapata, L., González Urquijo, J. and Ibáñez Estévez, J. J. (2000) Agricultura, alimentación y uso del combustible: Aplicación de modelos etnográficos en arqueobotánica. *Saguntum*, extra-3, 403–420.

Peña-Chocarro, L., Zapata, L., González Urquijo, J. E., Ibáñez Estévez, J. J. and Moreno García M. (2005) An ethnoarchaeological project in the Western Rif (Northern Morocco): First results. *Quaderni del Laboratorio di Archeobiologia* 1, 21–32.

Peña-Chocarro, L., Zapata, L., Garcia, J., Gonzalez, M., Sesma, J. and Strauss, L. G. (2005) The spread of agriculture in northern Iberia: new archaeobotanical data from El Mirón cave (Cantabria) and the open-air site of Los Cascajos (Navarra), *Vegetation History and Archaeobotany* 14 (4), 268–278.

Peña-Chocarro, L., Zapata, L., Gonzalez Urquijo, J. E. and Ibañez, J. J. (2009) Einkorn (*Triticum monococcum* L) cultivation in mountain communities of the western Rif (Morocco): an ethnoarchaeological project. In A. S. Fairnbairn and E. Weiss (eds.) *From Foragers to Farmers*. Gordon Hillman Festschrift. Oxbow Books, Oxford, pp. 103–111.

Peña-Chocarro, L. and Zapata, L. (2010) Neolithic agriculture in the southwestern Mediterranean region. In J. F. Gibaja and A. F. Carvalho (eds.) The last hunter-gatherers and the first farming communities in the South of the Iberian peninsula and North of Morocco. *Promontoria* 15, 191–198.

Peña-Chocarro, L. and Zapata, L. (2014) Versatile hulled wheats: farmers' traditional uses of three endangered crop species in the western Mediterranean. Chapter 6: A versatile world: examples of diversity in plant use. In A. Chevalier, E. Marinova and L. Peña- Chocarro (eds.) *Plants and People: Choices and Diversity through Time.* EARTH Series Volume 1. Oxbow Books, Oxford, pp. 276–281.

Perales, U. (2010) Aproximación a la cultura material del Mesoneolítico, nuevas perspectivas: aplicación del análisis funcional sobre las láminas del neolítico antiguo de Atxoste (Álava, País Vasco). *Estudios de cuaternario* 1, 153–169.

Perez, G. (2005) Nuevos datos paleocarpológicos en niveles neolíticos del Pais Valenciano. In P. Arias, R. Ontañón and C. García-Moncó (eds.): *Actas del III Congreso del Neolítico de la Península Ibérica.* Universidad de Cantabria, Servicio de Publicaciones, Santander, pp. 73–81.

Poplin, F. (2007) Des os supports à denter les faucilles, séance du 31 octobre 2007. *Bulletin de la Société Nationale des Antiquaires de France* 2007, 215–221.

Reigniez, P. (2002a) *L'outil agricole en France au moyen âge*, Errance, Paris.

Reigniez, P. (2002b) *Les baguettes à récolter les céréales en Géorgie (Caucase).* Méridiani, Tbilissi.

Reigniez, P. (2003) Les baguettes à récolter du Caucase. *Techniques & Culture* 41, 91–114.

Rodet-Belarbi, I., Esteban Nadal, M., Forest V., Moreno-Garcia M. and Pimenta C. (2007) Des aiguisoirs/polissoirs aux enclumes en os: historiographie des os piquetés, *Archéologie Médiévale* 37, 157–167.

Rodet-Belarbi, I., Forest, V., Gardel, M.-É and Ginouvez, O. (2002) Aiguisoirs, polissoirs médiévaux en os (VIIe–XIVe s.). Nouvelles données. *Archéologie du Midi Médiéval* 20, 149–153.

Rodríguez, A. C. (1999) Análisis funcional del instrumental lítico tallado del poblado de Cabecicos Negros. In M. D. Cámalich and D. Martín (eds.) *El territorio almeriense desde los inicios de la producción hasta fines de la Antigüedad : un modelo : la depresión de Vera y cuenca del Almanzora.* Consejería de Cultura, Sevilla, pp. 225–235.

Rodríguez, A. C. (2004) Análisis funcional de los instrumentos líticos tallados. In D. Martín *et al.* (ed.) *La Cueva de El Toro (Sierra de El Torcal, AntequeraMálaga): un modelo de ocupación ganadera en territorio andaluz entre el VI y II milenios A.N.E.* , Consejería de Cultura, Sevilla, pp. 135–160.

Rojo, M. A., Kunst, M., Garrido, R., Garcia, I. and Moran, G. (2008) *Paisajes de la Memoria. Asentamientos del Neolítico antiguo en el Valle de Ambrona (Soria).* Universidad de Valladolid, Valladolid.

Schreurs, J. and Brounen, F. T. S. (1998) Resten van een Michelsberg aardwerk op de Schelsberg te Heerlen. Een voorlopig bericht, *Archeologie in Limburg* 76, 21–32.

Semenov, S. A. (1964) *Prehistoric Technology.* Cory Adams & Nackay, London.

Sigaut, F. (1978) Identification des techniques de recolte des graines alimentaires. *Journal d'Agriculture Traditionelle et de Botanique Appliquée* 25 (3), 145–162.

Skakun, N. N. (1993) Agricultural implements in Neolithic and Eneolithic cultures of Bulgaria. In P. C. Anderson, S. Beyries, M. Otte and H. Plisson (eds.) *Traces et fonction: les gestes retrouvées*, Vol 2. Eraul 50. Liège, 361–368.

Skakun, N. N. (1999) Evolution of agricultural techniques in Eneolithic (Chalcolithic) Bulgaria: data from use-wear analysis. In P. C. Anderson (ed.) *Prehistory of Agriculture. New Experimental and Ethnographic Approaches.* Institute of Archaeology Monograph 40. University of California, Los Angeles, pp. 199–210.

Steensberg, A. (1943) *Ancient harvesting implements. A study in archaeology and human geography.* Copenhagen.

Stordeur, D. and Anderson-Gerfaud, P. (1985) Les omoplates encochees néolithiques de Ganj Dareh (Iran). Etude morphologique et fonctionnelle. *Cahiers de l'Euphrate* 4, 289–310.

Schreurs, J. (1992) The Michelsberg site Maastricht-Klinkers. A functional interpretation. *Analecta Praehistoria Leidensia* 25, 129–171.

Vanmontfort, B. (2004) *Converging worlds: the Neolithisation of the Scheldt basin during the late fifth and early fourth millennium cal. BC.* Unpublished Ph.D. thesis, KU Leuven.

Toffin, G. (1983) Moisson aux baguettes au Nepal central. *Objets et Mondes* 23 (3–4), 173–176.

Vavilov, N. F. (1926) Centers of origin of cultivated plants. *Bulletin of Applied Botany and plant breeding* 16 (2), 1–248

Willcox, G. (1998) Archaeobotanical Evidence for the Beginnings of Agriculture in Southwest Asia. In A. B. Damania, J. Valkoun, G. Willcox and C. O. Qualset (eds.) *The Origins of agriculture and crop domestication.* ICARDA/IPIGRI/Grcf/FAO, pp. 25–38.

Veiga de Oliveira, E., Galhano, F. and Pereira, B. (1983) Alfaia agrícola portuguesa. 2nd edition. Instituto Nacional de Investigação Científica, Centro de Estudos de Etnologia, Lisboa.

White, K. D. (1967) *Agricultural implements of the Roman world.* Cambridge University Press, Cambridge.

5 Threshing Processes and Tools

5.1. EXPLORING DIVERSITY IN THE PAST: AN INTRODUCTION

John C. Whittaker

Threshing is not merely a necessary step in processing the grain crops that are the foundation of most traditional agricultural foodways. Among the agricultural processes that transform crops into food, threshing has received a disproportionate amount of attention because it is an active process, of symbolic importance to many cultures, and involving distinctive artifacts and technologies that have attracted archaeological and ethnographical interest. Threshing and associated processes that transform crops into food and fibre show great diversity through time and space. The cultural and adaptive choices among different methods for doing similar things reflect webs of relationships among crops, tools, environmental conditions, and social contexts. Threshing usually involves more or less permanent facilities, minimally a clean, hard surface, but often a built structure. Threshing floors and related facilities are recognisable as 'passive' archaeological features on which the threshing with 'active' tools such as flails and threshing sledges takes place. The range of variability in the form of threshing floors is the result of variation in the type of threshing performed on them, the crops that are threshed, and other elements of the system. Threshing floors are made of local materials found in different landscapes, and their different layouts and forms also are due to different environmental conditions and cultural traditions.

The distribution of different types of threshing facilities, tools, and techniques in Crete, Cyprus, France, and Spain illustrates some of these points. Agricultural facilities express some aspects of the organisation of labour and social relationships within communities. Variation in the form and positioning of threshing floors in Cyprus relates to the social organisation of threshing and choices made by independent and cooperating families (Whittaker, this chapter). Patterns of social organisation and land ownership can also be seen in the examples from Spain and Portugal (Schippers, Mingote Calderón, this chapter).

However, the 'active' threshing tools have received more attention than the 'passive' floors they are used on. Different types of threshing tools represent alternative processes that are used to produce grain and chaff. The specific choices among tools and techniques reflect decisions that were made about what crops should be cultivated, the relative emphases placed on grain and chaff, and the possible uses of these products. These decisions about crops and products are also influenced by local environmental conditions, and by cultural preferences and traditions. In many areas, animals alone are used for threshing, harnessed in various combinations to trample the grain. The choice of animals over other threshing methods reflects decisions about which crops are better processed by trampling, and may also be affected by symbolic or ritual concerns.

Probably the most intensively studied threshing processes are those that rely on the *tribulum* or threshing sledge. The basic idea of a wooden sledge with cutting blades on the underside is very widespread, but the form and use of threshing sledges varies geographically and historically. Modern scholars, especially those of us who are not from agricultural backgrounds, tend to consider food for humans the primary product of agriculture. However, in many village economies, the use of threshing sledges often reflects an equal emphasis on the importance of chaff that was used as temper for ceramics, daub, mud bricks, and lime or clay plaster; for animal fodder, and as raw material for other processes. Crops which are intended to nourish stock rather than people were also threshed with the *tribulum*. In fact, one of the reasons threshing sledges are intriguing and good models for the understanding of any agricultural tool is that they solve several farming problems at the same time. They thresh the grain and cut up straw, and a person in charge of the animals can ride instead of walking during the threshing. These different aspects of the tool are hard to separate.

Some of the early variations on the threshing sledge, and evidence of its great time depth, can only be studied through archaeological methods. The earliest evidence for sledges appears in the Late Neolithic of the Middle East (Anderson 2003; 2006) in the form of distinctively worn flint blades, the cutting inserts on the underside of a *tribulum*. Ancient Mesopotamian threshing sledges have been reconstructed from texts and iconography, and these replicas have been tested in threshing experiments (Anderson 2006b). Archaeological evidence in the form of use-wear on stone sledge inserts allows us to examine the spread and timing of *tribulum* use in the Middle East and Europe. Archaeological analyses and material culture approaches such as replicative experiments are important because agricultural systems that still use threshing sledges have become rare. Through experimentation we can also begin to understand the engineering of this ancient implement, and the physical attributes that make it effective.

Threshing sledges use inserts of flint and other stone, and thus also represent a rare survival of Stone Age technology into the historic and even industrial eras. In the few places where threshing sledges have been used in modern times, the choice of insert material (stone, flint, metal) reflects local conditions, the relative efficiency of technological alternatives, and networks of production and exchange. Over the last half century, most traditional agricultural systems that used threshing sledges have abandoned them in favour of powered machinery, and only a few are still in use. Threshing sledges are thus a lens through which we can examine processes of obsolescence and survival, and the loss of technological skills such as flintknapping. The knappers who produced flints for sledges have been studied in Cyprus, Turkey and elsewhere (Whittaker, this chapter). They serve to remind us that all implements imply intimate connections between different craftsmen and technologies, from knappers and carpenters to ironworking and animal husbandry, as well as the social contexts of the technological processes. The use of any technology is affected by its social setting. In parts of Spain, aesthetic and personal choices can be seen in the retention of threshing sledges and the choice between animals or tractors to pull them (Schippers, this chapter). The use of sledges on threshing floors also has a ritual component in many parts of their circum-Mediterranean distribution. The abandonment of threshing sledges in recent times in favour of mechanised threshers (Fig. 5.1) also has implications for village social life and the surrounding landscape.

Technologies compete and coexist under different conditions. Crops may be threshed with rollers rather than with sledges, and different forms of rollers often coexist and compete with threshing sledges (Carpinschi and Whittaker, this chapter). French examples allow us to study rollers in terms of their distribution, relationships with other threshing techniques, choice of animal traction, and considerations of efficiency.

Flails may be regarded as a third class of threshing tools, perhaps the threshing implement most familiar to the public. They tend to be used in regions of moister climate, but the distribution of flails and flailing overlaps with other threshing methods in Europe. Flail construction can be fairly

Fig. 5.1. Mechanical thresher, Turkey, 2010. In the decades after 1950, mechanical threshers, powered by tractors and soon to be followed by ever more complex combines and other equipment, displaced traditional agricultural tools and techniques around the Mediterranean (Photo: J. C. Whittaker).

elaborate and show regional variation and individual style and identity marking (Mingote Calderón, this chapter), but in some areas, threshing is done very simply with sticks, often reflecting a need for rapid processing of small quantities, or poverty and small crop yields (Peña-Chocarro, this chapter).

Winnowing and other processes used to sort and separate grain and chaff have been less studied, but clearly relate to how crops were previously processed, and their intended uses. Deceptively simple tools such as baskets and trays may rely heavily on specialised skills for their effective use. An example from Crete provides a brief introduction to these issues (Beckmann, this chapter). Although necessarily partial, our survey of threshing and related grain crop processing illustrates some of the important dimensions of the variability in methods employed by traditional farmers. We can discuss the relationships of different kinds of tools and their social implications, some of the environmental constraints on technology, the choice of crops and products made from them, the relationships of manufacture and animal traction to harvest technology, and some of the processes of technological change. All of these are closely interconnected in complex ways.

5.2. THRESHING FLOORS IN CYPRUS

John C. Whittaker

Threshing of grain and other crops requires a clean, hard surface, and threshing floors are common features of the Mediterranean agricultural landscape. Although they are often mentioned in literature and folklore (Ure 1955; Stricker 1955), they are architecturally simple, often separated from habitations, and associated with few artefacts, so they have received little attention from anthropologists and archaeologists, beyond brief descriptions in reports of regional surveys (Lohman 1992, 42; Murray and Kardulias 1986; White 1970; Whitelaw 1991; Young 1956, 124). Nevertheless, they reflect complex relationships between crops, technology, local materials, and the social organisation of agriculture. Around the Mediterranean, threshing floors range in form from areas of packed earth or open spaces in villages, as is common in Turkey, to stone-walled spaces with flagstone or cement floors, typical in much of Greece.

Floors in the Paphos district of southwest Cyprus (Fig. 5.2) documented in 1995 (Whittaker 1999; 2000) represent some of the variation, interpreted through information from informants who remember their use before the 1950s. Threshing floors in Cyprus (*aloni*, plural *alonia* in local Greek), were mostly used with threshing sledges (*dhoukhanes*). In some other areas of the Mediterranean, such as parts of Greece, trampling by animals alone was the

Fig. 5.2. Map of the Aegean, with the islands of Crete, Cyprus, and Melos, and the region of Paphos on Cyprus, as mentioned in Chapters 5.2 and 5.15. Map: R. Lugon, J.-C. Loubier and A. Chevalier.

dominant threshing technique. Grain crops were the commonest crops threshed on floors, but legumes were also processed on *alonia* by trampling or with sledges. In Cyprus, most *alonia* are round, 15–25 m in diameter, floored with earth, lime plaster, or sometimes flagstones (Fig. 5.3). They are usually built on slopes or ridges, with limestone walls forming terraces and encircling them to confine animals and crops. Villagers said *alonia* need to be near both fields and villages for easy access, and exposed to prevailing winds for winnowing. Most of these small floors were owned by families, who might have more than one *aloni* depending on how their fields were distributed. Often *alonia* are clustered because threshing was a social occasion, and neighbours frequently cooperated, exchanging labour. A similar situation can be seen in Turkey, where threshing areas tend not to be visibly bounded, but families work their own crops side-by-side on large open floors (Bordaz 1965). Cyprus also had some much larger threshing floors, often rectangular, where teams of hired workers threshed for estates or wealthy land-owners.

Fig. 5.3. Cypriot threshing floor at Galataria, 1995. Recently abandoned and now used for storage, the flagstone surface is overgrown (Photo: J. C. Whittaker).

5.3. TRAMPLING THE CROP WITH ANIMALS

Patricia C. Anderson

A common threshing technique involves using walking or running animals to trample the crop, releasing the grain and breaking up the straw. Threshing by trampling is mentioned in 3rd millennium BCE Mesopotamian cuneiform texts (Grégoire 2010) and later, the Bible, but unfortunately the texts do not give enough information to determine which crops were processed by trampling, and which by using a threshing sledge. One particular cuneiform text says that trampling was carried out by four teams using five oxen led by two men, on a large round communal threshing floor. In order to reduce the amount of straw on the threshing floor and the subsequent waste after winnowing, cereal stems were cut just under the seed head, and the straw remaining in the fields was harvested later by pulling it up, and stored for use as animal bedding, roofing, and so on (Grégoire 2010). In ancient Egypt, threshing wheat by trampling on round floors is described (Erman and Ranke 1980; Hartmann 1923).

Trampling is still used in some western Mediterranean areas where threshing has not been mechanised, such as Morocco, most of southern France, and Italy, and where the threshing sledge is not known or has passed out of use. However, because trampling does not involve tools that might be preserved in archaeological sites, or cut the plant material in the characteristic ways that threshing sledge blades do, it is difficult to identify trampling using archaeological remains.

The threshing floors used for trampling the crop tend to be irregular in shape or rectangular, less frequently round, as we observed in the southeast of France, unless sledges are also used on the same floors, as we saw in Tunisia (Anderson 2006b). Animals are sometimes loaned between families, and people help one another during threshing time. In southern France in the last century, as in northern Tunisia today (Anderson 2006b), trampling the crop was said to be a joyous occasion involving family cooperation (Leroy 2003). In general, each family owned its own threshing floor, and one load of the harvested crop was threshed in one day. A similar process is used throughout the Mediterranean world to trample the crop, though the floors may be clay, stone-paved, or simply a cleared and packed open area. The crop, usually long stems with seed heads, but sometimes only the seed heads, is carried from the field to the threshing floor, usually on the backs of animals, stacked up in a heap, then spread out over the threshing floor. According to our observations in Tunisia, two to four animals (mules, sometimes with horses and donkeys) are tethered together, usually led by a person, and run over the crop, in a circle or spiral that narrows in circumference throughout the process (Fig. 5.4, 5.5). In southern France, a long cord attached animals to a central pole in the middle of the threshing floor, around which they would move in ever-narrowing circles until they reached the pole, which was then turned over, so the animals would unwind the cord as they moved back toward the edge of the threshing floor (Comet 1992). In another variant in the southern French Alps, oxen were used indoors in barns to trample the crop gradually during the winter after the harvest. Although their pace was slow, it was said that the heavy oxen could produce a fine grade of chopped straw (Leroy 2003; J-L. Domenge, personal communication 2002).

Trampling accomplishes two things: it either releases the grain from its enclosing pods (for legumes) or from the glumes (in the case of heads

Fig. 5.4. Trampling barley with four mules in northwestern Tunisia, 2006. The man with the fork is removing some of the longer straw so that threshing can progress. Another man is leading the animals, tethered together with rope, in a wide circle over a large threshing floor, about halfway through the work. Both the straw and the grain will be used as animal fodder, and some of the grain will be used for human consumption, in particular to make a ceremonial stew eaten during feasts, as at Ramadan (Photo: P. C. Anderson).

of 'free threshing cereals' such as oats) or it breaks the seed head up into spikelets or hulled grain (for cereals such as hulled barley). At the same time, the straw is broken up. In Tunisia, the longer pieces of straw are removed and set to the side of the threshing floor during the work, and a windowing fork or shovel is used to gradually separate the short straw and chaff from the grain (Fig. 5.5). As the plant material is flattened down as the work progresses, it is once again piled together on the threshing floor with threshing forks into a circle of lesser circumference than at the beginning, so there is always a certain thickness to the layer of plant material to be trodden. As this layer becomes thinner, more animals may be added.[1] Wind winnowing cleans the grain, with the fine chaff settling into one heap, and the heavier grain, into another. The broken straw is used for animal fodder, bedding, or as tempering material for clay bricks, ceramic, and lime or clay plaster. But, why cut or break the straw at all? In the Mediterranean world, whole stems are said to be hard for animals to eat. Indeed the stems contain more silica than in northern temperate or continental climates. This is principally due to higher plant transpiration occurring in hot, dry climates, which means that here, more silica in solution is carried up from the soil into plant stems and glumes (Parry and Smithson 1964). Chopped straw is better for fodder, as well as for tempering ceramics, mud plaster, lime plaster, and so on. Trampling is one way to chop straw quickly. Other cutting methods such as chopping or cutting bunches of straw with a knife or axe by hand are far slower, but may be used in wet

Fig. 5.5. Trampling the oat crop in northwestern Tunisia, 2008, near the end of the process, once the ring of crop being trampled had been reduced to a small circumference. Three mules are being pulled in a tight circle, in order to bring as many hooves to bear as possible on the final trampling of the grain. Note the winnowing shovel on the right, and also the two stacks of straw, the rough straw fraction just above the shovel, and the finer straw fraction to the left of it. This oat crop was cut ripe only to obtain grain to sow as animal fodder. Oats are not consumed by humans here, and they are consumed as green fodder by animals. (Photo: P. C. Anderson).

climates that are ill-adapted to an all-day, outdoor operation such as threshing on a threshing floor, which requires dry weather for the straw to break and the seeds to be released.

In fact, trampling plays a role in any threshing process where animals are used, including when they pull a threshing sledge or a roller. Our ethnographic observations in Tunisia and Syria (Anderson 2003; 2006) show that even when a threshing sledge is used, trampling is a part of the beginning of the process, as animals are run over the crop before the driver mounts the sledge. The role of trampling in these instances is to help tamp down the pile of harvested crop on the threshing floor, which averages a thickness of about 20–50 cm at the beginning, and to begin the release of grain, which filters down to the threshing floor surface, before the threshing sledge comes into play. Practical reasons for choosing trampling over threshing sledges can be seen in northwest Tunisia where both are in use today (Anderson 2003; 2006). The nature of the plant to be threshed (the perceived fragility or hardness of the grain, and the resistance of the stems), affects the choice, as does the form of the threshing sledges. In northwest Tunisia, where the threshing sledge had blades of flint (and now, metal blades), trampling is and was used for all crops, *i.e.* barley (*Hordeum vulgare* L.), oats (*Avena sativa* L.), chickpea (*Cicer arietinum* L.), broad bean (*Vicia faba* L.), except for wheat (here, *Triticum durum* Desf., durum or macaroni wheat). We are told that trampling is adequate to break up the stems and release the grain for crops other than wheat, and the threshing sledge is not used on these crops because, as informants say, it would cut the grains or crush them. On the other hand the sledge, not trampling, is said to be necessary for threshing durum wheat, because its stems are hard and its grain hard and resistant, and the blades are needed to cut the straw finely enough for animals.

In southern Syria, where trampling is not used, the threshing sledge was traditionally used to thresh all crops (Anderson 2003). Here, however, the inserts in the sledge are not sharp, but rather rounded basalt pieces, which would not risk damaging the grain. A Syrian family using the basalt studded sledge, said that bladed threshing sledges (which they had only heard about), were not good because they cut the grain, echoing the opinion of Tunisian sledge users. The choice of threshing method is part of specific agricultural *chaines opératoires*, or operational sequences, but often elusive in the archaeological record, and our knowledge of past and present threshing methods is still limited. We do not really know yet if the choice of trampling was due to conscious choice, to cultural habit, to constraints linked to plant types, climate, raw material availability for tools or merely a reflection of patterns of diffusion of the threshing sledge from its presumed origin in the eastern Mediterranean.

5.4. THE MANUFACTURE AND USE OF THRESHING SLEDGES

John C. Whittaker

The threshing sledge or *tribulum* was widely used in ethnographic times around the whole of the Mediterranean, from the Iberian peninsula to the Levant, the Balkan and Black Sea areas, parts of North Africa, northern Mesopotamia, and the Anatolian plateau (Anderson and Inizan 1994; Ataman 1999; Bordaz 1965; Darwish 1986; Fox 1984; Fujii 1986; Kardulias and Yerkes 1996; Whallon 1978; Whittaker 2000; 2003; Yerkes and Kardulias 1994). The threshing sledge is also of great antiquity, and throughout its distribution, the general form and use of the *tribulum* is remarkably consistent (Fig. 5.6). A heavy sledge made of wooden planks with an upturned front edge is slotted on the underside for the insertion of flint (or sometimes metal) blades (Fig. 5.7). Drawn by animals over a hard surfaced threshing floor, the *tribulum* separates the grain from the chaff and breaks the chaff up into short sections. Both of these functions are important.

Underlying a general consistency, there are important variations. Sledges vary in size and the number and kind of draft animals used. In some places, such as Palestine and Syria where flint may be lacking, lumps of basalt or other rock have been used as sledge inserts, and in recent times as the flint knapping trade declined, metal blades have replaced flint (Darwish 1986; Whittaker *et al.* 2009). Modern make-shift threshing sledges include sheets of metal with sharp-edged holes punched in them. In some areas such as Italy, southern France and the Dalmatian coastal area, we have almost no documentation of *tribulum* use, and in other areas, it appears as a late introduction (Beckmann, this

Fig. 5.6. Threshing with the *düven*, Eymür, Turkey, 2010.[2] One family still uses the threshing sledge occasionally. The wife is riding, guiding the pair of oxen with a stick, while one son leads them (Photo: J. Whittaker).

Fig. 5.7. The underside of the threshing sledge at Eymür. It is said to be 20–30 years old. The form is typical, including the addition of sheet metal under the front, but there is no longer a ready supply of flints and many have been lost (Photo: J. C. Whittaker).

chapter). Alternative threshing techniques like trampling with animals or flailing are preferred in many areas, and overlap with the *tribulum* in parts of its range, as do alternative methods that use rollers (Carpinschi and Whittaker, this chapter). In the large area of modern Turkey, EARTH members have observed regional *tribulum* styles varying in the size of the sledge, the number, kind, and arrangement of the inserts, the means of animal traction, and the crops processed. On Melos in the Greek Islands (Fig. 5.2), we were shown metal toothed *volosyroi*. These are short but very wide sledges, that were fastened to revolve around a central pole on a small paved threshing floor (Wagstaff and Augustson 1982, 120).

The threshing sledge is a pre-industrial tool composed of many parts, and locally manufactured, but often specialist craftsmen are involved in at least part of the production process. Threshing sledges thus serve as a fine example of the interconnections of agricultural technologies and the ways that labour may be organised. Threshing sledges have been of particular interest to archaeologists because their stone inserts are examples of a rare survival of Stone Age technology into modern times (Bordaz 1969; Runnels 1982). As they were abandoned in favour of modern machines in most areas in the last few decades, threshing sledges illustrate common processes of technological obsolescence and survival. Most threshing sledges have a wooden body armed on the underside with teeth of stone or metal. In recent times, the carpentry techniques used to make the body are common enough that they have attracted little attention, although in some areas there were carpenters who specialised in making sledges, or craftsmen who did the carpentry as well as the flint work. However, anthropological interest has focused on the production of the flints. The most intensively studied threshing sledge flint industry has been in Cyprus (Fox 1984; Kardulias and Yerkes 1996; Pearlman 1984; Whittaker 1996; Yerkes 2000; Yerkes and Kardulias 1994). Here the threshing sledge or *dhoukani* went out of general use around 1950, but a few *athkiajas* survived into the 1970s and 1980s to describe their craft to ethnoarchaeologists. The *athkiajas* or *athkiakadhes*, glossed as 'flint knapper', was a specialist craftsman who sometimes made entire sledges, but often only made the flint inserts and replaced them in the sledges as they became worn or were lost. Alphredhos Andreou was interviewed in 1995, at which time he considered himself the last of the Cypriot *athkiajas* (Fig. 5.8).

Fig. 5.8. Alphredhos Andreou demonstrating knapping, Paphos, Cyprus, 1995. His hammers are makeshift, not the proper tools which he has lost. He works large chert flakes (left) into discoidal cores, producing short blades (right) that will be trimmed to fit the chiseled slots in the sledge bottom (Photo: J. C. Whittaker).

He had learned as a boy, apprenticed to a 'master' who, like many of the Cypriot knappers, had a yearly round. The master lived in northern Cyprus, where many sledges were made because timber resources were plentiful. However, the flint sources were better in the south, so knappers would collect flint and make blades in the south in the spring, and then return home via a series of villages where they repaired sledges in preparation for the summer threshing season. During the other months, they built the sledges. Mr. Andreou simplified this system, remaining near his home in the south, making flints and traveling to repair sledges, but not doing the carpentry himself (Whittaker 1996). He was operating at the end of the industry, when the number of customers and the demand for new sledges must have been declining, but there were always a number of ways of subdividing the craft into different specialisations, as described for Bulgaria and Tunisia in following sections (Gurova; Anderson, this chapter).

In Turkey, the village of Çakmak (the word is used for flint and currently for cigarette lighters; Fig. 5.9) also specialised in the production of flint blades for

Fig. 5.9. Location of the towns of Çakmak and Gölcüğez (with the village of Eymür in its proximity) in Turkey, as mentioned in the text. Map: R. Lugon, J.-C. Loubier and A. Chevalier.

threshing sledges (Bordaz 1969; Whittaker, Kamp, and Yılmaz 2009). The wooden sledges themselves were made elsewhere, often by local carpenters, but the villagers of Çakmak added a further variant to the industry by mining the flint. Teams of men worked together to dig deep shafts in the limestone, extracting the flint from mines that were very similar to those of gunflint knappers in Brandon, England (Fig. 5.10), and the Neolithic stone workers who preceded them by several millennia (Skertchly 1879; Whittaker 2001). A Çakmak team included men who mostly dug, and those who mostly knapped. The knapper was the 'master' but all shared equally in the profits. The flints were mostly sold through merchants in local towns, and fitted to sledges by the villagers or the sledge carpenters, rather than by the knapping specialists. Two retired knappers interviewed in 2007 (Whittaker, Kamp, and Yılmaz 2009) agreed that this was a profitable industry, making Çakmak the richest village in the region. They claimed, as did Bordaz's (1969) informants, that Çakmak supplied flints to all of Turkey. Sledges observed in 2007 from all over Turkey did have distinctive greenish Çakmak flint, but it was also evident that other sources provided flints, and

Fig. 5.10. The location of Brandon, England. Map: R. Lugon, J.-C. Loubier and A. Chevalier.

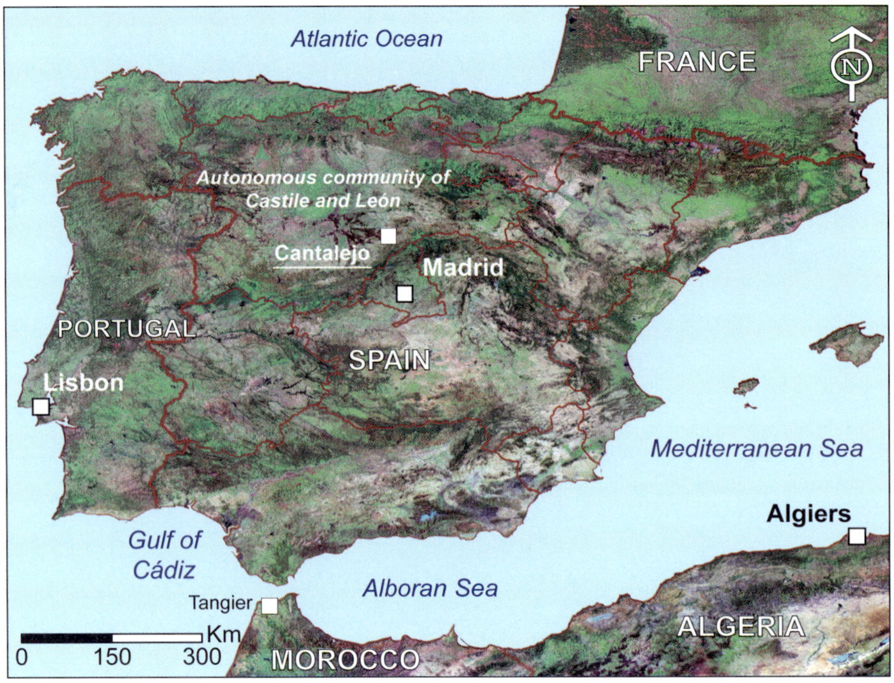

Fig. 5.11. Map of Spain with the mentioned city of Cantalejo in the autonomous community of Castile and León. Map: R. Lugon, J.-C. Loubier and A. Chevalier.

there must have been other knapping villages that have not yet been recorded. In 2010, Whittaker and Kamp paid a brief visit to the Turkish village of Gölcüğez, in the Kastamonu region near the Black Sea (Fig. 5.9). Although at some distance from Çakmak, a retired knapper at Gölcüğez told us that the industry was very similar. A quite different flint was extracted from limestone bedrock by deep shafts, worked on the spot, using similar tools and techniques, and bagged for sale around the region. We got the sense that the work was more seasonal, and farming more important at Gölcüğez than at Çakmak. At Çakmak, knappers used small stone huts with brush roofs to shelter knappers at the quarry; at Gölcüğez brush sufficed.

Threshing sledges were not always made by individual craftsmen or cottage industries. In Spain, where threshing sledges were still in common use into the twentieth century, small factories in Cantalejo (Fig. 5.11) and elsewhere produced sledges (Mingote Calderón 1990a). In Tunisia, Anderson (this chapter) has been working with carpenters who use electric band saws to cut wooden planks for *tribula*. The few in use today are armed with metal blades; the knapping industry has vanished.

5.5. ETHNOGRAPHIC THRESHING SLEDGE USE IN EASTERN EUROPE: EVIDENCE FROM BULGARIA

Maria Gurova

In the 1970s and 1980s, N. Skakun (1992, 1994, 1999, 2006) made the first ethnoarchaeological studies of the *tribulum* as a crucial element of traditional agriculture in Eastern Europe. Her work in Bulgaria included studies of Neolithic and Chalcolithic flint assemblages, ethnographic collections and archives, and experimentation. The author's interest was provoked by the availability of good ethnographic collections of *tribula* with flint inserts, and thus the opportunity to compare these artifacts with archaeological specimens. Using this approach to distinguish between sickle and *tribulum* flints in archaeological contexts improves our understanding of ancient agricultural practices (Gurova 2001; 2011; Chapters 5.9 and 10.2, this volume).

Bulgaria, a small country known for its traditional agriculture, is an example of the patchy use of the *tribulum* and our knowledge of it. In some corners of Bulgaria, cereals were threshed by trampling or flailing. Unfortunately, apart from some general studies of agricultural toolkits, the threshing sledge (*dikania* in Bulgarian) has not been subjected to a detailed study (Raichevski 1996; Vakarelski 1974).

The threshing sledges used in recent times by traditional patriarchal Bulgarian households are quite standardised in shape and dimensions, and so is the form of their flint inserts (Fig. 5.12). The traditional *dikania* consists of a platform of two to five suitable wooden planks hewn to shape, and 1.5–2.0 m long, held together by wooden crosspieces. Willow or poplar wood is preferred. The bottom of the planks contains rows of slots cut with a chisel. Flint inserts (Figs. 5.13 and 5.14) are driven into the slots using a wooden hammer. The flint flakes

Fig. 5.12. *Tribulum* from Brashlian village (southeast Bulgaria, Strandzha; Photo: M. Gurova).

Fig. 5.13. Detail of the flint inserts on the *tribulum* above (Photo: M. Gurova).

Fig. 5.14. *Tribulum* inserts from Yabalkovo village (South Bulgaria-Thrace; Photo: M. Gurova).

Fig. 5.15. Traditional use of *tribulum* in patriarchal Bulgaria. Photo from the Vakarelski archive (Skakun 1999).

(or more rarely blades), were carefully retouched to reduce their width and length and to obtain an ovoid form that facilitated insertion and threshing efficiency. The inserts were predominately made of high quality flint from Ludogorie in northeast Bulgaria. Commonly known as 'Dobrudžanski' flint, this material comes from quarries exploited in prehistoric times as well. Some old people from Dobrudža in Bulgaria, traditionally considered the most fertile cereal-producing land in the region, still remember the specialised craftsmen who knapped the local raw material into blade-like flakes and distributed the flint inserts widely around the country. The illustrated recent *tribulum* inserts are strikingly similar to their Chalcolithic predecessors in form and wear patterns (Gurova 2001). This suggests the stability and resilience of this technology, which dates back to the 5th millennium BC.

Threshing with the *dikania* was done at specially prepared threshing floors (Bulgarian *harman*) inside or in the vicinity of the villages. Often woman or children rode the *dikania* to guide the animals, and sometimes supplementary weights were added (Fig. 5.15).

> '..... the *harman* was soaked, then regularly strewn with clean straw, which was tamped down with a stone. Once the floor was well dried and hardened, it was carefully swept to become flat, smooth and durable. The wheat or barley brought from the field was scattered on the floor, with ears turned to the centre, then conscientiously trodden underfoot. Next oxen or horses were trotted in a circle. The sharp flint edges cut the straw and drove the grain from the seed head while a few women with special wooden forks shook the stems so that all ears could be husked and grain could fall from them' (Skakun 1999, 205).

In general, Bulgarian threshing sledges examined by the author were very carefully manufactured and maintained, with perfectly shaped and skilfully inserted flints. During the working life of a *tribulum* some elements unavoidably fall out and are replaced by other flints. In some cases when new flints were hard to obtain, the *tribulum* was repaired with different flints picked up from the fields. Especially as the flint knapping trade declined, *tribulum* owners around the Mediterranean began to replace the flints with metal blades. In Bulgaria, the *dikania* was certainly used until the 1950s, when they were rapidly replaced by industrial machines. Nevertheless many families still keep their threshing sledges as a mark of personal identity and nostalgia. Meanwhile the *tribulum* becomes attractive furniture for some modern rustic houses, and fortunately many survive for ethnographic study. Unfortunately the skills of flint knapping and *tribulum* use are irreversibly lost.

5.6. PERSISTENCE OF THE THRESHING SLEDGE: THE TUNISIAN *TRIBULUM*

Patricia C. Anderson

Introduction

Around Beja in northwestern Tunisia (Fig. 5.16), the *tribulum* is still made by a specialised carpenter and used by farmers. Wheat, the dominant crop, is traditionally the only crop treated using the threshing sledge. Local farmers say it is important to thresh it correctly, meaning with the sledge. The hard wheat (*Triticum durum*) grown here is considered to have resistant grain, and hard stems which must be finely cut to be consumed as animal fodder. The threshing sledge or *jarousha* is drawn by two or three mules or horses, as they are fast animals. The persistence of the threshing sledge in this region may have something to do with the fact that it was the 'granary of Rome' and is still a rich agricultural zone of non-mechanised farms in mountainous areas 292 to 510 m in altitude. Our study between 2005 and 2009 comes at a time of transition in this society, where commercial agriculture and mechanisation have become common in the past twenty years or so, alongside the traditional, non-mechanised system. The steep slopes and isolation of mountain farms contribute to the persistence of traditional tools. Most farms are subsistence operations, and tools, animals and products are shared or exchanged among family members and neighbours. Some surplus is sold immediately to local cooperatives or to small local private or government companies. The population is large enough to keep the non-mechanised agriculture going, among people united by kinship or community ties. Teens and young adults as well as older kin work in agriculture. A key factor is that animals are still kept and trained to work, and manual and animal-drawn tools are still made and maintained by local specialists in a nearby town, Beja (Fig. 5.16). Although in the 1980s and 1990s most of the mountain population moved to the nearby plain from their mud brick houses into new concrete ones, where running water and closer schools made life more convenient, they still

Fig. 5.16. Map of Tunisia with the towns of Beja, and Chemtou. Map: R. Lugon, J.-C. Loubier and A. Chevalier.

return to farm their mountain fields. In the plain, some now use farm machinery, but the families that stayed or maintain farms in the higher altitude areas continue to farm traditionally.

Although its origins in North Africa are not yet known, the threshing sledge or *jarousha* was observed over a large area of northern and central Tunisia in the first half of the 20th century (Lethielleux 1947; Luquet and Rivet 1933). Although traditional tools were more widespread forty or fifty years ago, the basic procedures described in the literature and by farmers over fifty years of age are still those used in our study area. An exception to this is the change to metal inserts from flint ones for the threshing sledges.

The Making of the Sledge

Threshing sledges in our study area are made by the same carpenter who makes ards. They are light and narrow, quite unlike the wide, heavy sledges used most recently in Spain, for example. The sledges we saw in use all measure between 40–60 cm in width, and 1.8–2 m in length, the size the sledge-maker said was necessary for proper function. The sledge weighs between 14 and 25 kilograms. The carpenter could not understand why in other countries one would use a larger or heavier sledge, or slow animals such as oxen and cows.

We watched him make sledges on several occasions. Using a form or pattern, the planks, which were up-curved at one end, were cut from a pine log (Fig. 5.17A). He made sledges of three planks, but two planks were used for sledges in the past, and we were told that the width of the planks and the number needed for each sledge depended on the diameter of the tree trunk. The carpenter assembled the planks using wooden pegs, and smoothed the sledge surfaces with a spokeshave. A taut string dipped in red paint was used to mark a pattern to guide the insertion of the cutting elements. Today these are of metal. Long arched pieces forged by a blacksmith go on the front of the sledge. For the rest of the underside, saw blades are bought in the market, and then cut into segments of approximately 2 cm. About ninety of these are pounded into the wood in a checkerboard pattern with the saw edge down, leaving smooth, canted cutting edges protruding from the wood (Fig. 5.17B). The carpenter said that the leading edge must be completely buried in the wood, and the actual direction of the slant

A

B

Fig. 5.17. A: The carpenter in northwest Tunisia about to cut a pine log into the three curved planks of a sledge. B: Tunisian threshing sledge with two types of metal inserts. On the left is another sledge prepared for us to insert the flint blades (Photos: P. C. Anderson).

is not important. Sometimes whole saw blades are also nailed along the edges of the sledge. None of the farms we visited had sledges with flint. The transition from flint to metal inserts seems to have occurred in the 1960s or 1970s, and we were told that this was because metal saw blades became easier to obtain. They are also easier to insert by pounding them into the wood, without the cutting of grooves needed for flint inserts. The carpenter had learned his trade from his late father, who was already using metal elements, but the father used to use flint inserts. We asked people in the region if they knew where the flint came from and who knapped it. Some claimed there was a local source, others said that the flint came from the south near the Sahara, and was sold by nomads in the local market. In either case, the carpenter would buy and insert the flints, but most informants agreed that the carpenter was not a flint knapper. Older sledge users identified fine-grained brown flint among flint samples we brought as the material most like flints formerly used in sledges. The local flint is coarse-grained, black, and too small to make blades. We are now receiving confirmation that the flint was indeed mined and knapped in the south.

A *jarousha* with flint inserts in the museum at Chemtou (Fig. 5.16) is labeled as coming from our study area. It has flint elements set in grooves. A few missing flint inserts have been replaced by pieces of pottery and glass, suggesting that no flints were available, but perhaps this was done at a time before metal pieces came into use. Ali ben Alaha, seventy-five years old, who examined the sledge in the museum with us, said these were indeed the kind he had always used up until the 1950s and 1960s. The transition to metal occurred progressively as mechanisation was introduced and metal became more readily available. Carpenters began to make sledges using metal blades, which were easier to insert, and people began to replace their sledges with metal-bladed sledges, rather than adding metal inserts as replacements to sledges with flint. Mr. ben Alaha said that although it was recognised that the flints cut better and lasted longer than the metal inserts, people welcomed the fact that the metal inserts did not tend to fall out of the sledge as readily, and the metal armed sledges began to be those in demand. The wooden part of the sledge (Fig. 5.18) remained the same, and the larger forged iron elements in the front of the sledge were already in use in older sledges. Our informant said these long elements begin cutting the long straw, whereas the flint behind them cut the straw finer. The forged iron elements were sometimes added along the sides, around the area of flint inserts. In this case, they would protrude slightly more than the flint elements; our informant said this was to protect the flint and add to the stability of the sledge (Fig. 5.17B).

Ethnographic Experiments

We wanted our Tunisian farmer informants to use and evaluate flint-armed sledges once more, after which we could examine microwear, a rare opportunity to experiment with experienced *tribulum* users, well-trained animals, and a sledge similar in size and inserts to our reconstructed Mesopotamian threshing sledge (see below, this chapter). John C. Whittaker and Carole McCartney provided knapped blades and we asked the carpenter to make the wooden part of the sledge as he always did and insert the flint blades. Our initial attempt was not fully satisfactory; the blades tended to fall out, but one of the two farmers who used it said it worked faster than his metal-armed sledge until it lost too many blades. For a second attempt, Whittaker made new blades that were more regular, with retouched backs. Based on his observations in Cyprus, he instructed us in a better way to cut standard grooves with a chisel and to insert blades securely without damaging them (Fig. 5.18A). We were told that the slightly thicker and bigger blades were easier to insert than smaller or thinner ones. Our farmer friend was delighted to receive a second sledge with flint, and used it daily for six weeks before the flint began to fall out. We provided him with a good number of replacement blades and the tools with which to insert them. Two years later he was still happily using the sledge and replacing the blades as needed, saying he was using it in preference to his metal-armed sledge, which worked more slowly. We are currently analysing the microwear on his blades for comparison to our archaeological samples.

The Threshing and Winnowing Procedure

Great care is taken in preparing floors for threshing the valuable wheat. Each year, our informant

Fig. 5.18. A: Inserting flint blades in the experimental sledge to reproduce the older form of Tunisian sledge. Note the three planks of the sledge, additional forged iron blades, and the grid lay-out for the flints. B: Farmers threshing with our experimental sledge with flint elements. The latter half of the work is shown here, and the circumference of the crop on the threshing floor has been decreased (Photos: P. C. Anderson).

removed any plant growth, and chopped straw and water were added to the floor, which was then trampled by the animals. The process was repeated until the clay floor was hard and as reflective as porcelain.

In the threshing we observed, wheat sheaves were left to dry next to the floor, then untied and spread on the floor in a layer approximately 50 cm deep.[3] First, the animals were run over the crop to tamp it down and begin the threshing process, then the threshing sledge was drawn by two or three mules walking or running around the 15–20 m diameter floor. The driver, male or female, stands on the sledge with a whip (Fig. 5.18B). The time needed to thresh each batch depends on the thickness of the plant layer, and a little on speed. The crop was turned with wooden forks to keep it an even depth, and to enhance the double process of separating the grain from the seed heads and stems, while cutting the stems into fine chaff. We are told that about halfway through the process, larger pieces of cut straw are removed from the threshing floor and set aside. The sledge then can work faster to finish the fine cutting and threshing. After several hours, usually near late afternoon, winnowing begins as the wind comes up. A fork and shovel are used to throw the threshed crop into the wind, separating it into grain, a larger straw fraction, and a finer straw fraction.[4] The fine fraction is sieved, to separate the grain, and the grain is cleaned using a large sieve or riddle. The chopped straw is piled carefully in the early morning next to the threshing floor on a base of barley straw, and gradually forms a house-shape (called a *nader*) as the threshing progresses. At the end of the season, this is covered with barley stems, then plastered with a mixture of clay and chopped straw to make a rain and wind-resistant cover. The *tibbn* (chopped straw) is then taken from the bottom of the structure first, to feed the animals when green fodder or grain is scarce (until the next harvest).

Perspectives

The farmers generally feel that rapidly fading traditional techniques should be maintained, and even returned to, particularly as they produce better-tasting and higher-quality crops. Farmers are proud of their animals and their use of the threshing sledge.

The Tunisian case is another example of traditional tools persisting under different conditions, and of the variable ways in which a technology can be developed and used. Traditional Tunisian sledges contrast with other recent European sledges which are usually drawn around the threshing floor more slowly (see Whittaker, this chapter, and Schippers, this chapter). It is hoped that microwear traces will reflect these differences in sledge use, and enable us to identify some of the variation in the use of ancient sledges.

5.7. THE CONTEMPORARY USE OF IBERIAN THRESHING SLEDGES: SOME ETHNOGRAPHIC OBSERVATIONS ABOUT AN OBSOLETE CHOICE

Thomas K. Schippers

The Iberian Peninsula and Spanish Canary Islands form the most western region of Europe where threshing sledges have been in use, in some cases until the present. During the 20th century they were extensively studied by Spanish and Portuguese ethnographers and historians, who have published numerous descriptions, photographs, maps and even films about this type of tool (see for example Veiga de Oliveira *et al.* 1995; Mingote Calderón 1988, 83–180 and 1990a, 7–72; Caro Baroja 1983). This region provides an interesting case study, as different threshing practices existed side by side.

The *tribulum* (threshing sledge) and the roller-equipped *plostellum* (threshing cart) were in use until quite recently, scattered over various parts of

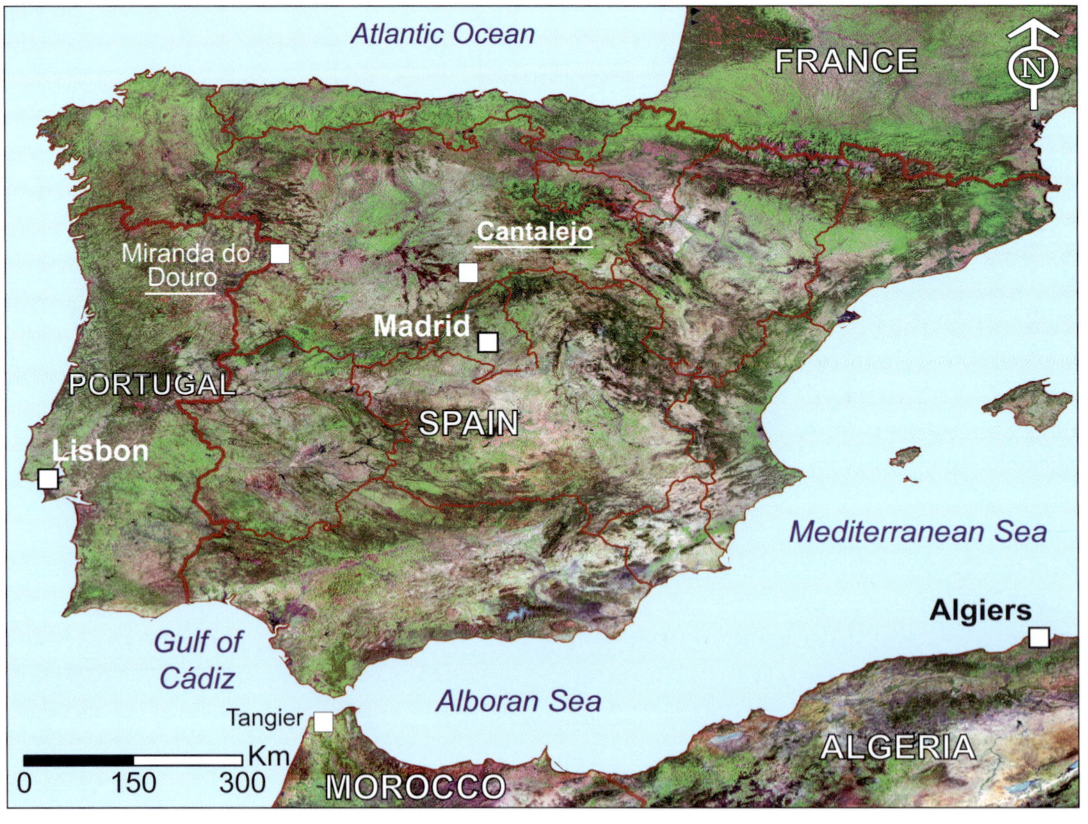

Fig. 5.19. Map of the Iberian Peninsula with the mentioned cities of Cantalejo in the autonomous community of Castile and León in Spain and of Miranda do Douro in the Bragança district in Portugal. Map: R. Lugon, J.-C. Loubier and A. Chevalier.

the peninsula. Threshing with flails and trampling with animals are other well known, sometimes co-occurring practices. During two exploratory ethnographic surveys in 2003 and 2004 on both sides of the Portuguese-Spanish border near the small town of Miranda do Douro in Portugal (Fig. 5.19), I encountered, on the Spanish side, several peasants busy threshing with wooden threshing sledges armed with flint inserts. The sledges were of a common type produced until the 1960s by the thousands in the numerous specialised workshops of the Cantalejo region (Fig. 5.19) and traded widely all around the Peninsula. These *in situ* observations of *tribula* at work brought up many ethnographic and technical questions. The first concerned the reasons why these tools were still in use here. As a matter of fact, it was possible to observe *simultaneously* a wide range of threshing techniques in action in this single village: 1) *tribula* drawn by horses and/or donkeys (rare, see Fig. 5.20A), 2) 'twinned' *tribula* drawn by tractors, and often combined with stationary mechanical winnowers (quite numerous, see Fig. 5.20B), 3) small tractor-driven stationary mechanical threshers combined with stationary mechanical winnowers (quite numerous, see Fig. 5.20C), 4) modern self-propelled combine harvesters with air-conditioned driver cabins (rather rare).

It proved quite instructive to listen to the various reasons mentioned by the users (all household-based small farmers), to explain their choice of threshing equipment. For many, hiring the services of a combine harvester was considered much too expensive in regard to the size and expected yield of their harvest (mainly barley and rye). Those using one or more *tribula,* drawn with a tractor or animals, claimed that the stationary thresher 'broke' the grains, rendering them unlikely to germinate when re-sown. Finally those using a *tribulum* with donkeys or horses were either too poor to have any tractor or mechanical thresher, or simply disliked the engine noise and smell while threshing. All the users stressed the importance of having nicely chopped straw of both rye and barley for cattle fodder. It clearly appeared that the persistence of the use of the *tribulum* as a threshing tool in this region was due as much to socio-economic reasons as to technical ones. Especially among the poorest users encountered, the lengthy (10–15 hours) trampling with donkeys and a *tribulum* produced for them a satisfying, low-cost result. Those using the *tribulum* with a tractor and a stationary winnower were often well aware of their 'double-bind' situation as they no longer had either horses or donkeys, nor did they have the means to access 'real modernity' represented by the services of a combine harvester. Apparently ashamed of their perceived 'backwardness', most refused to be photographed near their machines by the ethnographer, as did those using a stationary thresher combined with a stationary winnower.

Fig. 5.20. A: Transport of a heavy horse-drawn *tribulum* toward the threshing site (Photo: E. Salesse). B: Twinned *tribula* intended to be drawn by a tractor (Photo: T. K. Schippers). C: Small tractor-driven stationary mechanical thresher combined with stationary mechanical winnower neighbouring the *tribulum* threshing in the foreground (Photo: T. K. Schippers).

The observations in the field allowed me to collect some precise data on the use of a *tribulum* drawn by animals. Very low humidity of the straw seems crucial in order to have it finely chopped and to have the *tribulum* slide without creating 'bumps' or 'rolls' of straw. It is the texture of the layer of chopped straw that determines when some new sheaves of cereal have to be added during the process, and when everything is finished. It is the size of the initial stack that determines the quantity of cereal processed during one complete threshing cycle. During the threshing, the heavier grains tend to 'sink' toward the threshing floor, while the *tribulum* continues to chop the lighter material at the surface. The layer of straw is regularly turned over and pushed towards the centre during the trampling. Ant colonies established underneath the threshing floor tend to steal the grains from beneath and the farmers are keen to do the winnowing as soon as possible after the threshing is finished in order not to leave the threshed grain overnight on the floor. When the threshing is finished, the mixture of chopped straw and grains is swept and pushed into a diagonal swath which has to be perpendicular to the evening breeze in order to allow efficient winnowing.

Compared to the fast moving, lean Tunisian *tribula* observed by Anderson (this chapter), the Iberian ones are large and heavy, weighing about 30–50 kg, with approximately 1200 to 1600 flint inserts. They slide slowly at the pace of walking horses or donkeys. Some ethnographic films recorded in Miranda do Douro in the early 1960s, show the same *tribula* with running horses, but according to some witnesses this was due to the presence of the filmmakers. The users consider the weight of both the *tribulum* and the animals an important element; some add weight on the *tribulum* or ride on the smaller animals. Among the peasants, the importance of animal trampling in the threshing process is discounted by those who use two twinned *tribula* behind a tractor, but the action of the heavy tractor wheels and tires is also probably underestimated by these farmers. This seems surprising since elsewhere in Iberia and in other regions in southern Europe, animal trampling is practiced without any other threshing tool.

Only more precise experimental comparisons could offer data to evaluate the exact roles of the *tribulum*, the animal trampling, and in some cases the tractor tyres in the threshing process. This might reveal more precisely the various functions of a *tribulum* as a straw-chopper, a thresher and as a convenient transport device for the human operator. It could also provide some broader insights into the persistence of this apparently obsolete implement in this region until present times.

5.8. IS THE *TRIBULUM* TRADITIONAL IN CRETE? PROBLEMS OF HISTORICAL DOCUMENTATION

Sabine Beckmann

Establishing the history of agricultural techniques and tools is not always easy, especially in the gap between prehistoric archaeology and comprehensive historical documentation. The use of flint-studded threshing sledges in Crete is a case in point. Crete (Fig. 5.2), freed from Turkish rule at the beginning of the twentieth century,[5] has changed only recently from a poor traditional agricultural economy (Allbaugh 1953), to a rich European tourism-centred island. Now, at the beginning of the 21st century, threshing sledges have disappeared from agricultural use and are seen as old traditional farming tools suitable for wall decorations. But what exactly does 'old' or 'traditional' mean? Today, the function of the 'traditional' threshing sledges is known to older Cretan farmers who used them until the 1960s. The purpose of the sledges is also obvious from the finds of typical flint blades used as sledge inserts on or around threshing floors in Crete.

The locals call the threshing sledge a *volosyros*. The true *volosyros* (literally *sod-puller*) was a similar heavy board, but without the cutting flint or iron blades inserted in its base. The true *volosyros* was used like a harrow to break up chunks of soil after ploughing. There do not seem to be any indications of the use of flint or obsidian inserts in prehistoric Cretan threshing sledges, perhaps because of the lack of adequate local lithic raw material, and threshing sledges were not mentioned in Greek texts of the Classical period (Isager and Skydsgaard 1995, 53; Amouretti 1986, 108). The *tribulum* seems to have been introduced to Greece during Roman times under the name of '*tykane*' (Amouretti 1986), which corresponds to modern Greek and Cypriot '*doukane*' or '*adoukane*', but this word is not used in Crete. Even though we know of a few illustrations of threshing sledges from Byzantine Greece, it is not mentioned in medieval sources or later until the beginning of the 18th century, where an Anatolian-Greek example is known (engraving in Bryer 1986, 79).

Early travellers in Crete describe threshing reaped cereals by treading with animals (Sieber 1823, 54), confirming the impression that threshing sledges were not in use there until sometime in the middle of the 19th century (Hitier 1853, 593). That is when the first descriptions of threshing sledge use in Crete appear. A general use of the *volosyros* is documented from the beginning of the 20th century (*e.g.* Trevor-Battye 1919, travelling before 1909). According to surface finds (Blitzer in Watrous *et al.* 2004; Whittaker pers. comm.) and the testimony of local farmers, the flints used in Cretan sledges until *ca.* 1960 came from Turkey and were called '*tsakmakópetra*' (Çakmak-stone). Çakmak is the name of a Turkish village that produced widely used flints (Fig. 5.9; Bordaz 1965, 1969; Whittaker *et al.* 2009), and as in Turkey, *tsakmakópetra* referred to flint fire starters as well. Cretans also called these flint sledge inserts '*nychia*,' short for onyx, meaning 'finger-nail' because of the typical shape. Neither excavations nor surface surveys seem to have produced convincing examples of older chipped stone tools showing the use-wear typical in *tribula* inserts. Diamond (1975) identified some possible threshing flints by use-wear, but later work finds this doubtful (Kardulias and Yerkes 1996, 664), although there still seems to be an impression in the scholarly world that something similar to the *tribulum* must have existed already in Minoan times on Crete (Warren 1979, 560). However, archaeological, historical, and linguistic evidence all point to the same conclusion: even though threshing sledges with flints have been used in the eastern Mediterranean since Neolithic times (Anderson 1999a; 1999b) they probably were not used in early Crete – neither in the Minoan Bronze Age nor later Archaic or Classical times. Even after the Roman conquest of Greece, when a Greek name for a threshing sledge (*tykane*) appears, there seem to be no documented findings of flint sledge inserts in Crete (according to published surveys). It was not until the middle of the 19th century that threshing sledges were first introduced into Crete, presumably from Turkey, but sledges never became popular enough even to acquire a proper name in the local language.

It remains a mystery why the Turks who ruled Crete since the 17th century did not introduce threshing sledges on their farms, and it might be interesting to study this question in the future. The incomplete penetration of the threshing sledge into Crete may be due to several reasons. The finely cut chaff produced by threshing sledges was not seen as palatable by animals according to some (Halstead and Jone 1989). Under Turkish domination, agriculture in Crete remained underdeveloped, and it may be that farmers only became interested in threshing sledges after the island regained autonomy and self-confidence following the insurrections of the 19th century. It is also possible that a growing population from that time necessitated the accumulation of surplus as farmers were no longer working just for their own subsistence, and a more effective threshing tool became desirable. All in all, *volosyroi* obviously lasted just long enough to be seen by modern Cretans as traditional tools, even though this 'tradition' actually may not have lasted longer than four or five generations.

5.9. PREHISTORIC THRESHING SLEDGES: A CASE STUDY FROM BULGARIA

Maria Gurova

The origin of the prehistoric *tribulum* has only been explored in the last decade. P. C. Anderson and team, working in the Bronze age of the Near East, have identified Canaanean blades (very large, regular flint blades) as threshing sledge inserts (Anderson 2003, this chapter; Anderson and Chabot 2004; Anderson *et al.* 2004; Anderson *et al.* 2006; Chabot 2002; Chabot and Eid 2003; Chabot *et al.* 2007). Natalia Skakun and the author have studied Neolithic and Chalcolithic flint assemblages from Bulgaria, concluding that during the Chalcolithic (5th millennium BC) the use of the *tribulum* in prehistoric subsistence activity is indubitable (Gurova 2001; 2002; 2011; 2013; Skakun 1992; 1993a; 1993b; 1994; 1999; 2006).

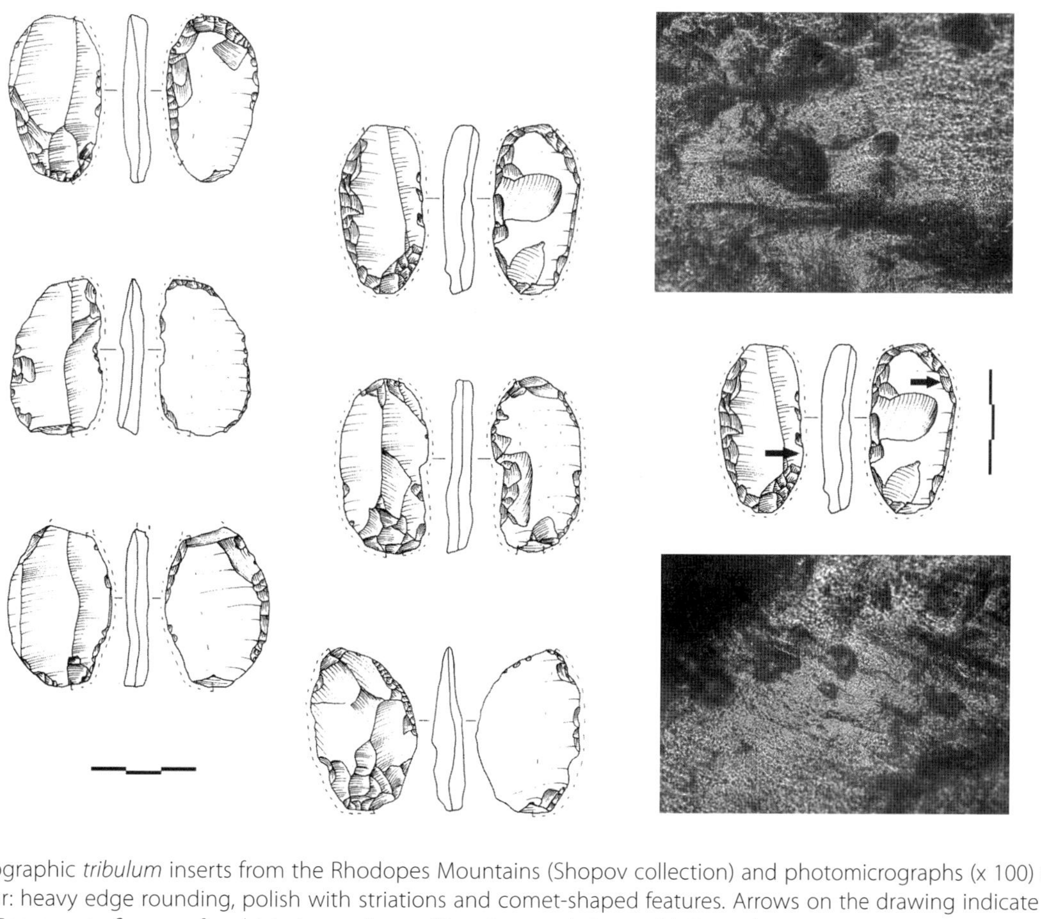

Fig. 5.21. Ethnographic *tribulum* inserts from the Rhodopes Mountains (Shopov collection) and photomicrographs (x 100) showing typical use-wear: heavy edge rounding, polish with striations and comet-shaped features. Arrows on the drawing indicate location of the photos. Compare to figures of prehistoric specimens (Drawings and photos: M. Gurova).

Archaeological *tribulum* inserts can be recognised because of their similarity to ethnographic examples in form and especially in their microscopic wear traces. *Tribulum* flints from Bulgaria, both ethnographic and prehistoric, have a distinctive elongated ovoid shape obtained by intentional retouch. It is not always evident whether they were made on flakes or elongated blades. Most of the pieces have been used on one edge, but some show bilateral traces of utilisation. The use-wear complex is very distinctive. Working edges are strongly smoothed and rounded bifacially. Under the microscope, a highly developed polish shows rough and opaque polish texture, with comet-shaped features and depressions (Fig. 5.21). All these features have been described repeatedly (Fujii 1986; Gurova 2001; Kardulias and Yerkes 1996; Whallon 1978; Yerkes 2000). No traces of bitumen, which is sometimes used to secure both sickle blades and *tribulum* inserts, have been discovered on the archaeological specimens from Bulgaria.

The problem with many of the early flint finds, for instance those described by Skakun (1993a; 1993b; 1999; 2006), is that there is little of the contextual data necessary for understanding the precise chronological and cultural position of the flints. The presence and use of *tribula* is affirmed by Skakun in a general way for a large area in northeastern Bulgaria (including the Varna culture area), at sporadic sites in south Bulgaria, and for late Neolithic sites in estern Bulgaria. Similarly, the author knows of about four archaeological artefacts coming from Late Neolithic contexts and has published one of them in the detailed study on the Tell Karanovo flint assemblage (Gurova 2001; 2002). Because very few late Neolithic flints identified as *tribulum* inserts come from secure stratigraphic contexts, the author preferred to consider them 'anecdotal' finds rather than unequivocal attestation of *tribulum* use throughout the Neolithic in Bulgaria.

Archaeological finds securely identified as early prehistoric *tribulum* inserts from Bulgaria come from excavations of the multilayer tell site of Drama-Merdžumekja[6] (Thrace). These are more useful for tracing the history and evolution of this agrarian tool and to evaluate the possibility of adequately

Fig. 5.22. Map covering south-eastern European to Middle Eastern countries, with the Bulgarian archaeological sites of 1) Drama-Kairiaka, Drama-Gerena, and Drama-Merdžumekja, 2) Durankulak, as well as the Bulgarian Ludogorie region, and the Iraqi archaeological site of 3) Tell Kutan Map: R. Lugon, J.-C. Loubier and A. Chevalier.

reconstructing the prehistoric threshing sledge in the Balkan region. Excavations in the Drama micro-region started in 1983 and continued under the direction of J. Lichardus from 1986 to 2003. The village of Drama is situated in the Tundza valley (Yambol district), in the eastern part of the Thracian plain (Fig. 5.22). The area is extremely rich in archaeological sites of the Neolithic and Chalcolithic periods, including the important sites of Drama-Gerena[7] (belonging to the late Neolithic – Karanovo III–IV periods), Drama-Merdžumekja (a tell with an impressive Chalcolithic sequence – Karanovo V and Karanovo VI[8]), and Drama-Kairiaka (a cult site from the Iron Age and cemetery from the Roman period) (Lichardus *et al.* 2001; 2002). The study of Drama-Gerena lithic assemblages produced two artefacts from the humus (*i.e.* not reliably stratified) that possessed morphometric and use-wear traits characteristic of *tribulum* inserts[9] (Gurova 2001). Microscopic study[10] of Chalcolithic assemblages from Drama-Merdžumekja permitted the identification of twelve items from good contexts (Fig. 5.23).

The most reliable pieces of evidence are the artefacts coming from well-defined and interpretable archaeological features ('objects') (see Lichardus *et al.* 2001):

- 'Object 360' is a large, almost circular ditch surrounding areas of Karanovo V and Karanovo VI settlements (the inhabited area).
- 'Object 101' is a house from the Karanovo V period.
- 'Object 825' is identified as a group of pits for clay extraction.
- 'Object 756' is unfortunately difficult to interpret now because of lack of sufficient information.

The stratum called 'humus' in the documentation of the German-Bulgarian team is not humus *senso stricto*, because the erosion on the tell in fact denuded the superficial sediments, creating deposits of settlement strata with some relics of a primary humus layer (Lichardus *et al.* 2001, 210). The so-called 'humus' is understood as belonging stratigraphically to the two main strata (Karanovo

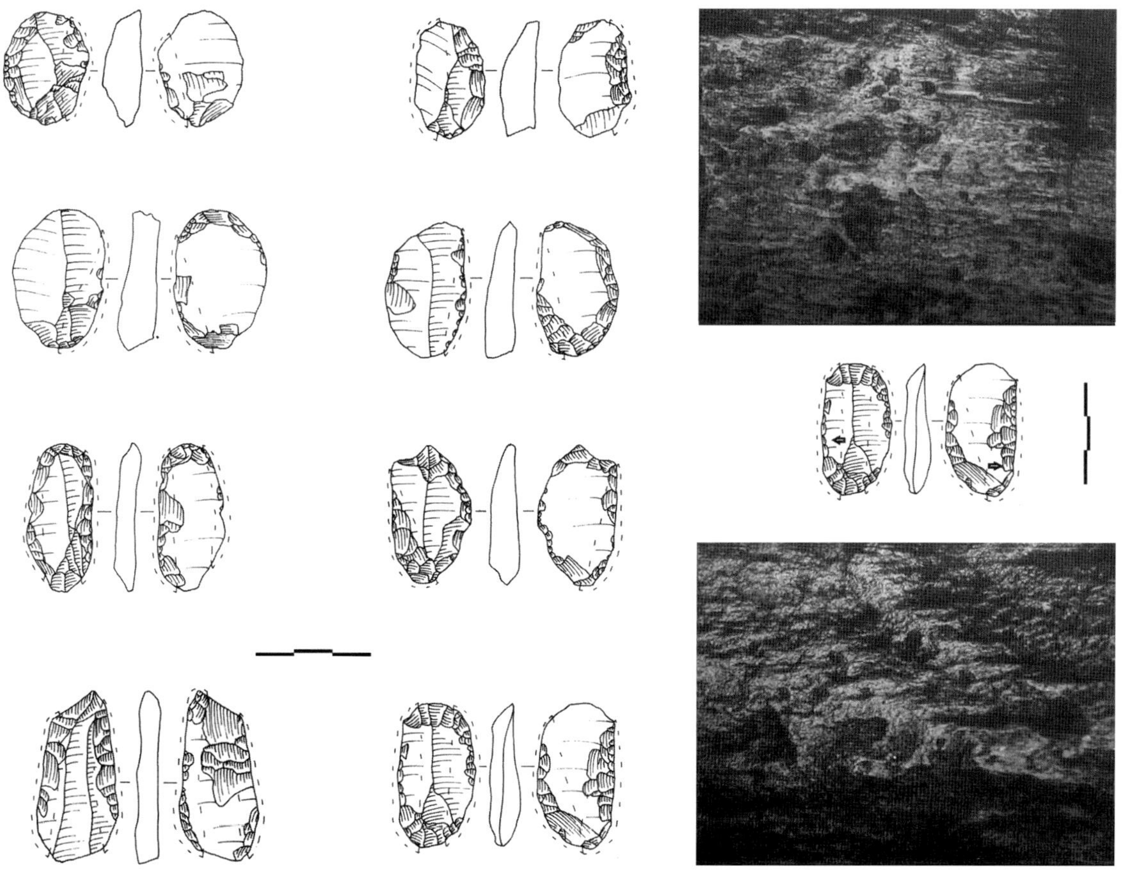

Fig. 5.23. Archaeological *tribulum* inserts from Drama-Merdžumekja, with photomicrograph (x 100) showing typical use-wear traces (Drawings and photos: M. Gurova).

V and Karanovo VI), but without precise distinction between them. Accordingly, only one piece with diagnostic *tribulum* use-wear lacks chronological placement. The spatial distribution of the threshing sledge inserts on the plans of the settlements of Karanovo V and VI shows that three of the *tribulum* inserts belonging to the 'humus' could be linked with object 360, the ditch (respectively from areas D6, G4 and F 16), while another three belong rather to the residential areas (F13 and H12). Although there are no ^{14}C dates from Drama-Merdžumekja, the Bulgarian Chalcolithic is firmly dated by materials from northern Bulgaria and the subdivision is as follows: Early Chalcolithic – 4900–4600; Middle – 4600–4500; Late – 4500–4100 cal BC (Boyadžiev 1995, 179). The relevant levels are cross-dated to the periods Karanovo V and VI on the basis of the material culture. The available dates from the Tell Karanovo sequence suggest the end of the Karanovo IV period ca. 5000 cal BC (Kutschera *et al.* 2005, 426). The Drama-Merdžumekja sequence (Karanovo V and VI) thus begins about 5000–4900 BC. Absolute ending dates are not available.

It is worth emphasising here that the careful use-wear and morphological analysis of ethnographic *tribulum* inserts allows unambiguous distinctions between different prehistoric items with cereal polishes. According to my observations, in Bulgaria, rectangular blade fragments do not seem to have been used as *tribulum* inserts and all flint artefacts with *tribulum* type use-wear are ovoid in shape. However, Skakun (1999, 202–204) has recently claimed that a subgroup of the fragmented large blades from northeastern Chalcolithic Bulgarian sites could be regarded as *tribulum* inserts, in some ways similar to Canaanean blades from Tell Kutan in Iraq (Fig. 5.22; Anderson and Inizan 1994). Anderson (2006a, 313–314), who has examined several of the blades in question, accepts Skakun's interpretation of the use-wear. Nonetheless, on the basis of Skakun's publications and my own studies of Chalcolithic flint assemblages, I find that the descriptions and drawings (Skakun 1999, 202–204) of these blades show that they are sickle blades typical of the Bulgarian Chalcolithic rather than *tribulum* inserts (Gurova 2007). The examples above, and a few other specimens coming from Bronze Age contexts at Durankulak (Fig. 5.22) and elsewhere, currently under study by the author (Gurova 2006, fig. 1), attest to the use of the *tribulum* in the Chalcolithic and Bronze Age of Bulgaria. While *tribula* are altogether missing in the archaeological record from Roman and Byzantine periods, there is evidence for *tribulum* use during Bulgarian Medieval and Ottoman periods, however not from domestic but only from funerary contexts.

5.10. BLADES, SICKLES, THRESHING SLEDGES AND EXPERIMENTAL ARCHAEOLOGY IN NORTHERN MESOPOTAMIA

Patricia C. Anderson and John C. Whittaker

Following the apparent collapse of the Uruk culture in the fourth millennium BCE, a new type of village life emerged at Amuq in southern Anatolia in present-day Turkey and northwest Iraq. During the third millennium BCE, in the Ninevite V period, new villages were constructed on virgin soil. Various suggestions have been made concerning the origins of these villages, and the economic and political organisation they represent (Butterlin 1998; 2003). In the Middle Khabur valley was a large concentration of these small, closely spaced village sites (Fig. 5.30), fairly similar in size and material remains (Schwartz and Klukas 1998; Schwartz 1987; Chabot 2002). Storage structures and macrobotanical (*i.e.* seed) remains found in many of the sites indicate they served principally for grain storage and treatment, while calculi, numeric tablets and seals in some of the sites suggest that this activity was under exterior administrative control (Fortin 1997; McCorriston 1998).

Combining the methods of experimental archaeology and microscopic use-wear analysis with ethnographic and archaeological data, we reconstruct a critical ancient agricultural tool and assess its wider implications for interpreting these Middle Khabur valley Ninevite V period sites. A characteristic lithic artefact found at these sites is the 'Canaanean' blade (Neuville 1930, 205–206; Cauvin 1962; Crowfoot-Payne 1960). The sample of Canaanean blades (Fig. 5.24A) we identified and studied includes approximately 800 Ninevite V implements from sites located in present-day northern Syria and Iraq and southern Turkey (Anderson *et al.* 2004). Canaanean blades are large, flat, and very regular. These blades contrast with the contemporary lithic industry of ad hoc flake tools made on local materials also associated with these sites. The Canaanean blades are usually broken intentionally into several straight segments, and often show macroscopically visible wear in the form of polish on both surfaces of one edge, commonly referred to as 'sickle sheen' (Fig. 5.24A). As Pélegrin (2002) has shown from experiments, technological features of the Canaanean blade platforms show that they were probably made using a long-lever pressure device, tipped with antler or copper (Pélegrin 2002; Chabot 1998; 1999; 2002).[11] The standardisation and regularity of these blades over a wide region and the difficulty of their manufacture indicates that they represent a specialised craft. Flint refuse from their manufacture and segmentation is not found in the Khabur valley sites, and raw material of the requisite size is found only along the modern Syrian border with Turkey (Anderson *et al.* 2004).

Canaanean blade segments are often identified as 'sickles' in the literature (*e.g.* Rosen 1996; 1997) because most display a faint to marked gloss on their lateral edges that is visible to the naked eye. However, a few researchers (*e.g.* Woolley 1956, 14) have suggested that they might be *tribulum* elements. Over the past 25 years, numerous studies, involving the examination of experimentally used tools at low and high magnifications, have shown that this glossy 'sickle sheen' wear on stone tools which appears the same to the naked eye, can in fact be produced by a variety of different uses. Cutting

Fig. 5.24. A: Canaanean blade segment from northern Syria, Bronze Age (about 5000–4000 BP), showing the appearance of the edge, and gloss from use in the threshing sledge (Photo: J.-D. Strich). B: Detail of the gloss on this tool seen at 100x magnification, showing microwear traces characteristic of use in a threshing sledge. Note large comet-shaped pits (Photo: P. C. Anderson).

and harvesting silica-rich plants (cereals and reeds) with stone sickles produces this gloss, but so does chopping plant material, cultivating (with hoes), and threshing. Glossy macrowear can also be produced on stone tools that were used to work wood or soft stone, and to scrape clay and hides (Anderson 1994b; Barkai and Yerkes 2008; Juel Jensen 1994; Van Gijn 1994; 1999; Yerkes *et al.* 2003). However, using incident or reflected light microscopes at magnifications of 100–200×, the surface of the gloss can be 'read', and microscopic attributes of its appearance (*e.g.* polish, morphology of abrasion, striae, pits, etc.) allow us to distinguish among different functions producing macroscopic gloss (Anderson 1994b; Juel Jensen 1994; Van Gijn 1994; 1999; Yerkes *et al.* 2003). Microscopic examination at magnifications between 100× and 200× revealed that the features of gloss on Canaanean blades did not resemble flints used in experimental sickles to harvest plants, but rather showed a remarkable similarity to tools known to us from our analyses of ethnoarchaeological *tribulum* flints from various countries where threshing sledges were used until recently (Anderson and Inizan 1994; Anderson 1994a; 2003). Microwear traces within the gloss on ethnographic threshing sledge inserts from Turkey (Ataman 1999; Whallon 1978), Cyprus (Kardulias and Yerkes 1996; Yerkes and Kardulias 1994) and Bulgaria (Skakun 1994; 1999; Gurova 2001 and this chapter) have been described by many others as well. On the Canaanean blades we observed wide linear traces, a dull and abraded surface, randomly oriented scratches and comet-shaped pits (Fig. 5.24 B), the same microscopic use features that we and others have observed on threshing sledge inserts removed from ethnographic sledges (Fig. 5.21).

Traces on threshing sledge blades are distinctive from other kinds of use in that they show the result of a continuous motion. The blades are inserted in a sledge frame and pressed down by the weight of the sledge (to which stones or a person often serve as additional ballast), then dragged through thick layers of plant material placed on a prepared surface, the threshing floor. This peculiar contact mode with plant material takes place in the presence of abrasive elements from the soil adhering to the plants. The combination of these factors accounts for the simultaneous presence of the distinctive features of the glossed micro-surface: smooth, bright areas interrupted by zones with rough, abraded and pitted topography, randomly oriented fine scratches, and characteristic linear features such as long troughs drawn from large irregular pits in the surface ('comet' features, see Fig. 5.24B). These features are all found on the Canaanean blades, as well as on the ethnographic threshing sledge inserts, and cover much greater areas on the surface than occurs for other uses. We explain these features by the fact there is a continuous motion of the blade when used in a threshing sledge, the blade penetrating the layer of plant material, with its entire working area in contact with the stems. The flow of silica-rich stems is usually parallel to the blade, but some material strikes the blade at various angles. This leaves long, continuous striations at varying angles, following the continuous motion of the sledge, helping to distinguish *tribulum* blades from sickles. On sickle blades, the shorter cutting motion leaves microwear polish and shorter striations that are oriented nearly parallel to one another and to the edge of the blade.

However, Canaanean blades also show some differences in wear from ethnographic *tribulum* blades.

They are larger, lack evidence of damage from pounding them into a plank-built sledge, and the corners do not show the extreme rounding wear of most modern *tribulum* inserts. Nearly all the Khabur blade segments had more or less abundant traces of bitumen residue on the backed or the non-active cutting edge, used as adhesive to fix the blades into an instrument. The bitumen from Khabur sites is of non-local origin (Fortin 1998). Bronze Age threshing sledges seem to have been somewhat different from their modern counterparts. Grégoire (in Anderson 2000; in Anderson and Inizan 1994) has pieced together the construction of the Bronze Age threshing sledge from brief contemporary references to the instrument, principally in the 'Farmer's Almanac' (in Anderson and Inizan 1994; Civil 1994; Littauer and Crouwel 1990) and various cuneiform administrative archives (*e.g.* Gelb 1955). These documents seem to describe a sledge made of wooden staves lashed together with leather strips. The blades were placed in the interstices between the staves and fixed to the structure using bitumen, heated and poured into the interstices of this 'raft-like' structure. In this manner a smooth, flat underside of the raft's working surface was obtained without use of a plank-type construction (Fig. 5.25 and 5.26). In the cuneiform inventories (*e.g.* Gelb 1955: 271, 273 tablet 33 [FM 229201 lines 31 and 35]), from 50 to 80 'teeth' were needed to fill the threshing sledges. One archaeological blade still has adhering bitumen with the imprint of the wooden staves.

Jacques Pélegrin produced the blades for our experiments, replicating the ancient Canaanean examples in dimensions and morphology. They were fixed in a raft-like structure comprising 10 wooden staves, each 3–5 cm in diameter (Fig. 5.25). Five sledges have been built and used with a wide variety of crops and surfaces in the course of 10 seasons of experiments (1995–2010), some since the original report (Anderson *et al.* 2004). We decided to use 50–80 blades, the minimum number deduced from the textual descriptions (Gelb 1955), to arm the undersurface of the sledges, to see whether they could function efficiently even at minimal size. Melted tar tempered with fine sand was spread between the staves as each blade was inserted. The blades were set to protrude according to the location of gloss and bitumen on the Bronze Age blades (Fig. 5.26). It was observed that standardisation of blades allowed for construction of a solid instrument. After

Fig. 5.25. Experimental threshing team, Seranon, France 2005, with the threshing sledge reconstructed in our latest experiments (Photo: A. Handley).

Fig. 5.26. The underside of the reconstructed Bronze Age threshing sledge, showing the insertion pattern and protrusion of the large, experimental blades like those from the Bronze Age (Photo: J. C. Whittaker).

the first year of experiments, 'skis' were added to the sides of the sledge, protruding 15 cm from the front of the frame, which helped the small *tribulum* glide over the thick plant layer, and raised the blades several millimetres off the ground, to help prevent contact with the hard threshing floor surface. Although not described in our ancient texts, such runners occur on some ethnographic *tribula* in Spain (J. J. Ibanez, pers. comm. 1994). We added them because the wear traces observed on Canaanean blades were evenly distributed over the blades' working edges and adjacent surfaces, unlike ethnographic sledge flints, which can have a blunted or facetted area on the very edge contact area, polished like a water-worn pebble (Whallon 1978). The experimental blades used in the first year touched the threshing floor, and the edges had begun to wear down more than the other contact surfaces. This pattern did not replicate the pattern seen on the archaeological blades (Anderson 1994a; 1999). However, subsequent experiments showed that the skis were impeding the smooth flow of the straw under the sledge (Anderson *et al.* 2006), and that the sledge did not need to be held above the level of the threshing floor, an observation made by both our collaborating engineers, and by a sledge maker in northwestern Tunisia, where a similar small sledge is used today (Anderson 2006b, and this chapter). Indeed, when the skis were removed, and the front of the sledge was curved upward by thinning, heating and bending the staves, the sledge functioned well, and the wear pattern on the blades was no longer concentrated on the very edge, but rather on the edges and the sides, as on the ancient blades. Our mistake had been putting too little plant material on the threshing floor in the early years of our experiments, causing the sledge to actually touch the threshing floor, which as we learned in observing sledges still in use in Spain, Syria, and Tunisia, never should occur.

Fig. 5.27. Reconstructed Mesopotamian threshing sledge in use on a threshing floor made in a field in Seranon, in the Alpes Maritimes region of southern France (Photo: P. C. Anderson).

Since 1995, we have experimented with these sledges on beaten and hardened clay threshing floors in southern France and in Spain, and on an ancient paved floor and on a surface of dry grass clipped short, in southern France. We have processed a variety of crops, including einkorn, hulled barley, and bread wheat, mostly cut close to the ground surface to leave long stems. The light sledge has been augmented by different amounts of weight, and pulled by horse, donkey and human traction (Fig. 5.27). The sledge appears to cut most efficiently after about 20 minutes of work, and a layer of sheaves 30 cm thick on a threshing floor 8–9 m in diameter is processed in about two hours. The sledge is remarkably efficient and chops the chaff without damaging the grain. Our results in weight of grain threshed in a given time were similar to those recorded today, for example in Tunisia (Anderson 2005) although in southern Syria, the basalt studded sledges used there took longer to obtain this result (Anderson 2003). Study of the experimental flint blades under the microscope showed that the resulting wear attributes corresponded to those seen on the Canaanean blades: a matt, metallic-appearing surface with white smooth areas, with large comet-shaped removals and drawn linear features extending well into the blade surface.

Engineers working at the École Centrale, Lyons, France, participated in our experiments and were able to carry out measurements that shed light upon the complex nature of the action of the Bronze Age Mediterranean threshing sledge (Vargiolu *et al.* 2003, 448–51; Anderson *et al.* 2004). One of the experimental sledges was equipped with instrumentation attached to one of the blades, and these recorded pressure on the blade in three directions, as well as any changes in temperature of the flint blade as it cut the straw. A hole was made in the sledge frame and covered with a transparent piece of plexi-glass, while a video camera mounted on the plexi-glass filmed the movement of the plant material and of the sledge as it worked.[12] Measurements over the first 45 minutes of work with the tool show that the friction is greatest during the first 15 minutes but then falls abruptly and stabilises. This conforms to our own field observations, and must be the result of a modification in the thickness of the bed of plant material. At first, the random orientation of the plants and the long stems forms a resistant layer. As work progresses, the straw is chopped and a new,

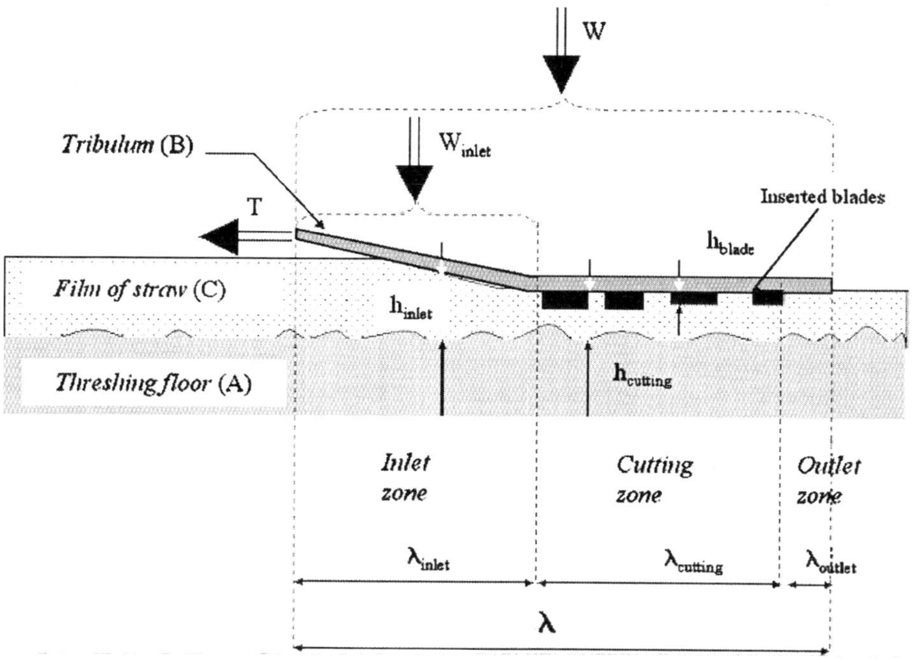

Fig. 5.28. Diagram showing optimum functioning of the Mesopotamian threshing sledge with large blade segments such as those from the Bronze Age. Engineers have emphasised the importance of the inlet area, the action of the film of straw on the blades in the cutting area, and the outlet zone, as well as interaction of all this with the surface of the threshing floor itself (J.-M. Georges, R. Vargiolu).

more elastic layer of straw forms. The temperature varies only minimally, indicating that there is little thermal change on the flint surface during use. The video of the straw being cut under the *tribulum* revealed that, after 15 minutes, the stems moved parallel to the sledge blades, whereas at the same time the straw was cut perpendicularly. The cutting mechanism involves a complex system with two layers of plant material: an upper one, made up of fragments of straw and seed heads already cut and being ejected from the back of the sledge; and a bottom layer, composed of plant material in the process of being cut. The action of the sledge serves as a comb, guiding the straw from the front to the back. The movement of this layer, which circulates like a fluid, causes the straw to be cut with great efficiency. Experiments and calculations (Anderson *et al.* 2006) show the great importance of the angle of curvature of the front of the sledge, creating a working inlet zone, followed by a cutting zone and finally, an outlet zone, where the cut straw is ejected (Fig. 5.28). This distinctive and complex movement on the threshing floor also explains the diversity in orientation of the distinctive traces that form on sledge blades. The video shows the progressive cutting of straw fragments: first the straw is scored many times as it moves against the sledge flints, making smooth, curved or straight patterns on the epidermal surface. It breaks apart progressively, following the pattern of the incisions; these patterns are reflected in the cut profiles of silica phytolith

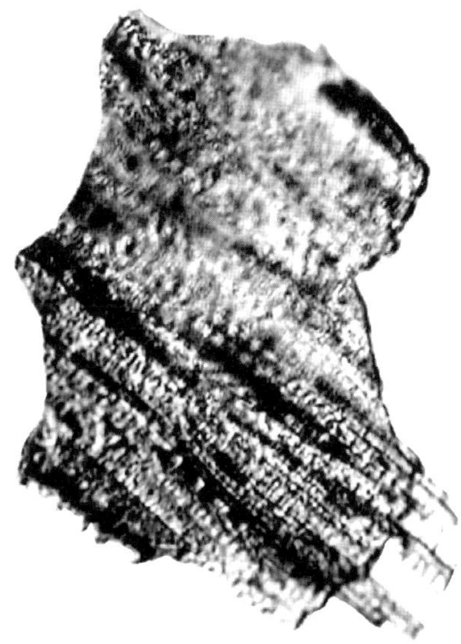

Fig. 5.29. Phytolith from a threshing floor in Spain, where the *tribulum* was used to work bread wheat. This phytolith is a minute fragment of the epidermis of the stem, shown at 400 x magnification, showing the two concave cuts on its edge to the left of the photograph, characteristic of cuts occurring when the straw is moving on a threshing floor under the threshing sledge (Photo: P. C. Anderson).

sheets (Fig. 5.29). Phytoliths are microscopic plant silica structures that survive in most environments in the soil after plants have decayed, or as a residue on tools; their shape or their imprint reflect characteristic forms of cells in the stems and chaff of cereals (and other plants) in which they form. These cut phytoliths are recognisable in archaeological contexts as a distinctive signature of *tribulum* use, while both harvesting with a sickle and using a blade to cut straw on the ground produces crushing and jagged cuts of the stem and its phytoliths.

Laboratory experiments with friction between flint and cereal stems, and field trials with the *tribulum*, also offer some insights into the formation of traces on threshing sledge blades. These results suggest why even dull sledge inserts can function, although slowly compared with the experimental *tribulum*, and also why dull sledge inserts of basalt, highly rounded areas of sledge blades, and sharper blades all show the same basic microscopic use features under the reflected-light microscope at 100–200x magnification. In fact, the 'wear' is not acting on the stone surface itself, but rather on a microscopic layer deposited on the working surface of the insert, probably derived from plant silica. This surface will be in similar contact with the straw whatever its shape. According to our discussions with present-day sledge users in Spain and Syria (Anderson 2003), all sledges need to have inserts projecting at least a few centimetres in order to function, but apparently whatever the nature of these projections, the sledge will work, and the distinctive flow of the plant material will occur on the threshing floor. We observed this for sledges with metal projections and basalt inserts in the Jebel Hauran area of Syria. When we asked the farmer if he were aware of stone-bladed threshing sledges, he said he knew of them, and they were the very best, far more efficient than his, but he could not obtain blades with which to make one, and the local volcanic bedrock does not yield stones which can be knapped into blades. When we showed him a photograph of the experimentally reconstructed sledge with large Canaanean-type segments, he greatly admired the fine blades, stating again that this sledge would work far more quickly and efficiently than any he knew of.

Although we modern folk think of threshing and winnowing as processes that produce grain, the production of chaff is equally important. Ethnographic observations show that traditional winnowing methods enable different size fractions of the chopped plant material to be separated, each fraction having particular uses: fuel, animal food, bedding material for humans and animals, temper for mudbrick and plaster, or for ceramic containers, such as granaries (Anderson 1998; 1999; 2000; 2003; Oates 1990, 389–390; Palmer 1998, 150; Whittaker 1999, 13; 2003, 381).

Details about the use of the *tribulum* in Early Bronze Age sites are revealed by the analysis of phytoliths. Cut straw from ethnographic threshing floors where sledges armed with blades were used, in Catalonia, Cyprus and Tunisia, was subjected to chemical digestion of its organic components, leaving the silica shell, to produce an ethnographic comparison. Some phytolith 'sheets' were broken in ways that can be seen in breakage patterns for any assemblage of decayed plants (Khedhaier *et al.* 2003), whereas other shapes reflected the action of the threshing sledge: the phytoliths, linked in sheets of cell imprints, were cut with peculiar smooth, 'razor-sharp' edges, and straight or curved complex patterns, first observed by Juan Tresseras (1997) then by Anderson (1998; 1999; 2000; 2003) and others (Khedaier *et al.* 2003; Cummings 2003). We confirmed this observation by analysing remains of cut cereal fragments gathered from threshing floors after our experimental sledge had been used. The phytolith damage produced by threshing sledges differs from ethnographic and experimental material harvested with a sickle or processed by animal trampling or other threshing techniques. The distinctive smooth-cut, complex phytolith profiles have been found in Bronze Age sites we studied, in levels which also had Canaanean blade segments showing traces of use as inserts in a threshing sledge. At Tell Leilan in northeastern Syria (Fig. 5.30A), samples came from ashy deposits in ovens, the remains of dung used as fuel. This finding provides some evidence for the use of chopped straw to feed domestic animals (Cummings 2003), and underlines the importance of dung fuel in this period (McCorriston 1998), as well as today throughout the Mediterranean region (Anderson 2003, 424–5; Anderson and Ertug-Yaras 1998). Ongoing analyses of phytoliths show that straw processed with the threshing sledge in use at this time at Tell 'Atij (Fig. 5.30B), also in northeastern Syria, was present in storage structures, and incorporated into mudbrick walls directly or as burnt material, probably from middens (McCorriston, pers. comm. 2001). Only phytoliths from chopped stems, not glumes, were found at Tell 'Atij, which may reflect the harvesting of long stems. This would support the results from studying macro-remains, which indicate that plants were harvested by pulling and that the cereal was a

hulled variety, such as hulled barley (McCorriston 1998) whose glumes would normally not be detached from the grain by the threshing sledge, and therefore not enter into the archaeobotanical remains at this stage. In addition, this indicates that the temper was from the threshing floor, and not a residue of the dehusking process (Procopiou 2003). The particular level showing these phytolith data at Tel 'Atij is the same one in which nearly 250 Canaanean blade segments were found, all having traces indicating that they functioned over long periods of time while inserted in a threshing sledge (Anderson and Chabot 2001).

In our Ninevite sites, threshing floors may have been relatively unmarked surfaces, as in many ethnographic areas, or may have been away from living and high-status areas usually excavated by archaeologists. Nevertheless, our studies may eventually allow them to be recognised. Preliminary phytolith studies of 5th to 3rd millennia sites in the Uvda Valley in the southern Sinai show phytolith remains from chopped straw preserved in the soil from two threshing floors. They present the complex types of cut produced by a threshing sledge (Avner *et al.* 2003, 470, Fig.15), which would concur with the presence of sledge inserts at the site.

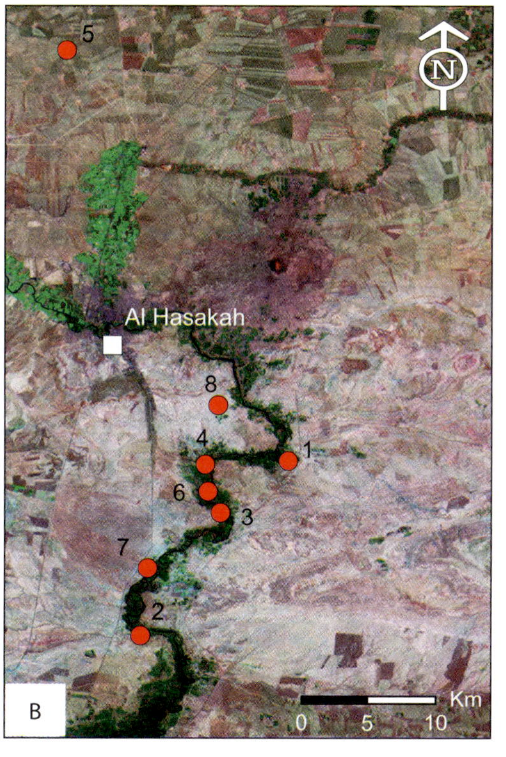

Fig. 5.30. A: Map of the Near East with the location of the archaeological site of 1) Tell Leilan in north-eastern Syria. B: Zoom on north-eastern Syria with the archaeological sites of 1) Tell 'Atij; 2) Bderi; 3) Gudeda; 4) Mulla Matar; 5) Nusstell; 6) Raqa'i; 7) Tell Melebiya; and 8) Tell Tuneinir. Map: R. Lugon, J.-C. Loubier and A. Chevalier.

Our reconstructions and experiments allow us to interpret the archaeological evidence to show the use of an important agricultural tool with extensive implications. The wear on a distinctive class of flint blades is now identifiable, and the presence of these blades explained. They were the product of specialist workshops engaged in a wide trade network. The blades were standardised to fit a common form of sledge, necessitating an unusual degree of specialised knapping skill, even greater than that of the ethnographic specialists in Turkey, Cyprus, and elsewhere, who mined flint and knapped blades with metal hammers, and also participated in complex economic relations (Bordaz 1965, 1969; Karimali 1994; Pearlman 1984; Whittaker 1999; 2003). The archaeological data do not yet show by whom and in what form the threshing sledge was distributed in Canaanean networks, nor if its production occurred in several stages and in different locations. The sledges efficiently processed the crop into valuable resources, grain and chopped straw chaff. The demand for chopped straw in turn implies harvest technique: some or much of the crop was probably harvested by uprooting (see Chapter 4), or at least cracking off the stems at their very base to obtain a maximum amount of straw. Both techniques are used in southern Syria today. In areas where long straw is used for thatching or basketry, sickle cutting reduces the straw length, but removes roots with contaminants such as soil and small stones, which are not a problem for crops threshed by a *tribulum* and winnowed.

As the Canaanean blades are recognised as *tribulum* elements, not sickles, the resulting apparent dearth of sickle blades in these sites is explained by the way the crop was processed. The Early Bronze Age of northern Mesopotamia (Ninevite V) is well known for the appearance of numerous villages built directly on virgin soil, and in the Middle Khabur valley of northeast Syria, many such sites have been recently excavated ('Atij, Gudeda, Bderi, Nusstell, Raqa'i, Mulla Matar, Melebiya, Tuneinir; Fig. 5.30B). It has been suggested that their inhabitants produced surplus grain (Fortin 1997), or 'staple finance' (Philip 2001, 167; Schwartz 1987). The use of a highly efficient threshing sledge in village sites throughout the region to produce large quantities of clean grain and finely chopped straw may well correspond to this concept. Philip (2001, 167; 2003) has also proposed a model of flexible, heterarchical corporate villages and middle-range societies rather than institutional power structures associated with city states, to describe Jordanian settlements during the Uruk period and the Early Bronze Age, emphasising the importance of resource storage at these sites. This model could apply to the northern Mesopotamian sites considered in this study. These villages could be seen as small homogeneous units ensuring the supply of primary products (*i.e.* grain, chopped straw, etc.), but dependent upon urban centres for specialised goods, including perhaps the Canaanean blades necessary for the threshing.

5.11. THE USE OF FLAILS FOR THRESHING CEREALS

José Luis Mingote Calderón

The flail is a relatively modern threshing tool in the history of agriculture. Other tools are much older, dating back to Neolithic times, but the flail has been attested in Europe only since the 4th century, and widespread use did not occur until the Middle Ages (Fig. 5.31). In spite of the importance of the flail in Europe, we should not forget that this tool is also known in places located very far away, like Japan, where it has a different history. Some researchers say that in Europe, the introduction and later expansion of rye cultivation is the main reason for the increasing use of flails for threshing. This could be true for medieval times, but in the 19th century and later this association is complex. For instance, in some Spanish areas such as the province of Leon, the flail and the *tribulum* coexisted in the Middle Ages. In this province, from the 19th century, both tools were used. Nevertheless, the situation depends on the particular area we are looking at. In some places both cereals are flailed; in other areas, rye is flailed and wheat is threshed with a *tribulum*; finally, there is another area where both are grown but only the *tribulum* is used (Mingote Calderón 1990a; 1990b).

As with many other agricultural tools, the flail has been studied from morphological and functional perspectives (Trotzig 1943; Herculano de Carvalho 1953; Ó Danachair 1970; Veiga de Oliveira *et al.* 1983; Mingote Calderón 1988; 1990a; 1990b; 2003). Some of these authors have looked at the relationship between the shape of the tool and the way it is used, proposing typological approaches to explain local diversity in their own countries and neighbouring regions. All these typologies tend to look at similar aspects, like the connection between the handle and the hammer of the flail. Different proposals reflect geographical diversity, as we can see through Swedish and Portuguese examples (Fig. 5.32). Apart from the way the hammer and the handle are put together, we need to focus our attention on their absolute and relative lengths, because, together with the first factor, the proportions of the flail condition the way the tool is used.

The work of flailing involves complex organisation. This complexity relates to all aspects of the flailing process, including the choice of flailers, preparation of meals and many other minor activities carried out during the threshing. Flailer teams were composed mostly of men, who used to take part in work-exchange systems. These groups tended to be

Fig. 5.31. Romanesque scene of flailing. Duomo of Modena, Italy (Photo: J. L. Mingote Calderón).

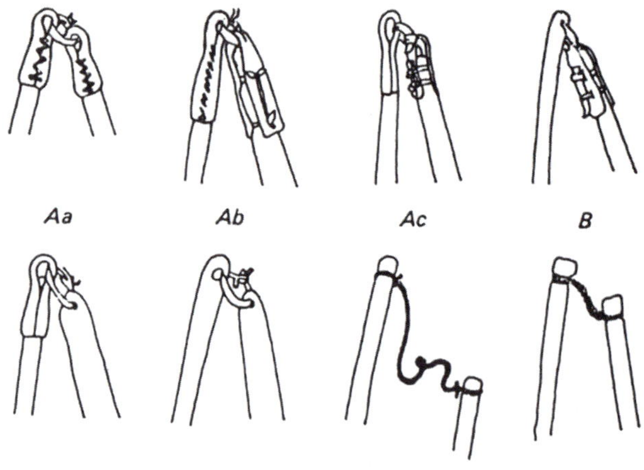

Fig. 5.32. Portuguese flail typology (from Veiga de Oliveira *et al.* 1983).

composed of men at more or less similar economic levels, that is, having a similar amount of crop to flail. The group may or may not include relatives of the owner of the crop. They used to be stable through time (Portela 1986). The important principle is that the reciprocal obligations be kept in balance; families tend to work with others of similar economic level so that a small farmer does not work many days for a larger farmer who will then need to work less on the small farmer's crop. Flailing was, in fact, unpaid work, but, in exchange for the labour, the host offered special meals during the whole threshing period. The flailing process was a very important social activity. Because threshing represented the culminating event in the entire agricultural cycle, and because exceptional meals were offered as part of the labour exchange, the flailing itself became something of a ludic activity, a party, and a feast.

It is convenient to stress the complexity of the spatial as well as the social organisation of flailers, and the social and functional importance of individual positions in the flailing team. A man's position depended mainly on his strength and ability using the tool. The best flailers would be in the middle of the row of flailers, while the weak or less expert tended to occupy the lateral positions. The most common spatial organisation was two rows of flailers, one facing the other. But sometimes we find more complex situations: at Celorico de Basto, Tecla (Portugal; Fig. 5.34), Veiga de Oliveira *et al.* (1983) described four rows of flailers organised in two groups, one in front of the other. Each group is composed of a variable number of flailers organised in two rows, one behind the other, flailing all together at the same time.

Hammers may be short (half the length of the handle, more or less) or long (similar to the handle or even longer). In the first case, the small hammer may be swung around over the heads of flailers who are arranged in tight rows. In the second case, the hammer turns around in a slightly different way: it passes by the right or the left side of the flailer, not over his head. In this case, the row of flailers has to be divided into pairs. As a consequence, the flailing teams tend to appear separated in small groups of four people, two opposite the other pair (Fig. 5.33). In the province of León (Fig. 5.34), these small groups received special names. They were called *perrona* or *perrina*: the first name applied to groups formed only by men; the second to feminine or mixed groups. *Perrona* and *perrina* were names derived from small coins, 10 and 5 cents of a *peseta*, respectively. Depending on the length of the

Fig. 5.33. A team of four persons flailing in Leon province. Spain (Photo: J. L. Mingote Calderón).

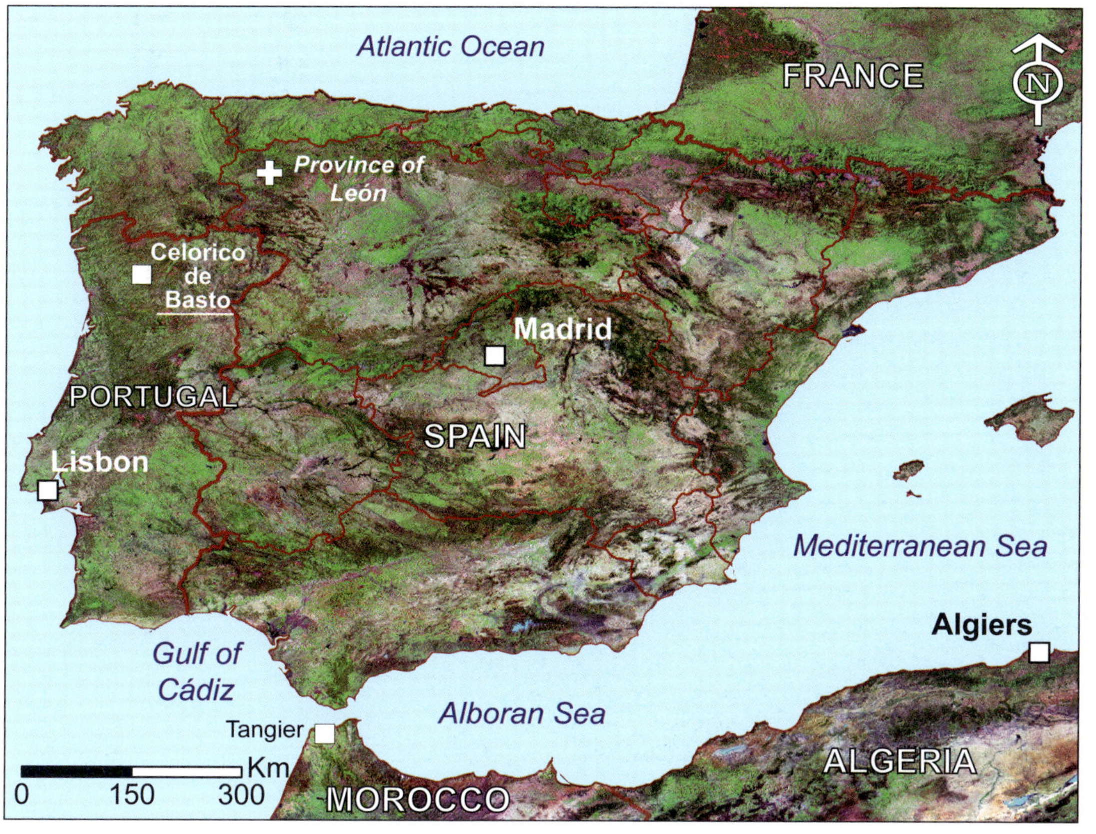

Fig. 5.34. Map of the Iberian Peninsula with the mentioned town of Celorico do Basto in the Braga district, Portugal, and the Province of León in Spain. Map: R. Lugon, J.-C. Loubier and A. Chevalier.

hammer, which affected the way of working, the men then needed to act as left or right-handed flailers. They hold the handle in different manners because of their different positions in the row. Right-handed people who, if needed, were able to work also as left-handed flailers, were highly valued. There are different kinds of flails, as well, depending on the way they turn around the handle: in some of them, the hammer makes a perpendicular revolution around the handle, changing completely the spatial distribution of flailers. This method of flailing is not well known yet, and has not been studied in detail.

From another point of view, flailing was organised in different ways according to the weather. In regions with hot summers, the whole crop is flailed together at the same time, immediately after the harvest. People living in rainier regions, where there was a chance of rain during the harvest, tended to stock the crop in a barn and did the flailing at different times. As mentioned, women may participate in flailing, but, in most of the regions, women's work was restricted to 'secondary' tasks. Women would commonly lay out the crop, hold a sheet next to the flailers to prevent the grain jumping outside the threshing pile, turn the ears before the second phase of flailing, shake the cereal bundles to free the last grains from the ears, gather the grain, etc. Apart from these tasks, women were in charge of everything related to the meals, a very relevant aspect because of the extraordinary significance and elaboration of those meals. Finally, we should mention the existence of some rituals and jokes associated with flailing that present sort of an 'archaic' aspect, that is, they seem to come from quite a long time ago, and appear to contradict the relatively modern development of the flail. These rituals have been mainly studied in the northern Portugal case, where they show a surprising complexity. Sometimes, these rituals and jokes imply parodies of religious rituals; often they involve peculiar songs, sometimes talking about sex or the role of the host's wife. There are even practices and beliefs relating to an animal hiding in the last sheave, something that we may compare with Roman or even pre-Roman traditions.

5.12. ALTERNATIVE THRESHING METHODS: LASHING AND BEATING WITH STICKS AND MALLETS IN THE WESTERN MEDITERRANEAN

Leonor Peña-Chocarro

In traditional Mediterranean agriculture, threshing was generally performed using threshing sledges, flails or animal trampling. However, these were not the only techniques available. There were also other threshing methods influenced by a number of factors, such as the final use of the straw. Variation in threshing methods in both Spain and Morocco seems to be very much related to the need to obtain un-chopped or uncrushed straw for thatching and crafts such as basketry or stuffing. In addition, the choice of threshing technique may depend on the type of crop. The methods of beating or lashing may be used to thresh hulled wheats (as opposed to the naked wheats more common today). Hulled wheats are tightly enclosed within their glumes but, what is more important when we discuss threshing, hulled wheats have a semi-fragile rachis which implies that when threshed, the ear breaks into spikelets

Fig. 5.35. Fig. 5.35. A: Farmer lashing an einkorn crop in the western Rif (Morocco), and B: The *porro*, a threshing tool used in Asturias (northern Spain) to break the ears of hulled wheats into spikelets (Photos: L. Peña-Chocarro).

leaving behind clean or almost clean stalks. The free-threshing naked wheat varieties differ in that their grains get separated from their husks during the threshing process, leaving the rachises attached to the stalk. So, when the full length of the straw is needed, and a free-threshing wheat is involved, beating or lashing are only applied when the amount to be threshed is small, whereas if a hulled wheat is threshed, and the full length of the straw is required, the same farmers may use lashing or beating as the threshing technique.

In southern Spain and northern Morocco where einkorn, a hulled wheat (*Triticum monococcum* L.) was grown until recently, farmers used two different techniques which allowed the ears to be broken into spikelets and thus separated from the stalks: a) lashing, *i.e.* beating of cereal sheaves against a fixed object (threshing sledge, rock, tree trunk, etc.) or the floor (Fig. 5.35A), or b) beating with sticks or specific tools so the ears get broken into spikelets. Einkorn ears could be beaten using simple sticks, or as in the case of northern Morocco, using specific tools such as wooden beaters with a carved surface and a short handle (see Peña-Chocarro, Chapter 7). These are called *rsama* or *marzab* and they were also used in the area for cloth washing, clay grinding and sometimes einkorn dehusking. In both these two areas, these simple threshing techniques were used because uncrushed long stalks were required for a number of different activities such as thatching, basketry, and stuffing saddles and mattresses.

In the most southern part of Asturias (northern Spain; Fig. 5.11), an area of hulled wheat (emmer and spelt) cultivation, old farmers still remember the use of wooden mallets to break cereal ears into smaller units or spikelets. The method is very similar to the one just described for the area of einkorn cultivation. Wooden mallets were used under different local names according to the different areas where the technique was applied: *mazos* or *porros*. The mallet is a quadrangular piece of wood with an undulated lower surface, attached to an angled handle (Fig. 5.35B). In other areas of the Iberian Peninsula, similar methods, using variations of the hammers and mallets just described, were applied to small quantities of cereal or legumes. For example, wooden hammers were used in Zamora province to thresh chickpeas (Peña-Chocarro 1999). Similar techniques are described for threshing gleanings or small quantities of cereals in southern Spain and Portugal (Veiga de Oliveira *et al.* 1983).

5.13. PRE-MECHANISED THRESHING SYSTEMS IN FRANCE

Carolina Carpinschi and John C. Whittaker

Introduction

Although mechanised agriculture expanded in France throughout the 19th and 20th centuries, traditional, non-mechanised techniques continued to be used at least into the 1950s. In the case of threshing, a variety of human- and animal-powered technologies held their own because they operated as a system, which was determined by socio-economic factors as well as local conditions and crops, and was capable of rapid responses to change (Castel Carpinschi 2003). The development of agricultural science and state-sponsored agronomy promoted mechanised technology, but also promoted some of the non-mechanised techniques. At the same time, the specific socioeconomic patterns that affected both were documented. The 19th century also saw the development of steam power and eventually other forms of mechanised energy, as well as state projects such as canals and road systems. Large agricultural companies, specialised and market-oriented, were quick to adopt new technologies. Smaller family operations, with limited resources, small land holdings, and an orientation toward polyculture still found it advantageous to maintain traditional practices, although these too changed in many ways.

C. Parain (1938) distinguished four areas of France in the 19th century where different modes of non-mechanised threshing were employed: 1) beating with a flail under shelter (in grange or barn), 2) beating with flail in the open, 3) trampling with animals, and 4) threshing with the *rouleau* or threshing roller. The use of traditional tools like the flail, or of new instruments, like the threshing roller, was determined by a variety of factors. These could be economic, climatic, or physical. They were also influenced by the choice of plants grown, and by local usages and customs. Despite the strong association of specific tools with particular crops, the non-mechanised techniques, far from being anachronisms, were organised and perfected according to logical properties. They formed what can be called 'systems of pre-mechanised threshing' that were associated with different communities and different assemblages of crops.

Diffusion of the Roller

During the first half of the 19th century, the large enterprises in the northern half of France, the area where the use of flails in barns was the traditional threshing technique, rapidly adopted the mechanical thresher. In the southern half of France, a warmer, drier climate had traditionally allowed farmers to thresh their crops by flailing in the open (Fig. 5.36) and trampling with animals. In the south of France (Fig. 5.37), the use of a new threshing implement expanded in the 19th century:

> 'a gunner from Haute Garonne built for M. Crouzet, farm owner and amateur agricultural enthusiast, the threshing roller of Italy, the '*ritolo*'... It was known to the Romans and used by the Carthaginians. One works as fast with the *rouleau* as one can thresh the wheat with horses, and one obtains as much grain as with the flail; it uses only labour that would be useless for threshing with flails. The straw is layered by the *rouleau*, the least possible is lost, and it is sought avidly by the beasts (Puymaurin 1809, 66, our translation).'

Although the *rouleau* or threshing roller was an ancient implement, there is little evidence of its

Fig. 5.36. Threshing oil rapeseed in open air with flails. The artist has depicted the orderly sequence of strikes used by an experienced threshing team (Mingote Calderon, this chapter). From Jacque (1835).

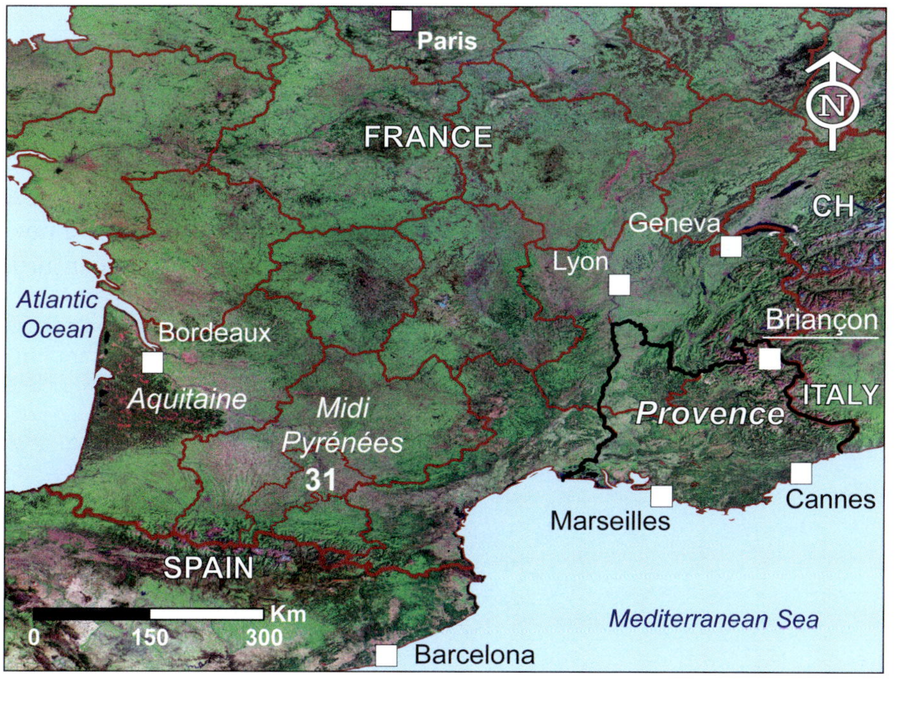

Fig. 5.37. Map of France with the Provence cultural and historical region that includes de Provence-Alpes-Côte d'Azur political region; the 31) Haute-Garonne Department; and the town of Briançon mentioned in Chapters 5.13 and 5.14. Map: R. Lugon, J.-C. Loubier and A. Chevalier.

use in France until the wooden *ritolo* described by Puymaurin was recognised in 1804 as an efficient innovation by the Royal Society of Agriculture (Dispan 1804, 3). The favourable report by the Royal Society on the advantages of the threshing roller marks a turn in the diffusion of this instrument, and encouraged the large farm holders of the *Midi* region in southern France to adopt and improve the *rouleau*.

The *rouleau* was at first a heavy log or cone of wood, held by an axle in a frame that could be drawn by animals (Fig. 5.40). Most had bars or teeth worked from the mass or added. After about 1830, threshing rollers made of stone or even iron appeared. They had smooth, toothed, or grooved surfaces and diameters from 35 to 100 cm. The forms varied depending on the region. The diffusion of the *rouleau*, like that of the threshing machine, was tied to the cultivation of wheat. Introduced first in Haut-Garonne (Fig. 5.37), the *rouleau* was rapidly adopted in the middle Garonne, and on the Mediterranean and Atlantic coasts. In the area of France where trampling the grain with horses was typical, the *rouleau* allowed middle-size operations to reduce

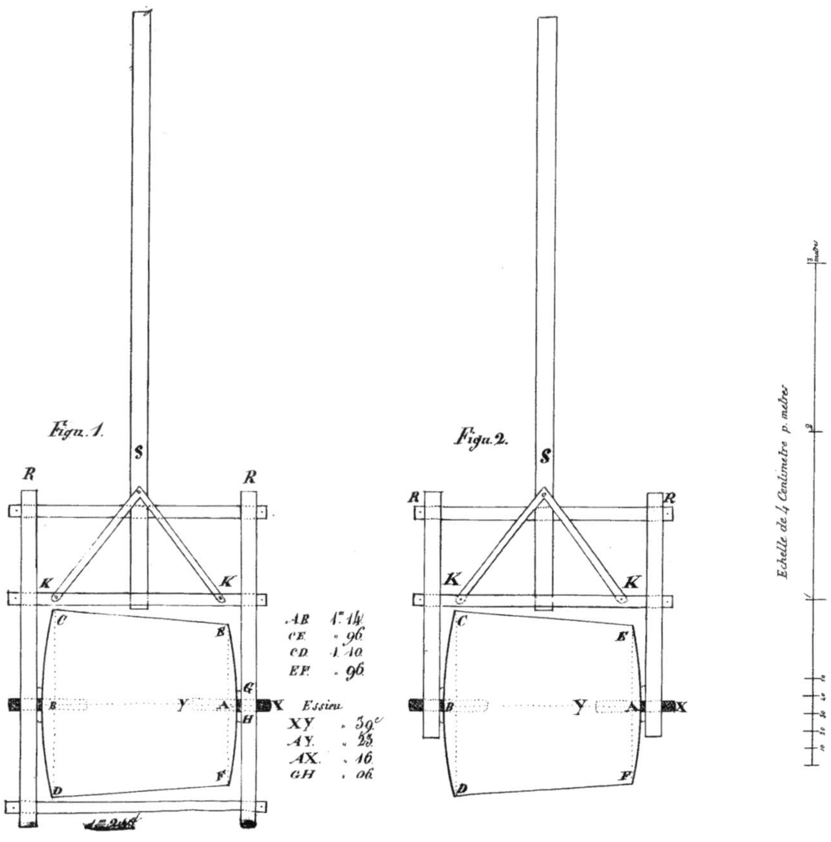

Fig. 5.38. The stone threshing roller of Mr. Duclos used with two mules (stone of Carcassonne), 1835, Gragnague near Toulouse, France, Haute-Garonne, from Duplan (1835).

Below right:
Fig. 5.39. Threshing beans with a horse-drawn stone roller, August 1942, Damvix, Vendée, France. The driver holds the reins and a stick; in front of him is a box for collecting droppings to prevent soiling the beans. Drawn from a photograph by G. Franceschi, archives of MuCEM (Musée des civilisations de l'Europe et de la Mediterrannée), C. Carpinschi 2011.

Below left:
Fig. 5.40. A wooden threshing roller with bars, probably end of the 19th century, Musée du Pays Brignolais, France, Brignoles, Var (Drawing: C. Carpinschi, 2001).

the number of horses needed, and thus work at less cost. On the Atlantic coast, the introduction of threshing rollers on a grand scale replaced the laborious operation of threshing by flail in the open, and reduced the use of oxen for traction. The *rouleau* also did not require specialised labour (as did flailing), accentuating the similarity between the ancient *rouleau* and the modern threshing machine. The *rouleau* developed under the influence of related horse-drawn forms, including the *traineau*, a sledge or beam of wood, sometimes in a frame with cart-wheels in front. By the 1840s, this was being combined with the *rouleau*, by building the roller into a heavy frame that dragged, adding the friction of the *traineau* to the compressive action of the *rouleau*. The frame could be weighted with stones or ridden by children, recalling the *tribulum*, but there is almost no evidence of the use of threshing sledges in France, although they were known to French agronomists in the 1770s and used across the Pyrenees in Spain (Castel-Carpinschi 2003; Comet 1992, 358; Mingote Calderón 1990a).

Despite the increasing use of horse-drawn implements as well as threshing machines, flailing remained popular in the 19th century and survived until the 1950s for threshing smaller quantities of cereal and legumes. French farmers used two different kinds of flails. Flails to be used for threshing in the open most of the time are composed of a handle and a striker of about the same length. These were common in areas of Mediterranean climate, warm and more or less dry, where the crop was usually threshed immediately after the harvest. The light striker allowed a team to keep the rhythm of work for several hours and over several days. This flail was typical of small operations with limited financial resources, using familial, non-specialised labour, male and female. In contrast, the flail for use in barns had a shorter striker, but of diameter and weight greater than the handle. It was used in winter, several months after harvest, almost exclusively by specialised workers. These could be hired by the day, or were part of the workers hired for the year, who must be occupied over winter, so they threshed during the slack season of the agricultural cycle. The difficult work of threshing in the closed barn was done by one or two men. Even in the region of indoor threshing, certain grains, such as buckwheat, were threshed in the open immediately after harvest, to avoid the risk of spoilage if they were stored too long. This flexible use of flails along with that of other non-mechanised techniques such as trampling (*foulage*) or lashing/beating off (*chaubage*), was able to adapt to the introduction of new crops such as oilseed rape, carnations, and clover, which developed in the 19th century.

The Renewal of Old Techniques with the Introduction of New Crops

Some ancient alternatives to flailing, threshing rollers, and mechanised threshers were readily adapted to processing small quantities of crops for special purposes, or by small labour pools, and were actually sustained by the introduction of new crops. *Chaubage*, for instance, meant beating a sheaf against a fixed object. *Chaubage* provided two advantages: it produced whole long straws for thatching from grains such as rye or wheat, and it allowed the farmer to select and recover the best grain for next year's sowing. *Chaubage* was as widespread as mechanical threshing or the *rouleau* in France. *Battage*, or beating with sticks, could be done over a container to catch the seeds. Both worked well for some of the alternate crops like poppy and carnation that thresh easily. As stock-raising grew in importance in the 19th century, feed crops were also processed with traditional techniques. Clover seed was collected for planting by dragging a sledge (*traineau*) or flailing, although larger operations preferred mills or machines. Maize was usually shelled from the cob by rubbing, but could also be threshed on a table with flails or sticks. Buckwheat was threshed with the *rouleau* or by beating with long sticks, a different form of *battage* (Castel-Carpinschi 2003).

Systems of Pre-Mechanical Threshing

The systems of pre-mechanised threshing in 19th century France show the complex relations among operations of the harvest, including threshing. The processes of threshing of course do not stand in isolation. The larger agricultural system includes different harvesting methods, and operations after threshing such as winnowing and sieving, and eventual storage. For instance, where long-stemmed crops are tangled by winds, a large sickle is needed to harvest the ears by cutting high on the stem. Given that particular cut, the short seed heads obtained were easier to work by trampling and *rouleau*, and these two techniques used in the open were favoured on the Mediterranean coast, where one harvested with the sickle.

The factors which condition the choice and evolution of elements of the threshing system are of two orders. On one hand, there are physical conditions to which the system must adapt, including the morphology of the plants, and climatic and physical conditions of the region. On the other hand are socio-economic and cultural aspects, which evolve from these. In turn, the cultural and economic context of an agricultural tool may trigger evolutionary effects with changes to the tool itself. The improvement and diffusion of the *rouleau* was promoted by the use of stone, obtained from mining industries that were already well-developed as the *rouleau* spread. In some areas, the availability of cheap mountain labour for the summer harvest put a brake on the need for mechanised threshing. When a technique had proven effective, it was likely to become fixed by custom, until the introduction of new factors changed the situation. The use of roof tiles, for example, led to the decline of thatched roofs and replaced the cultivation of rye with other crops, accompanied by other agricultural, social, and economic practices. In the case where rye continued to be cultivated for the grain, several solutions offered themselves to farmers. They could employ a different manual technique like flailing, invest in a threshing machine, or even hire threshing specialists. Any of these choices in turn affected the social order (Castel-Carpinschi 2003; Sigaut 1989).

5.14. THE INTERDEPENDENCE OF TIME, CROPS AND TECHNIQUES

Thomas K. Schippers

Among the less often analysed aspects of agro-technical operations are the time-related constraints to which they must adjust. An ethnographic example for the southern French Alps will show how the time spans allotted for different harvests – here hay and cereals – are mutually interdependent in family-based agro-pastoral farms and how quantitative changes in one crop may influence the technical choices in working with the others.

The point of departure of this study was research on the practice of 'lashing' (French: *chaubage*, Provençal: *esbarbo*) as a threshing technique in the southern French (as well as neighbouring Italian) Alps noticed by early 20th ethnographers (Parain 1937/1979, 17–27) as a possible 'primitive' survival. According to these early studies on threshing techniques in France, the region appeared geographically as a 'relict area' between the 'Mediterranean' trampling (French: *dépiquage*) and the 'Northern' and Alpine flailing zones. But more recent ethnographic research (Llaty 1997, 83–100; Schippers 2003, 357–362) revealed more complex factors involved in the choice of lashing as a threshing technique. Not only the quantities of cereals to be processed by a household or the use or non-use of the unbroken straw appeared as factors influencing the choice of lashing, but also the available time for threshing during the very busy and short Alpine summer. For those who processed only a small quantity of cereals (mainly rye or wheat) for domestic consumption, lashing was used as the only 'degraining' technique, rapidly performed either on a special stone or an inclined wooden board (Provençal: *esbarbouiro*) or even simply against the wall of a house. However, other farmers employed lashing as a first, intermediary form of threshing stack-ripened cereal in summer, followed by a second, more lengthy threshing with flails inside a barn during the less busy winter season.

According to my informants in the Briançon region (Fig. 5.37), this two-step threshing process was used to deal with the shortage of time in summer. Since hay harvesting in the higher Alps was a vital part of the household economy, and large quantities of hay were necessary to feed the domestic animals (cows and sheep) during the rather long winter period, there was less time for threshing in the summer. But when, from the 1950s on, the regional roads and transport conditions (especially for moving cattle by trucks and vans) improved, the local economy changed and farmers started to prefer to buy young animals in spring and sell them in autumn rather than to feed them inside during the winter. A consequence of this was a decrease in the summer hay harvest which in its turn 'liberated' time in summer to perform all of the cereal processing with flails or small, fixed mechanical threshers. Accordingly, the necessity of a first threshing of the cereals by lashing disappeared due to changes in the hay harvest. This example, chosen among many possible others, shows the dynamics of the relationships between many different elements of the agricultural system that result in successful farming. Changes in one element, here the local transport conditions, may result in other changes, often difficult to predict. A time-based approach to local agro-pastoral systems allows ethnographers to get a better insight into local complexities and their 'practical logics', which may prove thought-provoking for other disciplines.

5.15. BETWEEN THE THRESHING FLOOR AND THE OVEN: WINNOWING, CLEANING AND MILLING GRAIN IN CRETE

Sabine Beckmann

Whoever thought that grain would ready to be milled and baked as it came from the threshing floor is sadly mistaken. Between those two stations on the way from grain to bread, men and especially women have to spend a lot of time working with the harvest before it is ready to be made into bread. The process in Crete is representative of chores that were found throughout agricultural Europe. Winnowing is a necessary first step after threshing, requiring a set of specialised tools and skills. The winnowers heaped the cereals to be winnowed, still mixed with straw and chaff, on the lee side of the threshing floor, the *alóni*, (αλώνι). Using a spade-shaped wooden fork, the *thrináki* (θρινάκι), with three or four broad tines, they threw the mixture into the wind (Fig. 5.41) that had to be blowing on that day. The light straw and chaff was blown outside of the threshing floor, while the heavier grains collected in a heap at the winnower's feet. This process was repeated until the grains reached an acceptable stage of purity. After winnowing, the chaff was left next to the winnowing floor.

The grain still had a long and stony road to travel before it was ready to be milled and baked. Cleaning the grain was a women's chore in traditional Crete.[13] Sieves were the necessary tools, all handmade with a rim of bent wood with various diameters between 30 and 60 cm, and bottoms of either perforated sheet metal or different sizes of wire mesh. Still on the threshing floor, right after winnowing, the first sieve, called *bolistri* (βολίστρι), was used. It had large perforations and could be used for cereals. The grain passed through the mesh, and the straw which remained in the sieve was added to the chaff that had been removed earlier by winnowing. The same *bolistri* was also used after threshing broad beans, but in this case the beans stayed in the sieve while the chaff and other impurities fell out. Further steps of the grain-cleaning work were done later back at home. First came sieving with a finer sieve called *koskino* (κόσκινο). The grain would stay inside the sieve and smaller stones

Fig. 5.41. Winnowing wheat, Eymür village, Turkey 2010. The winnowing fork is essentially the same in Crete and around the Mediterranean (Photo: J. C. Whittaker).

and unwanted weed seeds would fall out. After this step the grain needed for next year's sowing was set aside. Then it was necessary to get rid of dust. For this the woman needed a low bucket, the *kouvás* (κουβάς) or comparable vessel. Using the *kouvás* in a process similar to winnowing – the grains were thrown lightly from the bucket into the air and caught again, working in the house's courtyard where a soft wind would carry off most of the dust. For more thorough cleaning the grain next needed to be washed. In the days before water was piped to houses (starting in the 1970s) women took the harvest to the village spring or any other water source, even to the sea, if no other possibility was available. There, with the help of a ceramic or wooden basin (*lekani* or *skafi* – λεκάνη/σκάφη) the harvest was washed thoroughly to remove the rest of the dust and fine straw particles that would float off on top of the water. A large water source was needed because the water had to be changed several times. Then the harvest was taken back to the homes and spread to dry in the sun on the roof of the house, which was thoroughly swept or covered with sheets. Drying usually took two days. To test if the grains were dry, the woman would take a few between her teeth to bite down on them. If they were still soft they needed more drying. Finally there would be one more step of finer sieving, so that in the end only the grain and some weed seeds (*e.g.*, *Agrostemma githago* L.: corncockle, Greek *gourdoúlakas* - γκουρντούλακας; *Lolium temulentum* L.: darnel, Greek *níra* – νήρα) of similar size remained. Then before milling (or right after the last sieving) the grain was spread on the table and the foreign weeds were picked out. Corncockle was easily recognisable, nearly pea-sized and black. Darnel was not quite so easily picked out, as it looks like small wheat seeds. Both are toxic, which is why they were painstakingly removed.[14] In this process all the female members of the family (small children, elder daughters and grandmothers) would sit together and help. Now the family's donkey could be loaded with the cleaned cereal in sacks for the trip to the mill.

Mills in Crete were mostly windmills, individually owned although they often appear in long rows on windy hills and ridges (Fig. 5.42). Much rarer (because of the sparseness of running water) were the traditional watermills built on steep slopes. A high chimney-like headrace collected enough water to power the mill. It directed the water to fall through the pipe (a cone wider on top than in the bottom). From the lower end of the pipe, water would spray out on the small horizontal mill-wheel and turn it and its perpendicular shaft (Nomikos 1997, 29). This shaft reached up through the floor of the milling room above and turned the upper of two millstones. Windmills had a slightly more complicated mechanism (Fig. 5.43), as the power had to be transmitted from the horizontally turning sails to the vertically set shaft. The stones for mills were often imported from Melos (Fig. 5.2) or the surrounding islands as their volcanic rock was especially hard and durable. Families preparing to bake bread would take their grain to a mill on

Fig. 5.42. Windmills on the hill. Lakonia, Crete, 2011 (Photo: S. Beckmann).

Fig. 5.43. The interior workings of a Cretan windmill. The vertical cogged wheel transfers the rotation of the vanes to the horizontal grindstone at the bottom (Photo: S. Beckmann).

their donkeys or mules, usually in loads weighing between 80 and 90 kg. The miller would grind it for a fee of two or three percent of the grain. Milling always took place shortly before baking, as otherwise the flour would undergo oxidation. (Whole grain flour from previously degermed grains keeps longer, but is a modern invention.) Family size and needs determined how often bread was made. Cretan bread was a kind of rusk or twice-baked bread called *paximádi* (παξιμάδι) or, in Cretan, *dáko* (ντάκο). This bread is dry and thus can be stored for a long time, an important quality for typical Cretan men engaged as shepherds or sailors. *Dáko* is much lighter to carry than fresh bread and keeps much longer. Before eating, it is slightly soaked in water and then tastes like fresh bread.

As some mills were very busy in regions where few were available, people would sometimes travel three to four hours to find the next mill where they would not have to wait. Regions that always had enough water to run the water-mill independently of wind would also attract people from many surrounding villages for milling. Depending on the kind of grain, flour would be ground in various coarsenesses, with hard wheat ground much finer than softer barley. This also reflected the fact that barley grains were hulled and a large amount of husks would be produced together with the flour. The work of cleaning was not done even after the grain had been ground. In the last step before finally making the bread, the flour was sieved again, with a fine sieve, called a *knisara* (κνισάρα or κνησάρα). In this process *pitira* (πίτυρα) bran (and in the case of barley husks, *skivala*, σκύβαλα) were sieved out to refine the flour. The bran and husks were used as animal fodder. From 50 kg of flour, *ca.* 2–3k g of bran would be sieved out if the flour was to be used for bread – or even more, with another go at an even finer sieve if the flour was to be used for cakes or cookies or the white church bread *artos* (άρτος).

To be ready for mixing and kneading dough, women would take a tiny bouquet of herbs to church to be specially blessed in the middle of the forty-day fast before Easter (Sunday of the Veneration of the Cross). This tiny bouquet would be carefully kept in a safe place until September, when on the day of the Exaltation of the Cross (14th) a sprig of basil would be blessed in church. This was soaked in water and the water, blessed also by the earlier bouquet, would be used for preparing the yeast sponge. After a few hours of waiting until the yeast sponge was ready, women would proceed to the actual baking process, preparing the dough for bread, which required another evening's and early morning's work in which other rituals had to be followed to make sure God's benevolence would secure the outcome (Chapter 1).

The introduction of modern harvester-threshers made a huge difference in Cretan agricultural life. Even then, not all the laborious process of cleaning was eliminated. Although mechanical threshers and mills have now taken over, women still need to sieve their flour depending on the coarseness of the bread they want to bake. When we consider the whole arduous process used in traditional Crete and similar societies, we can see why bread was everywhere esteemed as a precious thing.

5.16. CONCLUSIONS

John C. Whittaker

The preceding discussions of threshing and related agricultural activities have emphasised the tools and technologies. The tools used in any activity are of great interest to the people using them, and in studying the agricultural past, are often the best evidence available for testing hypotheses about ancient agricultural practices. However, we should remember that farmers think more about the activity than the technology, the doing rather than the implement. Implements do provide clues about some of the activities where they were used, because they are made to perform a particular function, and to be used with a distinctive motion. Each tool is also shaped by other tools and techniques used with it, and to make it. For example, the way threshing sledges are made and used varies from area to area, responding to different crops to be threshed, different raw materials available for their manufacture, and different choices about how to pull the sledges. However, they also vary on small, local scales, as individual craftsmen shape them, and individual farmers modify them to suit their uses.

A *tribulum* implies animal traction; oxen are used in some areas, but elsewhere donkeys or horses are preferred, and farmers in any one region may prefer (or can afford) different animals, or a modern tractor. While sledges in some areas are not intended to be ridden, and the animal is led instead, usually the animal or team is guided by a rider. This skill has to be learned: to balance on a moving board while wielding reins and a prod, and keeping an eye on the produce, the animals, and fellow workers. In some areas, a chair is preferred, or an individual farmer like Ismail Çamuşçu (Fig. 5.44) may add a seat to his *düven*, in an area of Turkey where he claimed this was an innovation. In 2010 he was an eccentric in any case, the last *tribulum* user in the region, enough of a novelty that when he gave us a demonstration, other villagers brought their children to ride the *düven* and 'see how we used to do it.'

Ismail's personal choice of implement was informed by many factors. If the crop was really good, it was worth using a neighbour's mechanical thresher. If there was not too much grain, it was more pleasant to use his friendly oxen and familiar *düven*. If the 'American scientists' wanted to see a *düven* in action, and the other villagers could make an event of it, he was pleased to have his family the centre of attention, and his use of a threshing sledge became a source of pride rather than a mark of poverty or backwardness. All technologies are associated with such emotional and symbolic meanings. All agricultural activities do more than process a product. Despite our bias toward the physical objects in this chapter, some of the immaterial aspects show through, and will become more focused in later sections of the book.

Fig. 5.44. Ismail Çamuşçu on his *düven* seat, Eymür village, Turkey, 2010 (Photo: J. C. Whittaker).

Chapter Notes

1 See film clip made by the author showing the trampling of broad bean and oats in Tunisia at https://vimeo.com/77179966.
2 http://vimeo.com/88645869 shows a film clip of threshing wheat with the düven (threshing sledge) in the village of Eymür, Kastamonu district, Turkey.
3 See film clip made by the author showing the process of use of this threshing sledge in Tunisia at https://vimeo.com/77179927.
4 See film clip on the use of another threshing sledge in Tunisia, which briefly shows winnowing, at https://vimeo.com/77179954.
5 Administrative autonomy from Ottoman rule 1899, liberation and union with Greece 1913.
6 Excavations were directed by J. Lichardus, Saarland University, Germany. I would like to thank my colleague Iliya Iliev (Yambol Museum) for his kind permission to publish this series of artefacts.
7 The flint assemblage from this site was prepared for publication quite a while ago, but still is not published.
8 The excavated area of Karanovo V is 14 000 m^2, and of Karanovo VI 10 000 square metres.
9 The results of my study have been prepared for publication and given to the Drama excavation team.
10 The use-wear study (completed but not yet published) was made using the microscope MBS 10 (x100) and Leitz Orthoplan (Carl Zeiss Germany – x200). The photomicrographs were taken by digital camera Olympus C 50-60, with magnification x100.
11 A film on this subject by the author can be viewed at https://vimeo.com/91603982 in French, and in English at http://www.archaeologychannel.org/video-guide/video-guide-menu/video-guide-summary/144-tools-techniques-and-tablets-retracing-ancient-agricultural-heritage. Alternatively, select the film "tools techniques and tablets: retracing ancient agricultural heritage" on The Archaeology Channel.
12 See https://vimeo.com/91603979 for a film clip by the author showing these experiments.
13 I am thankful to Kostis Bardas from Kroustas and Maria and Jorgos Brokos from Kritsa for their patience in explaining all the details to me. Note that the Greek terms are Cretan Greek and not necessarily similar to other Greek terms for the same items and plants.
14 Interestingly both these plants, that had migrated from southern to northern Europe a long time ago together with the cereals, are now, with modern agricultural techniques, close to extinction in central Europe.

Chapter Bibliography

Allbaugh, L. G., (1953). *Crete, a case study of an underdeveloped area.* Princeton University Press, Princeton.

Amouretti, M.-C. (1986). *Le pain et l'huile dans la Grèce antique. De l'araire au moulin.* Belles Lettres, Paris.

Anderson, P. C. (1994a) Interpreting traces of Near Eastern Neolithic craft activities: an ancestor of the threshing sledge for processing domestic crops? In A. L. van Gijn (ed.) *Enigmatic Wear-Traces on Stone Implements: Evidence for Handicraft in Prehistory.* Helinium 34(2), 281–293.

Anderson, P. C. (1994b) Reflections on the significance of two typological classes in the light of experimentation and microwear analysis: flint sickles and obsidian 'Cayönü tools'. In H. G. Gebel, and S. K. Koslowski (eds.) *Neolithic Chipped Stone Industries of the Fertile Crescent: Proceedings of the First Workshop on PPN Chipped Lithic Industries.* Ex Oriente, Berlin, pp. 61–82.

Anderson, P. C. (1998) The history of harvesting and threshing techniques for cereals in the prehistoric Near East. In A. B. Damania, J. Valkoun, G. Willcox and C. O. Qualset (eds.) *The Origins of Agriculture and Crop Domestication.* University of California, Genetic Resources Conservation Program, Report 21. ICARDA, Aleppo, Syria, pp. 141–155.

Anderson, P. C. (1999a) Experimental cultivation, harvest and threshing of wild cereals. In P. C. Anderson (ed.) *Prehistory of Agriculture: New Experimental and Ethnographic Approaches.* Institute of Archaeology Monograph 40. University of California, Los Angeles, pp. 104–118.

Anderson, P. C. (1999b) (ed.) *Prehistory of Agriculture, New Experimental and Ethnographic Approaches.* Institute of Archaeology Monograph 40. University of California, Los Angeles.

Anderson, P. C. (2000) *Tools, Techniques and Tablets:*

Retracing Ancient Agricultural Heritage. 26 minute film, with the participation of J. Chabot, J.-P. Grégoire and J. Pélegrin. VSP Production, Sophia Antipolis. Co-production of the EARTH project and the CEPAM, CNRS, Valbonne, France, and the CELAT, Quebec, Canada.

Anderson, P. C. (2003) Observations on the threshing sledge and its products in ancient and present-day Mesopotamia. In P. C. Anderson, L. S. Cummings, T. K. Schippers and B. Simonel (eds.) *Le traitement des récoltes: un regard sur la diversité du Néolithique au present*. XXIII Rencontres internationales d'archéologie et d'histoire d'Antibes. Editions APDCA, Antibes, pp. 417–438.

Anderson, P. C. (2006a) Premiers Tribulums, Premières Tractions Animales au Proche-Orient vers 8000-7500 BP? In P. Pétrequin, R.-M. Arbogast, A.-M. Pétrequin, S. van Willigen and M. Bailly (eds.) *Premiers chariots, premières araires. La diffusion de la traction animale en Europe pendant les IVe et IIIe millénaires avant notre ère*. CNRS Editions, Monographies du CRA 29, pp. 299-316.

Anderson, P. C. (2006b) Non-mechanised processing and storage of cereals, grasses and pulses used for fodder, fuel, food and crafts: Examples from N. Tunisia, Atlas Region, Northwestern Tell, In F. Ertuğ (ed.) *Proceedings of the IVth International Congress of Ethnobotany (ICEB 2005)*, Ege Yaynlar, pp. 223-231.

Anderson, P. C. and Chabot, J. (2004). Les lames cananéennes et la première machine agricole. *Les dossiers de l'archéologie* 290, 44–51.

Anderson, P. C. and Chabot, J. (2001) Functional analysis of glossed blades from northern Mesopotamia in the Early Bronze Age (3000–2500 BC): the case of Tell 'Atij. Recherches Archéométriques. *Cahiers Archéologiques du CELAT*, Université Laval, Québec, pp. 257–276.

Anderson, P. C., Chabot, J. and Van Gijn, A. L. (2004). The functional riddle of 'glossy' Canaanean blades and the Near Eastern threshing sledge. *Journal of Mediterranean Archaeology* 17, 87–130.

Anderson, P. C. and Inizan, M.-L. (1994) Utilisation du tribulum au début du IIIe millénaire: Des lames « cananéennes » lustrées a Kutan (Ninive V) dans la région de Mossoul, Iraq. *Paléorient* 20, 85–103.

Anderson, P. C., Georges, J.-M., Vargiolu, R. and Zahouani, H. (2006). Insight from a tribological analysis of the *tribulum*. *Journal of Archaeological Science* 33: 1559–1568.

Anderson, S. and Ertuğ-Yaras, F. (1998) Fuel, fodder and faeces: an ethnographic and botanical study of dung fuel use in Central Anatolia. *Environmental Archaeology* 1: 99–109.

Ataman, K. (1999) Threshing sledges and archaeology. In P. C. Anderson (ed.) *Prehistory of Agriculture: New Experimental and Ethnographic Approaches*. Institute of Archaeology Monograph 40. University of California, Los Angeles, pp. 211–222.

Avner, U., Anderson, P. C., Bui-thi-Mai, Chabot, J. and Cummings, L. S. (2003) Ancient threshing floors, threshing tools and plant remains in Uvda Valley, southern Negev desert, Israel. A preliminary report. In P. C. Anderson, L. S. Cummings, T. K. Schippers and B. Simonel (eds.) *Le traitement des récoltes: un regard sur la diversité du Néolithique au present*. XXIII Rencontres internationales d'archéologie et d'histoire d'Antibes. Editions APDCA, Antibes, pp. 455–476.

Barkai, R. and Yerkes, R. W. (2008) Stone axes as cultural markers: technological, functional and symbolic changes in bifacial tools during the transition from hunter-gatherers to sedentary agriculturalists in the Southern Levant. In L. Longo and N. Skakun (eds.) *'Prehistoric Technology' 40 Years Later: Functional Studies and the Russian Legacy*. BAR International Series 1783, Oxford, pp. 159–167.

Bordaz, J. (1965) The threshing sledge. *Natural History* 74 (4), 26–29.

Bordaz, J. (1969) Flint flaking in Turkey. *Natural History* 78(2), 73–79.

Boyadžiev, Y. (1995). Chronology of prehistoric cultures in Bulgaria. In D. Bailey and I. Panayotov (eds.) *Prehistoric Bulgaria*. Monographs in World Archaeology 22. Madison, pp. 149–191.

Bryer, A. (1986). Byzantine agricultural implements: the evidence of mediaeval illustrations of Hesiod's 'Works and Days'. *Annual of the British School at Athens* 8, 45–85.

Butterlin, P. (1998) Espaces urukéens en Syrie: problèmes de cartographie et de méthodologie. In M. Fortin and O. Aurenche (eds.) *Espace naturel, espace habité en Syrie du nord (10e-2e millénaires av. J.-C.)*. Bulletin 33, Travaux de la Maison de l'Orient 28. Canadian Society for Mesopotamian Studies, Québec and Lyon, pp. 149–166.

Caro Baroja, J. (1983) *Tecnologia popular española*, Editorial Nacional, Madrid.

Castel Carpinschi, C. (2003) Le choix de techniques d'egrenage 'pre-machinistes' en France (1786–1950). In P. C. Anderson, L. S. Cummings, T. K. Schippers and B. Simonel (eds.) *Le traitement des récoltes: un regard sur la diversité du Néolithique au present*. XXIII Rencontres internationales d'archéologie et d'histoire d'Antibes. Editions APDCA, Antibes, pp. 327–346.

Cauvin, J. (1962) Les industries lithiques de Tell Byblos (Liban). *L'Anthropologie* 66, 88–502.

Chabot, J. (1998) Analyse spatiale et stratigraphique des artefacts de pierre sur éclats de tell 'Atij-centre. In M. Fortin and O. Aurenche (eds.) *Espace naturel, espace habité en Syrie du nord (10e-2e millénaires av. J.-C.)* Bulletin 33, Travaux de la Maison de l'Orient 28, 257–69. Canadian Society for Mesopotamian Studies, Québec and Lyon.

Chabot, J. (1999) *Deux chaînes opératoires d'artefacts en pierre taillée sur un site de Mésopotamie septentrionale daté de l'Age du Bronze Ancien (Tell 'Atij)*. Philipp von Zabern, Mainz.

Chabot, J. (2002) *Tell' Atij et Tell Gudeda: industrie lithique.* Cahiers d'archéologie de CELAT 13, Série archéométrie 3. Quebec, Université Laval.

Chabot, J. and Eid, P. (2003) Le phénomène des lames « cananéennes » : état de la question en Mésopotamie du Nord et au Levant sud. In P. C. Anderson, L. S. Cummings, T. K. Schippers and B. Simonel (eds.) *Le traitement des récoltes: un regard sur la diversité du Néolithique au present.* XXIII Rencontres internationales d'archéologie et d'histoire d'Antibes. Editions APDCA, Antibes, pp. 401–416.

Chabot, J., Eid, P. and Anderson, P. C. (2007) Deux nouveaux indices de production agricole à Tell 'Acharneh, Syrie (région du Ghab). *Subartu* 18, 239–257.

Civil, M. (1994) *Farmer's Instructions: A Sumerian Agricultural Manual.* Aula Orientalis, Supplementa 5. Ausa, Barcelona.

Comet, G. (1992) *Le paysan et son outil, essai d'histoire technique de céréales.* Collection de l'École française de Rome 165, 1–711.

Crowfoot-Payne, J. (1960) Flint implements from Tell al-Judaidah. In R. J. Braidwood and L. S. Braidwood (eds.) *Excavations in the Plains of Antioch.* Oriental Institute Publications 61, 525–39. University of Chicago Press, Chicago.

Cummings, L. S. (2003) Tracing cereal grain threshing through time and space using cut phytoliths. In P. C. Anderson, L. S. Cummings, T. K. Schippers and B. Simonel (eds.) *Le traitement des récoltes: un regard sur la diversité du Néolithique au present.* XXIII Rencontres internationales d'archéologie et d'histoire d'Antibes. Editions APDCA, Antibes, pp. 507–514.

Darwish, I. R. (1986) *Threshing Sledge Use and Variablity in Palestine: an Ethnographic Classification and Use Wear Analysis.* Unpublished MA, University of Texas, Austin.

Diamond, G. P. (1975). A study of microscopic wear patterns on the chipped stone artefacts from the Neolithic and early Bronze Age levels of Knossos. Ph.D. Thesis, University of London.

Dispan, P. (1804) Notice sur une machine à battre les grains, publiée, sur les résultats de l'expérience, par la Société d'agriculture du département de la Haute-Garonne. Machines instruments aratoires, pp. 1–8.

Duplan (1835) Battage des blés. *Journal des propriétaires ruraux pour le Midi de la France.* Tome 31, 129–141.

Erman, A., and Ranke, H. (1980) *LaCivilisation Egyptienne.* Payot, Paris.

Fortin, M. (1997) Urbanisation et redistribution de surplus agricoles en Mésopotamie Septentrionale (3000–2500 av. J.-C.). In W. Aufrecht (ed.) *The Origins and Development of Urbanism in the Ancient Near East.* Journal for the Study of the Old Testament, Supplement Series 244, 50–81. Sheffield Academic Press, Sheffield.

Fortin, M. (1998) L'habitat de la station commerciale de Tell 'Atij, sur le moyen Khabour, au IIIe millénaire av. J.-C. In M. Fortin and O. Aurenche (eds.) *Espace naturel, espace habité en Syrie du nord (10e-2e millénaires av. J.-C.).* Bulletin 33, Travaux de la Maison de l'Orient 28, 229–242. Canadian Society for Mesopotamian Studies, Quebec.

Fox, W. A. (1984) *Dhoukani* flake blade production in Cyprus. *Lithic Technology* 13(2), 62–68.

Fujii, S. (1986) Criteria for the identification of threshing sledge blades. *Bulletin of the Okayama Orient Museum* 5, 1–34.

Gelb, I. J. (1955) Old Akkadian inscriptions in Chicago Natural History Museum. Texts of legal and business interest. *Fieldiana Anthropology* 44(2), 270–77.

Van Gijn, A. L. (1994) Introduction: there is more to life than butchering and harvesting. In A. L. van Gijn (ed.) *Enigmatic Wear Traces on Stone Implements: Evidence for Handicraft in Prehistory.* Helinium 34. Universa, Wetteren, pp. 176–185.

Van Gijn, A. L. (1999) The interpretation of sickles: a cautionary tale. In P. C. Anderson (ed.) *Prehistory of Agriculture: New Experimental and Ethnographic Approaches.* Institute of Archaeology Monograph 40. University of California, Los Angeles, pp. 254–259.

Grégoire J.-P. (2010) Le Système Après-Récolte dans l'Hydro-Agriculture Mésopotamienne à la Fin du IIIe Millénaireavant notre Ere, in *Actes de la Rencontre Assyriologique Internationale de Barcelone*, 56. (in press).

Gurova, M. (2001) Eléments de tribulum de la Bulgarie - références ethnographiques et contexte préhistorique. *Archaeologia Bulgarica* 5(1), 1–19.

Gurova, M. (2002) Feursteininventar aus Sondage O 19 in Tell Karanovo: typologische und funktionale Analyse. In S. Hiller and V. Nikolov (eds.) *Karanovo, II, De Ausgrabungen in O19.* Phoibos Verlag, Wien, pp. 149–175.

Gurova, M. (2006) Kremachen ansambal ot iamno svetilishte ot zheliaznata epoha i selishte ot rannata bronzova epoha pri Svilengrad. In V. Nikolov, G. Nehrizov and I. Tsvetkova (eds.) *Arheologicheski razkopki po traseto na zhelezopatnata linija Plovdiv-Svilengrad prez 2004.* V. Tarnovo, pp. 512–517 (in Bulgarian).

Gurova, M. (2007) N. N. Skakun. Orudia truda i hoziaistvo drevnezemledelcheskih plemen Iugo-vostochnoi Evropy v epohu eneolita. St.-Petersburg, 2006, Nestor-Istoria (in Russian). - Археология, 2007, 1–4, 207–209 (Book review).

Gurova, M. (2008) Kremachni artefakti ot Svilengrad. In V. Nikolov, G. Nehrizov and I. Tsvetkova (eds.) *Spasitelni arheologicheski razkopki po traseto na zhelezopatnata linija Plovdiv-Svilengrad prez 2005.* V. Tarnovo 2008, 527–533 (in Bulgarian).

Gurova, M. (2011) Etnografski i arheologicheski dikani: transregionalna perspektiva [Ethnographic and archaeological *tribula*: trans-regional perspective]. *Bulgarian e-Journal of Archaeology* 1 (1): 1–39. http://be-ja.org

Gurova, M. (2013) *Tribulum* inserts in ethnographic and

archaeological perspective: case studies from Bulgaria and Israel. *Lithic Technology* 38 (3): 179–201.

Halstead, P. and Jones, G. (1989) Agrarian ecology in the Greek islands: Time stress, scale and risk. *Journal of Hellenic Studies* 109, 41–55.

Hartmann, F. (1923) *L'Agriculture dansl'Ancienne Egypte.* Librairies-Imprimeries Réunies, Paris.

Herculano de Carvalho, Jose Gonçalo C. (1953) Coisas e palavras. Alguns problemas etnográficos e linguísticos relacionados com primitivos sistemas de debulha na Península Ibérica, *Biblos* XXIX, 1–365.

Hitier, M. (1853) Appendice sur l'etat de l'agriculture et les productions de l'ile de Crete. In L. Lacroix (ed.) *Les Isles de la Grèce.* Didot, Paris, pp. 589–610.

Jacque, M. (1835) Battage du colza. *L'Illustration* 1015 (9 August 1862), 88.

Juel Jensen, H. (1994) *Flint Tools and Plant Working: Hidden Traces of Stone Age Technology.* Aarhus University Press, Aarhus.

Kardulias, N. and Yerkes, R. (1996) Microwear and metric analysis of threshing sledge flints from Greece and Cyprus. *Journal of Archaeological Science* 23, 657–666.

Karimali, E. (1994) The Neolithic Mode of Production Reconsidered: Lithic Production and Exchange Patterns in Thessaly (LN–FN periods). Unpublished PhD dissertation, Boston University.

Khedhaier, R., Verdin, P., Furestier, R., Lemercier, O. and Müller, A. (2003) Dépiquage au tribulum au Néolithique final dans le Sud-Est de la France. Indices convergents de la tracéologie et de l'analyse des phytolithes. Le cas du site de Forcalquier-La Fare (Alpes de Haute-Provence). In P. C. Anderson, L. S. Cummings, T. K. Schippers and B. Simonel (eds.) *Le traitement des récoltes: un regard sur la diversité du Néolithique au present.* XXIII Rencontres internationales d'archéologie et d'histoire d'Antibes. Editions APDCA, Antibes, pp. 477–493.

Kutschera, W., Steier, P., Wild, E. M. and Hiller, S. (2005) Radiokarbondaten. In S. Hiller and V. Nikolov (eds.) *Karanovo, IV.1, Die Ausgrabungen im Nordsüd-Schnitt, 1993-1999.* Phoibos Verlag, Wien, pp. 421–426.

Leroy C. (2003) Les aires de battages dans la moyenne vallée du Verdon (Alpes-de-Haute-Provence). In P. C. Anderson, L. S. Cummings, T. K. Schippers and B. Simonel (eds.) *Le traitement des récoltes: un regard sur la diversité du Néolithique au present.* XXIII Rencontres internationales d'archéologie et d'histoire d'Antibes. Editions APDCA, Antibes, pp. 363–374.

Lethielleux, J. (1947) La moisson au Fezzan. *Institut des Belles Lettres Arabes (IBLA)* 39 (3), 243–269.

Lichardus, J., Fol, A., Getov, L., Bertemes, F., Echt, R., Katinčarov, R. and Iliev, I. (2001) *Izsledvanija v mikroregiona na s. Drama (Jugoiztochna Balgarija). Obobshtenie na osnovnite rezultati na Balgaro-germanskite razkopki ot 1983 do 999 g.* Universitetskf Izdatelstvo 'Sv. Kliment Ohridski', Sofia (in Bulgarian).

Lichardus, J., Iliev, I. and Christov, C. (2002) Die Karanovo I-IV-Perioden an der unteren Tundža und ihre chronologische Stellung zu den benachbarten Gebieten. In M. Lichardusitten, J. Lichardus and V. Nikolov (eds.) *Beiträge zu jungsteinzeitlichen Forschungen in Bulgarien.* Saarbrücker. Beitr. Altkde. 74. Dr. Rudolf Habelt GmbH, Bonn, pp. 325–410.

Littauer, M. A. and J. H. Crouwel (1990) Ceremonial threshing in the ancient Near East. *Iraq* 52, 19–23.

Llaty, C. (1997) Le fléau, la planche et le mulet, l'égrenage des céréales dans les Alpes du Sud. *Le Monde alpin et rhodanien* 25(1), 83–100.

Lohmann, H. (1992) Agriculture and country life in classical Attica. In B. Wells (ed.) *Agriculture in Ancient Greece.* Paul Astroms Forlag, Stockholm, pp. 29–60.

Luquet, G.-H. and Rivet, P. (1933). Sur le tribulum. *Mélanges offerts à Nicolas Iorga*, Librairie J. Gamber, Paris, pp. 613–638.

McCorriston, J. (1998) Landscape and human-environment interaction in the Middle Khabur drainage from the Neolithic period to the Bronze. In M. Fortin and O. Aurenche (eds.) *Espace naturel, espace habité en Syrie du nord (10e-2e millénaires av. J.-C.).* Bulletin 33, Travaux de la Maison de l'Orient 28, 43–53. Canadian Society for Mesopotamian Studies, Quebec.

Mingote Calderón, J. L. (1988) Mayales y trillos en España (1), *Anales del Museo del Pueblo Español* 2, 83–167.

Mingote Calderón, J. L. (1990a) Mayales y trillos en España (2), *Anales del Museo del Pueblo Español* 3, 7–72.

Mingote Calderón, J. L. (1990b) *Mayales y trillos en la provincia de León.* Diputación Provincial de León, León.

Mingote Calderón, J. L. (2003) Travail agricole et savoir-faire paysan. Quelques aspects de l'égrenage des céréales dans la péninsule Ibérique. In P. C. Anderson, L. S. Cummings, T. K. Schippers and B. Simonel (eds.) *Le traitement des récoltes: un regard sur la diversité du Néolithique au present.* XXIII Rencontres internationales d'archéologie et d'histoire d'Antibes. Editions APDCA, Antibes.

Murray, P. and Kardulias, P. N. (1986) A modern-site survey in the southern Argolid, Greece. *Journal of Field Archaeology* 13(1): 21–42.

Neuville, R. (1930) Notes de préhistoire palestinienne. I. La grotte d'El-Touamin. *Journal of the Palestine Oriental School*, 10, 64–75.

Nomikos, S. (1997) Water *Power in Pre-Industrial Greece.* ETBA cultural foundation, Athens.

Ó Danachair, C. (1970) The Flail in Ireland, *Ethnologia Europea* IV, 50–55.

Oates, D. (1990) Innovations in mud-brick: decorative and structural techniques in ancient Mesopotamia. *World Archaeology* 21, 388–406.

Palmer, C. (1998) 'Following the plough': The agricultural environment of northern Jordan. *Levant* 30, 129–165.

Parain, C. (1937/1979) 'Les anciens procédés de battage et de dépiquage en France', reprinted in: Parain, C.

(ed.) *Outils, ethnies et développement historique*, Éditions Sociales, Paris.

Parain, C. (1938) Les anciens procédés de battage et de dépiquage en France. *Folklore Paysan* 1(6), 89–91.

Parain, C. (1979) Les anciens procédés de battage et de dépiquage en France, *Outils, ethnies et developpement historique*. Éditions Sociales, Paris, pp. 17–27.

Parry, D. W. and Smithson, F. (1964) Types of opaline silica depositions in the leaves of British grasses. *Annals of Botany* 28, 169–185.

Pearlman, D. A. (1984) *Threshing Sledges in the East Mediterranean: Ethnoarchaeology with Chert Knappers and Dhoukanes in Cyprus.* Unpublished M.A. thesis, University of Minnesota, Minneapolis.

Pélegrin, J. (2002) Principes de la reconnaissance des méthodes et techniques de taille: annexe A. In J. Chabot (ed.) *Tell 'Atij et Tell Gudeda: Les artefacts en pierre taillée.* Cahiers d'archéologie du CELAT 13, série Archéométrie 3. CELAT, Quebec, pp. 215–26.

Peña-Chocarro, L. (1999). *Prehistoric agriculture in southern Spain during the Neolithic and the Bronze Age; the application of ethnographic models.* BAR International Series 818. Archaeopress, Oxford.

Philip, G. (2001) The Early Bronze I–III ages. In B. MacDonald, R. Adams and P. Bienkowski (eds.) *The Archaeology of Jordan.* Levantine Archaeology 1, 163–232. Sheffield Academic Press, Sheffield.

Philip, G. (2003) The Early Bronze Age of the southern Levant: a landscape approach. *Journal of Mediterranean Archaeology* 16, 103–32.

Portela, J. (1986) *Trabalho cooperativo em duas aldeias de Tras-os-Montes.* Edições Afrontamento, Porto.

Procopiou, H. (2003) Les techniques de décorticage dans le monde égéen: étude ethnoarchéologique dans les Cyclades. In P. C. Anderson, L. S. Cummings, T. K. Schippers and B. Simonel (eds.) *Le traitement des récoltes: un regard sur la diversité du Néolithique au present.* XXIII Rencontres internationales d'archéologie et d'histoire d'Antibes. Editions APDCA, Antibes, pp. 115–136.

Puymaurin, M. de (1809) Mémoire sure un nouveau rouleau à battre le blé. Annales de l'agriculture française 38, 64–80. Librairie de Mme Huzard, Paris.

Raichevski C. (1996) Zemedelie. V: Strandza. Materialna i duhovna kultura. BAN, Sofia (in Bulgarian).

Rosen, S. (1996) The decline and fall of flint. In G. H. Odell (ed.) *Theoretical Insights into Human Prehistory.* Plenum, New York, pp. 129–158.

Rosen S. (1997) *Lithics after the Stone Age: A Handbook of Stone Tools from the Levant.* Alta Mira Press, Walnut Creek.

Runnels, Curtis N. (1982) Flaked-stone Artifacts in Greece During the Historical Period. *Journal of Field Archaeology* 9(3), 363–374.

Schippers, T. K. (2003) 'Quand la paille se mêle au grain', note sur la complexité des chrono-logiques à l'œuvre dans des choix techniques dans les Alpes du Sud. In P. C. Anderson, L. S. Cummings, T. K. Schippers and B. Simonel (eds.) *Le traitement des récoltes: un regard sur la diversité du Néolithique au present.* XXIII Rencontres internationales d'archéologie et d'histoire d'Antibes. Editions APDCA, Antibes, pp. 357–361.

Schwartz, G. M. (1987) The Ninivite V Period and the development of complex society in northern Mesopotamia. *Paléorient* 13(2), 93–99.

Schwartz, G. M. and Klukas, E. E. (1998) Spatial analysis and social structure at Tell al Raqa'i. In M. Fortin and O. Aurenche (eds.) *Espace naturel, espace habité en Syrie du Nord (10e-2e millénaires av. J.-C.).* Bulletin 33 and Travaux de la Maison de l'Orient 28, 199–208. Canadian Society for Mesopotamian Studies, Quebec.

Sieber F. W. (1823) *Reise nach der Insel Kreta im griechischen Archipelagus im Jahre 1817.* Fleischer, Leipzig.

Sigaut, F. (1989) Storage and threshing in preindustrial Europe: additional notes. *Tools and Tillage* 6(2), 119–124.

Skakun, N. (1992) Evolution des techniques agricoles in Bulgarie chalcolithique (d'après les analyses tracéologiques). In P. C. Anderson (ed.) *Préhistoire de L'Agriculture: Nouvelles Approches Expérimentales et Ethnographiques.* Centre de Recherches Archéologiques Monograph 6. Centre National de la Recherche Scientifique, Paris, pp. 289–303.

Skakun, N. (1993a) Results of traseological examination of flint implements from Neolithic settlements in Western Bulgaria. In I. Gatsov (ed.) *Neolithic Chipped Stone Industries in Western Bulgaria.* Jagellonian University Press, Kraków, pp. 52–54.

Skakun, N. (1993b) Agricultural implements in the Neolithic and Eneolithic cultures of Bulgaria. In P. Anderson, S. Beyries, M. Otte, and H. Plisson (eds.) *Traces et fonctions: les gestes retrouvés.* Etudes et Recherches Archéologiques de l'Université de Liège 50(2). Liège, pp. 361–368.

Skakun, N. (1994) Agricultural implements and the problem of spreading of agriculture in Southeastern Europe. *Helinium* 34(2), 294–305.

Skakun, N. (1999) Evolution of agricultural techniques in Eneolithic (Chalkolithic) Bulgaria: Data from use-wear analysis. In P. C. Anderson (ed.) *Prehistory of Agriculture: New Experimental and Ethnographic Approaches.* Institute of Archaeology Monograph 40. University of California, Los Angeles, pp. 199–210.

Skakun, N. (2006) *Orudia truda i hoziaistvo drevnezemledel-cheskih plemen Iugo-vostochnoi Evropy v epohu eneolita.* Nestor-Istoria, St.-Petersburg. (in Russian).

Skertchly, S. B. J. (1879) *On the Manufacture of Gun-Fints, the Methods of Excavating for Flint, the Age of Palaeolithic Man, and the Connexion Between Neolithic Art and the Gun-Flint Trade.* Memoirs of the Geological Society of England and Wales, London. Reprinted 1984, Museum Restoration Service, Bloomfield, Ontario.

Stricker, B. H. (1955) The Origin of the Greek Theatre. *Journal of Egyptian Archaeology* 41, 34–47.

Trevor-Battye, A. (1919) Crete: Its Scenery and Natural Features. *The Geographical Journal*, 54(3), 137–153.

Trotzig, D. (1943) *Slagan och andra tröskredskap. En etnologisk undersökning med utängspunkt frän svenskt material.* Nosdiska Museets, Stockholm.

Ure, A. D. (1955) Threshing floor or vineyard? *Classical Quarterly* 5, 225–230.

Vakarelski, H. (1974) *Etnographija na Balgarija*. BAN, Sofia. (in Bulgarian).

Vargiolu, R., Zahouani, H. and Anderson, P. C. (2003) Etude tribologique du processus d'usure des lames de silex et fonctionnement du tribulum. In P. C. Anderson, L. S. Cummings, T. K. Schippers and B. Simonel (eds.) *Le traitement des récoltes: un regard sur la diversité du Néolithique au present.* XXIII Rencontres internationales d'archéologie et d'histoire d'Antibes. Editions APDCA, Antibes, pp. 439–454.

Veiga de Oliveira, E., Galhano, F. and Pereira, B. (1983) *Alfaia agrícola portuguesa.* Instituto Nacional de Investigação Científica. Centro de Estudos de Etnologia, Lisboa.

Veiga de Oliveira, E., Galhano, F. and Pereira, B. (1995) *Alfaia Agricola Portuguesa.* Dom Quixote, Lisbon.

Wagstaff, M. and Auguston, S. (1982) Traditional land use. In C. Renfrew and M. Wagstaff (eds.) *An Island Polity: The Archaeology of Exploitation in Melos.* Cambridge University Press, Cambridge, pp. 106–133.

Warren, P. M. (1979) The Unfinished Red Marble Jar at Akrotiri. In C. Doumas (ed.) *Thera and the Aegean World I.* Papers presented at the Second International Scientific Congress, Santorini, Greece, August 1978. The Thera Foundation, London, pp. 555–568.

Watrous, L. V., Hadzi-Vallianou, D. and Blitzer, H. (2004) *The Plain of Phaistos: Cycles of Social Complexity in the Mesara Region of Crete.* Monumenta Archaeologica 23. Cotsen Institute of Archaeology, Los Angeles.

Whallon, R.-J. (1978) Threshing sledge flints: a distinctive pattern of wear. *Paléorient* 4, 319–324.

White, K. D. (1970) *Roman Farming.* Cornell University Press, Ithaca.

Whitelaw, T. M. (1991) The ethnography of recent rural settlement and land use in northwest Keos. In J. F. Cherry, J. L. Davis and E. Mantzourani (eds.) *Landscape Archaeology as Long-Term History: Northern Keos in the Cycladic Islands from Earliest Settlement Until Modern Times.*UCLA Institute of Archaeology, Los Angeles, pp. 403–454.

Whittaker, J. C. (1996) Athkiajas: A Cypriot flintknapper and the threshing sledge industry. *Lithic Technology* 21(2), 108–120.

Whittaker, J. C. (1999) *Alonia*: the ethnoarchaeology of Cypriot threshing floors. *Journal of Mediterranean Archaeology* 12, 7–25.

Whittaker, J. C. (2000) *Alonia* and *dhoukanes*: The ethnoarchaeology of threshing in Cyprus. *Near Eastern Archaeology* 63(2), 62–69.

Whittaker, J. C. (2001) 'The Oldest British Industry:' Continuity and Obsolescence in a Flintknapper's Sample Set. *Antiquity* 75(288), 382–390.

Whittaker, J. C. (2003) Threshing sledges and threshing floors in Cyprus. In P. C. Anderson, L. S. Cummings, T. K. Schippers and B. Simonel (eds.) *Le traitement des récoltes: un regard sur la diversité du Néolithique au present.* XXIII Rencontres internationales d'archéologie et d'histoire d'Antibes. Editions APDCA, Antibes, pp. 375–387.

Whittaker, J. C., Kamp, K. and Yılmaz, E. (2009) Çakmak Revisited: Turkish Flintknappers Today. *Lithic Technology* 34(2), 93–110.

Woolley, L. (1956) *Ur Excavations: The Early Periods.* 4. University of Pennsylvania, Philadelphia, and the British Museum, London

Yerkes, R. W. (2000) Ethnoarchaeology in Central Cyprus: Interdisciplinary Studies of Ancient Population and Agriculture by the Athienou Archaeological Project. *Near Eastern Archaeology* 63, 20–34.

Yerkes, R. W., Barkai, R., Gopher, A. and O. Bar-Yosef (2003) Microwear Analysis of Early Neolithic (PPNA) Axes and Bifacial Tools from Netiv Hagdud in the Jordan Valley, Israel. *Journal of Archaeological Science* 30(8), 1051–1066.

Yerkes, R. and Kardulias, P. N. (1994) Microwear analysis of threshing sledge flints from Cyprus and Greece: Implications for the study of ancient agriculture. *Helinium* 34(2), 281–293.

Young, J. H. (1956) Studies in South Attica: Country estates at Sounion. *Hesperia* 25, 122–146.

6 Storage and Preservation

6.1. INTRODUCTION

Marie Russel, Pascal Verdin and François Sigaut (†)

Introduction

Storage aims to regulate the available quantity of food and is especially relevant in those situations where large amounts of food become available during a short period of the year, being unavailable the rest of the time. In areas with a distinct seasonality, storage is especially relevant and has been practised for millennia, long before the introduction of crop farming (Halstead 1981; Ingold 1983; Testart 1982). It is practised both on animal and plant foods. Meat and fish are dried, smoked, or mixed with other foodstuffs like in pemmican among the American Indians. Sometimes they are buried in caches, as is the case among the Inuit (Binford 1978). Wild plants which abound for a short time in the year, such as berries, are commonly dried and stored for future consumption. Because domesticated plants like cereals are also harvested seasonally, these too need to be stored (see contributions by Anderson, Smerdel, Kadim and Peña-Chocarro and Zapata in the following chapter). Most of the contributions in this chapter address the various ways in which domesticated plants can be stored, resulting in a great variety of structures, containers and other features across the world.

Storage is often the only guarantee of survival under severe conditions, such as poor harvests or wars (Kunz 2004). A secondary use of storage is to strengthen the established power of landlords or elites, sometimes using speculation or food reserves as a weapon or a means of pressure (Kunz 2004). In any case, storage is important at all levels of rural and urban societies, with purposes ranging from simple survival to power politics.

Storage can also be perilous, as stored crops may be lost to pests or climatic conditions. Nowadays, storage is the main cause for post-harvest losses in southern countries and represents, for instance, 13.5% of the total value of cereal production in sub-Saharan Africa (World Bank 2011). Studies of storage structures and techniques are still important as a contribution for achieving global food security (Redlingshöfer and Soyeux 2013), and are a temptation to raiders. The main point indeed is probably that without proper storage techniques, seasonal resources cannot be fully used. The consequences of this 'principle' are obvious, especially for sedentary societies, with purposes ranging from simple survival to power politics.

Storage is closely linked to preservation, various processes that transform a perishable food product into a more lasting one, such as the making of jams from strawberries. Often the harvest of large amounts of a specific resource is accompanied by such preservation measures. For example, during the Mesolithic in northwestern Europe, hazelnuts were available in profusion in the autumn. In order to preserve these for consumption at a later time, they were roasted in large roasting pits, found in huge quantities in settlement areas (Groenendijk 1987).

One more recent example of the close relationship between storage and transformation and the social implications thereof is wine- versus beer-making: beer can be made and drunk quickly, while wine requires storage to mature. In the making of most beers, the grain and the ferment (starter) are stored separately, so that beer-making can take place at any time of the year. The housewife can just go to the granary to fetch the necessary quantity of grain, large or small, according to the number of drinkers involved. It will take her two or three days to make the beer, which is not expected to keep and will be drunk as soon as it is ready. So, beer-making is a domestic activity, which as such is often in the hands of women. There were exceptions, especially in the palaces of kings, for example. But as a matter of course, beer-making was and remained women's work everywhere in the world until quite recently, except in continental Europe since medieval times, where beer-making developed into the men's craft which seems so natural to us today. In contrast, wine-making is a seasonal activity. Grapes are not stored (except by drying, which is another matter). The entire harvest of grapes has to be made into wine at once, and it is the wine that is stored, to be drunk in the course of the following year, not the grapes. This example shows how complex and how diverse the motivations for storage and its consequences on social organizations and developmental trajectories can be. It will not be developed further in this chapter, but research on this theme is just beginning (Jennings *et al.* 2005).

Many researchers from various backgrounds, using a wide range of methods, have addressed the question of storage and preservation. Agronomists, economists and biologists were the first and predominant scientists interested in storage, until the 1950s (Muntz 1881; Reneaume 1708; Serres 1600). Then archaeologists (Avrilleau 1975; Blanchet 1923; Mauny and Cordier 1967), ethnographers and historians (Gast and Sigaut 1979; Kunz 1951) became involved. Experimental archaeology (Currid and Navon 1989; Hill *et al.* 1983; Reynolds 1979), as well as laboratory analysis and physical measurements developed in the second half of the 20th century (Adrian *et al.* 1979; Reynolds 1974).

Since the 1980s, storage structures have been widely documented in archaeology (Bouby 2003; Buxó *et al.* 2000; Derreumaux *et al.* 2003; Gransar 2003), focusing on crop choice, harvesting techniques and on the study of silo location in the settlements. Interdisciplinary approaches are increasingly applied to this subject. Jacomet conducted an experiment on *in situ* carbonised wooden barrels at the bottom of which were pomegranates, carbonised and split into small pieces (Jacomet *et al.* 2002). In collaboration with two scientific laboratories (one specialised in model calculations, and the other one in engineering), they arrived at a model as to how the process of burning and crushing of the barrels and their contents could have taken place. Such an interdisciplinary approach leads to better interpretations.

The case studies presented in this chapter cover the approaches mentioned above, from ethnography to experimentation, often combining several disciplines. First, the chapter deals with massive long-term storage for community consumption, and ends with small-scale storage for domestic consumption. Three ethnographic examples are given from outside Europe (Morocco, Syria and Tunisia). Silo layouts, lining, flooring and wall strengthening with fire concerning long term storage technology for community consumption will be illustrated by case studies of Bronze Age, Iron Age and Upper Middle Age material from two sites in northern France. The principles that make storage pits work effectively will be studied through experiments and physical measurements, using the case of maize storage in simple pits in North America. The question of techniques to prevent losses during grain storage will be addressed by an agronomical approach to the structures for cereal storage in Morocco. The variability in storage structures (cereals, hay and fruit) in Slovenia will be explored through a review of ethnographic case studies. Household storage will be illustrated by ethnographic examples of jars, baskets, and sack management in Morocco. Patterns of storage of fodder and straw will be presented through ethnographic examples from northwestern Tunisia and southern Syria. This chapter will also deal with small-scale preservation techniques for domestic consumption by applying current knowledge of fruit biochemistry to textual data.

6.2. THE PREPARATION OF STORAGE PITS IN ANCIENT FRANCE: PHYTOLITH EVIDENCE

Pascal Verdin

Introduction

Several methods of storage pit preparation are known to improve the preservation of crops in underground silos. Proper preparation of the pit protects the products from the assault of humidity, insects, bacteria and moisture, avoiding the direct contact of crops with the earthen wall of the silo. Common techniques, lining and flooring, are cited by Varro (*R. R.*, I, 57), and other authors of Antiquity (Columella, Philo of Byzantium, Pliny, Xenophon). These techniques were the subject of experiments some years ago (Gast and Sigaut 1979) and have been documented by historical sources and ethnography (Kunz 2004). Drying and cleaning by burning has been observed archaeologically and studied experimentally (Firmin 1984), and has also been observed ethnographically (Beyer 2007).

Phytolith analysis (Piperno 2006), the study of silica micro-remains produced in some plants, is a tool that can be used in studying archaeological silos. The resistance of phytoliths to taphonomical processes[1] and physio-chemical alteration allows us to sample archaeological sediments to shed light on construction and preparation techniques. The analysis of pit fillings and use levels helps us to identify storage structures (in addition to their morphology), and to reveal how plants cultivated or gathered on a site were treated (Verdin 2002; Cammas *et al.* 2005). It can sometimes show storage levels, or plant materials comprising possible silo linings and floors, or drying by burning. These techniques are the subject of the study presented here. This kind of approach has already been successful in the study of the lining of some Iron Age silos in Turkey (Madella 2001).

Fig. 6.1. Location of the sites 1) Bussy-Saint-Georges; and 2) Saint-Denis (ZAC Nozal-Chaudron) in northern France. Map: R. Lugon, J.-C. Loubier and A. Chevalier.

This study considers phytolith analysis of structures on two sites. On the first site, situated in Bussy-Saint-Georges (La Manjoire; Fig 6.1) (Brunet and Sethian 2001), in the Paris region, samples were taken in the bottom of two silos dating from the Bronze Age and the Upper Middle Ages, respectively. On the second site, samples were taken from the bottom layers of two archaeological structures interpreted as silos; in addition, two samples from the silo walls of these structures in the natural sediment were taken as reference samples. These structures were part of a rural settlement of the second Iron Age (La Tène), situated in Saint-Denis (Nozal-Chaudron; Fig. 6.1), north of Paris, France (Leconte *et al.* 2002).

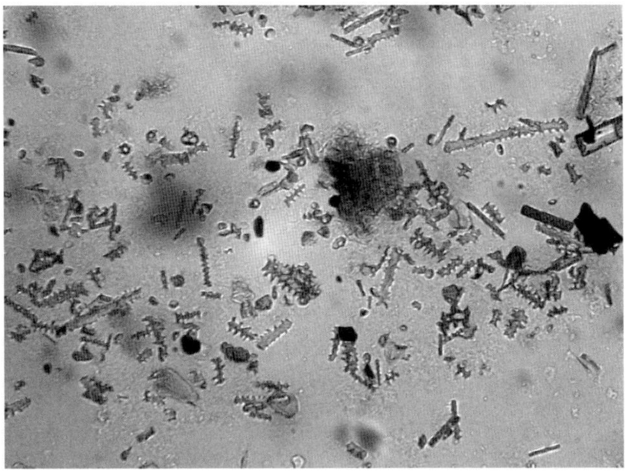

Fig. 6.2. St 155 – Microscopic panorama of a phytolith assemblage dominated by Poaceae glume phytoliths (spikey rods). 200x magnification (Image: P. Verdin).

Questions and Hypotheses

This study aimed at answering the following questions: do the samples from the bottom of these structures represent, entirely or partly, a primary deposit of storage, a level of secondary deposition, or remains of silo preparation? In these different cases, which form of storage, of preparation, or of discard are we seeing? Were remains of masses of cereal or other crops in loose storage left in the silo bottom, and in which form? Will the remains shed light on preparation of the silo bottom? Did this silo have a secondary function as a garbage pit before its final filling, and is this visible? To try to answer these questions, analyses were conducted using 40 g of dry sediment, according to the extraction method used by D. Pearsall (2001).[2] More than 100 phytoliths per slide were observed at 400x magnification.

The Bronze Age Silo at the Site of Bussy-Saint-Georges

This silo sample is characterised by a high proportion (57%) of glume phytoliths of Poaceae (the grass family, Fig. 6.2). There is thus little doubt that this level is of human origin. It could have been composed of hulled grains or spikelets that decayed on the spot. However, the phytoliths could also come from a level of husks of threshed or dehusked cereals that could be used as a protection for grain against pests and humidity. These phytoliths, sub-families of which are characterised by distinct sets of morpho-types, are produced in pooid (festucoid sub-family) Poaceae, including cereals such as wheat, barley, rye and oats, but excluding millet (panicoid sub-family). Such a process leads us to consider that this quartet of cereals was very likely to be involved. If wild Poaceae had been stored in this pit, this would have resulted in an assemblage displaying a wider variety of morpho-types. Experiments showed that complete separation of grain, glumes and culm is more difficult to accomplish (Anderson 1999).

So, in addition to the more obvious interpretation of this phytolith assemblage as a decayed hulled grain level, we can propose a second one: this glume level could be the remains of flooring composed of threshing by-products, such as the grain husks. This technique is known ethnographically in central Europe (Kunz 2004) and is mentioned in ancient texts.

The Upper Middle Ages Silo at the Site of Bussy-Saint-Georges

The silo dating to the Upper Middle Ages (Figs 6.3 and 6.4) showed a very different phytolith assemblage, dominated by dicotyledon phytoliths (not grasses or cereals). In addition to the phytoliths, a huge amount of micro-charcoal was observed (Figs 6.5 and 6.6). This phytolith spectrum could reflect a secondary filling made of dumped remains of a hearth. An alternative hypothesis can, however, be proposed: like the Bronze Age silo example, this spectrum can

Fig. 6.3. Silo 225 profile, showing carbonised remains, result of hearth dumping or disinfection by burning (Image: INRAP).

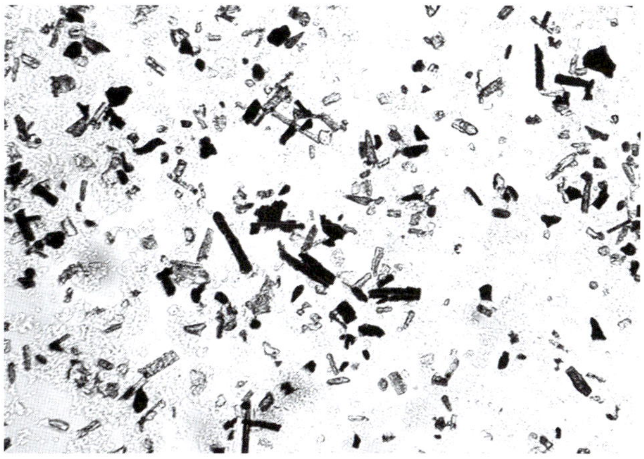

Fig. 6.5. Structure 225 – phytolith assemblage composed of varied taxa mixed with micro-charcoals. 200x magnification (Image: P. Verdin).

Fig. 6.4. Silo 225 – fully excavated, showing the typical bulb shape of Upper Middle Ages storage pits (Image: INRAP).

be interpreted as a preparation technique used in the silo, namely drying and disinfection by burning. This technique, documented by ethnography and experimentation, was used to purify and clean silo walls and bottoms, and to harden the walls (Sigaut 1975). The phytoliths could easily come from remains of diverse plants used as fuel to burn and clean the silo.

The Iron Age Silo at the Site of Saint-Denis

Four samples were taken from Iron Age silos at the site of Saint-Denis. One contained phytoliths, the other three did not. The positive sample came from the bottom layer of one of the silos (F 1130-us 1133). One of the negative samples originated from the wall of this same silo, whereas the other two negative samples derive from the other silo (F 1120). The abundance of phytoliths in the positive sample (more than 500), and the absence of phytoliths in a reference sample taken in a natural layer on the site, suggest the human origin of the deposit at the bottom of this structure; even if the second archaeological sample did not contain any phytoliths.

The particularity of this phytolith assemblage is the low variety of taxa (morpho-types), a result rarely observed in similar taphonomical and geomorphological contexts (personal observations on silts of the Paris Basin, on many reference samples). Poaceae produced 97% of all phytoliths in this sample (Fig. 6.7). A selection of plants of

	POACEAE		POACEAE Pooids			POACEAE Panicoids	DICOTYLEDONS	MICRO-CHARCOALS
	Hair	Long cells	Short cells	Long cells, glumes	Long cells, epidermis	Short cells	Cells	
SILOS								
St 155	0.6	16.5	23.4	58.9		0.6		
St 225	4.4	12.4	20.4	2.5	4	2.6	21.4	31.6

Fig. 6.6. Bussy-Saint-Georges – phytolith and micro-charcoal frequency (%).

	POACEAE/ CYPERACEAE			POACEAE pooids				POACEAE panicoids	PINACEAE	DICOTYLEDONS
	Hair	Bulliform cells	Long Cells	Short cells	Long cells, epidermis	Long cells, Glumes	Silica skeletons	Short cells	Epidermis	Cells
Structure										
F1130 US 1133	11.1	0.4	55.6	24	4.4	0.9	0.9	0.4	1.3	0.9

Fig. 6.7. Saint-Denis – phytolith frequency (%).

agricultural origin is highly probable, with a low number of different taxa with a high dominance of grasses, but with very different proportions of each morpho-type. Such data have been observed on several sites of the Iron Age in the Ile-de-France region, in silos in Rungis (Verdin 2002), or on the site of Le Plessis-Gassot (Cammas *et al.* 2005). Which plants and which plant parts were selected and ended up in the bottom of this silo on the site of Saint-Denis?

Such a high proportion of grasses (97%) in such a context suggests the hypothesis that these grasses are very likely to be cereals, or at least gathered or cultivated wild grasses. Among the morpho-types coming from these taxa, a few come from spikelets or ears (glume phytoliths). Long cell phytoliths (elongate psilate, Madella *et al.* 2005, 257–258) are strongly represented (55%). This proportion is rarely observed in natural contexts. This leads us to hypothesise that the remains of cereals left after winnowing were probably used as a pit bottom lining. The phytoliths suggest that these remains were in the form of culms or stems, probably with the leaves removed, since very few bulliform (bubble-shaped) cell phytoliths produced by leaves were found.

The presence of some silica skeletons (phytoliths in anatomical connexion), coming from the decay *in situ* of cereal straw, strengthens the hypothesis of remains of a primary deposit at the bottom of the silo, in the form of a straw lining and not in the form of edible products. These results can be compared to those obtained by Madella (2001) on Iron Age silos of the same type in Turkey: a low taxonomic variability, a large proportion of long cells, associated with Poaceae silica skeletons. His interpretation agrees with ours: these are the remains of silo linings.

Conclusion

This case study shows that it is possible, with the help of specific methods, to identify on archaeological sites agricultural methods such as pit lining and burning, as mentioned by ancient authors such as Varro, Columella, and Pliny, and observed in some ethnographical studies. It brings additional elements to the technical knowledge of storage structure construction and therefore their archaeological identification, their function, and their place in the agricultural process.

6.3. MAIZE STORAGE IN SIMPLE PITS

John C. Whittaker, Kathryn A. Kamp, Barrett Brenton, and Linda Scott Cummings

In both ethnographic and archaeological contexts, pits used to store grain and other crops are common, but the detailed workings of pit storage are little studied, although Reynolds (1974) performed pioneering experiments with pit storage of grain, and others have followed (Brenton 1989; Currid and Navon 1989; Hill *et al.* 1983). In North America, maize was stored in a wide variety of pit features, indoors or outside, lined or unlined, and ranging in size up to rooms large enough to stand in. We modelled our experiment on ethnographic Plains Indian practices in the Midwest (Wilson 1987), and tested small pits as maize storage facilities. The principles that make them work should be widely applicable to grain storage in pits anywhere.

In 1995, Whittaker, with Kamp, Brenton, and students dug four pits in an Iowa field. The pits were dug 1 or 1.5 m deep in hard clay, lined with native prairie grasses, and filled with maize, either shelled or on the cob. The 1 m deep pit held 273 l of maize kernels, and the 1.5 m pit held 477 l. Each pit was sealed with a layer of grass and earth. Such pits work because they maintain stable temperatures, and in a sealed environment the grain continues to transpire until high CO2 levels prevent pests and the grain becomes dormant and stable. A pipe to the interior of the pits allowed us to monitor temperature and CO2 throughout the winter.

The pits were filled in October and opened in May (Fig. 6.8). Winter temperatures included long periods below freezing, with the lowest recorded temperature at the pits -14 C. The temperature inside the pits stabilised at 1–7 degrees C, and the pits never froze. CO2 stabilised at 2–6% inside the pits. (Atmospheric CO2 is 0.33%; at 2.5% a flame is extinguished and at 5% human breathing is difficult). Three pits were opened May 4, 29 weeks after filling. After a normally wet and cold Iowa winter, both shelled and on-the-cob maize was dry and well preserved. Aside from a mouldy crust around the pit perimeter, germination rates on 100 random seeds from several parts of each pit were 90–100%. Aflatoxin tests to detect mould and bacterial action were negative. The corn was suitable for either food or seed.

Fig. 6.8. Pit filled with maize on the cob, showing grass lining and instrument pipe (Image: J. C. Whittaker).

Fig. 6.9. (above) Emptying pit, showing good condition of maize on the cob (Image: J. C. Whittaker).

Fig. 6.10. (right) Cross section of last pit, opened in July, showing pit structure and wet, packed shelled maize (Image: J. C. Whittaker).

The fourth pit was opened 2 months later (July 16) (Figs 6.9 and 6.10). The weather had been warmer and wetter. In this pit, the shelled maize smelled fermented, the seeds were swollen and wet, and the mouldy crust around the top and sides was much thicker. The rest of the corn was not mouldy, and no aflatoxin was detected. It could probably have been eaten, but none of it germinated, so it was unfit for seed corn.

In 2005, ten years later, we excavated one of the backfilled pits to see what archaeological indications of maize storage remained. Remnants of the grass lining were found, but no recognisable maize kernels. Cummings analysed samples from several locations in the pit and surrounding soil, and concluded that there were elevated levels of maize pollen and phytoliths in some areas, but because the pits were dug in a former maize field, the background micro-remains made conclusions uncertain.

Many variables in pit storage, such as size, lining material, and the type or condition of the crop could be manipulated to test for optimal performance. Pits could also be used to process food through fermentation. The ease and high success rates of a storage method with low investments in labour and technology explains the common use of pit storage.

6.4. A PARTICIPATORY APPROACH TO POST-HARVEST LOSS ASSESSMENT: UNDERGROUND AND OUTDOOR CEREAL STORAGE IN DOUKKALA, MOROCCO

Ouafaa Kadim

Introduction

Cereals constitute the base of Moroccan food. Their production occupies 4.5 million ha, or 85% of Moroccan farmland (Besri 1989). Except in some special cases, cereals are usually stored before consumption. Efforts to improve cereal production are not fruitful if the storage component is neglected. Storage plays major roles in protecting the harvested crop and its value, forming a strategic stockpile and increasing the capacity of cereal farmers to negotiate their economic interests. The present contribution has two objectives, first to analyse the performance and limits of the two principal techniques of storage in the area of Doukkala: in underground pits (silos) (Fig. 6.11) and in plastic bags, which were stored in open, lean-to sheds exposed to outdoor conditions. The

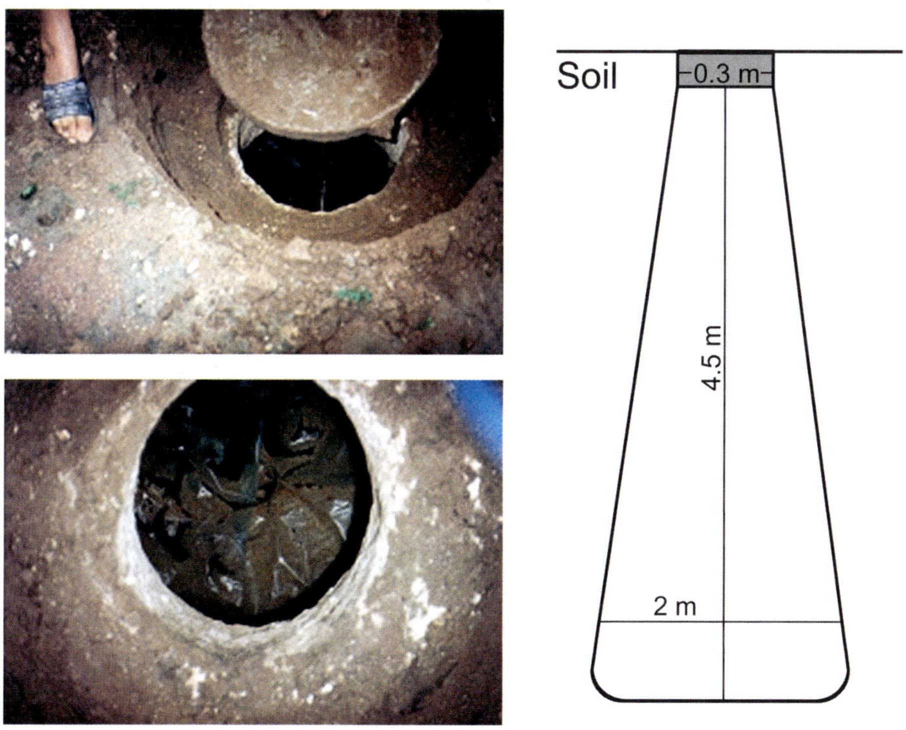

Fig. 6.11. Form and cross-section of underground pit in study area (Image: O. Kadim and A. G. Heiss).

second objective will be to evaluate the dry weight losses due to insect attack during cereal storage in these two types of storage. Before tackling the problem of the losses of stored cereals in the area of Doukkala, it is essential to introduce the geographical characteristics and socio-economic environment of this area.

Study Area

The area of Doukkala is situated in western Morocco, between latitude 32°15' and 33°15' north and longitude 7°55' and 9°15' west. The region of Doukkala is located between the rivers of Tensift and Oum Rbiaa and is 160 km long. According to the census of 1994, the density of population is 162 habitants per km^2; one of the highest densities in Morocco (Office Régional de Mise en Valeur Agricole D'El-Jadida). Precipitation averages about 400 mm per year on the coast to 300 mm inland. The region of Doukkala presents two distinct geographical units: the inland plain of Sidi Bennour and the sublittoral Sahel (Fig. 6.12).

Infestation of Wheat and Maize in the Study Area

Three species were considered in our study, durum wheat (*Triticum durum* L.), common wheat (*Triticum aestivum* L.) and maize (*Zea mays* L.). Insects which devastate stored cereals belong to two orders: Lepidoptera (butterflies / moths) and Coleoptera (beetles). They require precise ranges of temperature and moisture of the grain within the limits of which they proliferate (Ducom 1980). Lepidoptera are dangerous at the larval stage and cause important damage, whereas the Coleoptera are destructive from the larval stage to

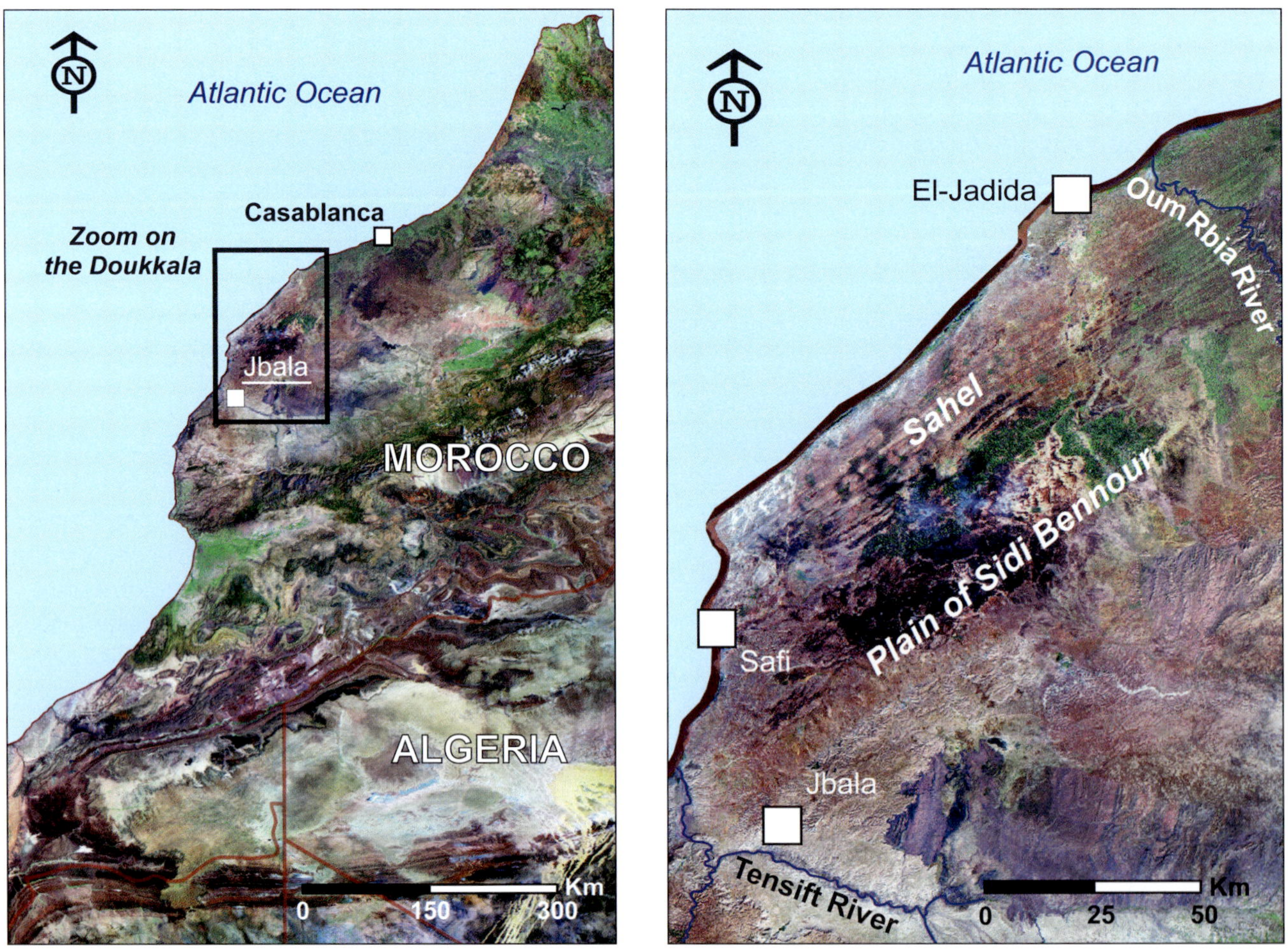

Fig. 6.12. Location of the Doukkala area in northern Morocco, with detailed view on the city of Jbala and the Sidi Bennour region located therein. Maps: R. Lugon, J.-C. Loubier and A. Chevalier.

the adult stage. For the two groups, Coleoptera and Lepidoptera, the optimal conditions are between 15°C and 30°C (Jazouli Belahboub 2000), to ensure high reproductive rates.

In grain stocks, the insects can cause massive damage by consuming the endosperm and sometimes the germ of the grains. Moreover, they contribute to the dissemination of mushroom spores by creating a favourable medium. The water content of a batch of wheat infested by weevil increased from 15% to 23% in three months, while a control batch which was not infested remained unchanged (Boyeldieu 1980).

Evaluation of the Losses: Technique of Sampling and Analysis

The post-harvest losses included: indirect losses, that is, any reduction in quality and edibility, and direct losses, all the damage caused by pests measured as dry weight loss. Here we limit our study to the direct losses and specifically the weight loss caused by the insects during storage. Estimating losses is an important element in any study aiming at the improvement of storage conditions.

On four occasions we took twenty samples of each grain species at the beginning of storage and after 2, 6 and 12 months from cereal stored in underground pits and plastic bags deposited in open lean-to sheds or stores. At each opening of the underground pits or the plastic bags, we took a sample of more than 1 g from the surface and the bottom of the stock. For each sample, the rate of insect attack and the percentage of weight loss were evaluated by counting and weighing methods recommended by the FAO (Pointel and Coquard 1977).

Results and Discussions

The observations made in our study show that over the 12-month period of storage, all the products except for durum wheat stored in the underground pit underwent attack by grain moths (genus *Sitotroga*) and grain weevils (genus *Sitophilus*). The results obtained for dry weight loss and rate of insect attack as a function of storage time indicated in Figs 6.13 to 6.16 show that the infestation had already started in the field for maize, whereas the common wheat and the durum wheat were not yet infested at the time they were placed in storage. The maize represents the highest rates of losses, in both modes of storage.

These results also show that the recorded rates of wheat loss in the case of storage in plastic bags are more significant than those recorded in underground storage.

This study revealed that the first factor responsible for the rates of dry weight loss and infestation during storage is excessively prolonged drying. That is, in the field – what is called drying on stem – or later when the cereals are spread out on the threshing floor, the court or over the roof of the house. Certain farmers in the study area have started a practice of drying which consists in leaving cereals on the stem for several weeks after maturity, to make sure they are fully dry. Certain peasants of the area wait up to one month after the physiological maturity of maize. In addition to the fact that it prolongs the duration of laying on the ground, this practice has the disadvantage of allowing attack by insects, birds and rodents. Prolonged drying after harvest also exposes cereals to insect attack.

The second factor causing weight loss is the fact that farmers of the area periodically extract cereals for family needs or for trade. In the case of underground pits, these must be ventilated to emit carbon dioxide accumulated inside before cereal extraction. Exposing the contents of the pits to the air will restart insect activity. Thus, it is recommended to entirely empty an opened silo, which obviously then poses other problems of management. Farmers will be obliged to find an alternative to this type of storage and, lacking an adequate infrastructure, could be constrained to sell these grains at a moment when the prices of cereal are low and to buy some later when the prices are higher.

The presence of insects after twelve months of storage of maize and wheat in the underground pits shows how important it is to seal these structures. In spite of the strong infestation of the maize grain at the time of silo filling, the rate of insect attack decreased from 1.30% at the beginning of storage to 0.07% after two months of storage. Whatever the choice of storage method, it is important to

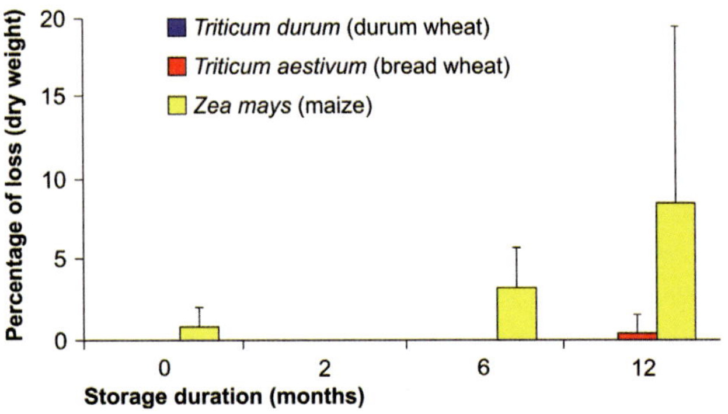

Fig. 6.13. Variation in cereal dry weight losses as a function of storage duration in the underground pit (Image: O. Kadim and A. G. Heiss).

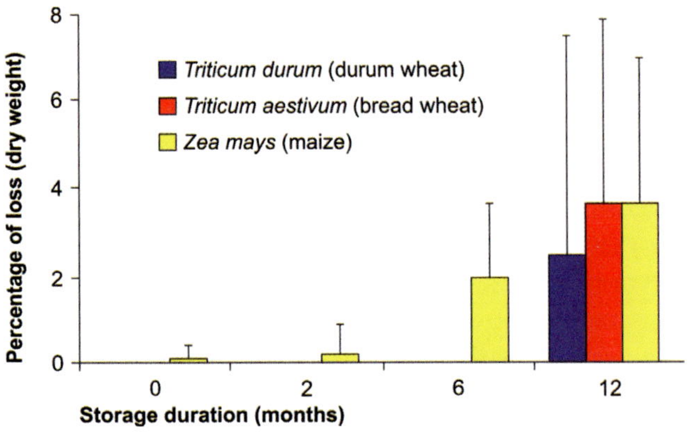

Fig. 6.14. Variation in cereal dry weight losses as a function of storage duration in plastic bags (Image: O. Kadim and A. G. Heiss).

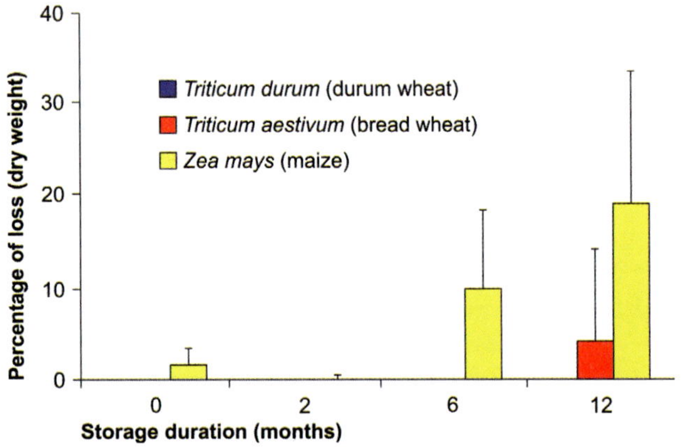

Fig. 6.15. Variation in insect attack rate as a function of storage duration in the underground pit (Image: O. Kadim and A. G. Heiss).

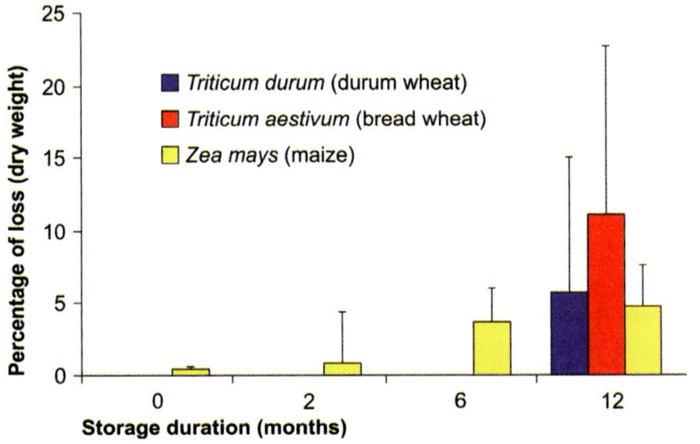

Fig. 6.16. Variation in insect attack rate as a function of storage duration in plastic bags (Image: O. Kadim and A. G. Heiss).

safeguard the quantitative integrity of the stored grains. To achieve these goals, the choice, the design and the maintenance of the structures of storage must meet a minimum of criteria:

- The reapers, the handling, the transport, and finally the state of cleanliness of the structures must all be checked.
- The site intended to receive the structure must be, preferably, a zone drained well or raised in the case of underground pits.
- Storage places should be inaccessible to pests.
- The design and the filling of the structures must facilitate the insecticidal treatments; in the case of grain stores, it is necessary to try to have surfaces relatively smooth, and easy to maintain.

Conclusion

This study shows that the importance of the damage caused by insects during storage depends on the intensity of the insect population present in the mass of cereals after harvest, the type of storage structure employed, the stored species, and on stock management. The length of the storage period may constitute an additional factor of risk. The highest level of loss was recorded in the case of maize after 12 months of storage in the two types of studied structures. The infestation of this species started in the field and deterioration continued in storage. The losses in dry weight of stored maize were greater when maize was stored in plastic bags exposed to the air than when it was stored in sealed underground pits. The farmer must intervene before and during harvest and during the phase of drying (on stem and/or terraces). This last is especially important to reduce the infestations during storage. Successful storage depends on the farmer's treatment of the grain, especially in not reopening storage structures and not drying the crop too long in the open air. The aim is to ensure a good conservation and to obtain a marketable product, whether the buyer is a storage organisation, an industrial operator, or simply the consumer. Any failure to reach optimum conditions for good conservation results in loss, which absorbs and sometimes exceeds the benefit to the farmer. Good storage structures for the harvest therefore constitute a valid investment, whether the products are intended for sale or personal consumption.

The control of storage techniques constitutes an essential link in mobilising resources for sustainable development. Finally, the evaluation of losses and the performance of a storage structure constitute a stage which is necessary for the development of any program of loss reduction. It will also serve as the basis for the final model of the decision-making, supported by the ratio of cost to profit.

6.5. EXPLAINING THE VARIABILITY IN STORAGE STRUCTURES IN SLOVENIA

Inja Smerdel

Introduction

The Slovene landscape still bears witness to the existence of a variety of storage structures and facilities, from simple poles and pits to more sophisticated buildings, some of which can be traced back at least five centuries (Figs 6.18 to 6.23).

The reasons for the various forms of storage are mainly functional, but also environmental (climatic), social, historical and cultural. Four culturally diverse regions can be recognised, reflecting centuries of entwined development at the junction of the Mediterranean, Alpine, Dinaric and Pannonian environments (Fig. 6.19). On the Smerdel basis of function we can distinguish structures or facilities developed for combined drying and preserving various crops. These include thin tree trunks with truncated branches (*ostrvi*) and poles for stacks of hay, grain sheaves and straw (*stogi*), and different types of hayracks (*kozolci*). There are other structures whose purpose is mainly storage, such as haylofts and barns (*svisli, seniki*), a variety of sheds (*lope*), including those for leaf litter, pits (*jame*), granaries (*kašče*), special maize granaries (*koruznjaki*), and cellars (*hrami, kleti*) for storing wine, dried meat, root crops and potatoes (Figs. 6.17 and 6.18). In the major part of the Slovene territory, especially in the mountainous and hilly areas, these outbuildings were mostly constructed near

Fig. 6.17. (left) The unmarried 'aunt' Luce, a domestic maid servant of the homestead by *Cendran* from the village of Selce in the Pivka region, in the shed carefully selecting potatoes, before storing them in the cellar (Image: I. Smerdel, 1981; SEM Archives).

Fig. 6.18. (right) A wickerwork maize granary with thatched roof in the village of Črensovci by Lendava in the Prekmurje region (Image: F. Šarf, 1963; SEM Archives).

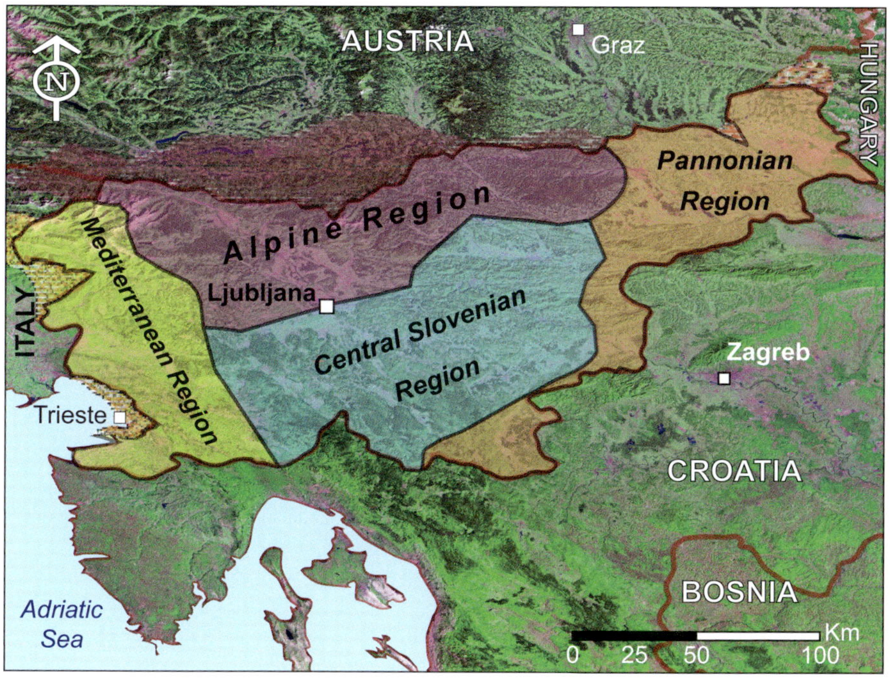

Fig. 6.19. Map of the four Slovenian ethnographical regions, defined and cartographically distinguished (though life in detail did not take its course exactly within these borders) on the basis of differences in essential cultural elements that reflect, historically and geographically, centuries of development in the Mediterranean, Alpine, Dinaric and Pannonian environments. These four regions – Mediterranean, Alpine, Central-Slovenian and Pannonian – virtually coincide with the macro geographical areas. Map: R. Lugon, J.-C. Loubier and A. Chevalier.

or around the farmhouse in a seemingly disorderly spread depending on factors like local topography, type of colonisation, individual farms' economic structure and so forth. However, in some western and eastern Slovene regions all storage functions were gathered under one roof.

Written evidence and pictorial sources on almost all such storage structures and facilities seem to be scarce (at least before the 18th century), and there are also only a few Slovene ethnological studies dealing with them, mainly on a rather general level. (On haylofts: Baš 1970; Cevc 1984. On granaries: Baš 1952; Baš 1970; Cevc 1988; Ložar 1944. On chests: Miklavčič Brezigar 2000. On maize granaries: Novak 1960; Šarf 1971). The only structures that have indeed been thoroughly researched so far are the hayracks, first by the geographer Melik (1931), then architects Mušič (1970) and Juvanec (2007) and ethnologists (Cevc and Čop 1993; Hazler 2004). Most recently the origin of hayracks was finely documented by the art historian and ethnologist Makarovič (2007).

Hayracks

Hayracks are 'freestanding, roofed, permanent buildings with a structural column-like head rail (or two head rails), into which in the same vertical plane horizontal poles are set at intervals equal to the thickness of a grain sheaf, and whose purpose is to dry and preserve various crops and fodder' (Makarovič 2007, 247). The need for hayracks presumably developed with the climate changes of the Late Middle Ages. In the wet and cold 'Little Ice Age' it was no longer possible to let grain sheaves dry in the field after the harvest, at least not in the Central and Alpine Slovenian regions. *Kozolci* first spread in Carinthia (Slovenska Koroška, Fig. 6.20; the earliest reference dates from 1558) and part of Carniola (Kranjska, Krain; Fig. 6.20) in the 16th century, were in use all over Carniola in the 17th, and became common also in Styria (Štajerska, Untersteiermark; Fig. 6.20) in the mid-18th century. From a social viewpoint, hayracks were first built on feudal estates, especially for drying moist tithe sheaves, but became common on many farms by the 17th century (Fig. 6.21).[3] The need for only drying declined in the 18th century, when the agrarian technical revolution generated new requirements, in particular the necessity of storing large quantities of fodder required by the then common practice of livestock breeding in stables, for sorting newly introduced field crops and for storing the increasing number of agricultural implements and vehicles (Makarovič 2007, 248). These needs then extended the function of hayracks and were indeed perfectly met by them (for example, the upper parts of double hayracks were constructed as haylofts). Their great variety (nine different types) and innovative

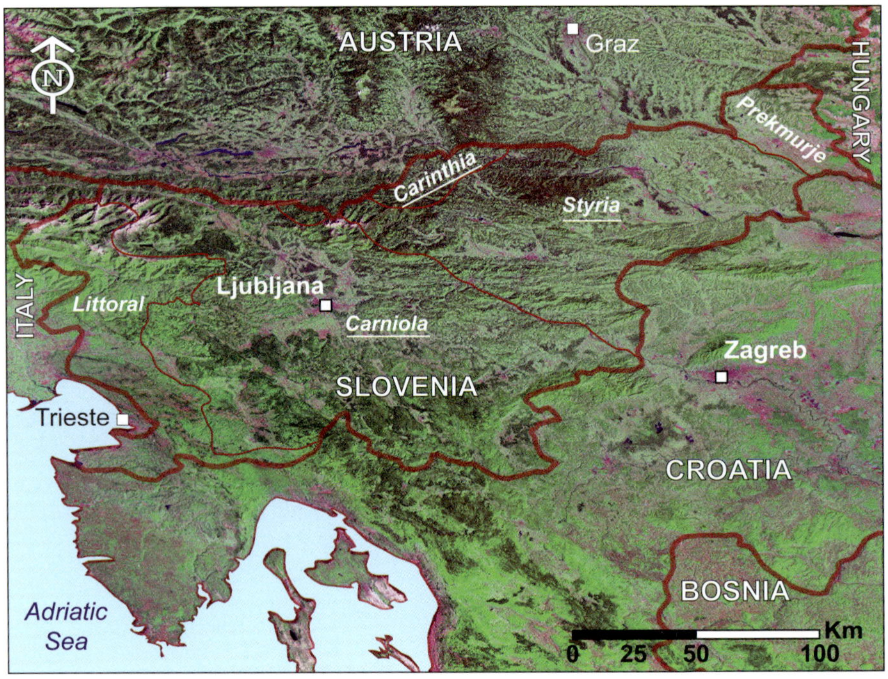

Fig. 6.20. The historical Slovenian regions of Carinthia, Carniola and Styria. Map: R. Lugon, J.-C. Loubier and A. Chevalier.

structures mainly date from the 19th century. *Kozolci* are essentially connected with the space in which they are erected. Their exterior reveals the impact of cultural and technical choices. Within a village - where it was said that there is no farm without a hayrack - they differed only in size (those with several *okna* or 'windows' indicated prosperity), and the quality of details and ornaments, thus revealing social stratification.

Granaries

Granaries seem to have appeared with advanced medieval carpentry, and developed as outbuildings. They were mainly built of wood, or with a masonry lower part, the cellar, and a wooden upper part, the granary (Fig. 6.22). When the buildings of homesteads concentrated closer to each other or even gathered more functions under one roof (from

A

B

Fig. 6.21. Hayracks of the elongated type: A) as depicted on a copperplate in Valvasor's *Die Ehre des Herzogthums Crain* (Valvasor 1689), and B) a specimen of the same type from the 1980s, in a meadow by the village of Šenčur near Kranj (Image: J. Klemenčič, 1981; SEM Archives).

Fig. 6.22. Two types of granaries. A) A stone (and brick) granary in Verd by Vrhnika in Central Slovenia (Image: R. Kregar; from the old SEM Archives, dating from the 1st half of the 20th century). B) An example of a half-stone, half-wooden structure, a granary with thatched roof in Obrh by Vinica in the Bela krajina region (Image: M. Kambič; from the old SEM Archives, dating from the first half of the 20th century) .

the second half of the 18th century onwards), the granaries as outbuildings gradually disappeared; after the introduction of chimneys the farmhouse lofts slowly took over their function (Baš 1970, 606). In both, grains (and also beans, long beans, and dried fruit) were kept either in built-in wooden chests with sections for wheat, oats, barley and other crops, in a number of plain hewed chests, or in wicker baskets (Figs 6.22 and 6.23).

Fig. 6.23. A watercolour of a granary interior from the Bela krajina region depicting hewn wheat chests. (SEM Archives).

Contemporary Use and Socio-Cultural Status

Where granaries are still preserved, people often use them as all sorts of old junk storage facilities. Among them there are some that became weekend houses. In the last decades of the past century it was for example popular to buy a hewed upper part of a granary, dismantle it, displace it to a chosen location, and reconstruct it for this new function. But the contemporary socio-cultural status of hayracks is even more complex. Picturesquely blended into the cultural landscape, the hayrack has always been a favourite motif for Slovene painters; it is also reflected in Slovenian architecture, in the design of urban furniture such as stands for advertising and local announcements, and in the production and marketing of ethnic souvenirs. Today, scaled down hayracks figure as ornaments in courtyards and linked hayracks even appear as garages (Smerdel 2004, 65). The intrinsic value of hayracks and their special meaning for the Slovenes thus also enables their interpretation as symbols of national identity.

6.6. STORAGE IN THE WESTERN RIF (MOROCCO): BASKETS AND CLAY/DUNG CONTAINERS

Leonor Peña-Chocarro and Lydia Zapata

This contribution focuses on two storage techniques documented in the western Rif (northern Morocco), in the Jbala area where ethnoarchaeological research was carried out between 1998 and 2002 (González Urquijo *et al.* 2001a, b; Ibáñez *et al.* 2001; Peña-Chocarro and Zapata 2003; Peña-Chocarro *et al.* 2005; 2009; Zapata *et al.* 2003a, b). Although different techniques were documented using various types of structures and materials such as canes, dung, clay, plant fibres, and cork, this paper focuses on the most common types: the *sulla*, large baskets made of cane (*Arundo donax*), and the clay or dung recipients called *tonna*.

The *sulla* (Fig. 6.24) is a large barrel-shaped basket woven with *Arundo donax* (giant reed) canes covered first by a coating of dung to avoid losses through the interwoven fibres and, then by a second layer of clay. Their height ranged from 1.20 to 1.50 cm which allowed the farmer to look into the container from above. The base of the container was around 50 cm wide. In the lower part of the container, the weaver opened a small hole through which the contents (wheat, barley, etc.) could be extracted. Every house used to have several of these baskets, one for each cereal crop. In the case of large grains or, for example, when einkorn spikelets were stored, there was no need to cover the basket with clay and/or dung as they were big enough to avoid passing through the small holes created by the interwoven fibres. The *sulla* were generally used for crops produced in large quantities, *e.g.* cereals, as opposed to *tonna* in which smaller productions, *e.g.* legumes, dry figs, einkorn were stored. The biggest baskets could contain up to 300 kg whereas the average ones contained around 150 kg.

Cane basketry was a specialised male craft, although in the past most men knew the techniques. The harvest used to take place in the autumn when the craftsman collected the canes. Only the shoots from the previous year were harvested as these were considered the best ones. The canes were peeled on site and then split into different lengths. Once transported to the craftsman's workshop, the canes were covered and wetted regularly (3–4 times a day), kept this way for 2–3 months, and then dried. Afterwards, the material was ready to be woven.

Fig. 6.24. Farmer showing different types of containers. The largest one is a *sulla*, a big basket made with woven canes. The medium-size container is a *tonna* made of cattle dung, and the smallest one is a *tabtoba* made also of cattle dung (Image: L. Peña-Chocarro).

The owner was in charge of covering the basket with either clay or dung. Craftsmen concentrated most of their work in the summer when the harvest took place and there was an increasing demand for storage baskets. Basket makers could either sell their products in the market or directly to the users. Many craftsmen were itinerant. They travelled to different villages where they lived for short periods producing baskets according to the particular needs of each family.

Apart from baskets, people from the Jbala used other types of containers such as the *tonna* made out of cattle dung (Fig. 6.25). In this region, dung has been traditionally used for different activities amongst which we highlight the making of storage containers (Ibáñez *et al.* 2001). The *tonna* had a barrel shape and were made out of clay and/or dung, usually covered by a white clay layer. As in the *sulla,* a small hole was cut at the lower part of the container (using a knife when the container was already dry) to allow extraction of the contents. The upper part was covered by a piece of cork or just an object (sieve) covering the aperture. The technique used for making this type of containers is one of the most commonly used in pottery making: a series of short coils joined and smoothed together.

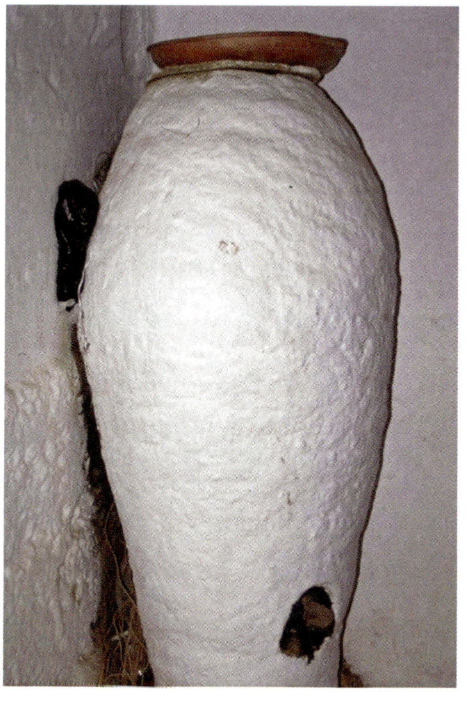

Fig. 6.25. Storage container *tonna* made of dung and/or clay. This type of container was used to store crops (generally large items such as einkorn spikelets or dried fruit) (Image: J. E. Gonzalez Urquijo).

A similar use of dung has been documented in other parts of the world (Ertuğ-Yaras 1997; Hillman 1984; Makal 1954). In northern Morocco, the *tonna* were made with three different types of material: cattle dung, unfired clay and a mixture of both clay and cattle dung. The choice of material depended very much on the availability and characteristics of the raw material. During the spring, for example, dung was generally too soft due to the abundance of fresh grass and, therefore, its consistency was improved by mixing it with clay. *Tonna* making was carried out during the summer when women could work outside, and drying was quicker. In addition, in this period they were able to calculate the number of containers needed according to the harvest volume, so they made enough for the winter.

Tonna containers varied in size from the biggest examples at *ca.* 200 kg to the smallest ones of *ca.* 50 kg. While the making of *sulla* was specialised work, the *tonna* were made by the women of the village. Each house used to have three to four *sulla* and five to seven *tonna*. Both types of containers were kept in a special room under the roof where all crops were stored. These containers were aligned along the walls together with oil containers. However, if there was not enough space in this area, they could be placed in any other room of the house. Other types of containers were also made of cattle dung such as the *tabtoba* used for placing the animal-fodder cereal meal.

Ethnographic research has shown the importance of dung in the Rifian society in the past. Dung has been an invaluable product that was rarely discarded in many traditional societies; its production was regular and predictable and, therefore, used for a wide variety of purposes, such as for manuring, as fuel, for tempering walls and surfaces (Zapata *et al.* 2003b). Dung's advantages as a material include its light weight and longevity, and insulating properties. Disadvantages include the unsuitability of dung containers for liquid storage and use as a cooking vessel.

Both dung and canes were common products available in the rural society of northern Morocco. Dung use for container-making has already disappeared from the traditions of the Jbala region, whereas the use of canes for basketry may still be in use in some isolated areas.

6.7. FODDER AND STRAW IN TUNISIA AND SYRIA: MODERN AND ARCHAEOLOGICAL CONTEXTS

Patricia C. Anderson

Production and use of straw and fodder are intertwined with virtually all agricultural and some artisanal activities. In particular, in order to have enough food for draft, milk and meat animals, it is important to supplement grazing and green fodder, which is not available during certain seasons of the year, with large quantities of dry fodder in the form of chopped straw from cereals and pulses (*tibbn*, in Arabic).

In southern Syria, in the Hauran, in the Jabal al-Druze mountainous region (Fig. 6.26), at approximately 1000m altitude, plants are always harvested by pulling up the whole plant by hand (see Chapter 4). This is an unusual practice for durum (hard) wheat (*Triticum durum* L.) as compared to elsewhere, but is done to obtain the entire stem, as the hard bases of the stems of hard (durum) wheat, called *casalls*, are winnowed out separately and stored, and used as fuel in this wood-poor region. Threshing sledges are still used today in Salim, near Sweida (Fig. 6.27), to separate out grain and break up the stems of cereals (wheat and barley), pulses, and other plants (vetch, alfalfa), for production on a household scale, albeit rarely nowadays. They are also used in Dama to thresh pulses (such as chick pea or faba bean). They are pulled on the threshing floor by donkeys or horses. Stems and glumes, the membranes surrounding the grains, are cut into fine pieces by the combined action of draft animal feet and the grinding of threshing sledge inserts. Here the inserts are not blades, but rather blunt ball-like pieces of obsidian, the only local stone. Alternatively, a peculiar form of threshing sledge is used, made from a sheet of metal with small cutting parts, such as seen in Salim (Fig. 6.27) (Anderson 2003). Then the threshed plant material is sorted into

Fig. 6.26. Map of Syria with the region of Jabal al-Druze (Djébel el-Druze) and the site of 1) Tell Halula. Map: R. Lugon, J.-C. Loubier and A. Chevalier.

Fig. 6.27. A farmer using a threshing sledge, made from a flattened oil drum with holes punched downwards in the metal sheet, creating rough edges serving as 'teeth', in Jabal al-Druze, near Sweida, Syria. It is pulled in a circle around a threshing floor by donkeys over sheaves of cereals or pulses, to chop straw and chaff and thresh out grain (Image: P. C. Anderson).

Fig. 6.29. Stores of finely chopped straw from the threshing floor, *tibbn*, piled up several metres high to the ceiling, in a ground-level cellar in Jabal al-Druze, Syria (Image: P. C. Anderson).

Fig. 6.28. Two clay bins for storage of cereal and pulse grain, in cellars in Jabal al-Druze, Syria. The grain is introduced at the top, and progressively removed from the hole near the base (Image: P. C. Anderson).

different fractions by wind-winnowing: grain, larger straw, and fine straw with, for hard wheat only, an additional winnowed fraction of the hard stem bases. All of these fractions are stored in different places.

Grain is stored in clay granaries above the ground, which are made either by building up clay coils (coiling) or by stacking up mudbricks (Fig. 6.28). Some older people had heard of silos, underground storage pits, but said these were only used during the Ottoman Turk occupation; these were used to hide grain, as it was taxed. The soft cut straw was (and still is) principally used for animal fodder in this region. It is also employed to temper the clay used to make grain storage containers, or mud plaster to spread on dry stone walls. The chopped *tibbn* produced (soft part of straw) and *cassals* (hard stem bases) are stored in sacks or piled up loose as high as the ceiling in ground-level cellars (Fig. 6.29). Different rooms or areas of rooms contain straw from different plants, that is, straw from faba or chickpea, and wheat straw and barley straw are each stored in separate areas. Straw was taken out of the store over the course of the year, with the year's harvest supplying enough chopped straw for one year. Storage of chopped straw was therefore not generally for over one year in duration at any given time.

Stores of chopped straw appear to be present in archaeological sites, such as at Halula, in the Euphrates Valley in northern Syria (Fig. 6.26; Molist 1998). Here a thick, ashy layer dating from a Late Neolithic level (about 7800 BP) was visible in a room or partition that was adjacent to a living structure. This could have been a compartment for storing chopped cereal straw. This hypothesis is supported by examination of this ash-like sediment under the transmitted and reflected light microscopes at 200× and 400× as well as the scanning electron microscope (SEM), which revealed the 'ash' was composed of thousands of finely cut phytoliths

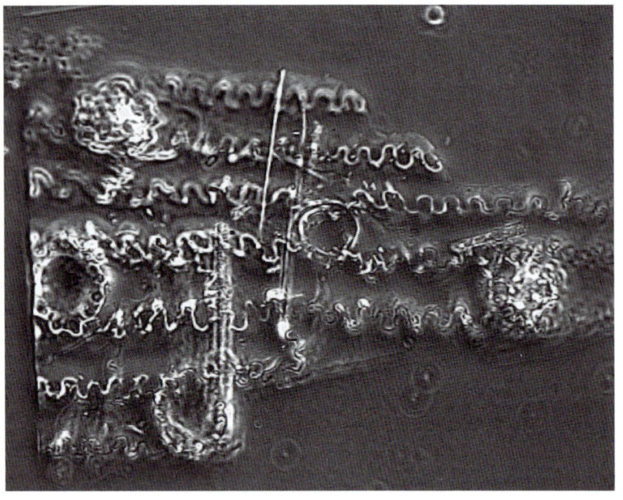

Fig. 6.30. Phytolith, silica cast of a fragment of the epidermis (outer membrane) of cut cereal straw, from the culm, showing several linked long and short cells (spodogram). The phytolith is from a thick, extensive ash-like deposit in a small room at Halula (Late Neolithic, Syria) which was most likely used for storage of the chopped straw fraction. 400x, transmitted light microscope. The smooth, straight cut at the left side of the phytolith, as well as the cut scars in the middle of the phytolith, suggests the straw was scored by the action of threshing sledge blades on a threshing floor, as in the Tunisian example (Image: P. C. Anderson).

from the epidermis (outer layer) of cereal/grass stems and glumes, the membranes enclosing the grain (Fig. 6.30). The straw would have been cut using a different threshing sledge from those used in the Jabal al-Druze region (Fig. 6.26; see above), where only basalt is available. Here, the threshing sledge would have been armed with flint blades, as this raw material is readily available and used both in the past, as evidenced from archaeological sites, as well as for threshing sledge inserts in the recent past. This impression was supported by the fact that the phytoliths formed 'sheets' (multi-cellular structures, that is, multi-celled or interconnected phytoliths) that were smoothly cut in regular patterns. Studies of material from ethnographic contexts and from experiments have shown that these patterns are not produced by other cutting techniques than that of the threshing sledge. Furthermore, flint blades with microwear patterns characteristic of use in threshing sledges, have been identified from this and other sites of this age (Anderson 2003; Anderson 2006a) (see Chapter 5).

A second case study takes us to the Atlas northwest tell area in northern Tunisia, near the town of Beja (Fig. 6.31), approximately 100 km to the west of Tunis. On farms at around 200–500 m in altitude, mules, horses, donkeys, and oxen used for draft and portage of people and goods, and sheep, goats and milk cows, are all highly valued animals kept by farmers. Cut stems and chaff of durum wheat, barley, vetch, oats, alfalfa, chick pea, broad bean and horsebean are used as dry fodder, as in Syria. However, cereals are harvested with a sickle, whereas pulses are pulled up by hand (see Chapter 4). It should be noted that the hard stem bases of durum wheat are left in the ground here, unlike in Syria, and the stubble is grazed. Cows are the first to pasture, breaking up these hard stem bases, allowing smaller animals to follow on the grazing. The harvested plants are processed (chopped) using the threshing sledge armed with cutting blades (Fig. 6.32) for durum (hard) wheat, or by trampling under animal feet (barley, vetch, oat, chick pea, broad bean, horse bean, see Chapter 5). This process produces clean grain after winnowing, which, according to an elderly farmer in Ain Sallem (Fig. 6.31), used to be stored in a room in the house in a large, circular basketry ring reaching to chest height. These basketry structures were woven from wheat straw, and measured about 5 m around and one metre and a half in height. They were said to contain grain to be used over the year, about 1200 kg. These are no longer in use, and there do not

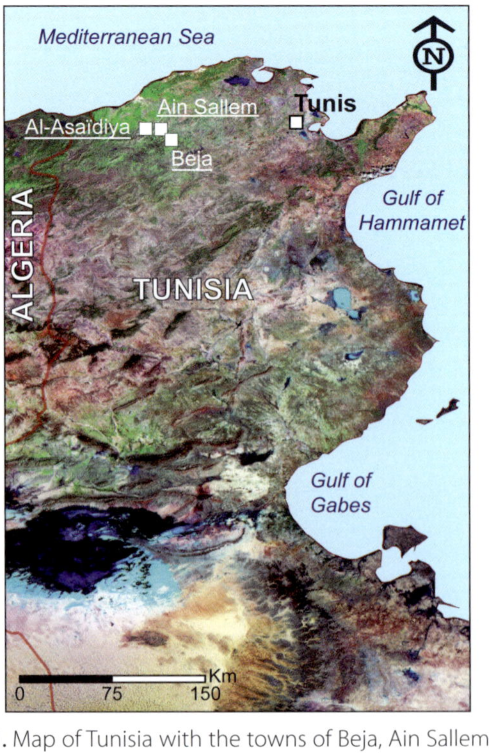

Fig. 6.31. Map of Tunisia with the towns of Beja, Ain Sallem and Al-Asaïdiya. Map: R. Lugon, J.-C. Loubier and A. Chevalier.

Fig. 6.32. The threshing sledge being driven over durum wheat in NW Tunisia, with the driver accompanied by his daughter, along for the fast ride. Mules pull this bladed sledge at high speed, rapidly producing chopped straw and chaff, and releasing grain (Image: P. C. Anderson).

Fig. 6.33. Storage structures (*nadir*) mainly for fodder, in NW Tunisia, built up from chopped straw and chaff, next to a field and threshing floor, seen in the autumn (Image: F. Ghozzi).

Fig. 6.34. Closer view of a *nadir*, showing its coating of mud and straw, protecting the *tibbn* beneath from rain and wind (Image: P. C. Anderson).

appear to be any remains or photographs of them. These would be very interesting to identify in archaeological contexts.

People several generations back talked about storage structures made of clay like those in the Jabal al-Druze in Syria (Fig. 6.26), described above, but these are no longer in use anywhere. Underground silos, called *matmoor* (like those described by Kadim, this chapter), are used in central Tunisia in the high steppes (see Anderson and M'hamdi, Chapter 4) but are unknown in the north. The *tibbn,* the secondary product of threshing, is stored in *nadir*, which are constructed outdoors, next to the threshing floor; straw is not stored indoors. The *nadir* is built by laying long barley stems down as a base, then piling up chopped straw as it comes off the threshing floor, until it is eventually built up into a longhouse-like structure, for example in Al-Asaïdiya (Fig. 6.33, Fig. 6.34). First the chopped legume straw (*tibbn*) is laid down, followed by cereal *tibbn*, following the seasonal order of harvesting and threshing. At the end of the threshing season, this structure is covered with stubble from the cereal field, mixed with mud and water (Fig. 6.33). The *tibbn* is progressively scooped out of the *nadir*, beginning at the bottom, for use throughout the year as dry fodder. The *tibbn* is also used to produce mud plaster and to mix with cow dung and clay each year to put a fresh, smooth finish on the threshing floors, where animals pulling the sledge are driven around at high speeds over the harvested durum wheat sheaves (Fig. 6.32) (Anderson 2006b).

This outdoor *tibbn* storage, for which we know of no archaeological equivalent, may be a response to climatic conditions in Tunisia, notably less windy than in the Jabal al-Druze, or may reflect the close proximity of houses and stables to the threshing floors there, but people from Syria in the Jabal al-Druze were not aware of the storage structures used in Tunisia. It is important to take into account these grain and straw stores – both on and 'off' site – when excavating archaeological sites and sampling for phytolith and archaeobotanical analysis. These provide us an avenue to exploring ancient harvest, threshing and storage techniques, as well as the significance of the use of chopped straw, which remains a little-known topic.

6.8. THE PRESERVATION OF QUINCE IN HONEY ACCORDING TO COLUMELLA: AN INTERPRETATION USING FOOD BIOCHEMISTRY

Marie Russel

Introduction

The study of archaeobotanical remains and ancient texts provides an overview of the fruit that was used for human consumption in Graeco-Roman times. Physiochemical analyses of organic remains of ancient vessel contents have been increasingly accurate in discriminating vegetal from animal remains, cooked from raw remains, and digested from non-digested remains (Rottländer 1990; Malainey *et al.* 1999). These techniques have recently been applied to data from the Graeco-Roman period (Kimpe *et al.* 2002; Colombini *et al.* 2004), but not yet to food remains from that period. The main sources showing preservation techniques for fruit are thus textual data: Greek theatre, agronomical texts, encyclopaedias, medical or philosophical texts, or texts on wars. Historical research, which has also focused on meat, reveals the central role of cereals, grapes and olives in the diet of Graeco-Roman times (Amouretti 1984; Jardé 1979). Apart from figs, well known as a nourishing fruit, and grapes, mentioned above, the role that fruit may have played in the ancient diet has not been studied yet from a nutritional point of view.

It can be questioned to what extent transformed (processed) fruit could have been used in a significant way in the daily diet in Classical times, and especially during food shortages. The Ancients indeed used to preserve fruit. Various techniques are described in many Latin texts (André 1981, 75–93). The wide range of fruit preservation techniques may mean that preserved fruit was indeed of great interest to Graeco-Roman populations. Some of these techniques described in the texts have been analysed in light of knowledge gleaned from food biochemistry and nutrition (Russel 2004, 55–85).

Applying Food Biochemistry to the Interpretation of Textual Data

The following case-study will deal with preserving fruit in honey, and will consider textual data in the light of current knowledge in food biochemistry. This method for preserving fruit is described by Columella (*R.R.* XII, 47, 2–4, translation by E. S. Forster and Edward H. Heffner):

> *'(2) Nevertheless we have not experienced any more sure and satisfactory method than, when the weather is calm and the moon waning, to pick quinces which are very ripe, sound and without blemish and, after wiping off the down which is upon them, to arrange them lightly and loosely, so that they may not be bruised, in a new flagon with a very wide mouth; then,* when *they have been stowed in up to the neck of the vessel, they should be confined with willow-twigs laid across them in such a way that they compress the fruit slightly and do not allow them to be lifted up when they have liquid poured upon them. Then the vessel should be filled up to the top with the very best and most liquid honey, so that the fruit is submerged. (3) This method not only preserves the fruit itself, but also provides a liquor which has the flavour of honey-water and can without danger be given at their meals to sufferers of fever. It is called* melomeli. *But care must be taken that the fruit which you wish to preserve in honey is not stored before it is ripe; for if it is picked when it is unripe, it becomes so hard as not to be fit for use. (4) But the practice of many people of cutting up the fruit with a bone knife and taking out the seeds, because they think that the fruit receives harm from these, is quite superfluous,*

but the method which I have just described is, indeed, so sure that even if there is a small worm in the fruits, they do not deteriorate any further once they have the liquid described above added to them; for such is the nature of honey that it checks any corruption and does not allow it to spread, and this is the reason too why it preserves a dead human body for very many years without decay'

Why can honey preserve fruit so well? Why does Columella seem so interested and worried by the phenomenon of maturation (XII, 10, 4; 14; 47, 3; 47, 6)? For instance, he says that quinces must be harvested very ripe, otherwise they neither mature nor give any flavour to honey.

Growth and Maturation of Fruits: The Quince

In order to understand this text, we need to focus on the chemical phenomena that occur in fruit. During growth, fruit breathes ('to respire' will be used in the following): it consumes oxygen while turning glucose into energy, and emits carbon dioxide. This respiration rate diminishes through time. In the case of some fruits, like apples, the respiration rate suddenly increases again, at the end of growth and the beginning of maturation. At this point, apples start emitting a gas called ethylene. In the case of other fruit, like tomatoes, maturation occurs without ethylene and without any change in the respiration rate. Maturation of these fruits can however be activated by ethylene.

Quince is the same kind of fruit as apple, but it starts maturing much later than an apple does: its respiration rate is much less than in apples, and increases far less rapidly. Quinces mature much more slowly than apples. At the end of growth, quinces are green, hard, acid and astringent. During maturation, due to chemical phenomena, they become less green, softer and juicier, less acidic and more sugary, and less astringent and more flavourful. These transformations are facilitated by ethylene.

The respiration rate during harvest is usually the main factor limiting preservation of fresh fruit. The stronger the respiration rate, the shorter the preservation period. Maturation may occur before or after harvest. Unripe fruit can be harvested, but preservation must then facilitate maturation.

As quince has a relatively low respiration rate, its maturation after harvest will be slow. Once a ripe quince is immersed in honey, it can remain ripe for a long while. If the quince is unripe when harvested, the technique for its preservation has to facilitate maturation, otherwise the quince will never ripen. As Columella notices, a green quince never matures in honey. In honey, an unripe quince cannot end its growth, thus cannot breathe. So honey prevents respiration by acting as a natural barrier to oxygen. This seems obvious. By analysing the end of the chemical process to see how honey protects fruit from oxygen, one can appreciate Columella's intuition and powers of observation.

Alteration of Fruit: The Role of Honey in the Preservation of Quince

In the natural process of aging in fruit, maturation is followed by senescence, a stage where all the components of the fruits are destroyed. To slow this process and keep the fruits safe as long as possible, it is necessary to reduce the respiration rate. This is why fruits are stored in cold places: cold slows down metabolism by lowering the speed of all the chemical reactions. For instance, it stiffens the lipids that are part of cell membranes. Membranes thus become more permeable to exchanges between cells, which limits the overall cellular activity.

The places in which the ancients used to store fruit were intended to preserve it in the best conditions. Varro (*R. R.* I, 59, 1–2; 3) explains that, for better aerated storage rooms, the windows should be on the north side. Pliny (*N. H.* XV, 59) adds that the windows of the north side should be opened when the weather is calm and that there should be no access on the south side. They describe rooms both cool and dry, *i.e.* convenient for preserving fruit. However, during storage, alterations may occur and lead to brown spots on the surface of the fruit. These alterations may be caused or accelerated by intense cold.

One enzyme linked to alteration is polyphenoloxidase (PPO). Together with oxygen, PPO transforms phenolic components into quinones. Quinones react with molecules like proteins and become melanins, polymers that can be seen by the naked eye. These molecules, muddled in a compact mass, protect the fruit from any attack by microorganisms. These chemical reactions belong to a process called enzymatic browning. The fruits survive thanks to these reactions, but, due to the loss of proteins, they also loose part of their nutritional value.

Present-day research proved that this process can be stopped at different steps. An acid environment transforms quinones into phenolic components, which is the opposite reaction to the first one described above; before this reaction occurs, the oxidative action of the enzymes can be stopped by reducing oxygen. Modern food industry experiments show that, for preserving apples, water containing 30% of sugar and 0.32% to 4% of calcium can prevent oxygen from getting near the surface of fruits (see for example patents EP95203393 and WO9907230, taken out at the French National Institute for Intellectual Property respectively in 1995 by Nestlé S. A. and in 1999 by Mantrose-Haeuser Co. Inc.).

Columella closely studied the preservation of quince. He noticed that honey 'checks any corruption and does not allow it to spread' (XII, 47, 4). The word 'corruption' can here be interpreted as the result of exposure to oxygen from the open air. Columella describes what he calls 'the nature of honey' as even able to prevent a small worm from deteriorating the fruits any further once the latter have honey added to them. The high concentration of sugar in honey eliminates oxygen and prevents oxidation.

Without oxygen, quince can neither breathe, nor mature. This is why Columella recommends choosing quinces that are already well ripened. Another reason is that honey is hygroscopic: its high sugar concentration attracts water, the water contained in the quinces migrates little by little into the honey, and the quinces become hard. Columella notes this point too.

If the quince is immersed in honey after its maturation has begun, its metabolism will be lowered significantly, not only because it is by nature slow, but also because honey prevents oxygen from reaching the surface of the quinces. But the barrier should not be completely airproof. Columella insists that the vessels should have a large neck, and he does not say to close them, whereas he clearly says to close vessels in other cases. Indeed, without oxygen, how could the quince keep on slowly maturing to gradually provide the honey with the flavour Columella enjoys so much? Honey lets just the right quantity of oxygen migrate from the open air to the surface of the quinces.

Conclusion

Even if the Ancients did not know about sulphates (used today in the food industry for preventing enzymatic browning), they had the idea of using honey and its properties for preserving fruit, as attested to by Columella. This practice is still used today by the food industry, and can be found in different patents that describe preservation of fruits in solutions of sugared water or of calcium. Columella had very good reasons to approve this method for preserving quinces and most other fruits in honey. Although the know-how of the Ancients was based on experience, this case-study shows that they were dealing with issues that are still relevant today.

6.9. FINAL REMARKS

Marie Russel, Pascal Verdin and François Sigaut (†)

Storage has been a decisive step in the process of plant food treatment since before the Neolithic, critical for subsistence and the economic development of mankind, especially in sedentary societies. Storage processes do not always occur at the same stage in the plant treatment process, between harvest and transformation. Food products may be preserved at any point from raw material to elaborately transformed products such as jam, bread, cheese or wine.

Through the different approaches of history, archaeology, experimentation, ethnography, and archaeobotanical analysis, it is possible to explore the rich variety of storage techniques and purposes. In spite of this variety, it is still useful to consider the subject of storage as a whole, and to approach storage through interdisciplinary analyses. This focuses researchers' attention on new ways of studying storage techniques, to utilise subtle archaeological traces of construction and maintenance, and to highlight the biochemical processes of preservation and transformation as well as the social elements affecting storage structures and the choice of agricultural treatment processes.

Chapter Acknowledgements

The work by Leonor Peña-Chocarro and Lydia Zapata (Chapter 6.6) was carried out within the projects AGRIWESTMED (ERC-AdG-230561), HAR2008-03976/HIST, HAR2008-01920/HIST and HAR2011-23716 *Nuevos cultivos, nuevos paisajes* and Programa Consolider TCP-CSD2007-00058 from the Spanish Government. Lydia Zapata is part of the Research Group of the Basque Government UPV/EHU IT-288-07 and UPV/EHU UFI UFI11/09. Patricia C. Anderson acknowledges support from the GDR 2517, CNRS, France.

Chapter Abbreviations

BASOR	*Bulletin of the American Schools of Oriental Research*
APDCA	Association pour la Promotion et la Diffusion des Connaissances Archéologiques
CNRS	Centre National de la Recherche Scientifique
DRAC	Direction Régionale des Affaires Culturelles
FAO	Food and Agriculture Organization
SRA	Service Régional de l'Archéologie
INRAP	Institut National de la Recherche en Archéologie Préventive
SEM	Slovene Ethnographic Museum
MASCA	Museum Applied Science Center for Archaeology

Chapter Notes

1 Taphonomy is the study of decaying or fossilising organisms, involving the processes prior to and after their deposition.

2 To extract phytoliths from the sediment, the latter was:
- Deflocculated in sodium hexametaphosphate
- Sieved through a 125µm sieve to eliminate the larger particles
- Decanted in distilled water baths to eliminate particles smaller than 5 µm
- Limestone was dissolved in hydrochloric acid
- The phosphates were eliminated with nitric acid
- Phytoliths were collected through levigation with heavy liquid at 2.4 g/ml density (sodium polytungstate)
- Extracted phytoliths were then preserved in ethanol, put on slides and observed at 400x.

3 These historical and nowadays unofficial (= non-administrative) regions are not to be confused with the names of the adjacent Austrian federal states of Styria (Steiermark) and Carinthia (Kärnten).

Chapter Bibiography

Adrian, J., Drapron, R. and Gast, M. (1979) Caractéristiques biochimiques d'un grain de sorgho conservé pendant cinq ans dans un silo souterrain, au Yémen. In M. Gast and F. Sigaut (eds.). Tome 1, pp. 41–47.

Amouretti, M.-C. (1984) *Le pain et l'huile dans la Grèce antique. De l'araire au moulin.* Les Belles Lettres, Paris.

Anderson, P. C. (1999) Postscript. Experimental cultivation, harvest and threshing of wild cereals.

In P. C. Anderson (ed.) *Prehistory of Agriculture. New Experimental and Ethnographic approaches.* Institute of Archaeology Monograph 40. University of California, Los Angeles, pp. 119–144.

Anderson, P. C. (2003) Observations on the threshing sledge and its products in ancient and present-day Mesopotamia. In P. C. Anderson, L. S. Cummings, T. K. Schippers and B. Simonel (eds.) *Le traitement des récoltes: un regard sur la diversité du Néolithique au présent.* APDCA, Antibes, pp. 417–38.

Anderson P. C. (2006a) Premiers tribulums, premières tractions animales au Proche-Orient vers 8000–7500 BP? In P. Pétrequin, R.-M. Arbogast, A.-M. Pétrequin, S. van Willigen and M. Bailly (eds.) *Premiers chariots, premières araires. La diffusion de la traction animale en Europe pendant les IVe et IIIe millénaires avant notre ère.* CNRS Editions, Monographies du CRA 29, Antibes, pp. 299–316.

Anderson, P. C. (2006b) Non-mechanised processing and storage of cereals, grasses and pulses used for fodder, fuel, food and crafts: Examples from N. Tunisia, Atlas Region, Northwestern Tell. In Z. F. Ertuğ (ed.) *Proceedings of the IVth International Congress of Ethnobotany (ICEB 2005)*, Ege Yaynlar, pp. 223-231.

André, J. (1981) *L'alimentation et la cuisine à Rome*, Klincksieck, Paris. Reprint of the 1st edition from 1961.

Avrilleau S. (1975) *Cluzeaux et souterrains en Périgord.* Archéologie 24, tome 1, Périgueux.

Baš, F. (1952) Kašte na Dobrovljah (Granaries on Dobrovlje). *Slovenski etnograf* 5, 18–43.

Baš, F. (1970) Gospodarska poslopja (Farm Outbuildings). In P. Blaznik, B. Grafenauer and S. Vilfan (eds.) *Gospodarska in družbena zgodovina Slovencev: Zgodovina agrarnih panog I.* Državna založba Slovenije, Ljubljana, pp. 595–607.

Besri, M. (1989) Etat sanitaire des semences de blé et d'orge utilisées au Maroc. In *Céréales en régions chaudes.* AUPELF-UREF, John Libbey Eurotext, Paris, pp. 85–94.

Beyer, L., Boyrie-Fénié, B. and Tucoo-Chala, J. (2007) *Le Paysan de la forêt dans les Landes de Gascogne. Habitation, travail, famille.* Éditions Cairn, Pau.

Binford, L. R. (1978) *Nunamiut ethnoarchaeology.* Academic Press, New York.

Blanchet, A. (1923) *Les souterrains refuges de la France.* Picard, Paris.

Bouby, L. (2003) De la récolte au stockage : éclairages carpologiques sur les opérations de traitement des céréales à l'âge du Bronze dans le sud de la France. In P. C. Anderson, L. S. Cummings, T. K. Schippers and B. Simonel (eds.) *Le Traitement des récoltes : un regard sur la diversité du Néolithique au présent.* APDCA, Antibes, pp. 21–46.

Boyeldieu, J. (1980) *Les cultures céréalières.* Hachette, Paris.

Brenton, B. P. (1989) *Rotten to the Core?: Experimental Archaeology of Iron-Age Type Food Storage Pits.* Unpublished manuscript. European Field Research Program, University of Massachussetts, Amherst.

Brunet, P. and Sethian, E. (2001) La fosse de l'étape 1 du Bronze final de Bussy-Saint-Georges, « la Manjoire ». *BGASM*, 39–42, 1998–2001, 23–29.

Buxó, R., Gransar, F., Matterne, V. and Pommepuy, C. (2000) Témoins archéologiques de la chaîne opératoire de traitement des céréales vêtues à l'Age du Fer dans le Nord de la France. *Els productes alimentaris d'origen vegetal a l'edat del ferro de l'Europa Occidental : de la producció al consum. Série monogràfica,* Girona, pp. 237–249.

Cammas, C., Marti, F. and Verdin, P. (2005) Aménagement et fonctionnement des structures de stockage de la fin de l'âge du Fer dans le Bassin Parisien. In *L'âge du Fer en Ile de France.* 26e supplément à la Revue Archéologique du Centre de la France, vol. 2, FERACF, Tours, pp. 33–54.

Cevc, T. (1984) *Arhitekturno izročilo pastirjev, drvarjev in oglarjev na Slovenskem (Architectural Tradition of Shepherds, Woodcutters, and Charcoal-burners on the Slovene Territory).* Državna založba Slovenije, Ljubljana.

Cevc, T. and Primožič, I. (1988) *Kmečke hiše v Karavankah (Farmhouses in Karavanke).* Drava, Celovec and Založništvo tržaškega tiska, Trst.

Cevc, T. and Čop, J. (1993) *Slovenski kozolec (The Slovene Hayrack).* Agens, Žirovnica.

Columella, *De Re Rustica* (cited *R. R.*), XII. E. S. Forster and E. H. Heffner (eds.). Harvard University Press, Cambridge Ma; Heinemann, London, 1955.

Columelle (cited Columella) *De l'agriculture,* livre XII. J. André (ed.). Les Belles Lettres, Paris, 1ère édition 1988, 2002.

Colombini, M. P., Giachi, G., Modugno, F. and Ribechini, E. (2004) Characterisation of organic residues in pottery vessels of the Roman age from Antinoe (Egypt). *Microchemistry Journal* 79, 83–90.

Currid, J. D. and Navon, A. (1989) Iron Age pits and the Lahav (Tell Halif) Grain Storage Project. *BASOR* 273, 67–78.

Derreumaux M., Malrain, F. and Matterne, V. (2003) Indices archéologiques et archéobotaniques du traitement des céréales du 2e âge du Fer à la fin de la période gallo-romaine en France septentrionale. In P. C. Anderson, L. S. Cummings, T. K. Schippers and B. Simonel (eds.) *Le traitement des récoltes: un regard sur la diversité du Néolithique au présent.* APDCA, Antibes, pp. 219–234.

Ducom, P. (1980) Eléments d'écologie des stocks et de lutte contre les ravageurs. *L'amélioration des systèmes post-récolte en Afrique de l'Ouest,* ACCT, Paris, pp. 65–83.

Ertuğ-Yaras, F. (1997) *An ethnoarchaeological study of subsistence and plant gathering in Central Anatolia.* PhD thesis, St. Louis, Missouri, Washington University.

Firmin, G. (1984) Paléoenvironnement et archéologie expérimentale. Agriculture expérimentale dans la vallée de l'Aisne, Amiens. *Revue Archéologique de Picardie* 1, 95–102.

Gast M. and Sigaut F. (1979) (eds.), *Les techniques de conservation des grains à long terme*. Vol. 1, CNRS, Paris.

González Urquijo, J. E., González Vázquez, A., Ibáñez Estévez, J. J., Moreno García, M., Peña-Chocarro, L., Ruiz Idarraga, R., Zapata Peña, L. and Gómez Pellón, E. (2001) Un proyecto etnoarqueológico y antropológico en el Rif occidental. Avance sobre los estudios del trabajo de campo del año 2000. *Edades. Revista de Historia* 8, 91–104.

González Urquijo, J. E., Ibáñez Estévez, J. J., Peña-Chocarro, L. and Zapata Peña, L. (2001) Estudio etnoarqueológico sobre la cerámica Gzaua (Marruecos). Técnica y contexto social de un artesanado arcaico. *Trabajos de Prehistoria* 58 (1), 5–27.

Gransar, F. (2003) L'Apport de l'étude du stockage à la réconciliation des systèmes agro-alimentaires de l'âge du Fer en France septentrionale. In P. C. Anderson, L. S. Cummings, T. K. Schippers and B. Simonel (eds.) *Le traitement des récoltes: un regard sur la diversité du Néolithique au présent*. APDCA, Antibes, pp. 201–217.

Groenendijk, H. A. (1987) Mesolithic hearth-pits in the Veenkoloniën (prov. Groningen, the Netherlands): defining a specific use of fire in the Mesolithic. *Palaeohistoria, 29*, 85–102.

Halstead, P. (1981). From determinism to uncertainty. Social storage and the rise of the Minoan palace. In A. Sheridan and G. Bailey (eds.) *Economic archaeology* Oxford, Archaeopress, pp. 187–214.

Hazler, V. (2004) *Kozolci na Slovenskem* (*Hayracks on the Slovene Territory*). Kmečki glas, Ljubljana.

Hill, R. A., Lacey, J. and Reynolds, P. J. (1983) Storage of Barley Grain in Iron Age Type Underground Pits. *Journal of Stored Products Research* 19 (4), 163–171.

Hillman, G. C. (1984) Traditional husbandry and processing of archaic cereals in recent times: the operations, products and equipment which might feature in Sumerian texts. Part I: the glume wheats. *Bulletin of Sumerian Agriculture* 1, 114–152.

Ibáñez Estévez, J. J, Peña-Chocarro, L., Zapata, L. and González Urquijo, J. E. (2001) Les récipients en bouse de vache et argile non cuite (*tabtoba* et *tonna*) dans la région Jebala (Maroc). Exemple d'un processus technique de type domestique. *Techniques et Culture* 38, 175–194.

Ingold, T. (1983) The significance of storage in hunting societies. *Man, 18*, 553–571.

Jacomet, S., Kučan, D., Ritter, A., Suter, G. and Hagendorn, A. (2002) *Punica granatum* L. (pomegranates) from early Roman contexts in Vindonissa (Switzerland). *Vegetation History and Archaeobotany* 11, 79–92.

Jardé, A. (1979) *Les Céréales dans l'Antiquité grecque : la production*. De Boccard, Paris.

Jazouli Belhaboub, H. (2000) *Ravageurs des céréales stockées* (Formation of the retailers of pesticides, organized by the Département Provincial d'Agriculture de Meknes), Meknes, pp. 1–8.

Jennings, J., Antrobus, K. L., Atencio, S. J., Glavich, E., Johnson, R., Loffler G. and Luu C. (2005) Drinking Beer in a Blissful Mood. Alcohol Production, Operational Chains, and Feasting in the Ancient World. *Current Anthropology* 46 (2), 275–303.

Juvanec, B. (2007) *Kozolec* (*The Hayrack*). Fakulteta za arhitekturo, Ljubljana.

Kimpe, K., Jacobs, P. A. and Waelkens, M. (2002) Mass spectrometric methods prove the use of beeswax and ruminant fat in late Roman cooking pots, *Journal of Chromatography* A, 151–160.

Kunz, L. (2004) *Grain pits*. Wallachian open-air museum, vol. 1 Roznov pod Radhostem, 180 p. on CD.

Kunz, L. (1951) 'Doły' na Záhoří. *Časopis Moravského Muzea* 36,182–188.

Kunz, L. (2004) DVD

Leconte, L., Auxiette, G., Boyer, J.-F. and Giraud, P. (2002) Rapport de fouille préventive du site de Saint-Denis-ZAC Nozal-Chaudron. DRAC-SRA-INRAP, Saint-Denis.

Ložar, R. (1944) Pridobivanje hrane in gospodarstvo (Obtaining food and agriculture). In R. Ložar (ed.) *Narodopisje Slovencev* (Ethnography of the Slovenes). Založba Klas, Ljubljana, pp. 98–192.

Madella, M. (2001) Understanding archaeological structures by means of phytolith analysis: a test from the Iron Age site Kilise Tepe-Turkey. In J. D Meunier and F. Colin (eds.), *Phytoliths: Applications in Earth Sciences and Human History*. Balkema Publishers, Netherlands, pp. 173–182.

Madella, M., Alexandre, A. and Ball, T. (2005) International Code for Phytolith Nomenclature 1.0. *Annals of Botany* 96, 253–260.

Makal, M. (1954) *A Village in Anatolia*. Vallentine, Mitchell and Co., London.

Makarovič, G. (2007) Kdaj so nastali kozolci? (When did hayracks first appear?). *Etnolog* 17, 209–248.

Malainey, M. E., Przybylski, R. and Sherriff, B. L. (1999) The effects of thermal and oxidative degradation on the fatty acid composition of food plants and animals of Western Canada: implications for the identification of archaeological vessel residues. *Journal of Archaeological Science* 26(1), 95–113.

Mauny R. and Cordier G. (1967) Souterrains refuges, caves-fortes et 'hypogées' de Touraine. *Bulletin de la Société des Amis du Vieux Chinon* VII (1), 13–95.

Melik, A. (1931) *Kozolec na Slovenskem* (*Hayracks on the Slovene Territory*). Znanstveno društvo, Ljubljana.

Miklavčič Brezigar, I. (2000) *Ljudske skrinje na Goriškem* (*Folk Chests in Goriško Region*). Goriški muzej, Nova Gorica.

Molist, M. (1998) Espace collectif et espace domestique dans le Néolithique des ixe et viiie millénaires BP au nord de la Syrie : apports du site de Tell Halula (vallée de l'Euphrate). In M. Fortin and O. Aurenche (eds.), *Espace naturel, espace habité en Syrie du Nord (10e–2e millénaires av. J.-C.)*, TMO 28, Lyon and BCMS 33, Québec, pp. 115–130.

Muntz A. (1881) Etudes sur la conservation des grains par l'ensilage. *Annales de l'Institut National Agronomique* 3, 19–73.

Mušič, M. (1970) *Arhitektura slovenskega kozolca (Architecture of the Slovene Hayrack)*. Cankarjeva založba, Ljubljana.

Novak, V. (1960) *Slovenska ljudska kultura (Slovene Folk Culture)*. Državna založba Slovenije, Ljubljana.

Pearsall, D. (2001) *Paleoethnobotany: A Handbook of Procedures.* 2nd edition. Academic Press, San Diego (CA).

Peña-Chocarro, L. and Zapata, L. (2003) Post-harvest processing of hulled wheats. An ethnoarchaeological approach. In P. C. Anderson, L. S. Cummings, T. K. Schippers and B. Simonel (eds.) *Le traitement des récoltes: un regard sur la diversité du Néolithique au présent.* APDCA, Antibes, pp. 99–114.

Peña-Chocarro, L., Zapata, L., González Urquijo, J. E., Ibáñez Estévez, J. J. and Moreno García, M. (2005) An ethnoarchaeological project in the Western Rif (northern Morocco): First results. *Quaderni del Laboratorio di Archeobiologia* 1, 21–32.

Peña-Chocarro, L., Zapata, L., González Urquijo, J.E. and Ibáñez, J. J. (2009). Einkorn (*Triticum monococcum* L.) cultivation in mountain communities of the western Rif (Morocco): an ethnoarchaeological project. In A. S. Fairnbairn and E. Weiss (eds.) *From foragers to farmers. Gordon Hillman Festschrift.* Oxbow Books, Oxford, pp. 103–111.

Piperno, D. R. (2006) *Phytoliths. A Comprehensive Guide for Archaeologists and Paleoecologists.* AltaMira Press (Rowman & Littlefield), Lanham, New York, Toronto, Oxford.

Pline l'Ancien (cited Pliny) *Histoire naturelle* (cited *N. H.*), livre XV. J. André (ed.). Les Belles Lettres, Paris, 1960.

Pointel, J. C. and Coquard, J. (1977) (eds.) *Le pourcentage de pertes en poids et la perte spécifique, critères d'évaluation des dégâts causés par les insectes dans les céréales et les légumineuses stockés.* Agro Trop. Paris, pp. 377–381.

Redlingshöfer, B. and Soyeux, A. (2013) Losses and wastage. In C. Esnouf, M. Russel and N. Bricas (eds.) *Food System Sustainability. Insights From duALIne.* Cambridge University Press, Cambridge, pp. 136–157..

Reneaume, (1708) Sur la manière de conserver les grains. *Histoire de l'Académie royale des sciences, Mémoires de mathématiques et de physique.* Paris, pp. 63–86.

Reynolds, P. J. (1974) Experimental Iron age storage Pits. An interim Report. *Proceedings Prehistoric Society* 40, 118–131.

Reynolds, P. J. (1979) A general report of underground grain storage experiments at the Butser Ancient Farm research project. In M. Gast and F. Sigaut (eds.) *Les techniques de conservation des grains à long terme*, vol. 1. CNRS, Paris, pp. 70–80.

Rottländer, R. C. A. (1990) Lipid analysis in the identification of vessel contents. *MASCA Research Papers in Science and Archaeology* 7, 37–40.

Russel, M. (2004) *Le rôle alimentaire et la conservation des fruits dans l'antiquité grecque et romaine. Étude historique et expérimentale.* Unpublished PhD thesis, University of Provence.

Šarf, F. (1971) *Kmečka hiša v slovenskem panonskem svetu (The Farmhouse in the Slovene Pannonian World).* Slovenski etnografski muzej, Ljubljana.

Serres, O. de (1600) *Le théâtre d'agriculture et de mesnage des champs.* Paris.

Sigaut, F. (1975) *L'agriculture et le feu. Rôle et place du feu dans les techniques de préparation du champ de l'ancienne agriculture européenne.* Mouton, Paris.

Smerdel, I. (2004) Slovenska atenska akropola (Slovenia's Very Own Acropolis). In M. Schindler (ed.) *15 + 10 European Identities.* Österreichisches Museum für Volkskunde, Vienna, pp. 63–65.

Testart, A. (1982) The significance of food storage among hunter-gatherers: residence patterns, population densities and social inequalities. *Current Anthropology,* 23, 523–538.

Varron (cited Varro), *Économie rurale* (cited *R. R.*), I, livre I. J. Heurgon (ed.). Les Belles Lettres, Paris, 1978.

Verdin, P. (2002) Analyse de phytolithes des structures archéologiques du site de Rungis-Les Antes. In *Néolithique et protohistoire du site de Rungis-Les Antes.* Artcom, Paris, pp. 41–47.

Wilson, G. L. (1987) *Buffalo Bird Woman's Garden: Agriculture of the Hidatsa Indians.* Minnesota Historical Society Press, St Paul.

World Bank (2011) *Missing food: the case of post-harvest grain losses in Sub-Saharan Africa.* Report No. 60371-AFR. World Bank, Washington D.C.

Zapata, L., Ibáñez, J. J., González Urquijo, J. E. and Peña-Chocarro, L. (2003a) Las casas de cubierta vegetal del Rif occidental: aspectos arquitectónicos, simbólicos y sociales. *El Pajar. Cuaderno de Etnografía Canaria* 14, 137–143.

Zapata, L., Peña-Chocarro, L., Ibáñez Estévez, J. J. and González Urquijo, J. E. (2003b) Ethoarchaeology in the Moroccan Jebala (Western Rif): Wood and Dung as Fuel. *Africa Praehistorica* 15, 163–175.

7 Cereal Processing and Cooking: Techniques, Tools and Gestures

7.1. INTRODUCTION

Hara Procopiou

Techniques employed in plant processing, prior to food preparation, are strong indicators of dietary habits. Each type of processing modifies the preservation of the plant, determines its nutritional value and modifies the texture and the flavour of the meal. Through case studies, this chapter presents a selection of plant processing techniques in preindustrial societies, although it does not aim at being exhaustive. Emphasis is put on the main staple food of Mediterranean and western European societies, cereals. More particularly, contributions deal with dehusking, parching and milling techniques, as well as with culinary preparations.

Different methodological approaches will be combined: archaeobotany, phytolith and residue analysis and use-wear studies. Confrontation of archaeological data with ethnographic, textual, and iconographic data reveals fundamental insights. Other approaches, such as palaeopathology (Molleson 1994) and dental use-wear, not considered in this chapter, are also essential for reconstructing cereal processing and dietary habits (Eshed *et al.* 2006), but fall outside the scope of this book.

Based on ethnographic data and ancient textual evidence, Leonor Peña-Chocarro proposes an overview of hulled wheat processing in Chapter 7.2. Although hulled wheats are nowadays in the process of extinction, they were among the earliest domesticated plants, spread over the Mediterranean and western Europe during the Neolithic, and they remained staple crops for many millennia (Perrino *et al.* 1996; Nesbitt and Samuel 1996; Colledge and Conolly 2007). Why were hulled cereals favoured crops, even though they require an extra step after threshing, in order to break the spikelets and release the grain? Their resistance to disease, their capacity to grow on poor soils, but also their particular taste, so appreciated by Greek and Roman authors (see for example Pliny's *Natural History*, XVIII, 7), are the essential factors influencing this choice (Nesbitt and Samuel 1996, 70). Peña-Chocarro also reveals the variability of tools (querns, mills, mortars, etc.) employed for dehusking, across space and time. Her research shows that techniques adopted are intimately related to the framework of production. For example, simple tools such as the mallet are utilised in the context of everyday life for dealing with small amounts of grains. Finally, this paper, using experimental and ethnographic data, shows that parching, although mentioned in ancient literature, was not a necessary operation. Its adoption seems to be related to specific cultural or technical preferences and contexts.

In Chapter 7.3, Pascal Verdin deals with such a specific context, the Gallo-Roman site of Lieusaint-La Pyramide in northern France. A phytolith analysis on several samples from a structure related to grain drying is presented. Even though several hypotheses concerning the purpose of this roasting have been considered, the low temperatures attested strongly suggest parching for malt production.

Chapters 7.4 and 7.5 present plant processing tools, essentially querns and handstones, and deal with tool evolution and variability. Grinding stones, often associated with the emergence of agriculture, came into light in several Upper Palaeolithic and Mesolithic sites in the Near East and Europe (Wright 1994; Arraguren *et al.* 2007). During these early phases, as Van Gijn points out for the Mesolithic of the Lower Rhine Basin, problems of recognition of these unmodified slabs and pebbles can occur. It is only recently that morphological studies (De Beaune 2000), use-wear and residue analysis (Piperno *et al.* 2004), as well as ethnographic data on hunter-gatherer societies (Stahl 1989) have revealed the role of such simple unmodified tools in plant processing (wild seeds, tubers, etc.). Distribution of wear on quern surfaces shows an evolution from a prominent circular to a more linear back-and-forth movement, which was to be generalised during the Neolithic period (De Beaune 2000).

Querns and handstones would persist through the Bronze and Iron Ages. It is only during the 5th century BCE that the Olynthus mill or 'hopper rubber' appears in the eastern Mediterranean and is diffused westward by Greek settlers (Moritz 1958; Frankel 2003). In this same period, the manual rotary mill appears in the northeastern Iberian Peninsula and is diffused eastward during the 4th and 3d centuries BCE (Alonso 1999). Finally, during the 2nd century BCE, the Roman biconical rotary querns appear. The more relevant type, the Pompeian mill, was animal-powered. Adoption of these more efficient grain mills, employed essentially in workshops, did not eliminate simple grinding slabs, which remain in use even nowadays in North Africa (Gast 1968) or Anatolia (Ertuğ-Yaras 2002), often in parallel with small domestic rotary querns. Although the history of milling tools is marked by several technical innovations, the persistence of simple types through the millennia reveals the tenacity of technical traditions in everyday life.

Case studies presented in Chapters 7.2 and 7.3 deal with Neolithic and Bronze Age grinding tools. The methodology adopted follows the recent development of low- and high-power use-wear analysis of grinding stones, which defined diagnostic wear traces related to the raw material of the querns and handstones, as well as to the composition of the substances ground (Procopiou 1998; Adams 2002; Delgado 2003; Dubreuil 2004; Van Gijn and Houkes 2006; Verbaas and Van Gijn 2007; Hamon 2006; Adams *et al.* 2009). Scientific advances in residue analysis of chemical components, phytoliths and starch grains have also been considered (Fullagar 1998, 2006; Procopiou *et al.* 2002; Fullagar *et al.* 2008). This methodology enables Van Gijn to show that cereals were processed by the Neolithic wetland inhabitants of the Lower Rhine delta (Chapter 7.2). Patterns of wear distribution have shown that querns were used in a longitudinal fashion and not a rotary one, even during the Middle Neolithic, when the querns became smaller and shorter than the early Neolithic ones. Persistence of the longitudinal movement, non-adapted to these small querns, suggests that this gesture was part of the cognitive baggage of Neolithic farmers.

A similar functional approach is applied by Procopiou to several sites in Bronze Age Crete, (Chapter 7.5) showing that, despite an apparent homogeneity, several changes occur in quern production processes and function. The choice of raw material suggests a preference for stones requiring low maintenance, at least in palatial contexts. In contrast, an increased investment in manufacture is attested. Patterns of distribution of mineral and cereal processing within settlements show a dispersed production and the absence of milling workshops. As a final point, the pronounced gloss on several querns led the authors to suppose that grains were ground without being systematically dehusked, thus indicating the consumption of a whole-grain meal.

Finally, both papers examine the context of production and the distribution of grinding activities. Data at our disposal do not permit the authors to deal with the social and sexual division of tasks, aspects analysed by several anthropological, ethnographic or ethno-archaeological studies (Roux 1985; Hayden 1987; David 1999; Katz 2003). Osteological deformations of Mesolithic female skeletons in the Near East are indeed associated with body position and forces applied during grinding (Mollesson 1994).

Mycenean texts of the Aegean Late Bronze Age attest to a female labour force (*me-re-ti-ri-ja*) responsible for cereal milling within palaces (Palmer 1992). As a laborious everyday task for Penelope's servants in the *Odyssey* (Song U, 118–119), grinding is intimately related, according to ethnographic data, to the female condition. It is frequently accompanied by songs, in which women express the wretchedness of their condition (Poitevin 1997), even sometimes by 'funeral dirges' (Pasayiannis 1928). Grinding can sometimes be performed by males, but in that case, as in Pharaonic Egypt and the Sumerian Near East, it is a task carried out by servants (Grégoire 1992). Concerning Classical Antiquity, the adoption of the Olynthus mill is accompanied by the development of workshops mobilising slaves, a practice which was to become more generalised in the Roman world (Moritz 1958). Nomads (Gast 2003) or soldiers can also undertake the task. Reducing a man to doing a woman's work, for instance milling, is also a severe sanction or humiliation, illustrated, among other examples, by Sampson's punishment (Judges 16, 22). All these exceptions establish that grinding is perceived as a female task par excellence.

Chapter 7.6 deals with recipes and taste, providing an overview of the processing of barley and pulse meal. In order to interpret the variability of culinary preparations observed, Procopiou considers the parameters governing the adoption of specific techniques. If scholars must take into account environmental and climatic factors, the context of production and consumption must also be evaluated. Culinary preparation can be carried out in domestic contexts but also in workshops. Meal or flour can be consumed on an everyday basis, but also during banquets, civic or religious feasts. Recent psychological, sociological and anthropological studies show the importance of dietary habits for constructing the sense of belonging and for creating cultural, ethnic, social or religious identities (Beardsworth and Keil 1997; Le Breton 2006). More particularly the perception of smells and flavours informs relations between the sexes, between classes and ethnic groups (Classen *et al.* 1994; Le Guérer 2002). Textual evidence from Classical Antiquity reveals recurrent oppositions: civilised vs. savage food, Greek vs. barbarian food, town-dwellers' vs. peasant food, citizen's vs. slave's food, food for the young vs. food for the old, food for the living, for the dead, for the gods. In the same period, the suffix '*-phagoi*' *(-eaters)* designates regional or ethnic identity: Boeotians are *kollikophagoi* (eaters of a barley bread roll), Thracians are referred to as *ospreophagoi* (pulse-eaters), whereas the Romans are called *pultiphagoi* (pulse-eaters, referring to a semi-liquid preparation of pulse flour and milk).

In order to cast light on these ancient '*phagoi*', a global approach is necessary. Archaeological and ethnographic studies of cooking pots, tools and structures implied in the processing, as well as botanical, textual and iconographical evidence can, by identifying specific techniques and recipes, evoke the smells, the flavours and the tastes of the past.

7.2. PARCHING AND DEHUSKING HULLED WHEATS

Leonor Peña-Chocarro and Lydia Zapata

Introduction

Einkorn (*Triticum monococcum*), emmer (*Triticum dicoccum*) and spelt (*Triticum spelta*) belong to the group of the so-called hulled wheats characterised by the presence of persistent tough glumes that surround their grains. In other words, and contrary to free-threshing wheats whose grain and chaff separates during threshing, the tough glumes of these species remain, thus tightly enclosing the grain after threshing. Therefore the spikelets, originated during threshing by the breaking of the cereal ear, need to be dehusked. This additional operation makes it possible to free the grain from the chaff prior to its consumption.

This contribution explores the range of steps involved in processing hulled wheats based on ethnographic information and data from Classical sources. The ethnographic data comes from both the authors' work in the Mediterranean area (Iberian Peninsula and Morocco) which has been largely published (Peña-Chocarro 1996; 1999; Peña-Chocarro and Zapata 2003; Peña-Chocarro *et al.* 2005; 2009) and also from other scholars' research (D'Andrea and Haile 2002). This contribution does not intend to provide brand new data on parching or dehusking, but to summarise otherwise scattered information and thus provide to the general public and researchers in their early career stages a more complete insight into two, often unknown, steps in the hulled wheat processing sequence, so that a more comprehensive understanding of the two operations can be gained.

When discussing hulled wheat processing, it has been commonly assumed that parching was a necessary operation prior to dehusking in order to render the chaff brittle and so free the grain from the husks (see examples in Nesbitt and Samuel 1996). This idea was partly based on the references found in some Classical authors who described contact with fire in the processing of some hulled wheats. However, as discussed by Nesbitt and Samuel, the different concepts and meanings of parching used by the various authors, as well as the translations made of the original sources have caused some controversy. In fact, when examining the various examples, it is often unclear which process is discussed in each case, *i.e.* roasting, drying, parching, etc., and at which point in the processing sequence this contact took place. A proper definition of parching is given by Nesbitt and Samuel, who point out that it is a type of cooking which implies a structural transformation of the chaff and sometimes of the grain as well. Drying, on the contrary, does not involve any structural transformation of the chaff. Lüning and Meurers-Balke (1980), based on experimental work, have suggested that temperatures up to 100°C refer to drying, whereas over 150°C, it is possible to talk of parching (Nesbitt and Samuel 1996).

Parching, Roasting, Drying: Historical and Ethnographic Sources

Various Classical authors (Varro, Pliny and Ovid) mentioned activities in which hulled wheats (in particular emmer) are involved. References to roasting, parching, or drying are commonly quoted, although it often remains unclear which activity they were actually recording.

Varro (116–27 BCE) in his *Res Rusticae* (1.63) in his discussion of cereal storage, explained that during the winter, emmer ears ('spelt' in the English translation, *far* in the Latin version) should be ground (*pisetur*) on the mill and parched (*torreatur*) (Hooper and Ash 1994). The use of *pisetur* and *torreatur* could perhaps be better translated as pounded and roasted. In a later passage (1.69), he indicated that in winter, emmer for consumption had to be taken to the mill to be roasted (*torrendum*). Although there are no further details, it is worth noting that in the processing sequence described, pounding comes prior to roasting, as implying that the contact with fire came as a later stage. This would automatically exclude the utilisation of fire for facilitating the pounding. Although Varro does not include details on the possible reasons for emmer roasting, this may be related to rendering it more wholesome. This is a practice encountered in traditional communities in order to add flavour to flour and meals. Amongst others, it has been described in Ethiopia (Lyons and D'Andrea 2003; Lyons 2007) and the authors have also documented the roasting of barley grain in Morocco, so Varro's description might be referring to a similar custom. This practice has also been recorded in other Classical authors such as Pliny (23–79 CE) (*Natural History* XVIII, 7–8) who recorded the Fornacalia Festival (Feast of the Ovens) as a festivity when the roasting of emmer took place in order to render it more wholesome. According to Pliny, Numa established the worship of the gods with an offering of wheat and obtained their favour with a salted cake which, following the Roman annalist Hemina, was of roasted emmer, as in this way it was more wholesome. This passage seems to refer to an activity aiming to improve the taste of the grain and not to a process of facilitating dehusking.

In another passage (XVIII, 97–98) focusing on the processing of cereals, Pliny says that in Etruria, emmer was roasted after it had been pounded. Roasting here could be related to two different cases, the drying of pounded spikelets or the removal of ears, as will be explained later. The description of pounding by this author includes a reference to a method described by the Carthaginian Mago (4th century BCE) which consisted of steeping the wheat in water before pounding, then shelling the spikelets, drying them in the sun and, finally, pounding them in a mortar. Experiments carried out by D. Samuel (1994; 2010) on ancient Egyptian emmer processing have tested this practice. In fact, Samuel has shown that dampening the spikelets facilitates pounding, at least when shallow mortars are used, whereas experiments with tall, narrow mortars did not require dampening. Similar results have been obtained by experiments on hulled barley processing in Bronze Age Crete (Procopiou 1998; 2003). Drying the damp mixture could be easily carried out outdoors under the sun, but sun-drying may have been difficult in wet areas where farmers could have been forced to use fire.

This might explain Varro's and Pliny's texts above. However, as mentioned earlier, the contact with fire after pounding may also concern other activities such as the removal of cereal awns. Pliny (XVIII, 299) says that emmer is stored in its chaff and only has the straw and the beard removed. No explanations are given on how this process took place, but it reminds us very much of a custom still practiced in Asturias (northern Spain; Fig. 7.1) where emmer and spelt ears are singed in the fire with the aim of removing the awns (Peña-Chocarro 1999) (Fig. 7.2).

Another important Classical author, Ovid (43 BCE–17 CE), mentioned that emmer 'shall bear twice the fire' (*Fasti* Book I, 690). In addition, his Book II records the roasting/toasting of emmer and the losses people may incur when doing so, as fire may spread to their huts. In trying to identify the reasons for this double contact with fire, one should also refer to a passage found in his Book I, where Ovid refers to the establishment of a shrine where the sacrificial spelt (emmer) would be burnt.

Consequently, bearing in mind the information already analysed above and contained in Pliny's text description of roasting emmer, one could assume that the burning (roasting) of emmer was part of a ritual which may have involved the roasting of grains to either improve their taste or stress the purity of the offering. The reasons for emmer to be put in contact with fire a second time are, however, still unknown, although the ethnographic record may provide some possible explanations.

Accounts of hulled wheat cultivation from various geographical areas seem to indicate that, if any contact with fire occurred during the processing of these species, this was not related to rendering the chaff brittle and therefore making pounding easier, but was instead connected with various other factors (see examples in Bennet and Elton 1898; Fenton 1978; 1991; Hillman 1981; 1982; 1984a;

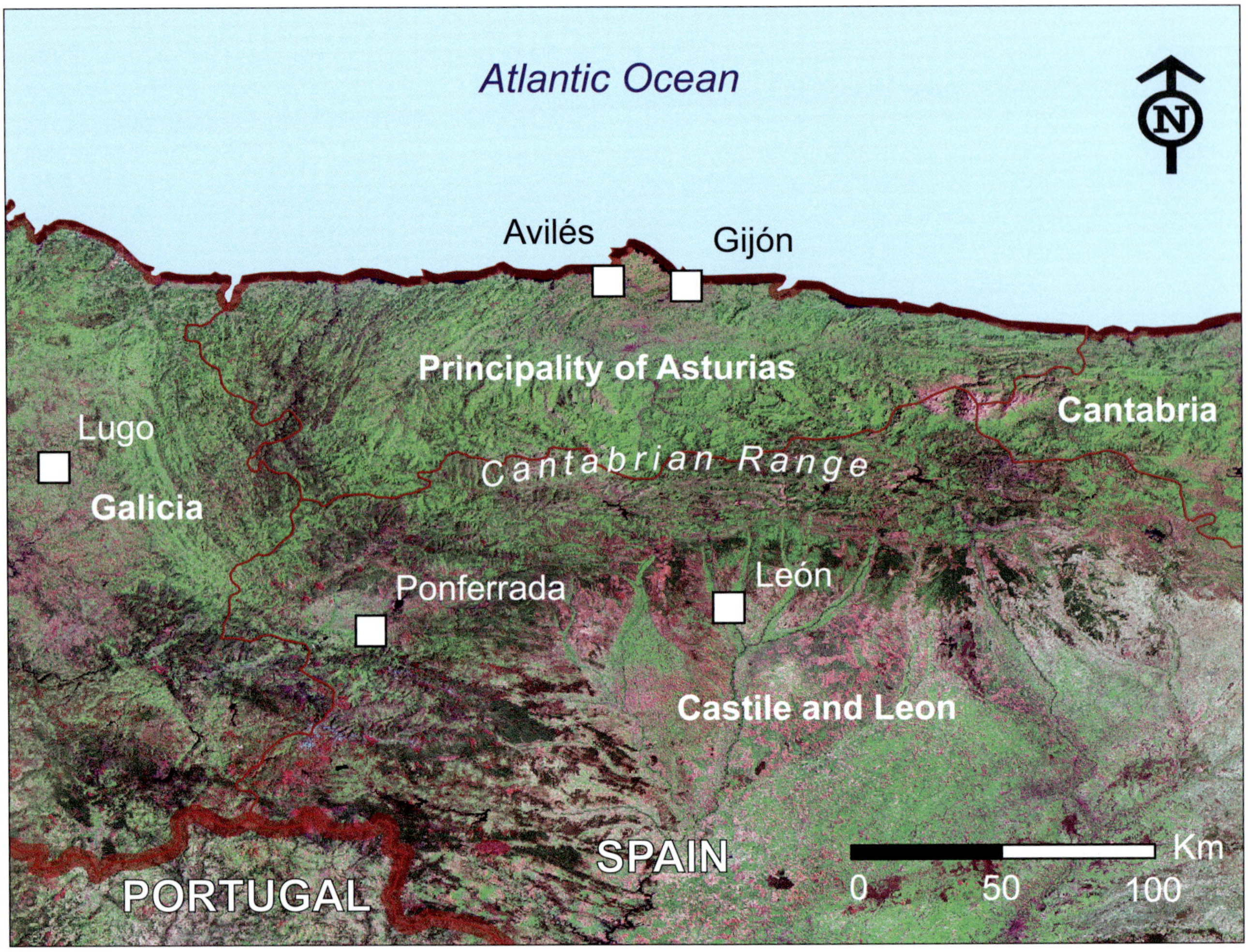

Fig. 7.1. The province of Asturias in northern Spain. Map: J. C. Loubier and A. Chevalier.

Fig. 7.2. Emmer ears are singed in order to remove the awns (Asturias, northern Spain).

1984b; 1985; Toffin 1983; Körber-Grohne 1987; Peña-Chocarro 1994; 1996; 1999; Peña-Chocarro and Zapata 2003; Van der Veen 1989). Moreover, experimental work on kiln-drying/parching prior to dehusking carried out by Meurers-Balke and Lüning (1992) has shown that previous parching produced very little effect on dehusking.

Factors that explain why hulled wheats may have come in contact with fire can be classified into three main groups: environmental conditions, cultural preferences (taste) and technical factors. Environmental conditions and in particular short and wet summers typical of many areas of northwestern Europe may determine the need to dry the crop by heating. Drying is necessary when crops are harvested still unripe due to excessive environmental humidity and before they are bulk-stored. This pertains to both free-threshing and hulled wheats. There are many ethnographic accounts of this practice using different devices (kilns, nets, pots) from Scotland and Ireland (Bennet and Elton 1898; Fenton 1978; 1991; 1992), the Faroe islands (Hillman 1982) or Iceland (Guðmundsson 1991). The use of perforated wooden boards placed on top of fires is also well documented in northern Spain (Asturias) for drying emmer and spelt ears, in Norway (Fenton 1991), and in Nepal for wheat and barley ears (Toffin 1983). Finally, Van der Veen (1989) provides numerous examples of corn-drying and malt production where permeable surfaces made of different materials (straw, wood, iron, etc.) were used to dry the grain, although in these cases, the main aim of drying/heating was malt production.

Cultural preferences and, in particular, taste have been among the factors accounting for the use of fire when processing cereals. Hulled and free-threshing grains at different ripening stages have traditionally been roasted in many parts of the world. References to hulled wheats are found in Turkey (Hillman 1981; 1982; 1985), Slovakia (Markus 1989), Germany (Körbe-Grohne 1989; Sigaut 1989), whereas for barley, the work by Lyons and D'Andrea (2003; Lyons 2007) is of particular interest. There are also many references from Europe and the Near East as well as from North Africa regarding free-threshing cereals. In fact, rye (Gunda 1983) and oats (Bennett and Elton 1898; Fenton 1978; 1991; 1992) amongst others are some of the crops involved.

Finally, technical factors may have led to the use of fire to facilitate hulled wheat processing. A clear example comes from northern Spain (Asturias; Fig. 7.1) where farmers use fire to remove the robust awns of emmer and spelt, as these hamper the dehusking process and sowing (Buxò 1989; Peña-Chocarro 1999; Peña-Chocarro and Zapata 2003). The operation is performed using a fork and scooping up ears from a cereal pile and shaking them over a fire very quickly so only the awns burn away while leaving the grain unaltered. Some farmers indeed only apply this procedure to the part of the crop to be sown. Of course free-threshing wheats may also be put in contact with fire due to technical reasons such as drying grains before milling to harden them or to reduce humidity in sprouted grains. Despite all this information, Hillman (1981; 1982; 1984a; 1984b; 1985 and pers. comm.) provided an example of parching ears to make them more easily 'breakable' in two Turkish villages, located in an area of high rainfall, but this could be more related to drying damp spikelets (Hillman pers. comm.).

Combining both the experimental and the ethnographic evidence, it is possible to conclude that, at some point in the crop processing sequence, the crop may enter in contact with fire, but the reasons appear related to drying or roasting for improving taste or to technical reasons which have little to do with rendering the chaff brittle and making dehusking easier.

Dehusking: Historical and Ethnographic Sources

The next issue to be taken up in this contribution is dehusking, a type of activity traditionally performed either by pounding or using water mills or hand mills. As in the section above, the aim is to gather information scattered in different works.

Classical references mentioned these practices in Roman times. For instance, Pliny in his description of dehusking methods (*Natural History* XVIII, 23) for

Fig. 7.3. The Rif region, northern Morocco. Map: R. Lugon, J.-C. Loubier and A. Chevalier.

Fig. 7.4. (left) Pounding spikelets in a mortar excavated into the soil (Rif region, northern Morocco).

Fig. 7.5. (right) Rubbing spikelets against a basket with a circular piece of cork (Rif region, northern Morocco).

emmer, included the use of pestles fitted with iron, and dehusking was also carried out in water-powered wheels, millstones and handmills. Cato (*De Re Rustica* 10.3), listing the farming equipment of an olive yard, mentions a pestle for emmer (*pilum farrearium*).

An additional source of information comes from the ethnographic record from those areas where hulled wheats are still cultivated and, more importantly, used (at least partially) for human consumption. This is the case in the western Rif region of northern Morocco (Fig. 7.3; Peña-Chocarro and Zapata 2003; Peña-Chocarro *et al.* 2005; 2009). In this region, different techniques were used depending on the amount of crop to be processed. If large quantities were available, and only in exceptional cases, olive

Fig. 7.6. Location of Transylvania, Romania. Map: R. Lugon, J.-C. Loubier and A. Chevalier.

Fig. 7.7. Rif region (northern Morocco). Rubbing spikelets with a mallet.

mills were used. However, the commonest method for dehusking einkorn has been the pounding of spikelets in sunken mortars excavated into the soil and, sometimes, plastered with clay (Fig. 7.4). Rotary querns were also used, often with a piece of cork in between the two stones which helped to protect the grain from breaking up by creating a soft surface for the grain to rub against. Rotary querns were also described for Turkey (Hillman 1984b) and Slovakia (Gunda 1983). In addition, there were also alternative techniques applied when the amount of cereal to be dehusked was small. In those cases, Rifian farmers used a circular piece of cork (Fig. 7.5) for rubbing the spikelets against a basket. Rubbing is also mentioned in Transylvania (Fig. 7.6; Gunda 1983). Another alternative method for Moroccan farmers was the use of a kind of pestle or mallet (Fig. 7.7 and Fig. 7.8).

Fig. 7.8. Rif region (northern Morocco). Detail of the mallet used to rub spikelets.

In the north of the Iberian Peninsula, in Asturias, spelt (and until recently also emmer) is still used for human consumption. Nowadays, dehusking in this area is carried out in water mills, as is also described for Turkey (Hillman 1984b; Nesbitt and Samuel 1996). Literary sources, however, point to the use of stone mortars and wooden pestles until the beginning of the 20th century. The device described consisted of two pieces: a stone mortar, sometimes sunken in the ground, and a proper pestle made of a pivoting wooden beam or board with a mallet at one end. The teeterboard was driven up and down by stepping on one end so the mallet pounded the cereal contained in the mortar. The continuous pounding broke the spikelets and dehusked the grain. References to the use of this type of dehusking method are found in one of the first accounts of the cultivation of hulled wheats in Asturias (Alvargonzález 1908) and its use is still within living memory. Mortars and pestles were also widely used in Europe and Asia for millet and flax processing (Bennett and Elton 1898; Dias 1950; Gunda 1983; Markus 1989).

Conclusions

Although parching was in the past acknowledged as an important stage in the hulled wheat dehusking process, experimental work and ethnographic information have demonstrated that parching was not a necessary operation. The examination of some of the most quoted Classical sources to support parching show that some of the processes described and their real meaning are still not fully understood. What is clear is that some contact with fire occurred during processing, but the ethnographic record suggests that this may have been related to environmental, cultural or technical factors more than to the actual need to render the chaff brittle.

Dehusking, on the other hand, appears to be characterised by some variability within the two main methods described and documented: the use of mortars and pestles and the use of mills in their multiple forms. Some alternative methods, practiced at a smaller scale and limited to specific regions, have been also documented. This is the case of the rubbing system used for dehusking small quantities of spikelets.

The ethnographic information has turned out to be an invaluable tool for exploring in depth the variability of techniques used for dehusking hulled wheats.

7.3. CROP DRYING AND ROASTING IN THE ROMAN PERIOD (NORTHERN FRANCE). IDENTIFYING A DRYING STRUCTURE BY PHYTOLITH ANALYSIS

Pascal Verdin

The archaeological site of Lieusaint-La Pyramide, dated to the Gallo-Roman period (northern France; Fig. 7.9), was studied in 2004, disclosing a structure measuring 2m × 3 m. The walls were made of stone, and tiles (*teguleae*) covered the floor. The structure was situated near pits and silos, located at various distances from the main settlement. The layout and location of this structure identified it as a grain-drying structure, a common type of building in Roman Britain (Van der Veen 1989), but only rarely occurring in northern France.

According to M. van der Veen, who proposed a synthesis on Roman grain-drying structures, six functions can be considered:

- Plant drying after a wet or rainy summer
- Drying for more efficient threshing
- Drying to more easily separate the glumes before winnowing
- Drying to better preserve the grain in the domestic unit
- Drying to prevent humid grain from sticking to grindstones before grinding
- Roasting to stop the germination process, for the production of malt, to make beer.

Fig. 7.9. The site of 1) Lieusaint-La Pyramide in northern France. Map: R. Lugon, J.-C. Loubier and A. Chevalier.

In order to elucidate the function of the structure of Lieusaint-La Pyramide, we conducted a phytolith analysis.[1] Grain dryers in Britain usually have a T-shaped morphology, but some are rectangular. The floor is laid with tiles. The presence of charred grains due to accidental combustion permitted a study of the macro-remains in these British structures (Van der Veen 1989). For our case study, archaeological levels did not furnish combustion traces. Detection of potential residues, on a macroscopic scale, motivated the phytolith analysis.

Four samples were analysed. The first sample was taken between the tiles, where grain residues could have been left after the abandonment of the structure. The second one was taken on top of the tiles, as a comparative sample. The third one was made in the soil under the tiles and the last one outside the structure, in the natural sediment, as a control sample.

Samples coming from natural sediment as well as the one taken under the tiles proved to be negative. However, both samples coming from the structure itself contained phytoliths, more than 95% from the grass family (Poaceae). The high proportion of short cell phytoliths (more than 55%) indicates the probable presence of cereal spikelets, spikes or hulled grains (upper part of the plant) (Anderson 1999). Furthermore, numerous micro-charcoals were found (50% of total micro-remains observed, cf. Fig. 7.10 and Fig. 7.11).

Structures	Phytoliths	Micro-Charcoals
St 1278, sediment between tiles	40.6	59.4
St 1278, sediment on tiles	42.1	57.9

Fig. 7.10. Percentages of phytoliths and microcharcoals from Lieusaint-La Pyramide.

One or more accidental combustions in the storage structure could have led to this type of remains. However, the lack of traces of fire on the tiles, to be expected in an accidental fire, makes this explanation unlikely. We can thus propose that this building was a drying structure. The building, which is well-preserved, probably did not possess any combustion structure (hearth or oven).

However, as Van der Veen (1989) has shown, cereal roasting to make malt does not require a sophisticated combustion system: a simple fire set on the tile level or on another type of soil could have been used, without necessarily leaving combustion traces. In fact, *Poaceae* do not leave large fragments of charcoal, but only micro-charcoals or powdery gray ashes, their components invisible to the naked eye. Combustion can also start directly from the top of the grain mound. The absence of phytoliths of glumes (grain envelopes) must also be noted. We can propose the process as follows: either the grains dried or roasted were naked or they were dehusked, but they could still germinate. The low temperature produced by this specific structure strongly suggests parching for malt production, but the hypothesis of a drying process still remains relevant. Each process could have been carried out at a different time. Finally, we can suggest that parching and drying were carried out successively.

	POACEAE/CYPERACEAE			POACEAE festucoids		POACEAE panicoids	DICOTS	SPONGE SPICULES
	Hairs	Bulliform cells	Long cells	Short cells	Long cells	Short cells	Cells	
St 1278, sediment between tiles	2.5	1.3	29.1	55.7	7.6	–	2.5	1.3
St 1278, sediment on tiles	7	–	22.3	55.3	9.4	1.2	4.7	–
Natural soil, inside	–	–	–	–	–	–	–	–
Natural soil, outside	–	–	–	–	–	–	–	–

Fig. 7.11. Phytolith composition (percentages) at Lieusaint-La Pyramide.

7.4. DIVERSITY IN QUERN SHAPE AND USE IN THE NEOLITHIC OF THE LOWER RHINE BASIN

Annelou van Gijn

The first farmers, of the Early Neolithic Linearbandkeramik culture (LBK), appear in the southern loess zones of the present-day Netherlands around 5300 cal BCE (Van Gijn and Louwe Kooijmans 2005). Among their characteristic artefacts are querns (Fig. 7.12). Remarkably enough, querns or quern-like implements do not seem to be present in the late Mesolithic context, even though plant-processing was probably quite important for these communities (Zvelebil 1994). Elsewhere, querns certainly are not solely associated with agricultural communities. For example, as early as 35,000 years ago, the first querns appear in Australia, associated with wild seed processing (Fullagar and Field 1997). Also in the Near East, at the Upper Palaeolithic site of Ohalo II, querns used for wild cereal processing have been attested by the presence of starch grains (Piperno *et al.* 2004). It may well be that we do not recognise the late Mesolithic querns in the Lower Rhine Basin, as they may not have been modified intentionally into recognisable 'tools'.

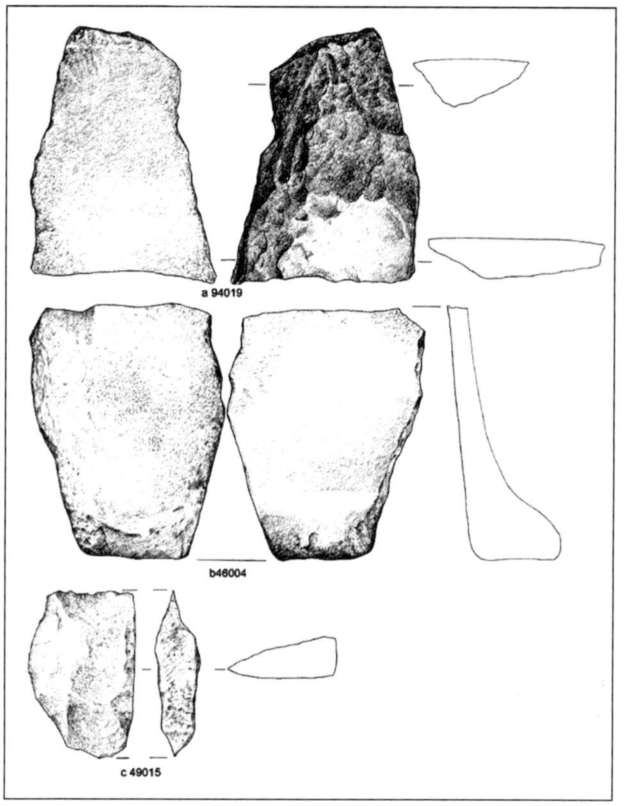

Fig. 7.12. Two querns from the LBK site of Geleen-Janskamperveld (drawing E. van Driel, from Verbaas and Van Gijn 2007, p. 195).

A group of querns from the LBK site of Geleen-Janskamperveld (Fig. 7.13) were studied to understand their biography. The querns have an elongated shape with a flat or concave surface. They are made of sandstone or quartzitic sandstone cobbles that can be collected along the Meuse or in nearby brook valleys. The presence of production flakes created by hard hammer percussion within the confines of the settlements shows that the manufacturing of the querns took place there (Verbaas and Van Gijn 2007). The flattened surface was then pecked by means of percussion, creating a roughened flat surface that was ready for the grinding process. It is likely that they used discarded blade cores of flint for this purpose. Experiments show these objects to be very suitable for this purpose (Van Gijn 1990; Verbaas and Van Gijn 2007).

Microscopic analysis of the querns shows a polish that closely resembles the wear obtained experimentally when milling cereals. The implements were used in a longitudinal fashion, as inferred from the

Fig. 7.13. The archaeological site of 1) Geleen-Janskamperveld in the Dutch province of Limburg. Map: R. Lugon, J.-C. Loubier and A. Chevalier.

predominant directionality in the use-wear polish. The heaviest pressure in such a longitudinal movement is exerted on the central part of the quern, wearing this section out most. It is also this part that has to be rejuvenated most frequently. The combined effect of the use and rejuvenation predominantly affects the central part of the quern, resulting in a saddle shape with a concave working surface (Fig. 7.14). The querns also display a lot of wear on their bottom side. This polish displays similarities to the type of wear seen on the actual grinding surface, but, in addition, a hide-like polish is visible, suggesting that the querns were placed on a piece of skin or hide to catch the spilt-over seeds and ground flour. The querns seem to be discarded at different stages in their use lives and underwent a special, presumably ritual, treatment upon discard (see Chapter 9).

Another thousand years passed by before the first farming communities left their traces in the wetlands of the western and northern Netherlands. Around 3700 cal BCE, small communities of the Hazendonk group settled on the dunes directly behind the sea in the area just south and west of the present-day town of The Hague. Their material culture bears similarities to that of the agricultural people of the contemporaneous Michelsberg culture of the Rhineland and surroundings. Much to the surprise of the archaeologists investigating these sites, they all provided evidence for the local cultivation of emmer and naked barley (Kubiak-Martens 2006). This was corroborated by the finds of a small number

Fig. 7.14. Photograph of a quern from the Middle Neolithic site of Schipluiden (photo: B. Grishaver, Leiden).

Fig. 7.15. Use-wear polish observed on the quern from Schipluiden (original magnification 200x) (photo: Laboratory for Artefact Studies, Leiden University).

Fig. 7.16. Phytolith on the quern from Fig. 7.3: worn dendriform of Poacaeae (grass family; orig. magnif. 200x) (photograph Laboratory for Artefact Studies, Leiden University).

of querns, both at Schipluiden (Van Gijn and Houkes 2006) and at Ypenburg (Houkes 2008; Van Gijn and Verbaas 2008). All of the stone must have been brought to the site from a considerable distance, as the area was devoid of stones.

The querns from these Middle Neolithic sites display a very different shape compared to the earlier LBK querns. In contrast to the elongated LBK querns, these Middle Neolithic ones are small and short (Fig. 7.14) and display a use-wear polish associated with the milling of cereals (Fig. 7.15). Phytolith research by Channah Nieuwenhuis showed them to have worn dendriforms of *Poaceae*, another indication for the actual milling of siliceous plants (Fig. 7.16) (Van Gijn and Houkes 2006). The querns were used in a longitudinal fashion and not a rotary one. Considering their limited size, this implies that the length of the strokes must have been relatively short. This would have increased the risk of 'slipping off' the surface of the stone and must have inhibited to some extent the gestures of the millers, and therefore the efficacy of the milling process. These milling stones must have been less 'comfortable' than the LBK querns. Most likely the inhabitants of the wetlands had limited access to large slabs of stone. Nevertheless they were 'copying' the milling movement customary amongst the people living in the Pleistocene uplands to the southeast, despite the fact that the stones available to them would have been more easily employed in a rotary motion.

7.5. GRINDING AND POUNDING TOOLS IN BRONZE AGE CRETE

Hara Procopiou

The beginning of the second millennium in Crete (Minoan Period) is characterised by the emergence of the palatial system, which was maintained in the Aegean world until the end of the 13th century BCE. This phenomenon was accompanied by the development of a self-sufficient subsistence agriculture based on cereals and pulses, vine and olive. Within the palaces, but also within urban or rural settlements, provision was made for the storage of agricultural surpluses. Grinding and pounding tools were very abundant and were used not only for processing agricultural products, but also for mineral resources, indicating the development of specialised crafts (pottery, metallurgy, faience, etc.). We have studied raw material, techniques of manufacture and use (by means of residue and use-wear analysis) of grinding and pounding tools from several Bronze Age sites: the administrative centre of Quartier Mu at Malia (Poursat 1996) and the palatial centre of Monastiraki in the Amari valley (Kanta and Marazzi 2006) both dated to the Middle Minoan Period (First Palace Period, *ca.* 1900–1700 BCE); the palace of Malia and the Sanctuary of Syme (Fig. 7.17; Lebessi and Muhly 1990) dated to the beginning of the Late Minoan period (Second Palace Period, *ca.* 1700–1600 BCE); finally, the settlements of Quartier Nu at Malia (Fig. 7.17; Driessen and Farnoux 1994) and Sybrita (Rocchetti 1994) dated to the Late Minoan III (Post-Palace Period, *ca.* 1400–1100 BCE). This study, as well as evidence from other Minoan sites, showed that despite an apparent homogeneity, several changes occur in tool production processes and function.

Fig. 7.17. Map of the Aegean Sea with the archaeological sites of 1) Akrotiri; 2) Galatas; 3) Malia; 4) the sanctuary of Syme; 5) Archondiko; and 6) Mesimeriani Toumba, as mentioned in Chapters 7.5 and 7.6. Map: R. Lugon, J.-C. Loubier and A. Chevalier.

The Raw Material Choice

Several parameters are involved in the selection of the raw material: availability and possibilities of procurement and how the material is anticipated to 'react' during manufacture and use (Horsfal 1987). During the Middle Minoan Period (First Palace Period, *ca.* 1900–1700 BCE) local, often quartzitic, sandstones, predominate. Outcrops were located either in the immediate vicinity of the settlements or in a larger zone which could stretch over a distance of 10–15 km. At the beginning of the Late Minoan Period (New Palace Period, *ca.*1700–1600 BCE), sandstones continue to predominate, but local porous and friable grainstones[2] were also introduced. It is only during the Late Minoan III Period (Post-Palace Period, *ca.*1400–1100 BCE), that volcanic rocks, such as andesite, were imported, probably from the Cyclades, for the manufacture of querns. However, local stones, previously exploited, continued to be used. During the Bronze Age, raw material procurement concerns the gathering of natural blocks, as there is no evidence for quarrying.

The introduction of friable grainstones during the New Palace phase, the apex period of Minoan civilisation, is difficult to explain from a utilitarian point of view, as this type of stone, at first sight, seems less suitable for cereal processing. We have attempted to shed light on this issue through an analysis of the mechanical characteristics of the stones and by means of experiments with the manufacture and use of the querns. This enabled us to show that the sandstones were resistant to wear, resulting in a low contamination of grit in the flour, but needed frequent re-roughening by pecking. In contrast, grainstones were automatically rejuvenated during use and hence did not require investment in maintenance. However, the resulting flour is contaminated with an abundant amount of calcareous grit. Assuming that procurement constraints played no role, this suggests that prehistoric craftsmen changed their choice of raw materials during the Minoan period, opting for different desired characteristics.

Data on 19th-century milling show that the quality of milling stones depends on the desired use, the milling technique, as well as on the individual consumer's appreciation (Sigaut 2002). Several studies of prehistoric stone selection stress the quality and efficiency of stones chosen (Runnels 1981; Schneider 2002; Santallier *et al.* 2002) and associated the use of poor-quality material to a low social value of grinding activities (Delgado *et al.* 2009). The case study of Galatas palace (Fig. 7.17), where the introduction of friable stones coincides with an intensification of cereal processing associated with feasting practices (Rethemiotakis 1999), suggests, on the contrary, that despite their friable nature, grainstones could have been perceived as raw materials of choice because they required no maintenance.

Morphology and Techniques of Manufacture through Time

General morphology of querns and handstones hardly changes during the Bronze Age: it concerns oval or rectangular slabs or rocks, with an active surface which often became concave from use. However, the techniques of manufacture vary. During the Old Palace Period, a low investment in manufacture predominates: only the active surface is transformed by pecking. In some case, the contours are regularised by flaking and pecking, but without substantial modification of the natural morphology of the stone slab (Fig. 7.18). It is during this same period that more time and effort was invested in the production of grinding stones, as inferred from some querns at the Quartier Mu at Malia, where flaking was aimed at producing a conical base. Here the contour was also regularised by pecking with stone pounders and, in some cases, with bronze tools.

The increased investment in production became the norm during the New Palace Period. For example, at the Syme Sanctuary or at the palace of Malia, a real modification of the stone's initial form is attested. For the Post-Palace Period, in contrast, a technological regression is evident because in most cases only the active surface is modified. If the general morphology of querns seems unchanged during the Minoan period, the investment in manufacture changes considerably. It is during the Palace Period, and within the palaces and sanctuaries, that we find the grinding stones in which the most time and effort was invested.

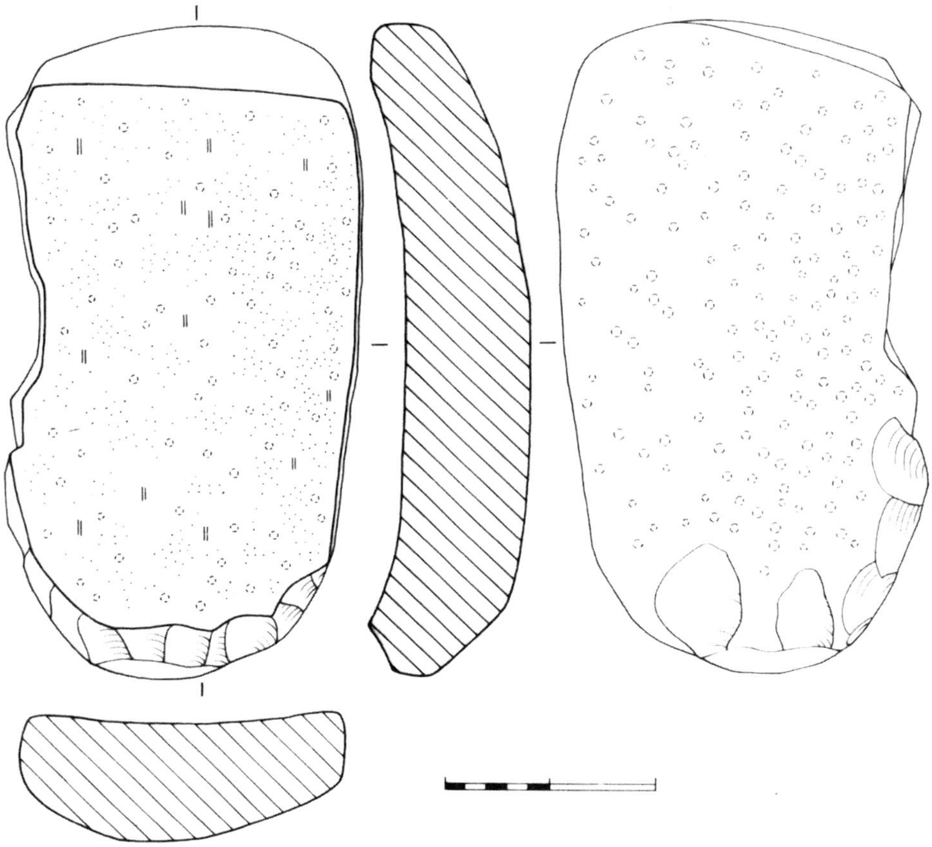

Fig. 7.18. Middle Minoan quern from the Quartier Mu at Malia (Crete). The slab is regularised by flaking and pecking.

Function and Processing Techniques

Use-wear studies (with low magnification and interferometry) and residue analysis (phytolith, starch grain and chemical analysis) permitted detection of diagnostic use-wear traces and identification of patterns of distribution of mineral and cereal processing within settlements. More particularly, a use-wear polish was associated with the grinding of hulled cereals. Indeed, experimental data show that even after dehusking, glume remains give a distinctive gloss (Procopiou 1998) to the quern surface (Fig. 7.19). On some querns, phytoliths, probably from glumes, were also preserved (Fig. 7.20), as well as starch grains and lipids detected by gas chromatography/mass spectrometry (Formenti and Procopiou 1998). On the other hand, mineral processing attacks the inter-granular matrix of sedimentary stones and creates a diagnostic rough surface (Procopiou 2004).

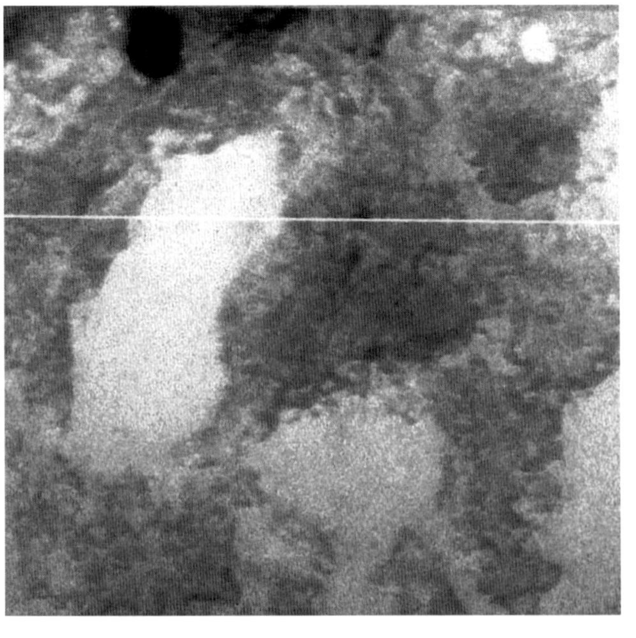

Fig. 7.19. Use-wear polish observed on a Middle Minoan quern from the Quartier Mu at Malia. 2D view obtained by interferometry (250 x 250 μm) at the LTDS, Laboratoire de Tribologie et Dynamique des Systèmes; UMR 5513, CNRS).

Observation of wear distribution also enabled us to reconstruct gesture. During the Middle

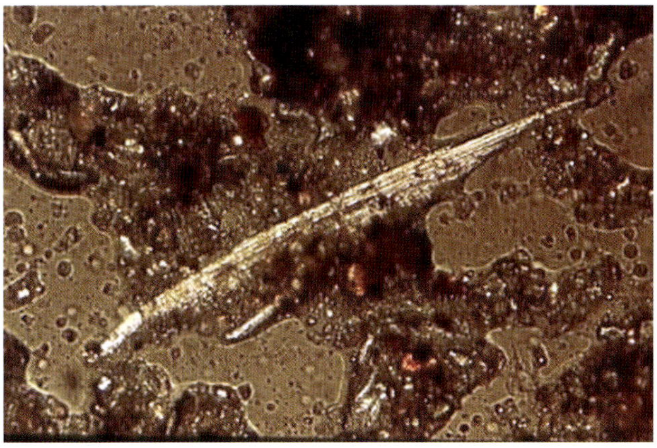

Fig. 7.20. Phytolith from Minoan quern (orig. magnif. 200x).

Minoan period, the back-and-forth movement predominated, sometimes accompanied by a circular one. During the Old Palace Period, the length of the handstones was smaller than the width of the quern surface, resulting in a distinctive wear distribution. The back-and-forth movement becomes systematic during the New Palace Period and the volume of handstones tends to increase. During the Post-Palace Period, the volume of querns and handstones increased considerably and cereal processing, as shown by the site of Quartier Nu at Malia, was demonstrated by a single quern within the settlement (Fig. 7.21).

In the Near East and in Egypt during the Bronze Age, bread production is often centralised. For example, at Ebla (Fig. 7.22), a grinding room with several querns laid out on benches was discovered (Matthiae 1982). In Crete, on the contrary, cereal processing took place essentially in a domestic context. Even within the palaces, where flour production could have been controlled by palace authorities, quern distribution showed a dispersed production and milling workshops were rare. Only at the palace of Phaistos (Pernier et Banti 1951, Fig. 88) was a quern built in a bench of the Neopalatial Period. A bench with three querns *in situ* was also discovered in Room 3a of the West House of Akrotiri (Fig. 7.17; Moundrea-Agrafioti 2007, 82–83). Finally, distribution of querns with traces of mineral processing for specialised crafts shows that this activity was carried on mainly within the Minoan workshops (Procopiou, forthcoming).

Concerning pounding tools, pestles and mortars, a variability could be observed within settlements. Hollow stone mortars, manufactured in homogeneous calcareous materials, were used to pulverise substances such as spices, but deep mortars suitable for cereal dehusking are rare. At Malia, they are absent in the Quartier Mu and Nu but present in the palace. On the contrary, a great variety of deep

Fig. 7.21. Settlement of Quartier Nu at Malia (Crete): Late Minoan quern *in situ*.

Fig. 7.22. Location of Ebla, Syria. Map: R. Lugon, J.-C. Loubier and A. Chevalier.

stone mortars (25–30 cm) adequate for dehusking, came to light at the palace of Galatas (Fig. 7.17).

When stone mortars are absent, can we presume the existence of wooden mortars, the use of other dehusking techniques or even that dehusking was not systematic? Association of two observations, the presence of gloss on querns and the absence of mortars – at least in the Quartier Mu – suggests no or partial dehusking, using alternative techniques (dehusking with querns, with mallets or by rubbing, see Chapter 7.2.).

From the choice of the raw material, to their manufacture and use, grinding tools are set into a long tradition that goes back to the early Minoan period. Closer analysis, however, has shown developments in raw material procurement, investment in manufacture, gestures and grinding techniques. Theses changes reflect distinct technical choices, between palaces and urban settlements, and suggest that the cereal meal consumed varied within these different contexts.

7.6. BARLEY MEAL PROCESSING IN THE AEGEAN WORLD: A LOOK AT DIVERSITY

Hara Procopiou

The Aegean Bronze Age is characterised by the genesis of Mediterranean subsistence agriculture based on the triad of cereals, olive, and grape. These species, despite the constant introduction of new products, formed the basic ingredients of the 'Aegean cuisine' till the mid-20th century. The emergence of a traditional Mediterranean diet, as defined by modern nutritionists (Hubert 1998), which would have survived by intergenerational transmission, is often argued in archaeological publications (Riley 1999). However, using the same ingredients does not imply the same type of processing, the same type of cooking, the same nutritional value, texture, smell and flavour of the meal. Technical choices concerning cereal meal preparation are strong indicators of dietary habits (Sigaut 2010 and also see Chapter 7.1). A brief overview of barley meal and bread production in the Aegean world will enable us, on the one hand, to consider the 'relationship that diet carries on with myth, culture, and social structure' (Flandrin 1996, 13) and, on the other hand, to deal with continuity or discontinuity of culinary practices.

Hulled barley is a predominant crop during the Aegean Bronze Age and Classical Antiquity. The diversity of processing methods for hulled cereals is attested by archaeological, textual, and ethnographic data (see Chapter 7.2). Greek and Roman textual sources show a large variety in barley meal: whole grain or gruel soups, beverages and decoctions, bread from flours of different coarseness and texture, and 'pancakes' from roasted flour. These culinary preparations imply different processing and consumption strategies.

Processing of Cereal Meal

In order to reconstruct Bronze Age cereal processing we have first analysed processing tools (mortars and querns) from several Middle (*ca.* 1900–1700 BCE) and Late Bronze Age (1600–1050 BCE) sites in Crete for the Minoan period (see Chapter 7.5), as well as botanical evidence from these periods. Taking the example of the Quartier Mu at Malia (Middle Minoan II Period) or the Syme sanctuary (Late Minoan Period), a use-wear analysis of querns revealed polish associated with the grinding of hulled cereals (see Chapter 7.5). This study was completed by the analysis of flour samples from the Late Bronze Age I city of Akrotiri in Thera (Cyclades; Fig. 7.17), a Minoan 'colony' destroyed by a volcanic eruption, which led to exceptional conservation of organic material (Sarpaki 1992).

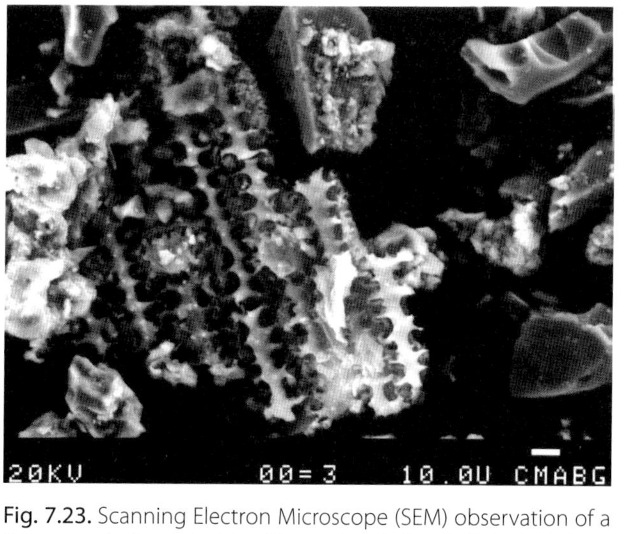

Fig. 7.23. Scanning Electron Microscope (SEM) observation of a flour sample from the Late Bronze Age I city of Akrotiri in Thera (Cyclades): einkorn glume phytolith.

A scanning electron microscope (SEM) analysis of einkorn and hulled barley flours (Procopiou *et al.* 2002) revealed a very high percentage of glume fragments (phytoliths) (Fig. 7.23), suggesting that dehusking of grains was not exhaustive and that no systematic sieving was subsequently practised. The absence of systematic sieving is also argued by D. Samuel (1993, 282) for Dynastic Egypt, where several loaves contain fragments of husks (light chaff or thick and robust pieces).

Akrotiri flour samples were compared with experimental ones and with ethnographic data (Procopiou 2003) from the island of Melos, where traditional barley rusks (chunks of hard bread) are still produced by a few families. At Melos, the dehusking sequence is omitted: barley is ground on the rotary quern, after which the glumes and chaff are partially eliminated by successive sieving of the flour (Fig. 7.24). The degree of sieving varies from one family to another, but all producers attest that the presence of chaff guarantees the taste of the bread (Fig. 7.25). The comparison of archaeological and ethnographic flours (microscopic and SEM analysis) and the study of the proportions and fractions of the phytoliths, shows that the grains from Akrotiri, like the ethnographic ones, were ground without prior dehusking.

Can we conclude that this technique was generalised in the Cyclades and Crete during the Bronze Age? On the contrary, the use-wear analysis of querns from the palaces of Malia and Galatas (Late Minoan I Period; Fig. 7.17), which showed a less pronounced polish, and the presence at these sites of deep stone mortars, suggest that grains were dehusked and

Fig. 7.24. Barley processing at Zefyria, Melos island (Cyclades). Barley is ground on the rotary quern (left), after which the glumes and chaff are partially eliminated by sieving (right).

Fig. 7.25. Left: Traditional barley rusks from Melos island (Cyclades); right: Detail of rusks with fragments of glumes and rachis.

sieved before grinding. A variability of processing is thus suggested, related respectively to the domestic and palatial context of production and consumption.

Textual and Archaeobotanical Evidence: A Variability of Recipes

This same variability is attested in Classical Antiquity. Hippocrates points out that barley is laxative by 'the juice of the glumes' (*On Regimen in Acute Diseases* II, XL, 1), which can be employed for therapeutic purposes (*On Regimen in Acute Diseases* II, XLII, 1). Several texts by Hippocrates, Aristotle, Galen or Athenaeus point out that bread acts differently in the human body according to the composition, coarseness, and texture of the flour. This perception seems to have evolved over time. For Hippocrates, wholemeal flour is less nutritive and laxative (*On Regimen in Acute Diseases* II, XL, 1) while for Athenaeus (*Deipnosophistes*, III, 115 e) wholemeal bread is more nutritious. This textual evidence on wholemeal flours shows, as M. C. Amouretti (1986, 129) had already pointed out, that dehusking hulled barley was not practised systematically during Classical Antiquity. The context of production and regional or social factors determine this variability. For example, Athenaeus mentions that some breads (*Deipnosophistes* III, 114 e) are consumed only within the family, suggesting implicitly that others were shared with members of the community.

From the Ptolemaic period on, but mainly during the Roman period, the value placed on the nutritional value of barley declines, and barley is often associated with slave food or fodder (Andorlini and Marconi 1999, 330). Barley bread consumption, however, does not disappear. During the Byzantine period, barley bread from non-sifted flours, thus containing a high percentage of chaff, is associated with poverty (Koukoulès 1952, 21), with monks, travellers, soldiers and peasants (Karapli 1994). The French traveller Thevenot (16th century) was astonished by the darkness and hardness of barley breads consumed by the inhabitants of Thera. Nowadays, barley bread has been rehabilitated (Procopiou 2010). As a symbol of Cretan or Cycladic cuisine, valorised by modern nutritionists, it is widespread in restaurants, bakeries, supermarkets, organic groceries, even on the web, where internet users from all over the world can buy traditional Greek barley rusks.

The coarseness and texture of flour, the gestures and tools employed during kneading of the dough, the ingredients added in the dough and cooking structures are essential factors in determining bread variability.

Except for some bread-like fragments from Akrotiri, bread samples have not been found in the Aegean World. On the basis of the low gluten content of the cereals found in Akrotiri (barley, einkorn and emmer) and the use of small portable ovens at the site, the production of an unleavened pita-type bread has been proposed (Sarpaki 2001, 40). However, barley flour was found associated with flour from pulses (6 samples of pulses versus 45 barley samples) (Sarpaki 2001, 34). Following Pliny (Natural History XVIII, 103), pulse flour (*ervi aut cicerculae farina*) was employed to leaven barley bread but is also attested in ethnographic data from Crete and the Cyclades. Could we presume that pulse flour at Akrotiri had this function? The strong contacts attested between the Late Bronze Age Aegean and Egypt, where leavened bread is well documented, imply that the inhabitants of Akrotiri not only knew of the existence of this type of bread but possessed at least the necessary raw material to produce it. Can we presume an opposition of leavened (*zymites*) versus unleavened (*azymos*) bread as a sign of identity within the societies of the Bronze Age eastern Mediterranean? Symbolic and religious prohibitions related to fermentation are widespread in different cultural contexts. For example, Greeks using leavened bread for the consecration, were in opposition with the unleavened bread of the Latin Church, and stigmatised as *fermentarians* or *prozymites*. During Antiquity, both types of breads coexisted, but leavened bread seems rarer and often associated with feasting (Amouretti 1986, 128) or religious festivities (Kerns 1994).

Other culinary preparations, based on whole or coarsely ground cereals are also attested. We can presume the consumption of whole boiled cereals, as the modern Greek *kollyva* consumed in honour of the dead (Eliade 1978, 63). Consumption of cracked barley was found at Akrotiri (Fig. 7.17; Sarpaki 2001), and cracked wheat on two Bronze Age sites in northern Greece, Mesimeriani Toumba and Archondiko (Fig. 7.17; Valamoti 2002). A SEM analysis of modified starch granules (Samuel 2006) conserved in these remains showed that grains were previously processed, probably cooked (Valamoti *et al.* 2008). Indeed, ground grain can reflect different transformations: grinding green or mature grains, with or without preliminary processing. Processing can include roasting or boiling, but also the addition of other ingredients (milk, sour milk, yoghurt) as the *kišk* in the Near East (Kanafani-Zahar 1994) or *xynochondros* or *trachanas*[3] in Greece (Valamoti and Anastasaki 2007). The ancient *tragos* probably required the same type of preparation. Hesiod's mention in his *Maza amolgaia* (*Works and Days*, 590) of barley meal made with goat's milk (Brumfield 1997, 153) is probably the oldest reference to this culinary preparation.

The Hellenic national meal, the *maza*, is prepared from *alphita*, a coarse flour deriving from the grinding of roasted grains (wheat and barley in the Homeric period, exclusively barley in the Classical period). The necessity of parching to process hulled cereals has recently been reconsidered (see Chapter 7.2. and bibliography). In the case of *alphita,* roasting aims at producing a 'pre-cooked' flour, which can be kept for several months. Two principle types of *maza* have been attested: firstly, there is the *tripte maza* (kneaded *maza)* prepared from kneading *alphita* with a liquid (water, milk, wine, oil). This kneaded 'cake' is consumed without being cooked. The *atripte maza* (unkneaded *maza)* is prepared by mixing *alphita* with a liquid (water, milk, wine, oil) and could take on the appearance of a beverage or a soup. The *cyceon* of the Eleusinian Mysteries is in fact a variety of this beverage, made with *alphita,* water and mint (*Homeric Hymn to Demeter*). This type of preparation began in the Mycenaean period (Sacconi 2001), when a meal of barley (*pa-ta*), probably a type of *cyceon*, was consumed during religious festivities dedicated to Mother Earth, Zeus and Kore.

During the Roman period (Pliny, Plautus), this same barley meal, associated with Greek identity, was called *polenta*. In Ptolemaic Egypt, it was a new term, *athara*, of Egyptian origin (*teret*) which referred to the *atripte maza* eaten by the Greek community. During the Byzantine period (Hesychius, Sofronius), different recipes of *maza* are documented (Koukoules 1952). The vocabulary is specialised with the occurrence of distinct names for each different recipe. The term *maza* is then reduced to a cake from *alfita* kneaded with water and oil. A beverage of *alphita* and sage is associated with the monasteries. At the end of the Byzantine period, a cream dish made with *alphita*, intended for consumption by children, appears (Migne 2012, 87, 3604). Present-day ethnographic data from Crete, Gavdos, Kythera, Karpathos and Pontos (Psilakis and Psilaki 2001, 327) call different recipes of *maza* by yet another name, *pachountas*.

On the basis of these data it is clear that even though the continuity of some recipes is attested, their social and symbolic connotations constantly varied. One particular recipe, the Greek *chidra,* a soup of cooked barley (whole or coarsely ground), reflects this evolution. A peasant's meal during the Classical period (Aristophanes *Pax*, 1304), it became a highly valued meal, associated with Greek identity in Ptolemaic Egypt (the mid-third century BCE Zenon Papyrus, see Orrieux 1983). Indeed, for immigrants, who are conservative in their dietary practices, continued attachment to rural dishes links them to their roots. Imported *Lathyrus clymenum* seeds, a specific Aegean pulse (Sarpaki and Jones, 1990) at Tel Nami, Tel Batash and Tel Miqne in the Near East (2nd millennium) (Kislev *et al.*, 1993; Mahler-Slasky and Kislev 2010) suggest that these ingredients accompanied Aegean migrants, for preparing their own cuisine. These Aegean dietary traditions persisted until the end of the Iron Age II Period, as indicated by finds from Ashkelon (*ca.* 604 BC).

From prehistory to the present day, the consumption of barley in the Aegean world reflects diversity. From the *cyceon* of the goddess Demeter to the Byzantine 'baby cream', from the monks' or peasants' bread, to the modern enthusiast of organic cuisine, the perception of barley constantly changes. Meanwhile, recipes persist.

7.7. CONCLUSIONS

Hara Procopiou

From parching and dehusking in the ethnographic record and during the Roman period, to milling and preparing cereals in the Neolithic Netherlands or Bronze Age Crete, these case studies have focused on the diversity of cereal processing techniques and on the economic, social and symbolic parameters governing technological change.

Dietary habits, strong indicators of identity and a sense of group belonging, also make it possible to deal with the issue of continuity versus discontinuity. If some processes, which are intimately related to the nature of cereals, to climatic and environmental factors, converge and do persist in some areas, nonetheless, consumption as also involving the gustatory and symbolic perception of meals develops over time. Daily bread is a relevant example of this evolution. Wholemeal bread, the devalorised 'muddy peasant's bread' at the beginning of the 20th century, has nowadays become the symbol of a balanced low-fat and fiber-rich diet that can save us 'from the diseases of civilisation' (Succar 2008).

Chapter Acknowledgements

The work by Leonor Peña-Chocarro and Lydia Zapata (Chapter 7.2.) was carried out within the Research Group of the Basque Government UPV/EHU IT-288-07, the UPV/EHU UFI UFI11/09 and the projects AGRIWESTMED (ERC-AdG-230561), HAR2008-03976/HIST, HAR2008-01920/HIST and HAR2011-23716 *Nuevos cultivos, nuevos paisajes* and Programa Consolider TCP-CSD2007-00058 from the Spanish Government. The authors are grateful to M. J. Torrens (CSIC) for helpful comments.

Chapter Notes

1 Phytoliths are microscopic siliceous particles formed in and on the cell walls of plants. They are highly resistant to decomposition and serve as a source of information in soil samples.
2 Following Dunham's classification, grainstones are grain-supported carbonate sedimentary rocks, that contain no micrite.
3 Cracked wheat boiled in milk, dried in the sun and stored for soup-making in winter.

Chapter Bibliography

Adams, J. (2002) *Ground stone analysis: A technological approach.* University of Utah Press, Salt Lake City.

Adams, J., Delgado, S., Dubreuil, L., Hamon, C. and Plisson, H. (2009) Functional analysis of macro-lithic artefacts. In F. Sternke, L. Eigeland and L. Costa (eds.) *Non-flint Raw Material Use in Prehistory Old Prejudices and New Direction.* 15th UISPP Congress, Lisbon – September 2006. BAR International Series 1939. Archaeopress, Oxford, pp. 43–66.

Alonso, N. (1999) *De la llavor a la farina: els processos agrícoles* protohistòrics a la Catalunya Occidental. MAM 4, Lattes.

Alvargonzález, C. (1908) *La escanda. Su origen. Su cultivo.* Noroeste, Gijón.

Amouretti, M.- C. (1986) *Le pain et l'huile dans la Grèce antique, de l'araire au moulin.* Les Belles Lettres, Paris.

Amouretti, M.-C. (1996) Villes et campagnes grecques. In J.-L. Flandrin and M. Montanari (eds.) *Histoire de l'alimentation*, Fayard, Paris, pp. 133–150.

Anderson, P. C. (1999) Postscript. Experimental cultivation, harvest and threshing of wild cereals. In P. C. Anderson (ed.) *Prehistory of Agriculture. New Experimental and Ethnographic approaches.* Institute of Archaeology Monograph 40. University of California, Los Angeles, pp. 119–144.

Andorlini, I. and Marcone, A. (1999) L'orzo nell' Egitto greco-romano. In D. V. Vera (ed.) *Demografia, sistemi agrari, regimi alimentari nel mondo antico.* Atti del Convegno Internazionale di Studi (Parma 17–19 ottobre 1997). Edipuglia, Bari, pp. 325–344.

Aranguren, B, Becattini, R, Mariotti Lippi M., Revedin, A. (2007) Grinding flour in Upper Paleolithic Europe (25 000 year bp). *Antiquity* 81, 845–855.

Beaune, S. A. de (2000) *Pour une archéologie du geste.* CNRS Editions, Paris.

Bennett, Ri and Elton, J. (1898) *History of corn milling. Handstones, slave and cattle mills.* Simpkin, Marshal and Company Ltd, London.

Beardsworth, A. and Keil, T. (1997) *Sociology on the menu. An invitation to the study of food and society.* Routledge, London and New York.

Brumfield, A. (1997) Cakes in the Liknon: Votives from the Sanctuary of Demeter and Kore on Acrocorinth. *Hesperia* 66 (1), 147–172.

Buxò, R. (1989) La présence de l'épeautre (*Triticum spelta* L.) dans l'alimentation en Espagne : les temps anciens et les temps modernes. In J. P. Devroey and J. J. van Mol, (eds.) *L'épeautre (Triticum spelta), histoire et ethnologie*. Editions Dire, Treignes, pp. 107–122.

Classen, C., Howes, D. and Synnott, A. (1994) *Aroma: the cultural history of smell*. Routledge, London/New York.

David, N. (1999) The ethnoarchaeology and field archaeology of grinding at Sukur, Adamawa State, Nigeria. *African Archaeological Review* 15 (1), 13–63.

Delgado-Raack, S., Gómez-Gras, D. and Risch, R. (2009) The mechanical properties of macrolithic artifacts: a methodological background for functional analysis. *Journal of Archaeological Science*, 1823–1831.

Dias, J. (1950) O pio de piar os milhos, *Trabalhos de Antropologia e Etnografia*, 12 (3), 323–344.

Driessen, J. and Farnoux, A. (1994) Malia, *Bulletin de correspondance hellénique* 118 (2), 471–477.

Eliade, M. (1978) *Histoire des croyances et des idées religieuses. 1. De l'âge de la pierre aux mystères d'Eleusis*. Payot, Paris.

Eshed, V., Gopher, A. and Hershkovitz, I. (2006) Tooth wear and dental pathology at the advent of agriculture: New evidence from the Levant. *American Journal of Physical Anthropology* 130, 145–159.

Ertuğ-Yaras, F. (2002) Pounders and Grinders in a Modem Central Anatolian Village. In H. Procopiou and R. Treuil (eds.) *Moudre et broyer. L'identification fonctionnelle de l'outillage de mouture dans la Préhistoire et l'Antiquité, I. Méthodes*. CTHS, Paris, pp. 211–226.

Fenton, A. (1978) *The Northern Isles: Orkney and Shetland*. J. Donald, Edinburgh.

Fenton, A. (1991) The processing of cereal grains in Scotland and around. *PACT* 26, 99–108.

Fenton, A. (1992) Small scale techniques in processing cereals for food. In M. Mesnil (ed.) *Du grain au pain. Symboles, savoirs, pratiques* 2. Institut de Sociologie. Université Libre de Bruxelles, Bruxelles, pp. 83–94.

Flandrin, J.-L. (1996) Introduction. In J.-L. Flandrin and M. Montanari (dir.) *Histoire de l'alimentation*. Fayard, Paris, pp. 7–15.

Formenti, F. and Procopiou, H. (1998) Analyse chromatographique des acides gras sur l'outillage de mouture et de broyage : apport et limites. *Cahiers de l'Euphrate* 8, 151–176.

Frankel, R. 2003, The Olynthus Mill, Its Origin, and Diffusion: Typology and Distribution, *American Journal of Archaeology* 107 (1), 1–21.

Fullagar, R. and Field, J. (1997) Pleistocene seed-grinding implements from the Australian arid zone. *Antiquity* 71, 300–307.

Fullagar, R. (ed.) (1998) *A Closer Look : Recent Australian Studies of Stone Tools*. Sydney University Press, Sydney.

Fullagar, R. (2006). Starch on artifacts. In R. Torrence and H. Barton (eds.) *Ancient Starch Research*. Left Coast Press California, pp. 177–204.

Fullagar, R., Field, J. and Kealhofer, L. (2008) Grinding stones and seeds of change: starch and phytoliths as evidence of plant food processing. In Y. M Rowan and J. R. Ebeling (eds.) *New Approaches to Old Stones: Recent Studies of Ground Stone Artifacts*. Equinox Press, London, pp. 159–172.

Gast, M. (1968) *Alimentations des populations en Ahaggar*. Mémoire du Centre de Recherches Anthropologiques, Préhistoriques et Ethnographiques VIII, Arts et Métiers Graphiques, Paris.

Gast, M. (2003) Meules et molettes sahariennes. In M. Barboff, C. Griffin-Kremer, F. Sigaut and R. Kremer (eds.) *Meules à grains, Actes du colloque international, La Ferté-sous-Jouarre, 16-19 mai 2002*. Ibis Press and Maison des Sciences de l'Homme, Paris, pp. 61–66.

Van Gijn, A. L. (1990) *The wear and tear of flint. Principles of functional analysis applied to Dutch Neolithic assemblages*. Analecta Praehistorica Leidensia 22, Leiden.

Van Gijn, A. L. and Houkes, R. (2006) Stone procurement and use. In L. P. Louwe Kooijmans and P. F. B. Jongste (eds.) *Schipluiden - Harnaschpolder. A Middle Neolithic site on the Dutch North Sea coast c. 3800 cal BC*. Analecta Praehistorica Leidensia 37/38, Leiden, pp. 167–194.

Van Gijn, A. L. and Louwe Kooijmans, L. P. (2005) The first farmers. Introduction: 5300–2900 BC. In L. P. Louwe Kooijmans, P. W. Van den Broeke, H. Fokkens and A. L. Van Gijn (eds.) *The Prehistory of the Netherlands*. Amsterdam University Press, Amsterdam, pp. 203–219.

Van Gijn, A. L. and Verbaas A. (2008) Het technologische systeem van Ypenburg. Een gebruikssporenanalyse van diverse werktuigtypen. In H. Koot, L. Bruning and R. A. Houkes (eds.) *Ypenburg-lokatie 4. Een nederzetting met grafveld uit het Midden-Neolithicum in het West-Nederlandse kustgebied*. Leiden, pp. 289–314.

Grégoire, J. P. (1992) Les grandes unités de transformation des céréales: l'exemple des minoteries de la Mesopotamie du Sud à la fin du IIIe millénaire avant notre ère. In P. C. Anderson (ed.) *Préhistoire de l'agriculture. Nouvelles approches experimentales et ethnographiques*. Monographies du CRA 6. Editions du CNRS, Paris, pp. 321–339.

Guðmundsson, G. (1996) Gathering and processing of lyme-grass (*Elymus arenarius* L.). *Vegetation History and Archaeobotany* 5 (1-2), 13–23.

Gunda, B. (1983) Cultural ecology of old cultivated plants in the Carpathian area. *Ethnologia Europea* 13 (2), 145–179.

Hamon, C. (2006) *Broyage et abrasion au Néolithique ancien. Caractérisation technique et fonctionnelle des outillages en grès du Bassin parisien*. BAR International Series 1551 Archaeopress, Oxford.

Hayden, B. (ed.) 1987 *Lithic studies among the contemporary Highland Maya*. University of Arizona Press, Tucson.

Hillman, G. C. (1981) Reconstructing crop husbandry practices from charred remains of crops. In R. Mercer (ed.) *Farming practice in prehistoric Britain*. Edinburgh University Press, Edinburgh, pp. 123–162.

Hillman, G. C. (1982) Evidence for spelt malt. In R. Leech (ed.) *Excavations at Castgore 1970-1973. A Romano-British village*, 2. Western Archaeological Trust Excavation Monograph 2, Bristol, pp. 137–141.

Hillman, G. C. (1984a) Interpretation of archaeological plant remains: the application of ethnographic models from Turkey, In W. van Zeist and W. Casparie (eds.) *Plants and ancient man: studies in palaeoethnobotany.* Balkema, Rotterdam, pp. 1–41.

Hillman, G. C. (1984b) Traditional husbandry and processing of archaic cereals in recent times: the operations, products and equipment which might feature in Sumerian texts, part I: the glume wheats. *Bulletin of Sumerian Agriculture* 1, 114–152.

Hillman, G. C. (1985) Traditional husbandry and processing of archaic cereals in recent times: the operations, products and equipment which might feature in Sumerian texts, part II: the free-threshing wheats. *Bulletin of Sumerian Agriculture* 2, 1–31.

Horsfal, G. (1987) A design theory perspective on variability in grinding stones. In B. Hayden (ed.) *Lithic studies among the contemporary Highland Maya.* University of Arizona Press, Tucson, pp. 332–377.

Houkes, R. A. (2008) Natuursteen. In H. Koot, L. Bruning and R. A. Houkes (eds.) *Ypenburg-lokatie 4. Een nederzetting met grafveld uit het Midden-Neolithicum in het West-Nederlandse kustgebied.* Leiden, pp. 247–262.

Hubert, A. (1998) Autour d'un concept: 'l'alimentation méditerranéenne', *Techniques et Culture* 31–32, 153–160.

Kanafani-Zahar, A. (1994) *Mûne, la conservation alimentaire traditionnelle au Liban.* Maison des Sciences de l'Homme, Paris.

Kanta, A. and Marazzi, M. (eds.) (2006) *Monastiraki 1, Campagne 2002/2004,* Università degli studî Suor Orsola Benincasa, Napoli.

Karapli, K. (1994) *To sitari, to psomi kai o Byzantinos stratos.* In *O artos imon. Apo to sitari sto psomi.* ETBA Editions, Athens, pp. 110.

Katz, E. (2003) Le metate, meule dormante du Mexique. In H. Franconie, M. Chastanet and F. Sigaut (eds.) *Couscous, boulgour et polenta: transformer et consommer les céréales dans le monde.* Editions Karthala, Paris, pp. 32–50.

Kearns, E. (1994) Cakes in Greek Sacrifice Regulations. In R. Hägg (ed.) *Ancient Greek Cult Practice from the Epigraphical Evidence.* Paul Aströms Förlag, Stockholm, pp. 65–70.

Kislev M. E., Artzy, M. and Marcus, E. (1993) Import of an Aegean Food Plant to a Middle Bronze IIA Coastal Site in Israel. *Levant* 25, 145–54.

Körber-Grohne, U. (1987) The history of spelt (*Triticum spelta*) on the basis of archaeobotanical findings from neolithic to medieval times, and the data by written sources until today. In J. P. Devroey and J. J. van Mol (eds.) *L'épeautre (Triticum spelta), histoire et ethnologie.* Editions Dire, Treignes, pp. 51–60.

Koukoulès, P. (1952) *Byzantinon Bios kai politismos,* vol. 5. Institut français d'Athènes, Athens.

Kubiak-Martens, L. (2006) Botanical remains and plant food subsistence. In L. P. Louwe Kooijmans and P. F. B. Jongste (eds.) *Schipluiden. A Neolithic settlement on the Dutch North Sea coast c. 3500 cal BC.* Analecta Praehistorica Leidensia 37/38, Leiden, pp. 317–338.

Le Breton, D. (2006) *La saveur du Monde. Une anthropologie des sens.* Métaillé, Paris.

Lebessi, A. and Muhly, P. (1990) Aspects of Minoan cult enclosures. The evidence of the Syme sanctuary (Crete), *Archäologischer Anzeiger* 1990, 315–336.

Le Guérer, A. (2002) *Le pouvoir des odeurs.* Odile Jacob, Paris.

Lüning, J. and Meurers-Balke, J. (1980) Experimenteller Getreindeanbaun im Hambacher Forst, Gemeinde Elsdorf, Kr. Bergheim/Rheinland, *Bonner Jahrbücher* 180, 305–344.

Lyons, D. and D'Andrea, A. C. (2003) Griddles, Ovens, and Agricultural Origins: An Ethnoarchaeological Study of Bread Baking in Highland Ethiopia, *American Anthropologist* 105 (3), 515–30.

Lyons, D. (2007) Integrating African cuisines: Rural cuisine and identity in Tigray, highland Ethiopia. *Journal of Social Archaeology* 7 (3), 346–371

Mahler-Slasky, Y. and Kislev, M. E. (2010) Lathyrus consumption in late Bronze and iron age sites in Israel: an Aegean affinity. *Journal of Archaeological Science* 37 (15), 2477–2485.

Markus, M. (1989) La presence de l'epeautre dans l'espace carpatique. In J. P. Devroey and J. J. van Mol (eds.) *L'épeautre (Triticum spelta), histoire et ethnologie.* Editions Dire, Treignes, pp. 123–127.

Matthiae, P. (1982) Die Fürstengräber des Palastes Q in Ebla. *Antike Welt* 13 (1), 3–14.

Migne, J.-P. (2012) *Patrologia Graeca.* Website. URL: *http://www.ellopos.net/elpenor/greek-texts/fathers/migne-patrologia-graeca.asp*

Molleson, T. (1994) The Eloquent Bones of Abu Hureyra. *Scientific American* 271 (2), 70–75.

Moundra-Agrafioti H. A. 2007, Lithina ergaleia. In Ch. Doumas (ed.) *Akrotiri Thiras (Dytiki Oikia),* Kathimerini, Athens, pp. 71–102. (in Greek)

Nesbitt, M. and Samuel, D. (1996) From staple crop to extinction? The archaeology and history of hulled wheats. In S. Padulosi, K. Hammer and J. Heller (eds.) *Hulled Wheats. Promoting the conservation and use of underutilized and neglected crops.* Proceedings of the First International Workshop on Hulled Wheats, July 1995. IPGRI, Rome, pp. 41–100.

Orrieux, C. (1983) *Les Papyrus de Zénon, l'horizon d'un grec en Égypte au IIIe siècle av. J.-C.* Deucalion, Paris.

Palmer, R. (1992) Wheat and barley in Mycenean Society. In Olivier J. P. (ed.), *Mykenaïka, Bulletin de Correspondance Héllenique Supplément* XXV, pp. 175–497.

Pasayiannis, K. (1928) *Maniatika Moiroloyia.* Sideris, Athens. (in greek)

Peña-Chocarro, L. (1994) Los modelos etnográficos en Arqueobotánica: los cereales vestidos. *I Jornadas*

Internacionales sobre Tecnología Agraria Tradicional 1992. Museo Nacional del Pueblo Español, Madrid, pp. 21–29.

Peña-Chocarro, L. (1996) In-situ conservation of hulled-wheat species: the case of Spain, In S. Padulosi, K. Hammer and J. Heller (eds.) *Hulled Wheats. Promoting the conservation and use of underutilized and neglected crops.* Proceedings of the First International Workshop on Hulled Wheats, July 1995. IPGRI, Rome, pp. 129–146.

Peña-Chocarro, L. (1999) *Prehistoric Agriculture in Southern Spain during the Neolithic and the Bronze Age. The application of ethnographic models.* BAR International Series 818. Archaeopress, Oxford.

Peña-Chocarro, L. and Zapata, L. (2003) Post-harvesting processing of hulled wheats. An ethnoarchaeological approach. In P. C. Anderson, L. S. Cummings, T. K. Schippers, and B. Simonel (eds.) *Le traitement des récoltes: un regard sur la diversité, du Néolithique au présent.* Actes des XXIIIe Rencontres Internationales d'Archéologie et d'Histoire d'Antibes, 17–19 octobre 2002, Éditions APDCA, Antibes, pp. 99–113.

Peña-Chocarro, L., Zapata, L., González Urquijo, J. E., Ibáñez Estévez, J. J. and Moreno García M. (2005) An ethnoarchaeological project in the Western Rif (northern Morocco): First results. *Quaderni del Laboratorio di Archeobiologia* 1, 21–32.

Peña-Chocarro, L., Zapata, L., González Urquijo, J. E. and Ibáñez, J. J. (2009) Einkorn (*Triticum monococcum* L.) cultivation in mountain communities of the western Rif (Morocco): an ethnoarchaeological project. In A. S. Fairnbairn and E. Weiss (eds.) *From foragers to farmers. Gordon Hillman Festschrift.* Oxbow Books, Oxford, pp. 103–111.

Pernier, L. and Banti, L. (1951) *Palazzo Minoico di Festôs, II. Seconda Palazzo.* Roma.

Perrino, P. Laghetti, G., D'Antuono, L. F., Al Ajloun, Mi, Kanbertay M. and Szabó A. T. and Hammer, K. (1996) Ecogeographical distribution of hulled wheat species. In S. Padulosi, K. Hammer and J. Heller (eds.) *Hulled Wheats. Promoting the conservation and use of underutilized and neglected crops.* Proceedings of the First International Workshop on Hulled Wheats, July 1995. IPGRI, Rome, pp. 100–118.

Piperno, D. R., Weiss, E., Holst, I. and Nadel, D. (2004) Processing of wild cereal grains in the Upper Palaeolithic revealed by starch grain analysis. *Nature* 430, 670–673.

Poitevin, G. (1997) *Le chant des meules. De la piété de paysannes à la philosophie de swamis.* Editions Kailash, Paris-Pondicherry.

Poursat, J.-C. (1996) *Fouilles exécutées à Malia. Le Quartier Mu III. Artisans minoens: les maisons-ateliers du Quartier Mu*, Etudes Crétoises 32, Ecole Française d'Athènes, Athènes.

Procopiou, H. (1998) *L'outillage de mouture et de broyage en Crète Minoenne,* Unpublished Ph.D. thesis, Université de Paris 1.

Procopiou, H. (2003) Les techniques de décorticage dans le monde égéen : étude ethno-archéologique dans les Cyclades. In P. C. Anderson, L. S. Cummings, T. K. Schippers, and B. Simonel (eds.) *Le traitement des récoltes: un regard sur la diversité, du Néolithique au présent.* Actes des XXIIIe Rencontres Internationales d'Archéologie et d'Histoire d'Antibes, 17–19 octobre 2002, Éditions APDCA, Antibes, pp. 115–136.

Procopiou, H. (2004) La transformation des matières minérales, *Dossiers d'Archéologie* 290, 58–61.

Procopiou, H. (2010) L'archéologie de la continuité :la perception du monde rural en protohistoire égéenne. In S. de Beaune (dir.) *Écrire le passé. La fabrique de la préhistoire et de l'histoire à travers les siècles*, CNRS Editions, Paris, pp. 149–161.

Procopiou, H. (forthcoming). Les outils lithiques. In J.-C. Poursat (ed.) *Le Quartier Mu V. Vie quotidienne et techniques au MMII*, Etudes Crétoises, Ecole Française d'Athènes, Athènes.

Procopiou, H., Anderson, P., Formenti, F. and Juan-Tresseras, J. (2002) Étude des matériaux transformés sur des outils de mouture : identification des résidus par analyse chimique et par observation en microscopie optique et électronique. In H. Procopiou and R. Treuil (eds.) *Moudre et broyer. L'identification fonctionnelle de l'outillage de mouture dans la Préhistoire et l'Antiquité, I. Méthodes.* CTHS, Paris, pp. 111–127.

Psilakis, N. and Psilakis, M. (2001) *To psomi ton Ellinon kai ta glykismata tis laikis mas paradosis.*Karmanor, Heraklion (in greek).

Rethemiotakis, G. (1999) The Hearths of the Minoan Palace at Galatas. In P. Betancourt, V. Karageorghis, R. Laffineur and W.-D. Niemeier (eds.) *Meletemata: Studies in Aegean Archaeology Presented to Malcolm H. Wiener as He Enters His 65th Year. Vol. III.* Aegeum 20. Peeters Publishers, Leuven, pp. 721–728.

Riley, F. R. (1999) *The Role of the Traditional Mediterranean Diet in the Development of Minoan Crete.* BAR International Series 810. Archaeopress, Oxford.

Rocchetti, L. 1994 (ed.) *Sybrita. La valle di Amari. Era Bronzo et Fero*, Incunabula Graeca XCVI, Roma.

Roux, V. (1985) *Le matériel de broyage. Etude ethno-archéologique à Tichiff (R. Mauritanie).* Editions Recherche sur les civilisations, Mémoire 58, Paris.

Runnels, C. N. (1981) *A diachronic study and economic analysis of millstones from the Argolid, Greece*, Ph.D. Dissertation, Indiana University, Bloomington.

Santallier, D., Caron, V., Gisclon, J.-L., Jautée, E. and Rantsordas, S. (2002) Réflexions préliminaires sur les qualités mécaniques des matériaux lithiques utilisés pour la confection du matériel de broyage et de mouture, In H. Procopiou and R. Treuil (eds.) *Moudre et Broyer. L'identification fonctionnelle de l'outillage de mouture dans la Préhistoire et l'Antiquité*, Vol. 1. Méthodes, Paris, pp. 15–29.

Sacconi, A. (2001) Les repas sacrés dans les textes

mycéniens. In R. Laffineur and R. Hägg (eds.) *POTNIA. Dieties and Religion in the Aegean Bronze Age*. Aegaeum 22. Peeters Publishers, Leuven, pp. 467–470.

Samuel, D. (1993) Ancient Egyptian cereal processing: the artistic record, *Cambridge Archaeological Journal* 3 (2), 276–283.

Samuel, D. (1994) *An archaeological study of baking and bread in New Kingdom Egypt*. Unpublished Ph.D. thesis, University of Cambridge.

Samuel, D. (2006) Modified starch. In R. Torrence and H. Barton (eds.) *Ancient starch research*. Left Coast Press, Walnut Creek, pp. 205–216.

Samuel, D. (2010) Experimental grinding and Ancient Egyptian flour production. In S. Ikram and A. Dodson (eds.) *Beyond the Horizon: Studies in Egyptian Art, Archaeology and History in Honour of Barry J. Kemp*. American University in Cairo Press, Cairo, pp. 456–477

Sarpaki, A. and Jones, G. (1990) Ancient and Modern Cultivation of *Lathyrus Clymenum L.* in the Greek Islands, *Annual of the British School at Athens* 85, 363–368.

Sarpaki, A. (1992) A palaeoethnobotanical study of the West House, Akrotiri, Thera, *Annual of the British School at Athens* 87, 220–230.

Sarpaki, A. (2001) Processed cereals and pulses from the late Bronze Age Site of Akrotiri, Thera. Preparations prior the consumption: a preliminary approach to their study, *Annual of the British School at Athens*, 96, pp. 27–40.

Schneider, J. S. (2002) Milling tool design, stone textures, and function. In H. Procopiou and R. Treuil (eds.) *Moudre et broyer. L'identification fonctionnelle de l'outillage de mouture dans la Préhistoire et l'Antiquité, I. Méthodes*. CTHS, Paris, pp. 31–53.

Sigaut, F. (1989) Les spécificités de l'épeautre et l'évolution des techniques. In J. P. Devroey and J. J. van Mol (eds.) *L'épeautre (Triticum spelta), histoire et ethnologie*. Editions Dire, Treignes, pp. 29–49.

Sigaut, F. (2002) Sortir de l'oubli. In AGAPAIN (eds.) *Les Meuliers. Meules et Pierres Meulières dans le Bassin parisien*. Presses du village, Étrépilly, pp. 9–20.

Sigaut, F. (2010) Qu'est-ce qu'on mange au juste? In H. Franconie, M. Chastanet and F. Sigaut (eds.) *Couscous, boulgour et polenta: transformer et consommer les céréales dans le monde*. Editions Karthala, Paris, pp. 13–29.

Souccar, T. (2006) *Le régime préhistorique. Comment l'alimentation des origines peut nous sauver des maladies de civilisation*. Indigènes, Barcelone.

Stahl, A. (1989) Plant food processing: implications for dietary quality. In D. R. Harris and G. C. Hillman (eds.) *Foraging and farming. The evolution of plant exploitation*. Unwin Hyman, London. pp. 171–194.

Toffin, G. (1983) Moisson aux baguettes au Nepal central. *Objets et Mondes* 23 (3–4), 173–176.

Valamoti, S. M. (2002) Food remains from Bronze Age Archondiko and Mesimeriani Toumba in northern Greece. *Vegetation History and Archaeobotany* 11, 17–22.

Valamoti, S. M. and Anastasaki, S. (2007) A Daily Bread – Prepared but Once a Year. *Petits Propos Culinaires* 84, 75–100.

Valamoti, S. M., Samuel, D., Bayram, M. and Marinova, E. (2008). Prehistoric cereal foods from Greece and Bulgaria: investigation of starch microstructure in experimental and archaeological charred remains. *Vegetation History and Archaeobotany* 17 (Suppl. 1), 265–276.

Van der Veen, M. (1989) Charred Grain Assemblages from Roman-Period Corn Driers in Britain, *Archaeological Journal* 146, 302–319.

Verbaas, A. and Van Gijn, A. L. (2007) Querns and other hard stone tools. In P. Van de Velde (ed.) *Excavations at Geleen-Janskamperveld 1990/1991*. Analecta Praehistorica Leidensia, Leiden, pp. 191–204.

Wright, K. I. (1994) Ground stone tools and hunter-gatherer subsistence in southwest Asia: implications for the transition to farming. *American Antiquity* 59(2), 238–263.

Zvelebil, M. (1994) Plant use in the Mesolithic and its role in the transition to farming. *Proceedings of the Prehistoric Society* 60, 35–74.

SECTION 3

The Social Context of Agricultural Technology

8 Aquiring Skills and the Transmission of Knowledge

8.1. INTRODUCTION

Inja Smerdel

From times immemorial people had to be resilient and innovative to adapt to a vast variety of natural environments and circumstances. Recent decades (in Europe at least from the 1960's onwards) have witnessed a loss of skills in most of the agricultural techniques, while in other domains new technologies, knowledge and skills have arisen. But the generations that were still 'touching' the past centuries, working with their hands, are slowly and irrevocably leaving us.

The knowledge and the skills people acquired, accumulated, further developed and passed on to new generations, were essential for their livelihood. The introduction of new knowledge or skills was not a rare event, confined to the elite classes, but because it was so valuable, rulers frequently claimed the credit. Agriculture, 'the largest industry of pre-industrial Europe' (Braudel 1988, 143) was fundamental not just for the livelihood of individuals but to the economy of the whole state. Let us introduce two examples.

The introduction of the shoulder yoke in Ireland by the mythical king Eochaid comes from an early Irish tale, *The Wooing of Étaíne* (Bergin and Best 1977, in Bell 1985, 73): 'Now until that night the men of Ireland used to put the strain on the foreheads of oxen, (but) it was seen that the folk of the elfmounds were putting it on their shoulders. Eochaid did the same, hence he is called Eochaid Airem (*i.e.* ploughman), for he was the first of the men of Ireland to put a yoke upon the necks of oxen.' The deed hypothetically dates from the beginning of the Christian era (Griffin-Kremer 2006, 65), but has an intriguing wider context of the transmission of innovation from one 'ethnic' group to another. In fact, it is a timeless 'spy story', the theft of a better tool that happens in the darkness of the night. King Eochaid sends out a spy, his superintendent, who hides in the darkness to secretly watch the hard work of road construction easily executed by King Midir's elfmound folk and their oxen, harnessed in shoulder yokes. Thus Eochaid, through industrial espionage, introduces a technical innovation to benefit his people (Griffin-Kremer 2006, 62–63, 66).

The second example, the ploughing of the Austrian emperor Joseph II is a historical fact, richly described in a book on his life and deeds (Bermann 1881). Joseph was advocating 'strong' agriculture, founding numerous Agricultural Societies (Vacha *et al.* 1994, 324). When in August 1769 he travelled under cover to visit the Prussian king Friedrich II, the axle of his carriage broke near the Slovak village Slavikovice. Coming out of the carriage he saw a farmer, Anton Trnka, ploughing his field before spring sowing. The emperor asked

Trnka to let him plough, but the farmer argued that ploughing is farmer's work, not gentlemen's because for a gentleman it would be a shame. To which the emperor replied: 'No work is shameful, particularly not the farmer's that gives us our daily bread.' He then successfully ploughed, and at the end smilingly said to his travel companions: 'See, my gentlemen, a ruler must master any work!' (Bermann 1881, 818–819). The prestigious example of the 'ploughing emperor' was imitated by the Russian emperor Alexander I (in 1815) and by the French governor of Algeria, marshal Bugeaud, in 1841 (Bermann 1881, 824).

Acquiring the special skills of agriculture has thus always been and still is so important and essential, that from ancient times to our contemporary industrial world, it was not uncommonly a matter of prestige, espionage or even minor wars. Contemporary events in pharmacological enterprises can be clearly seen as illustrative recent global examples. But there is more to the mythical rivalry between Midir and Eochaid. Technical solutions and the necessary skills to perform them properly are also markers of cultural boundaries (Barth 1969, cf. Schippers, this chapter).

When we deal with agriculture, trying to reveal knowledge and explain skills that were common among the majority of ethnic groups and their cultures, our knowledge on the subject is substantial but mostly fragmented; it is scattered through an immense number of studies, written by researchers of various scientific backgrounds. Acquiring skills, mastering various techniques, gestures, and the transmission of diverse knowledge is a topic suitable for diversified research methods and methodological approaches. Nevertheless, we can discern two foci of analysis and deliberation. One approaches knowledge and skills as the guaranty of continuity and stability, handed down in the environment of 'historical processes of adaptations to local agro-ecological conditions of slowly spreading 'tendencies'' (Leroi-Gourhan 1943/1971, cf. Schippers, this chapter). The other is more interested in the elements that 'arose inside' or 'came into' a community as an element of change, of necessary innovation, the proverbial 'daughters of need'.

In the first case – the predominant focus of this chapter – we are dealing with a process of imitative learning by observing and doing. The farmer, whether a traditional Asian rice grower or a modern American rancher, gains immense knowledge of his close environment, and 'his way of functioning was intimately related to the capabilities of his body, of which even the simplest tool became a prolongation with a great variety of possible actions...' (Steensberg 1993, 12). Children were crucial actors in this dynamic socio-historical game, and the awareness of their indispensable role in the process of knowledge transmission should be a relevant component, if not the central topic of individual research on the theme (as in Chapter 8.2, for example). Children absorb knowledge and acquire various skills by 'rehearsive play' (Cameron 1996, 22), with miniaturised tools and by rehearsing adult roles in play, by years of long apprenticeship in a variety of farming tasks and from oral tradition. Their 'first classroom' is usually nature, the animals and plants in the fields, and their teachers and 'employers', their elder kin, assigning them jobs according to age and gender. The social environment was a recognisable frame for these processes. Besides the initial role of an individual's family, there was the dominating, influential role of specific groups – ethnic or local and regional communities who taught 'their ways' and their knowhow in agricultural operations, either because of religious belief or strategic adaptation to circumstances.

In the second case – when we deal with the transmission of knowledge as an element of change and of innovation or independent invention – the central question is the following: who were the people introducing these new cultural elements, including tools, knowledge and skills? The range of possibilities extends from individuals to states. Individual farm-hands and seasonal workers, mowers, threshers, and the like can be the bearers of knowledge and skills from elsewhere. We may, for example, trace a broad European context in this aspect, bearing in mind long-term population movements (both medieval and modern times), also labour migrations of agrarian day labourers, and historical awareness of interdependence between these movements and technology transfer (see *e.g.* Bruneel *et al.* 1990, 48). Wandering craftsmen also bring knowledge and skills even from other countries. Representatives of a village community go to other villages to bring home useful innovations, like that group of 'wise men' from the Slovenian mountain village Jamnik on the steep slopes of Mount Jelovica

(Fig. 8.1), who went to a neighbouring valley to see a special ploughing technique they afterwards applied at home. Monastic orders and missionaries disseminated knowledge from their centres. The central state – with its various laws, agricultural societies and schools fostering knowledge – cannot be forgotten. In Europe, these were an influential factor, especially during the Enlightenment period with its interest in economy and agriculture.

But there are some other questions that were posed and are at least briefly answered in this chapter's contributions: What was the response of individuals or of a community to the introduction of cultural changes such as new or improved tools, or different techniques? Were there also cases where new skills and knowledge were rejected? What do these reveal and what were the reasons for such rejections? (See Chapter 8.4.) Were there any other relevant motives for acquiring various skills and gaining knowledge besides subsistence? (See Chapter 8.3. and 8.5.)

Trying to elucidate at least some of the possible topics and answer at least some of the questions posed, we introduce but a few examples. None of them was initiated in the frame of any ongoing debates on the cultural aspects of technology transfer. Our contributions merely introduce or discuss thematically relevant findings from our individual (and consequently geographically determined) past, long-term or current wider research: on the man-ox relationship, hay harvesting tools, technological choices and cultural boundaries, and on stone technology. Although they vary not only in theme and narration, and are diversified also in regard to research methods, our texts can nevertheless offer a handful of answers concerning the processes of knowledge transmission and acquisition of skills, thus contributing to recent debates on the topic. The study of skills should imply cross-examination of different relevant sources, for example comparison of oral testimonies with primary material sources like tools marked by traces of use, and with available pictorial and written historical sources.

In the initial contribution, Inja Smerdel addresses the apprenticeship of children in daily work with oxen; the mutual relationship between young animals and 'little humans', and the parallels in their training. Her findings are based mostly on ethnographic fieldwork in selected Slovenian regions and on pictorial, literary and other written sources. In her second text on the art of a mower, she introduces research on primary material sources, a Slovenian collection of whetstone holders, eloquent witnesses to some of the most valued and socially recognised agricultural skills, those of the hay harvest and scythe mowing. Among other issues, she strives to explain mower's skills as masculinity markers. Thomas Schippers addresses skills as identity markers, again on the basis of ethnographic fieldwork conducted in two southern Alpine valleys. Through two different techniques – threshing and carrying of loads – proper to two local (and national) groups, he reveals a cultural boundary. In the closing contribution, Helena Knutsson analyses archaeological sources, seeing blades as messengers of agriculture at the beginning of the Neolithic in Scandinavia. These objects, which bear traces of production and use, can further explain the skills of prehistoric knappers and harvesters, learned or reinvented experimentally by meticulous observation of archaeological tools.

The collected contributions hopefully prove that, although we are currently witnessing de-skilling processes in various fields of human activities, scientific tools and methods exist that allow us to detect, document, preserve, or even revive past skills and knowledge and help us reveal their historical, social, and environmental contexts.

8.2. 'TRAINING OXEN MEANT TRAINING FOR THE CHILDREN'

Inja Smerdel

Introduction: Investigating the Relationship between Children and Working Oxen

'Training oxen meant training for the children', said my informant Angela (from Blanca near Sevnica in the Posavje region of Slovenia; Fig. 8.1) in 2003, eloquently introducing the notion of an important parallel between training children for daily and seasonal agricultural tasks, and training oxen for carrying out all the work needed on an individual farm. The jobs related to oxen that also involved children were indeed 'the school of life' for them, as another informant, Cenco from Ošlje on the edge of the Karst region (Fig. 8.1), agreed.

The apprenticeship of children in animal husbandry is one of the prime insights of my ongoing research into the relationship between men and working oxen; on the mutual relations that are bonds (cf. Digard 2005, 38, 29) – with those animals of the farm that were 'like family' to their people and constituted 'a part of the human society' (Lévi-Strauss 1962, 272). The case-study thus contributes some relevant issues to the central focus of this chapter, related with children as crucial actors in the dynamic processes of knowledge transmission and acquisition of skills.

In 1961, Vilko Novak, one of the founders of Slovene ethnology, commented on stockbreeding that 'Working daily with livestock of all sorts shapes and defines a peasant's work, thoughts and speech much more than cultivation does...' (Novak 1961, in Smerdel 2005, 342). Contemporary anthropology sees similar themes in studies of animal domestication. Central cultural values have been recognised in attitudes to livestock, giving rise to expressions such as 'the African cattle complex' (Campbell 2005, 79; cf. Evans-Pritchard 1940, 1950). But oxen have been given little of the attention they certainly deserve. In European history, with a few French exceptions (Dalin 1999; Rognon and Lefèvre 2007; Sigaut *et al.* 1997), 'The ox, in spite of its historical importance which was in no way lesser than that of horse – has disappeared without any mention at all'... (Sigaut 1998, 1082). The ox has been inevitably present throughout the centuries, so commonplace that it was hardly noticed, yet proof of the importance and the intrinsic value of oxen in daily life and the significance of the knowledge and skills needed to work with them can be found everywhere. Oxen hitched to implements indeed deserve to be called those 'formidable machines which transformed the surface of the earth' (Griffin-Kremer 2007, 66). They were, and in some places still are, among the most common working animals worldwide. In places oxen, are so indispensable that Harris (1989, 16) explains the protection of the Indian holy cows more as a realistic rather than a religious taboo partly because the mother cow is the necessary 'factory for making oxen'.

Oxen predominated as draught animals in most parts of Slovenia (especially in the mountainous areas) and on most farms (in particular the prevalent medium-sized ones) until the 1970s. However, the limited scholarship on oxen is scattered in other work on stock-keeping (Stanonik 1987; Makarovič 1982; 1985; 1994) and a few more specialised studies (Smerdel 2005; 2007; 2009).

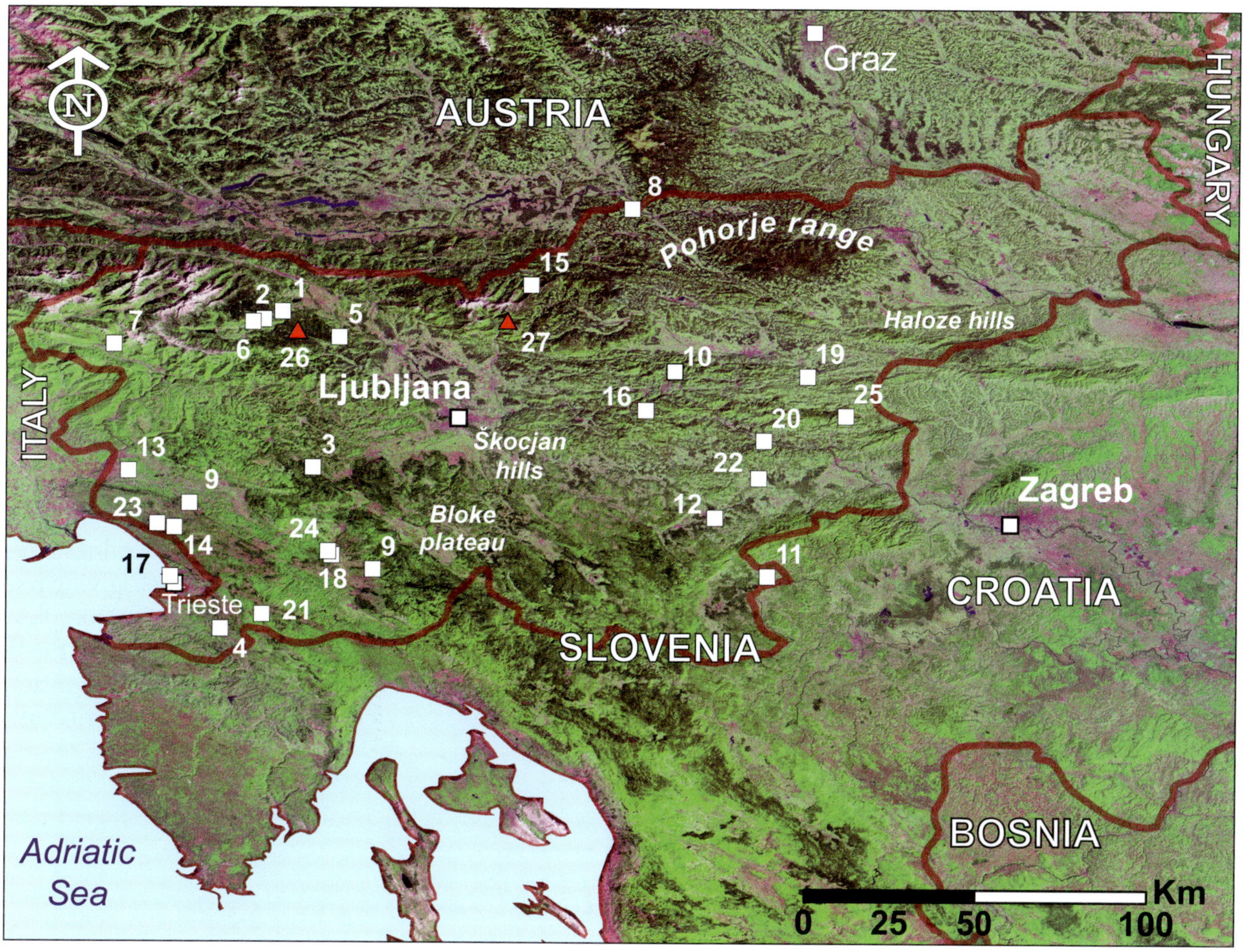

Fig. 8.1. Map of Slovenia with the settlements (□) of 1) Bohinjska Bela; 2) Gorjuše; 3) Hotedršica; 4) Hrastovlje; 5) Jamnik; 6) Koprivnik v Bohinju; 7) Koseč; 8) Libeliška gora; 9) Lukovec; 10) Marija Reka; 11) Metlika; 12) Novo Mesto; 13) Ošlje; 14) Pliskovica; 15) Robanov kot; 16) Rodež; 17) Ščedenj; 18) Selce; 19) Sele; 20) Sevnica; 21) Skadanščina; 22) Jurišče; 23) Škofi; 24) Slavina; 25) Vetrnik; the mountains (▲) of 26) Mount Jelovica and 27) Mount Velika planina, the Bloke plateau, the Škocjan hills, the Haloze hills, and the Pohorje range mentioned in Chapters 8.1, 8.2 and 8.3. Map: R. Lugon, J.-C. Loubier and A. Chevalier.

Oxen were the indispensable working companions of most Slovene peasants and, at the same time, highly appreciated assets whose labour earned the farm some money. 'The oxen were gold' for the farmers because 'they were there for everything: for dung carting, for ploughing, for hay and fodder, for work in the woods...' Their value even became a unit of money and a measure for a farmer's property (Smerdel 2005, 375). No wonder then that the farmers feared cattle diseases more than their own (Smerdel 2007, 147). A 'village poet' from Pliskovica in the Karst region (Fig. 8.1) memorialised the deep relationship between man and working oxen in the Slovene countryside in the inscription under a statue of an ox in 2006: 'Once we ploughed with oxen here about, yoked they carted heavy loads for us. Without them we would not survive, that's why we loved them all their life.'

The more important agricultural processes were for the livelihood of a society, the more fundamental were the knowledge and skills involved. As I became aware of the 'humble indispensability' of oxen and their meanings in the structures of rural life, I was led into research to reveal the range of interdependence in the relationship between man and ox and to record the intangible heritage of knowledge and wisdom about breeding, care, treatment, training, communicating and working with oxen. I was interested in the place of individuals with special skills in communicating with draught oxen, their roles in village and family labour and specialisation and, secondly, transmission of knowledge and skill through the socialisation and training of children (Smerdel 2005, 344). In agricultural techniques using limited tools, including intangible tools such as the human voice, abstract processes and practical skills and knowledge are based on observation (cf. Digard 1988, 144–147); on 'close attention to the movements, sounds and gestures of animals' (Ingold 2000, 25). Children are naturally around in the courtyards, fields and pastures, determined to learn how to deal with cattle and oxen by observing, and gradually by doing.

Although the number of anthropological and archaeological studies of children is growing (*e.g.* Arden 2006; Baxter 2005; Kamp 2001; 2002; Lillehammer 1989; Ramšak 2007; Sofaer 2000), until recently children's work has been almost as poorly investigated as the man-oxen relationship. Children's lives were rarely treated as a historical subject that deserves attention (Inkret 1982, 9). Looking for Slovene ethnological sources referring to children's agricultural work was 'looking for a needle in a haystack'. Those found mainly provide data on the ages at which children started to work, sometimes mentioning agricultural operations in which they were first engaged. For example, Marija Makarovič (1994, 228) recorded that children growing up in the village of Sele in Carinthia (Fig. 8.1) between the two World Wars started to work when they were about six years old. Between six and nine they mostly herded, gathered potatoes and led the oxen at ploughing. By the age of fifteen or sixteen years they already worked like adults. Gorazd Makarovič, synthesising Slovene attitudes toward time, argued that incorporating children in work

Fig. 8.2. A: Rock carvings with ploughing scenes from Mont-Bégo in the French Alps, dating from the Early Bronze Age (2300–1800 BCE). On all of them, smaller persons (children?) leading ox-teams are clearly visible. (From Rinieri-Villain 1997, fig. 8). B: The fresco painting from 1490, from the filial church of Hrastovlje (Slovene Istria), illustration of ploughing from the Labours of the months cycle. (Photo by M. Habič, 2006; SEM Archives). C: The picture of a Slovene farmer's homestead in Styria, made by Ferdinand Würst for the book '*Österreichische-Ungarische Monarchie in Wort und Bild*' (Wien 1890, page 211). A child stands there – a son observing father yoking their oxen... D: The little *Sadar's* boy Poldek from Male Češnjice (Dolenjska region) on pasture. (Photo: M. Matičetov, 1950; SEM Archives).

created two age-stages that became pronounced around the middle of the 18th century. At this time, changes in agriculture such as the introduction of a crop rotation regime, fostering of indoor rearing methods, and others which required more work, caused changes in attitude toward work. The first stage was rather early: by five or six years old a child must already attend to various tasks. The second age, introducing more demanding work, started at nine to twelve years (G. Makarovič 1995, 132).

So far the only in-depth study of childhood and education of the 19th century Slovenes is that of Alenka Puhar (1982), who tried to reconstruct children's daily lives, using scarce documents, especially memoirs. On the basis of these, she recognised (cf. Makarovič 1994) that the two most significant children's obligations were tending cattle and looking after younger children (Puhar 1982, 315 and Fig. 8.2D). Older children, eight to ten years old, might leave home for wage labour at similar occupations as nannies and herders (Puhar 1982, 312). Concerning the relationship between children and oxen, she notes that 19th-century parents – strongly influenced by Catholic doctrine, particularly in rural milieus – believed that 'stick and punishment bring wisdom', and that 'the best is for man to carry the yoke from his childhood'. It was also believed that children are by nature inclined to sin, and that nothing but the cane can cure that. To train children, to make them humble and obedient, applying physical punishment was considered the only and best method (Puhar 1982, 136–139). The parallel to oxen can be seen in a novel by Stanko Vuk (1998, 103) as he describes life in Karst in the early decades of the 20th century: 'He saw a plough and two pairs of yoked oxen. /.../ Plough, farmer, oxen and a boy, driving them with a thin rod...' Or in the juvenile memories of George Sand from 19th century France: 'A child of six to seven years, /.../ walked in the furrow, parallel to the plough, and kept on stinging the oxen's flank with a long and light cane, armed with a metal point. Proud animals trembled under the child's tiny hand...' (Sand 1962, 19). 'Oxen are educated like children.' (Versini 1997, 81).

Indeed some of the most telling descriptions of children working with oxen were written by various novelists (Blixen 1937/2004), authors of memoirs or realistic peasant stories, sensible and perceptive observers of daily life. In Rebula's autobiographic story (1957, 35, 111), a farmer grandfather trains his grandsons in yoking, unyoking and other work with oxen by making them a cart-toy and toy tools, and dividing chores among them by judging their capacities and preferences. 'Not yet four years old Štefan ran over the doorstep, carrying a chair to reach the beam above oxen's horns and take out the peg. All the rest can be done by Janko (the elder one) /.../ Mohor (the grandfather) can already watch the fruits of his training.' Prežihov (1969, 52, 10) describes children tending oxen, becoming little oxmen leading them at work, the process of training bullocks, and the bonds formed between children and oxen. 'As children, we early started to work with cattle. I did not yet go to school when I already drove oxen for ploughing and pulling, thus helping father.' 'The little herdsman silently cried /.../ When he stayed alone with cattle he snuggled to the oxen's bodies, seeking warmth....'

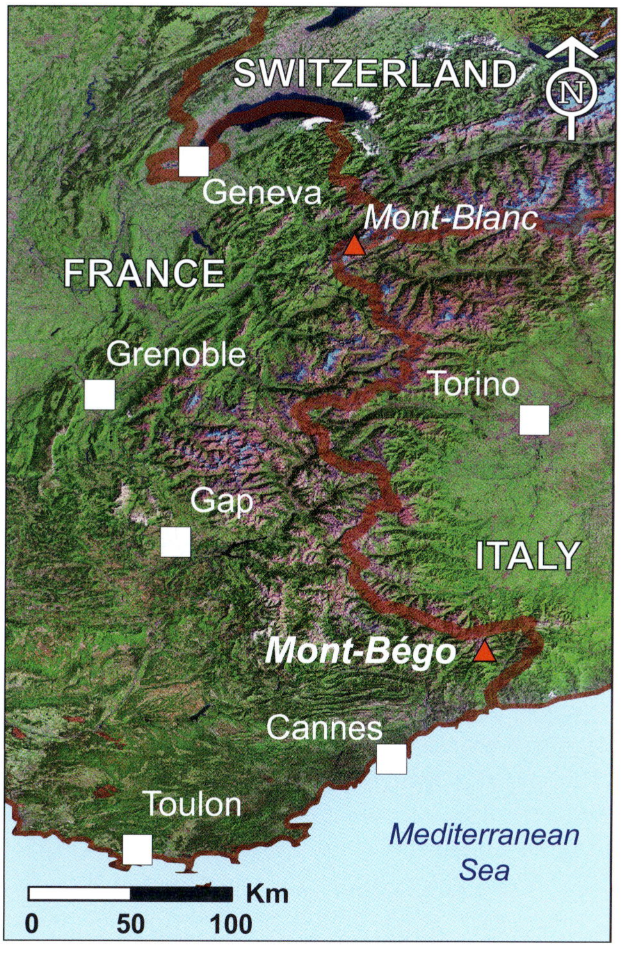

Fig. 8.3. Mont-Bégo in the southern Alps. Map: R. Lugon, J.-C. Loubier and A. Chevalier.

There do not seem to be many available pictorial sources, with the exception of 20th-century photographs of diverse provenance (Figs. 8.2D, 8.6). Yet there are some very early images depicting agricultural work with ox-teams in Europe. At Mont-Bégo in the French Alps (Fig. 8.3), rock carvings dating from the Early Bronze Age (2300–1800 BCE) reveal the importance of the ard and oxen (Rinieri-Villain 1997, 8–10). In some of the scenes, the human figures in front of ox-teams are evidently much smaller than the figures of the ploughmen behind them (Fig. 8.2A), and can plausibly be interpreted as children leading the oxen, although other interpretations are possible. Much later a fresco painting from 1490 provides a mediaeval illustration of ploughing from the Labours of the Months cycle in the filial church of Hrastovlje (Slovene Istria, Fig. 8.1). Here the two human figures are also of different size, and differently dressed (Fig. 8.2B), inviting us to assume that the working companionship of oxen and children, supported by numerous contemporary analogies, goes far back in time.

Research into Training Oxen and Children in Slovenia: Geographical, Historical and Social Contexts

Let us now move from more or less general reflections to focus on field research data from Slovenia on training oxen and children, and on their mutual relationship. I recorded them between 2001 and 2007 in four Slovenian ethnological regions which coincide closely with major geographical areas. These are 1. Mediterranean (in Pivka, Karst, Čičarija, and Šavrini in the Slovene Istria), 2. Alpine (on Pokljuka in Gorenjska, and in Carinthia), 3. Central-Slovenian (in part of Posavje, and in Zasavje region) and 4. Pannonian (in Kozjansko, and part of Posavje region). In these four regions, different bovine breeds were used, as well as varying forms of yoke which mark cultural boundaries. I chose research sites also based on the presence of previous ethnographic studies and the availability of farmers who currently or until recently worked with oxen. Depending on the age of my informants, their testimonies are relevant for the decades between the two World Wars, for the third quarter of the 20th century, and in isolated cases also for the present time.

In my research areas, among farmers who relied on oxen, the smallholders usually worked with one ox, and the owners of the most numerous, medium-sized farms used two animals. The wealthy farmers, particularly the owners of big lonely mountain farms in Carinthia, who usually owned two or three pairs of oxen, worked with one or two teams and kept a pair of young animals 'waiting behind'. In some other areas, for example in the lowland villages of Pivka region, the 'rich ones' had horses for ploughing, while 'average people' had oxen, and the poor ploughed with cows (Smerdel 2005, 357). In the needy villages of the Slovene Istria, where most people lived in 'miseria' (poverty), cultivating their hard clayey soil, which could not be ploughed with horses, the wealthy farmers owned an ox-team, 'some others maybe a donkey', and the poorest inhabitants of those villages did not even have a donkey to tether.

But on the farms where oxen were used, children and oxen were trained for their working obligations almost regardless of social context, whether a child drove his family's single ox (perhaps harnessed together with a neighbour's ox or even a cow), or proudly led the family's ox-team. Similarly, a child's training for managing the oxen usually began as he tended his homestead's cattle on pasture or herded for another farmer at the beginning of his 'serving career'. Training of oxen was mainly done by farmers themselves (the oxen were the subject of their chief care and pride), who selected, bought or sold, trained, and harnessed them, worked, carted, and ploughed with them. On some wealthy Carinthian farms, there were also farmhands called 'ox-men' who worked with cattle and were entrusted with training young oxen (Makarovič 1978, in Smerdel 2005, 359). Although oxen were men's work, some of the care of them was nevertheless occasionally carried out by other members of the family, commonly the children, if they were capable and old enough, or women, who might harness or lead oxen. However, the sons who were designated to become future masters of the farm were most often engaged in working with the oxen; they simply, naturally matured into real farmers in working companionship with their beasts.

I will focus here on the results of my research in the Pivka region, and conclude with comparative findings from other regions to reveal patterns of diversity in unity and the parallels between children and oxen; their interdependence or their mutual 'becoming with' (Haraway 2008, 19). The following issues dominate: how children grow from tending cattle to leading the oxen during the years of their apprenticeship in animal husbandry, and how 'the chosen ones' among them are recognised. The parallel issues for oxen include the choice of male calves to become bullocks, and the training of young oxen.

Pivka, at an altitude of 500 to 600 metres, is a transitional region on the passage from the Dinaric plateaus to the Karst and the Brkini regions, linking the Mediterranean and Central-Slovenian areas, with natural and socio-cultural influences from both sides. Agriculture in the villages of Selce, Slavina and Jurišče (Fig. 8.1) where I worked in the Upper Pivka was until the mid-20th century highly fragmented and poly-cultural, and aimed at feeding the farmer's household and his livestock. Owners of a 'half farm' (from 10 to 11 ha of land) and those with 'a quarter of a farm' (from 4 to 4.6 ha of land) dominated in the village's social structure, raising two to five head of beef cattle, with at most two pigs, and until the late 19th century a handful of sheep (Smerdel 2005, 353–357). The owners of the 'half a farm' yoked their own pairs of working oxen; the others, who often had only an ox and a cow, harnessed the two together for light jobs. For heavy work like ploughing they depended on mutual assistance, composing a team with a neighbour's ox. The demise of draught animals in Pivka started in the 1960s (as elsewhere in Europe), although when the first farmer of Selce bought a tractor in 1967, the people, suspicious of the novelty, used to say 'He won't plough over my land.'

Acquiring Skills and the Transmission of Knowledge: The Role of Folk Tales and 'Rehearsive Play'

A charmingly appropriate Slovene proverb of the 19th century reflects the notion of 'imitative learning' and the parallel between children and oxen: 'The little ox learns from the big ox' (Majer 1848, 118), as a son learns from his father (Fig. 8.4). As well as apprenticeship, 'observing and doing' in daily life, there existed two other ways of acquiring skills and transmitting knowledge related to working with oxen: oral traditions and the narration of folk tales, and 'rehearsive play'.

Folk tales whose content carries warnings about natural dangers are well known in the Karst region, in particular stories about the disappearance of young girls or boys, together with ox-teams, always

Fig. 8.4. A photograph of harrowing by *Gubenšek* in the village of Gubno (Kozjansko) before World War II, a proper illustration of the proverb: 'The little ox learns from the big ox'... (SEM Archives).

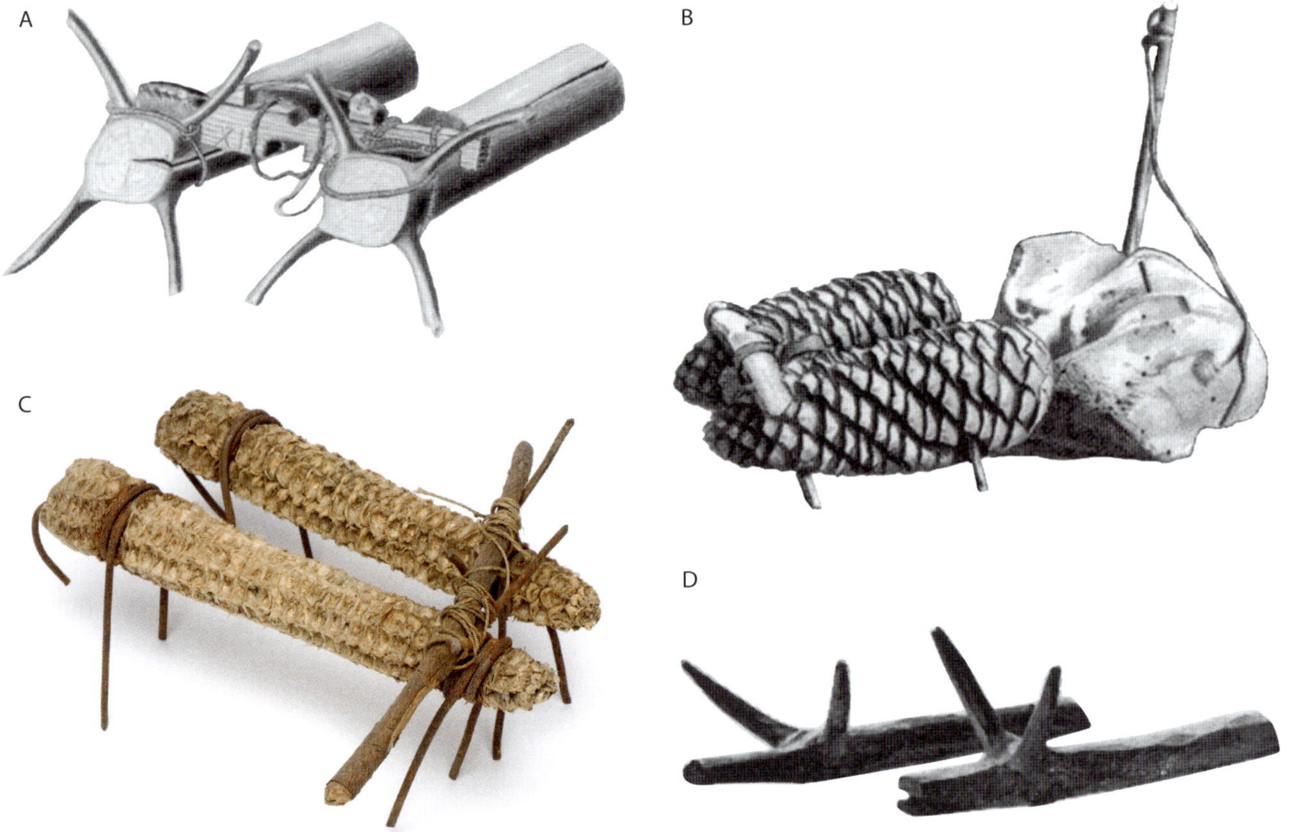

Fig. 8.5. A, B: Toy ox-teams – one (a) made of a small forked branch, the other one (b) of pine cones – from the collection of Eugenie Goldstern. (From Grieshofer 2004, page 142; nos. 15 and 31 from *Tafel 1* from the *Wiener Zeitschrift für Volkskunde*, 1924). C: Toy ox-team made of corn-cobs, from Los Llanos, Arenas de Iguña (Cantabria). From the catalogue of the exhibition Animalario (García-Hoz Rosales 2005, page 166). D: *Buše*, toy cows made of forked branches from Velika planina mountain. (From Cevc 1979, page 69).

in connection with karst watering places or abysses, after which the child's hair, or the oxen's horns and yoke, or both reappeared with water somewhere in the valley (Mencej 2006, 68). In the villages of the Pivka region, where the old people blamed witches for all animal diseases, tales transmitted the proper treatment: '... they could be chased away only with blessed things' (Žele 1996, in Smerdel 2005, 374).

I found no examples in my research area of children rehearsing adult roles through play with toy ox-teams, as are known elsewhere in Europe (Figs. 8.5A–C) (Grieshofer 2004, 142; Garcia-Hoz Rosales 2005, 166). And yet such toys are reported from the Velika planina mountain (Fig. 8.1, 8.5D), where in the 1930's some herdsmen still made wooden cows (called *buše*), and children gave them names like real ones (Cevc 1979, 69–70). Nevertheless, a picturesque story was recently told to me in Koprivnik on the Pokljuka plateau (Fig. 8.1), about a playful rehearsal for the adult roles of both children and oxen. In wintertime the boys used to harness their bullock to a small sledge, thus training him to pull a less heavy load and to obey the commands, improving their skills in managing draught animals, and merrily enjoying the ride.

From Pasturing Cattle to Leading Oxen: Peasant Children's Apprenticeship in Pivka

'I drove the oxen when I was just a kid,' remembers Urban from Selce. His father put him in charge of leading the oxen when he was just ten years old, while his younger brother was to be a joiner. 'I was just ten when I drove out and pastured the oxen – and I used to stand before them and wait while dad was seeing a merchant.' Grgáč from Slavina started working in much the same way. When he was ten or twelve, he and his older sister drove the oxen. While they were too young, their mother did, and the children's first training for working with the oxen was when they accompanied their father taking the oxen to water. Cendran from Selce taught his oldest and only son all there was to know about oxen, but after the lad served his term in the army, he moved to Germany and the women on the farm had to take over his duties. Of Cendan's three daughters, sister and wife, only the oldest daughter was suitable for the job. 'Joža knows how and when to crack the whip, you don't,' he told his wife Mima when their daughter started leading the oxen to plough at the age of thirteen or fourteen. 'Only she knows how to lead the oxen; she knows when to hit them and when to talk to them.' She acquired the skills of handling animals gradually, through constant contact with them, 'I must have gone for it myself, because I liked the work. I was very close to them. I just loved being in the stable to feed the calves, caress them and even shovel out the dung; I also milked the cows. That's how you learn the job, if you're interested, then it's like every other job – you accept it and you do it.'

Constant contact with the animals, as part of the process of learning to work with them, was as necessary as loving to be with them. The Grgač family children in Slavina remembered that after playing outside in winter and getting so wet that they did not dare enter the house, they went to warm their frozen hands between the front legs and body of the oxen and cows. Cunar from Jurišče who helped his father with the oxen from the age of seven or eight, tells a similar story. 'It just was my nature to be around the animals all the time. I felt at home in the stables even before I started attending primary school. If nothing else, I'd pasture them. Once you get familiar with the cattle, you can't get them out of your skin... Here's how it was with the oxen: one of us would lead the ox up front or have a rope tied around its horns – and we, the young, small kids just walked along with the ox and the ox followed our movements; at the back, dad led the plough when we were ploughing, or then when we were hilling up the potato plants.' The communication between man and oxen was most direct and important in leading them in work, and leading oxen was the most frequent and the most responsible task of children in the division of labour within the family. It was most important for boys, the future masters, to learn this, but there was no time for gender restrictions, especially if girls were the only children.

Choosing and Training Oxen

In the selection of suitable oxen, their character was as important as their physical properties. 'You can see right away what its nature is like; people always made sure that the ox had a wide breast, an open chest as they called it. They used to say that a low ox with an open chest would pull his weight,' said Cunar from Jurišče. Matija Urbanov from Selce had this to add: 'We looked at his legs, how they were shaped... Front and rear legs; the legs should be straight for a steady walk.' In selecting oxen that were mainly intended for carting, particular attention was given to the feet. 'The feet had to be hard for carting or he would be in pain. And if the feet weren't hard, the ox was no good – because it had to be used every day.' For this reason, farmers selecting oxen in Pivka after the Second World War preferred grey or grey-brown cattle of the Montafon breed and less often opted for Simmenthal oxen of the two traditional Slovenian breeds. The latter had soft feet; when someone did not want to work, it was said that 'he pulled the sad face of a spotted [Simmenthal] ox.'

In most Pivka villages except Jurišče, where working oxen were bred on the farm, the draught oxen were purchased elsewhere, in Čufarija, Dolenjska, and fairs in Novo Mesto and Metlika (Fig. 8.1). The farmers of Jurišče worked their oxen eight to ten years (compared with five years in Kozjansko, and up to twenty years in Istria) and they always had 'two young oxen in line who were already being trained.' The same was true in other places around Slovenia. A future working ox was chosen from among the farm's calves in view of its descent, its behaviour and its body build. Once chosen, it was castrated. Castration was performed on eight- to ten-week-old bull calves, usually in winter, when there were no flies that might cause an infection.

If the castrated bull calf was not 'right, did not fit, did not want to pull or was wild,' it was fattened for slaughter. Such an animal also was not given a name. But if it 'fitted in and was not easily scared, it was good for drawing and carting, it stayed on the farm.' And it was given a proper name, usually by the master of the farm or by his children. Names generally referred to colour and other body features or personality characteristics; they were indispensable markers of identity for working communication and companionship (Smerdel 2009,

66). As Cunar said, 'such an ox was like a member of the family; there wasn't much difference because the ox-team was the farmer's main aid'.

Training started when the castrated bull calf was about one year old and weighed around 300 kg. It was usually teamed with an older ox without a yoke or harness; the only purpose was to have the young ox walk alongside an experienced one to prepare it for work. Occasionally, it was urgent to prepare and train new oxen in winter time, before March, when the oxen had to be ready to cart and plough. At first 'a single yoke was put on his neck – in the stables – to make it aware of the burden. Later a small beam was put on and the ox was driven around when there was snow. Then the weight was slowly increased, with a half-full trough, a half-full basket... until it was trained; because if you trained an ox who was too young or put on too heavy loads, he'd think twice before moving...'

In Selce and Slavina the system was quite different from that in Jurišče. In Selce and Slavina and probably also in other villages of Pivka, perhaps because these villages were closer to traditional trade routes, oxen were exchanged, rotated every year or at most every two years, and the purpose was 'to make it pay off, make some money.' Every year between November and February, before prices went up in March, farmers went to buy oxen at fairs. The purpose was to make money from the animals' increase in weight. The farmers would buy lean oxen, two years old, and weighing 450 to 600 kg. They trained them if necessary, worked them for a year or two, then fattened them up to eight to nine hundred kg and sold them for meat. The proceeds were used at fairs to buy a new pair of oxen and what was left was spent for the household: 'something for the kids – a dress, sandals – and for taxes and insurance.' Taxes were the annually recurring obligation that made the farmers practise this system.

Farmers usually bought oxen that were already trained to plough, in particular in spring, because they had to take them to the fields as early as possible. But they often still had to teach them a lot: 'you'd put the yoke on him and he'd just stare at you – didn't know a thing,' because oxen were often bought from areas where they were traditionally led from the right side. In Pivka, however, they were 'from always' led from the left side 'as it was common to turn left when ploughing' (but cf. Smerdel, Chapter 2.5). Young, untrained bull calves were harnessed to a single yoke, and trained alongside an older ox on the latter's right side. In this socio-cultural system, in contrast to that of Jurišče, the bonds between men and their draught animals might seem to have been less strong. And yet the bonds between man and his oxen were nevertheless cherished, as they were essential for harmonious common work.

Conclusions

A few accounts from other areas than Pivka support the patterns there, and highlight the points I wish to make about training children and oxen. The parallels and analogies were widely recognised: 'You have all kinds of animals, just like us people,' stated Tinček from Koprivnik. Selections of oxen and their leaders were made on the basis of character as well as physical traits, and both oxen and children were recognised as individuals. For young oxen farmers ordered or themselves made small single yokes: 'like shoes for people; each must have his own ones...' (Jeglič from Lukovec, Karst, Fig. 8.1). Most children had to work with oxen and thus trained for all the necessary jobs, not just the sons who were by inheritance designated to be their fathers' successors on the farm. Nevertheless, their elders were constantly observing children, judging their character, physical characteristics and skilfulness, recognising 'the chosen ones' who would be best to work with oxen, in much the same way as they chose the proper bull calves to become working oxen. But what most marked 'the chosen ones' was their special relationship to oxen – their affection, their loving to be with them.

The special relationship between some children and oxen, so crucial for their working companionship, was begun early in life. As I was told by Ida Vankova from Škofi (Fig. 8.1), 'children were in the cattle-shed all the time, from their early childhood onwards. If a calf was born – they were there; they wanted to see and to touch the little one immediately. Thus they got used to each other; the animals knew the children.' At the age of six years or even earlier children 'drove animals to pasture on the commons; they had to get to know them.' They chased away the flies, accompanied their fathers while watering the animals, and helped by yoking and unyoking.

Fig. 8.6. Children – boys of different ages – just standing in front of an ox-team, driving oxen, leading the animals in front of a plough or a cart, pushing the forecarriage... A: Ploughing on Blečji vrh (Dolenjska region). (Photo by M. Matičetov, 1949; SEM Archives). B: Pulling firewood with sledge in Sedlašek (Haloze region). (Photo by M. Bras, 1959; SEM Archives). C: Carting dung in Nabrežina/Aurisina near Trieste. (Photo by M. Magajna, 1947; NŠK Archives). D: Carting wood near Hom (Dolenjska region). (Photo by F. Šarf, 1961; SEM Archives)

Children might begin driving oxen and leading the animals in front of a plough or a cart at eight or nine years of age, or as early as five or six years, depending on acquired skills, and the needs of a farm (Fig. 8.6).

Children's apprenticeship in animal husbandry was a school of life, required by fundamental socio-economic necessity. Children served an apprenticeship, beginning by listening to the wisdom of their elders in folk tales and family conversation, and observing them in the field. As Brijevc from Vetrnik (Fig. 8.1) said, 'I have been a little child, and I watched the old ones work with the oxen and care for them. Thus I learned. I remembered what I saw...' Both oxen and children acquired knowledge and skills for their adult roles step by step, moving from observation to practice. As the child might play with toy oxen, the young ox began his training with light loads. Farmers often first taught their animals to pull a branch or something similarly light and silent, and the humans involved in this training phase were often the children. Gradually the work increased in difficulty and seriousness. Young oxen had to get accustomed to pulling a harrow, and at last a plough or a cart. As their master acquired skills and knowledge for a certain job 'he moved forward to do something else, and the next one, the younger took it over' (Kasarnov from Skadanščina, Čičarija; Fig. 8.1). As Borišk from Rodež (Fig. 8.1) remembered about ploughing: 'The youngest of us was with the oxen, leading them, the next one by age pushed the forecarriage and father was ploughing. I started in front and then slowly moved backwards, from oxen to plough.'

As children learned by apprenticeship, working beside their elders, a young ox 'got a small single yoke around his neck, and a rope around his muzzle, and then you took him to the old one' (Kasarnov from Skadanščina). 'The old one taught him into the right direction' (Jezernik from Libeliška gora, Carinthia; Fig. 8.1). 'It was far easier to train him like that because the old one pulled harder' (Borišk from Rodež), and was 'so smart that he kept the young one aside – to prevent him running' (Ida from Škofi, Karst). If the young one was already big enough, it was preferable to train them harnessed in a double bow yoke. 'The young one depended on the old one, and thus he learned' (Kondi from Marija Reka, Zasavje; Fig. 8.1).

The training for both oxen and children was rigorous and could be harsh. Children were thought to be born sinful and punishment was necessary to rear a virtuous adult. If young oxen 'jumped too much' and did not respond to orders, they had to periodically experience the pain of whip or stick. 'You had to hold and pull him with that rope when he didn't want to move – like a child that doesn't obey...' (Ida from Škofi). However, stern training was not without care, respect, and love. 'You had to train oxen calmly, with reason', stated Brijevc from Vetrnik. And 'you had to do it slowly, and with love', added Jazbernik from Gorjuše (Fig. 8.1). After all, children and oxen were the most valuable possessions of the farmer, the future of the farm and the family.

Parallelism and analogies between training oxen and children can be traced throughout the processes of knowledge transmission and acquisition of skills. The equation between men (children) and oxen is made elsewhere in Slovenia, in the health care of working oxen (Smerdel 2007), and in the culture of communication with working oxen (Smerdel 2009). What I call the 'philosophy of everyday reality', which can be detected in the man-ox relationship, blurs the academic distinction between man and animal, and contributes some eloquent evidence, supporting the philosophical notion of 'companion species' (Haraway 2008, 16–19).

8.3. ART OF A MOWER: SKILLS WITH A SCENT OF SEXUALITY

Inja Smerdel

Introduction: The Eloquence of an Auxiliary Tool

Boys from various Slovenian regions learned the skill of mowing from their childhood. Why was this particular skill so important for the young men, who collaborated in the majority of working processes on their homesteads? Was it just because of the intrinsic economic value of hay as fodder? Were there tools in a mower's toolkit that bore some extra, multi-fold socio-cultural meanings, and what were their messages?

Some possible answers can be seen in the collection of whetstone holders from the Slovene Ethnographic Museum, which I studied in the 1990s (Smerdel 1994). Everybody knows what a scythe is, but what exactly is a whetstone holder? A riddle from the late 19th century (Erjavec 1883, 331) reveals the tools which are closely connected with the scythe: 'An iron mistress, a stone maid and a wooden well.' The metaphorical trio is a scythe, a whetstone and a whetstone holder. The principal implement, the mistress, is assisted by a maid, the whetstone, and by the source of the latter's effectiveness, a whetstone holder, *i.e.* a wooden vessel containing

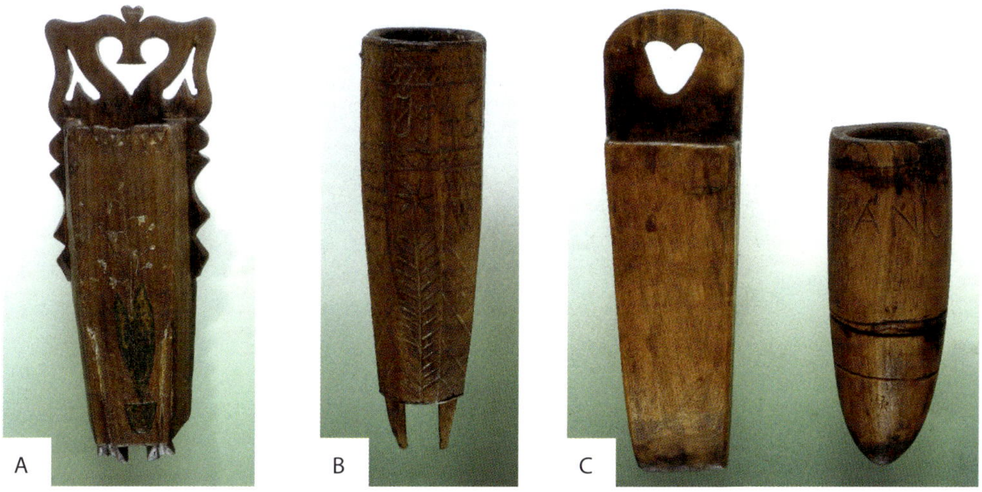

D

Fig. 8.7. Various whetstone holders – with carved flowers, initials, name, heart... A: The box-shaped example from the Upper Savinja valley decorated with a realistic flower motif of a flower pot with a lily of the valley in blossom. (From Smerdel 1994, cat. no. 39). B: Whetstone holder from Gradež (Dolenjska) decorated with carved geometric ornament and stylised plants. On the upper third the year and initials are carved. (From Smerdel 1994, cat. no. 30). C: The example from Luče (Upper Savinja valley) with a motif of a hart, and the one with unknown provenance with the carved name **FRANC** (probably manufacturer's as well as mower's). (From Smerdel 1994, cat. nos. 23, 61). D: Whetstone holder with the carved words **PIKČU SEM** (I'm Pikec's) from the Tolmin Museum collection. (From Smerdel 1994, fig. 41).

water. The whetstone, called an *osla*, is a longish piece of natural or artificial stone, used to sharpen primarily scythes, but also sickles. Because of its frequent use, the whetstone must be at hand to the mower. He therefore carries it close to his body, in a horn or wooden (and in modern times sheet-metal or plastic) container, filled with water, and suspended from his belt. This container is an *oselnik*, a whetstone holder (Fig. 8.7). Notwithstanding the mechanisation of modern agriculture, they are still in use in many places throughout Slovenia.

The physical contact between man and whetstone holder was a source of reflection on the multi-fold significance of whetstone holders. The whetstone holder is not just an economic and cultural element to be studied through the history and geographical distribution of types, and the evidence of manufacture, design and use. This tool bears hidden messages about the relationships between man and tool, and between the different social environments of the user. Visible aspects of whetstone holders include their shape, decoration, carved initials and names, and traces of maintenance, repair, and use. But these raise questions about other, hidden aspects we can merely suspect: erotic and symbolic meanings, personal attachments to a tool and expressions of identity. Such interpretations may be complicated and masked by prosaic, materialist factors: an old whetstone holder may be retained not for symbolic reasons, but for lack of money or time to buy or make a new one.

By researching the working context of whetstone holders, the labour and festivity of the hay harvest and the knowledge and skills of the mower, we can understand some of the deeper stories of the whetstone holders. From 1990 to 1993 I thus researched hay harvesting, in the same regions where museum specimens of whetstone holders originated, and in regions where they were still being made at that time.

Hay Harvest: Labour and Festivity

There is no lack of written sources on the hay harvest in Slovenia in the second half of the 19th and the first half of the 20th century. The whetstone holders in museum collections come from the same period. The middle of the 19th century saw the first stages of the introduction of livestock-breeding in stables and a consequentially increased need for fodder (Smerdel 1994, 85, 86). Cattle-breeding, which had been relatively limited in the first half of the 19th century, became the most important animal husbandry branch in the same period, and was directed at fattening oxen for sale and at dairy farming (Smerdel 1991, 40–43). There is little doubt that these changes substantially increased the importance of fodder. Hay became the mainstay of cattle-breeding and elevated the importance of the hay harvest, indeed of all efforts aimed at preparing as much hay as possible to feed the domestic cattle. How important hay was in some places is exemplified by an old proverb from Bohinjska Bela (Fig. 8.1), where a farmer was valued by his hay crop. 'A cart of hay on the first day of May, that's a fine farmer, I need to say' was said of farmers who still had a cart of the previous year's hay at the beginning of May (Šarf 1974, 134, 135).

Every calendar of farming activities features several weeks dedicated to the hay harvest. In the plains, some started as early as St. Vidus's Day (June 15th), in the hills on St. Peter and Paul (June 29th), but in general, harvesting was started in July or even later (Kuret 1989, 438) and it ended in September or 'when the whetstone froze' as they said about the second hay harvest in Koseč near Kobarid (Fig. 8.1; Smerdel 1994, 106).

Besides the fact that harvesting was very important in a farm's economy, and that it was considered to be one of the most trying farm jobs, something else was characteristic of the hay-making season: its festive excitement. Special dishes were served and people, attired in festive working clothes, sang and danced and joked. The festive mood perhaps had its roots in both the economic importance and the hard labour; but it may also have been enhanced by the particular reputation of those who did the job so well, the mowers, and by the fact that the harvest brought together so many people, men and women, boys and girls (Fig. 8.8). There are descriptions of the hay-making season changing the ambience of entire villages – with people asking each other for help, seasonal mowers hiring themselves out (like in some other European countries, *e.g.* in the Netherlands; see Lucassen 1990, 26–27), and with women joining the mowers in the meadows to rake up the hay (Smerdel 1994, 106; after Mrkun 1943; Koprivec 1939; Kuhar 1972, FN 1993). Quite

Fig. 8.8. Mowers and women rakers in a painting by Ivan Grohar, entitled Raking women (1902, oil canvas, 133.5 x 107 cm). Kept by the Ljubljana City Museum.

different are the descriptions of the hay harvest on individual, remote farms (for example, in Robanov kot; Fig. 8.1), where whole families tackled the job and older children as well as women had to mow. Things were also different in remote mountain villages where the men were absent and working in faraway places, and the women had to do the mowing when there was no alternative (Smerdel 1994, 106).

It is highly probable that the reason whetstone holders were decorated and why they embody so many different meanings can be traced to the festive mood in the hay-making season and to everything it brought with it. Stylised flowers, a common motif on decorated Slovene whetstone holders (Figs. 8.7A and B) indicate festivities and accentuate the special importance of the decorated object (Makarovič 1973, 34). And it also explains why some of them are not decorated: their owners thought of them only as mere utensils (cf. Smerdel 1994, 107), where the hay harvest was seen as a 'peasant's job' or lost its prestige in the course of time.

Many of the meanings a whetstone holder may communicate are implied by those aspects of the hay harvest that made it a job and a festivity all in one. Firstly, there is the division of labour: men mow, women rake. Second, there is the mowers' reputed skill and, in some places, the arrival of strangers, the seasonal mowers to whom girls customarily presented bunches of flowers. Finally, in such a context there was constant joking and fooling, often full with erotic implications.

Mower's Knowledge and Skills

Although women might mow when necessary, mowing was essentially a man's job (as it was in other European countries; see *e.g.* Fel and Hofer 1974, 212). The Slovene language has no feminine form for the word *kosec*, mower (though a masculine form for a woman reaper, for *žanjica*, exists: namely *žanjec*). Women and girls tossed, raked and turned over the hay. They were called *grabljice*, *grablojce* and the like, all words indicating women who rake. Their tools were wooden hay-forks and rakes. The men who came on the meadows each carried a scythe, a whetstone holder and whetstone. Some of them also brought along a straightener, a needle-like 'flattening iron' (German *Streicheisen*) that was of great importance on all meadows where the mower's scythe was likely to hit stones (Smerdel 1994, 104). And, of course, a good mower also had to be good at whetting his scythe. Whetting scythes and all other tools used to cut crops was considered a male job. Mowers generally carried their whetstone holder stuck behind their belt and took it off only for resting and eating. It was then stuck in the ground to prevent the water from spilling. Some preferred to carry it on their belly, but it was mostly worn on the right side of one's back (Fig. 8.9A).

To make the scythe cut better and the whetstone whet well, a little vinegar was poured in the whetstone holder in many places, but only when the whetstone was too hard. 'Vinegar kills the stone's hardness', people used to say. It had to be added only 'if the whetstone wasn't effective enough'. The water in the whetstone holder had to be cold (wooden whetstone holders kept it that way) and in some places soft rain-water was preferred. (Water with a high percentage of calcium corrodes) (Smerdel 1994, 108). And there also exists an old superstition in many places in Slovenia that a scythe cuts best if a snake's head is put in the whetstone holder's water. Did people believe that the sharpness of a snake's teeth or tongue was transmitted to their scythes? Perhaps. But competitive attacks on the pride of good mowers, real men with sharp scythes (metaphorically considered their 'tools'), are also documented (FN 2007). In the mountains of Zasavje region some mowers carried copper bits in their pockets when mowing, striving to catch a moment to rub other mowers' scythe blade (or whetstone) with them, thus making it loose its sharpness. To prevent such 'humiliation', the mowers were accustomed to hiding their scythes and their whetstone holders every evening of the hay harvest time. They indeed needed the above mentioned knowledge to master their job, successfully compete with other mowers and guard their reputation.

Experienced mowers of course knew when it was time to apply their whetstone. When the weather was very dry and the grass harder, they had to whet

Fig. 8.9. A: A mower from Selce (Pivka region) carrying his whetstone holder on his belly while mowing (Photo: I. Smerdel, 1982). B: Farmer Jofil from Grant in Baška grapa sharpening his scythe. By hammering the tempered cutting edge of a scythe – placing it between hammer and anvil (thus producing heat) – the blade regains its sharpness (Photo: I. Smerdel, 1993). C: Farmer Danilo from Škofi (Karst region) made for his ten-year old grandson a real, but shorter, smaller scythe. Note also the rake, another important tool in haying (Photo: I. Smerdel, 2008).

Fig. 8.10. 'Seventeen mowers stand in line and they start swinging their scythes at the same time...' Mowers in Lancova vas near Ptuj, in the 1960s. (Photo by L. Bras, 1962; SEM archives, Archives of the Ptuj Regional Museum).

their scythes more often. If even after whetting the blade did not cut well enough (its sharpness was tried out on a finger-nail: a properly whetted blade cut off a very thin slice), it was time to sharpen it by hammering (Fig. 8.9B). But sharpening was not often needed when fresh, wet grass was mown. And a tuft of grass, wet with dew, was the most efficient help to clean a scythe's blade before whetting it (Smerdel 1994, 109).

Harvesting usually started before day-break. Mowers were ranked and placed in the line according to their social status, except that when it came to mowing, the experience and skill of a good mower were even more important than his social position. It was of course evident who did well and who did not once they started mowing (Fig. 8.10). Everybody wanted to do a good job because those who did not were teased: 'What are you doing, man? Grazing perhaps?' was a common way to scorn the slower ones in Koseč (Smerdel 1994, 109). There are many versions of a song with the following refrain 'and all I do is drowse and mow at the back' (Štrekelj 1908–1923, 231–235). In the same song the mower who is last in line blames the bad service: 'The maid brings wine, but she doesn't bring me any. The way I'm served, that's the way my scythe cuts.' Or he will blame the conditions: 'Your grass is too high and my scythe rusty.' A version from Pohorje (Fig. 8.1) gives a pointed counter: 'A scythe is as good or as bad as the man who swings it.' After all, only two things really mattered: skill at mowing and whetting; everything else was of secondary importance. A native from Haloze hills (Fig. 8.1) recollects that soon after the Second World War her father had 'to go and beg mowers,' but that 'he waited and waited until, in the end, he got real men' (Smerdel 1994, 110). Being a good mower indeed meant being a real man.

Boys learned the skill from childhood. In Slovenske gorice 'young boys, who weren't ready for mowing yet' came to the meadows together with the girls to bring in the hay (Koprivec 1939, 163). Things seem to have been different in mountain villages. In Koprivnik, two grandfathers made little scythes for their respective grandsons: the first made a wooden one with a wooden whetstone and straightener in a little plastic whetstone holder (Figs. 8.11A,B),

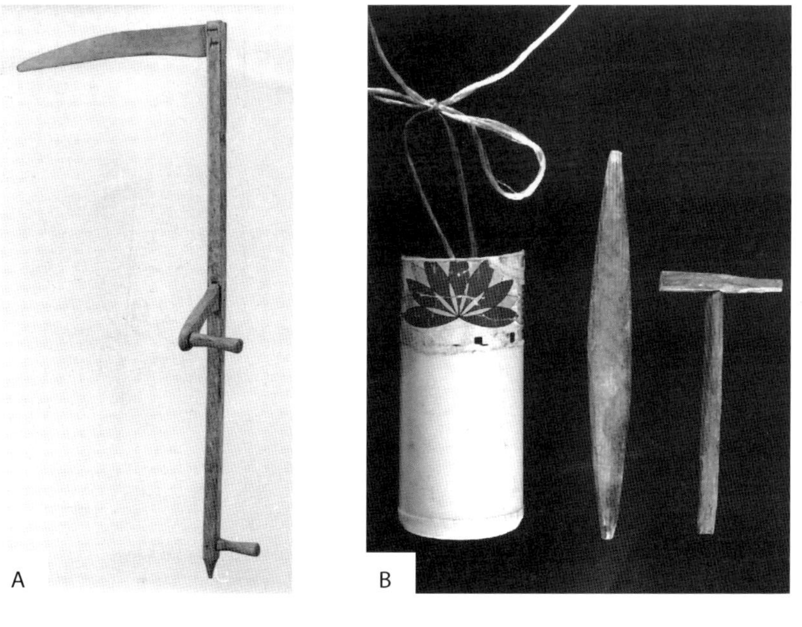

Fig. 8.11. A: The wooden scythe (length of snath 99 cm, length of blade 38 cm) a farmer from Koprivnik (Pokljuka) made for his five-year old grandson. B: To make a whetstone holder he cut off part of a plastic box of liquid fertiliser, taking care to preserve the flower-sticker! He also made him a wooden whetstone and straightener. (Photo by C. Narobe, 1993; SEM Archives)

Fig. 8.12. Location of Atany in Hungary. Map: R. Lugon, J.-C. Loubier and A. Chevalier.

and the second a real, though short scythe. The boys could easily imitate them and learn how to swing a scythe, and the other work of the mower. Boys as young as ten might be expected to actually mow (Smerdel 1994, 110), although in Hungarian Atany for example (Fig. 8.12), boys replaced their toy scythes with real ones at the age of fourteen or fifteen (Fel and Hofer 1974, 212).

Especially those lads and men who travelled around to mow as day-labourers had to prove themselves constantly. Farmers certainly did not like to hire inefficient labourers, and competed to get the best ones. Judging from the various versions of a folk song written down in Bloke near Cerknica and in Ščedenj (Fig. 8.1), mowers who were aware of their value decided which farmer could hire them: 'Oh no, I won't mow, I won't mow for forty five; but for forty five and tobacco, oh yes, there must be tobacco, then I'll mow!' (Štrekelj 1908–1923, 235). Childhood recollections of a native of Hotedršica in Notranjska region (Fig. 8.1) confirm that day-labourers liked to work only on those farms where the housewives served them well, not just with food but specifically with tobacco: '... mother sent me out to observe the cottagers at home, to get to know in what kind of smoking pleasure those indulged who would come to mow for us' (Albreht 1945, 55). Tobacco, at that time an exclusively male attribute, completed the manly image of a real mower.

Conclusion: Eroticism of the Hay-Making Time and Tools

The weeks of the hay-making season saw a lot of coming and going in some regions, a time when boys and girls, men and women, acquaintances and strangers all met at work, at a well-spread table, and in amusing themselves. There were the mowers with their reputation of being 'real men' and the women raking up the hay who 'the younger and prettier they were, the better', and so these were indeed weeks bursting with sensuality (cf. Smerdel 1994, 111). In different forms, eroticism is reported from so many places that it was certainly a pretty outspoken element of the hay harvest in Slovenia. One aspect of this involved the bunches of flowers customarily given to all mowers. Their individual message, communicated by the 'flower language', often was an essentially amorous one, as a bunch of cultivated flowers was (at least from the 18th century onwards) a token of love (Makarovič 1973, 30, 31; cf. Smerdel 1994, 111). In the hills of Škocjan (Fig. 8.1), girls used 'to chose their bridegrooms-to-be during the hay harvest' (Kuhar 1972, 115) and in Hotedršica, having a roll in the hay often ended by 'the banns being put up early in the following year and the couple being wed soon afterwards' (Albreht 1945, 55). The oral tradition of songs about the hay (and grain) harvest has preserved an evidently sound warning: 'Every girl is her own wreath's (=virginity) keeper!' (Štrekelj 1908–1923, 230, 231).

In this blend of explicit and implicit elements of the culture of hay-making, of life during the hay harvest, some whetstone holders contributed their own, visible erotic messages. Some were decorated with a heart (Fig. 8.7C). But the shape of whetstone holders and whetstones in general made them into perhaps less visible, but certainly outspoken, expressions of the sensuality which was in the air during the harvest. The eroticism may have been hidden in people's laughter and sparkling glances, but it certainly was explicit when the men were bragging. 'Take out your whetstone!' they would tease and offer their own 'tool': 'We've got a woman here, need a whetstone?' (Smerdel 1994, 112).

'*Foeniseca immittit cotem in cotarium sicut iuvenis penem suum in cunnum puellae*' or: 'A mower enters his whetstone into the whetstone holder like a boy enters his penis into a girl's vagina,' writes Koštial (1937–1939, 23) who saw the sexual symbolism of whetstones and their holders in their names in different languages. Leopold Schmidt (1952, 66, 67) established that handling the tool (whetting with a whetstone, since a good mower necessarily is a good whetter too) becomes part of the spoken language in describing a man's sexual activities; he assumes that this is the very reason why whetting tools became 'humanised' in speech. The whetstone's shape and function associated it with the male sexual organ, and the whetstone holder – its depository, a moist vessel that made it whet (function) well – was identified with the vagina. Direct evidence of this interpretation in Slovenia is articulated by a piece of superstition from Dolenjska, dating from the seventies of the 19th century (Trdina 1987, 373): 'To father a son, put on your whetstone holder *in actu*.'

In summary, we can say that the hay harvest was an eminently social moment in the structures of rural life in Slovenia from at least the middle of the 19th century to the sixties of the 20th century. The eroticism associated with mowers, symbolic expressions of the sexual role of 'real men', had its significance only within the social group in which the opposite sex was also represented through a clear gender division of labour. As both skilled labourers and courting males, mowers took great pains to be distinct individuals. Visibly distinguishing each of them from all the others in a line of mowers was the whetstone holder, a most personal tool, the external token and reflection of a mower's consciousness of his own personality and of his wish to be recognised and valued as an individual (Fig. 8.10). Learning the knowledge and acquiring the skills of a mower thus, among other things, signified learning to become an individual, a personality, a real man.

8.4. SKILLS AS IDENTITY MARKERS

Thomas Schippers

Introduction

Specific tools and implements used in pre-industrial rural societies have often been considered as the results of either historical processes of adaptations to local agro-ecological conditions of slowly spreading 'tendencies' (Leroi-Gourhan 1943/1971, 27–42) or of local innovations. During the 20th century, several European countries have launched large scale surveys into the spatial distribution of agricultural implements, as well as of other ethnographic material such as rural dwellings and houses, dialects or folklore (for an overview see Bromberger, Dossetto, Schippers, 1982–83, 15–40). By mapping out these data, at times contrasting geographic 'boundaries' between two technical solutions for a similar task have appeared, as between cereal threshing with flails and by animal trampling, for example.

I have been conducting ethnographic fieldwork in the southern Alpine Ubaye and Stura valleys (Schippers, 1993, 171–178; Llaty 1997) situated at one of these agro-technological boundaries located in the 1930s by Charles Parain (1937/1997) (which also corresponds here with the borderland between Italy and France), in order to collect information from local informants about their technological choices and their perception of the other, neighbouring solutions. Two different techniques have been precisely documented, as they were considered by the locals as clear markers of difference: threshing and the carrying of loads. Methodologically we differentiate 'indicators' which are objective qualitative or quantitative indices established by the researcher, 'markers' which are (a choice of) indicators recognised by the local population, and 'stereotypes' which are locally acknowledged indices of differentiation without any objective existence but sometimes high emblematic value.

Two Examples

In the French Ubaye valley, threshing has mainly been done with the use of animal trampling or by lashing in the case of very small harvests, while in the neighbouring Italian Stura valley (Fig. 8.13) lashing and flail threshing have been the usual ways to proceed. When in the 1920s and 1930s in the Ubaye peasant youngsters increasingly abandoned farming, some migrant Italian families took over local farms but also brought their flails to do the threshing. When we asked Ubaye farmers if they had tried or even adopted this flail threshing, the answers were not only negative but they added this technique was impossible for them to learn or to perform as *'we are not Italians...'* A very similar discourse was collected to 'explain' in the Ubaye valley why local people would only transport loads in hand-held baskets or by mules or donkeys, while on the Italian side the most common way to transport stones, manure, potatoes etc. is the *gerla*, a long conical basket with two shoulder straps, carried like a backpack. These different techniques were clearly perceived as proper to each group and inappropriate for the other.

These small examples, similar to many ethnographic cases worldwide, indicate that technical solutions and the necessary skills to perform them are often closely related to specific groups. This points not only to the importance of sociality (Carrithers 1992, 34–75) in the acquisition, exercise and transmission

Fig. 8.13. Location of the Stura valley in the southern Alps, at the border between France and Italy. Map: R. Lugon, J.-C. Loubier and A. Chevalier.

of specific skills but also to their role in marking cultural boundaries (Barth 1969, 9–38). The latter can prevent the use of certain skills or tools – not because they are ignored, but because they are explicitly refused and considered inappropriate by certain groups.

So, certain tools, skills and forms of know-how can enter the arena of inter-group relations as recognised social and cultural markers. Here they can be either valued cultural practices or negative attributes. In the same way, skills and tools are generally also quite strictly divided according to gender and even sometimes to an age-class (Tabet 1998). Performing a 'socially inappropriate' skill among members of one's own group can underline the 'abnormality' of an event. The Carnival period in European societies or certain episodes in Classic mythologies provide examples as gender roles are (symbolically) inverted (Leach 1961/2004, 132–144). On the other hand, socially recognised 'excellence' in certain agricultural skills, like ploughing or scythe mowing, can be a source for individual pride but also 'ethnic' prestige and social respect. Since the 19th century, rural competitions in various agricultural skills have been organised all over Europe by regional, national or even international authorities. Migrant workers and craftsmen have also used their regional identities to promote their skills and the quality of their products in rural societies.

Conclusion

To conclude, one can consider skills not only as technical gestures and know-how aimed at achieving certain goals, but also as indicators of individual and social belonging. As such they contribute to produce 'we-groups' as well as socio-cultural boundaries. Especially when studying the diffusion (or not) of skills and related tools, this dimension is important to take into consideration. Styles and fashions in skills also allow groups to recognise and differentiate each other.

8.5. BLADES AS MESSENGERS OF AGRICULTURE: A CASE STUDY FROM SCANDINAVIA[1]

Helena Knutsson

Introduction

In Scandinavia, sites from the period when farming was introduced are still comparatively few and collections from these times comprise stone, some pottery and a few eco-facts. So there is little material to use for the study of such complicated processes as the transmission of knowledge. We can be sure that the transfer of agricultural know-how was indispensable for survival, but farming became a full-time occupation at a time when knowledge was stored and transmitted through human memories, without a written language.

Even today, memory can be improved by storing information in material objects and their treatment. The objects used in farming *could be* loaded with commemorative meanings and produced, transported and treated meaningfully in work and ritual. This study investigates flint blades from the Mesolithic and the onset of the Early Neolithic in Sweden. It is suggested that they carried mythical and symbolic meanings related to agricultural practices, and as such were one of the channels through which knowledge and know-how in farming practices were transmitted. This is the best explanation for the sudden change in the production of blades in the periods of early farming. In the same way as many material objects of merely symbolic qualities in present-day cultural contexts, the blades and connected production in and around the farming cultures of the Old World seem to have been chapters in a 'handbook of farming practices'. Therefore the understanding of their role is an important issue in reading that 'handbook'.

Blades

From the beginning of archaeological research, the stone blade was recognised as a special artefact. As a need arose to make finer chronologies of the Stone Age and data were quantified, blades were defined as flakes at least twice as long as their width. Later a morphological criterion was added, defining the length and straightness of the dorsal scars (Malmer 1962, 530ff). Blade technology is often used as a chronological marker in Scandinavia: unifacial bipolar (or dual platform) blade-cores are associated with the Mesolithic Maglemose Culture, micro-blade handle core technology with the Middle Mesolithic and the Kongemose Culture, and cylindrical core technology with the Neolithic Pitted Ware Culture. This was followed by its integration into the functional logic of the New Archaeology. Later it was understood that blades could be a basis for a more value-oriented understanding of Stone Age life (Knutsson 1995, 2012), which in turn led to the realisation that these simple definitions and classifications needed rethinking. The most important consequence of this change of focus decisive for this short contribution is an insight into the complexity of functional, symbolic and memorial meaning of seemingly practical objects. In prehistory, these aspects are presumed to be intertwined in similarly complicated ways as we can observe in modern societies. More specific arguments about my opinion are developed elsewhere (Knutsson 1995).

Production and distribution of blades is one of the most persistent phenomena throughout the Stone and Bronze Age in Scandinavia. Through time they varied in size, and, blade production leaves characteristic waste in the form of flakes, core rejuvenation flakes and cores of about ten different shapes. Our understanding of the production process is based on modern experiments, which began as early as 1920 in Sweden (Klein 1920, 103; Malmer 1962, 353) and expanded into the 1960s with the work of experimenters like Sergej Semenov in Russia, François Bordes in France, and Don Crabtree in the United States. Blades from several periods are considered ideal products of complex technologies by recent replicators like Jacques Pélegrin, Espen Kannegård, Mikkel Sørensen and Erret Callahan, and in particular important criteria for distinguishing blade production techniques have been established by Pélegrin (2006), Callahan (1995), Knutsson (1995), and Sørensen (2006). Our knowledge of how blades were used is also based on experiments, now through the reproduction of and identification of use-wear (Pélegrin 2006; Sørensen 2006; Ballin 1999; Juel Jensen 1986; Knutsson 2001, among others).

The analysis of blades should be addressed on a continental level, since changes in operational chains seem to represent broader aspects of life than pure technology. Some of these changes are strongly related to the appearance of domesticates and agricultural structures. This paper is therefore to be seen as a suggestion for further work. In the examples that follow I intend to illustrate the needs and habits of the blade using Neolithic societies. Production and blade-using is compared to the preceding Mesolithic period and the Middle Neolithic period in the Scandinavian and a few Europen continental assemblages (cf. Knutsson 1995, 2000, 2001, 2012).

Scandinavian Blade Assemblages

In Mesolithic contexts, the blades seem to have been produced on a large scale within the settlements. These Mesolithic blades are relatively short, normally 4-8 cm, but not longer than 12 cm. The short blade tradition continues into the EN (TRB), but the production methods change during this period to different preparation of the core platforms (only a few of these cores are preserved) and probably the strategy could be related to pronounced use of antler punches as knapping tools in an indirect blade detachment technique.

Blades in the Mesolithic sample are regularly occurring at most sites such as settlements, graves and occasional hoards. In the Neolithic contexts, with one exception [sites attributed to the Middle Neolithic Pitted Ware Culture (PWC)], blades are rare; instead the assemblages contain few, but longer blades. In the Early Neolithic sites of *south* Scandinavia only 14 such longer blades are found. These blades are well made and occur only in graves or in hoards. They have the most reliable contextual association in my sample, but it has not yet been discovered where they were produced.

Imported *flint* blades and blade-like flakes are found along with thin-butted polished flint axes and other objects already occurring in the older TRB sites in *central* Sweden, where there is no natural occurrence of flint. These blades often show qualities, which represent a level of production skill achieved only through a lot of initial learning and constant practice. Since blades in other raw materials are common in the Mesolithic settlements of this area, the question is for which reason the *flint* blades were imported, and, whether blades were produced, used and deposited as part of cultural-ritual practices already in the Early Neolithic.

The burial assemblages of Middle Neolithic Scandinavian Corded Ware Culture (CWC) in the region also contain blades produced by skilled flintknappers and used as harvesting implements. A closer look at them suggests that the blades played an important role as symbols of the harvest and the idea of 'The Farmer' in the burials (Figs. 8.14–8.16; see also Pélegrin 2006; Callahan 1995; Knutsson 1995; Sørensen 2006).

Since other remains of production of the longer blades from TRB and CWC are missing, my conclusion is that they were for some reasons distributed as ready-made products. They have to be seen as either imported artefacts or, if produced on site, made by itinerant skilled specialists. This is, interestingly, a pattern documented from the beginning of the Neolithic in most parts of Europe, *i.e.* in sites with domesticates and structures related to processing and storing them (for example Perlès 1992; Gronenborn 1994).

The relation between the sophisticated craftsmanship of long blade production in several places of Europe and Asia (cf. Pélegrin 2006; Migal 2006) and

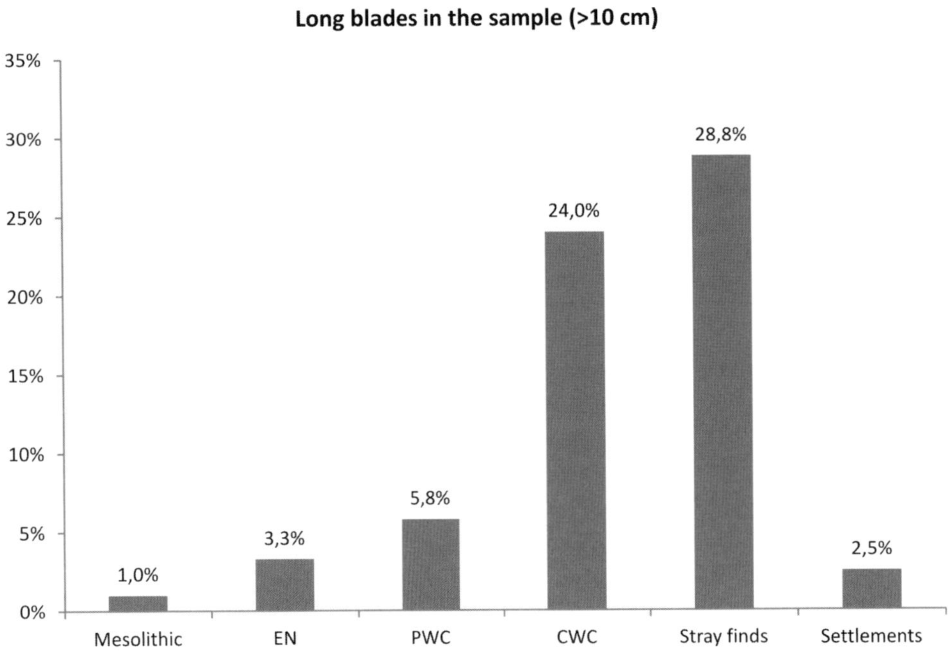

Fig. 8.14. Long blade (10–20 cm) deposits in Swedish Stone Age contexts. Frequencies of long blades increase through time. They are overrepresented in the stray find collections, probably due to better visibility ('larger' physical attraction), but partly also due to the fact that they are common in Corded Ware Culture graves, which tend to be more exposed by modern agricultural practices. Note that 10 cm is a classificatory category in this diagram and not a natural limit in production practices. The most interesting group of blades, that between 12 and 19 cm is very small, comprising only finds from the Neolithic sites. In general the Neolithic blades are longer then the Mesolithic ones.

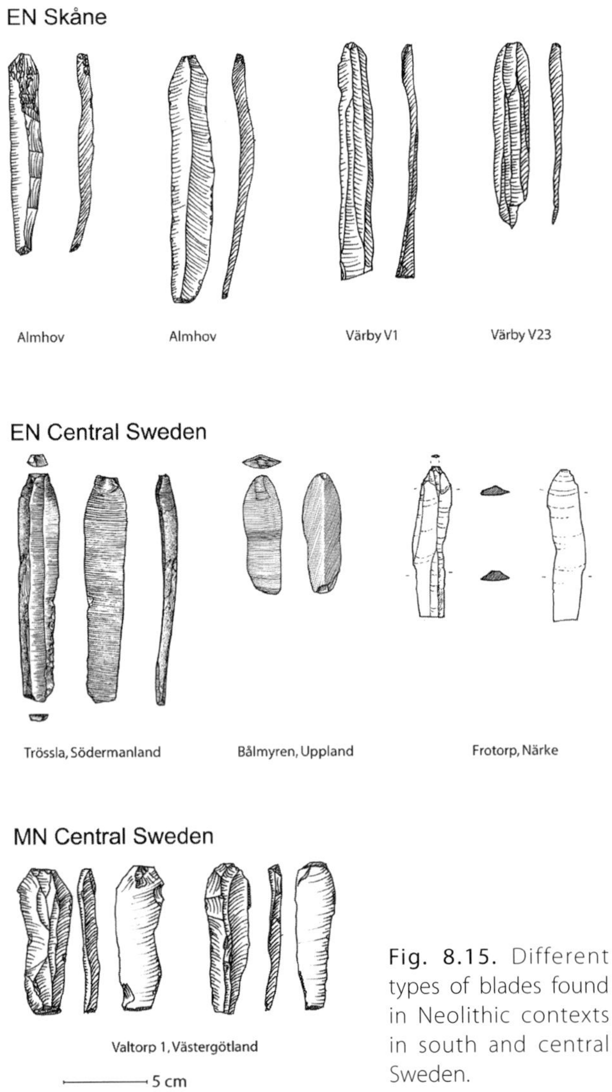

Fig. 8.15. Different types of blades found in Neolithic contexts in south and central Sweden.

the overall production of blades for more or less normal household use at the same time (this means between 7000–2000 BCE) has to be considered. Otte and Behm-Blancke (1992, 165ff) suggested that similar methods have been used in the same geographical areas, for both the sophisticated and the everyday products, related and reproduced via copying. There has obviously been a functional division of production, in terms of uses, status and distribution already at the time of the introduction of agriculture in the 'Old World'.

The occurrence of longer blades in the Neolithic materials in Scandinavia indicates that, as in continental Europe, during the EN an interest developed in such kinds of products. They are few in number, but they occur on EN sites across the entire distribution area. In the early TRB sites they are normally found in pits, sometimes sealed by other EN layers. In later contexts they are found in graves and hoards interpreted as ritual depositions. To understand how different the long blades are from the earlier blades, it is useful to look for the blade cores and other production waste.

Production Remains

As already stated, cores are rarely found. The cores occasionally found on EN sites do not conform to

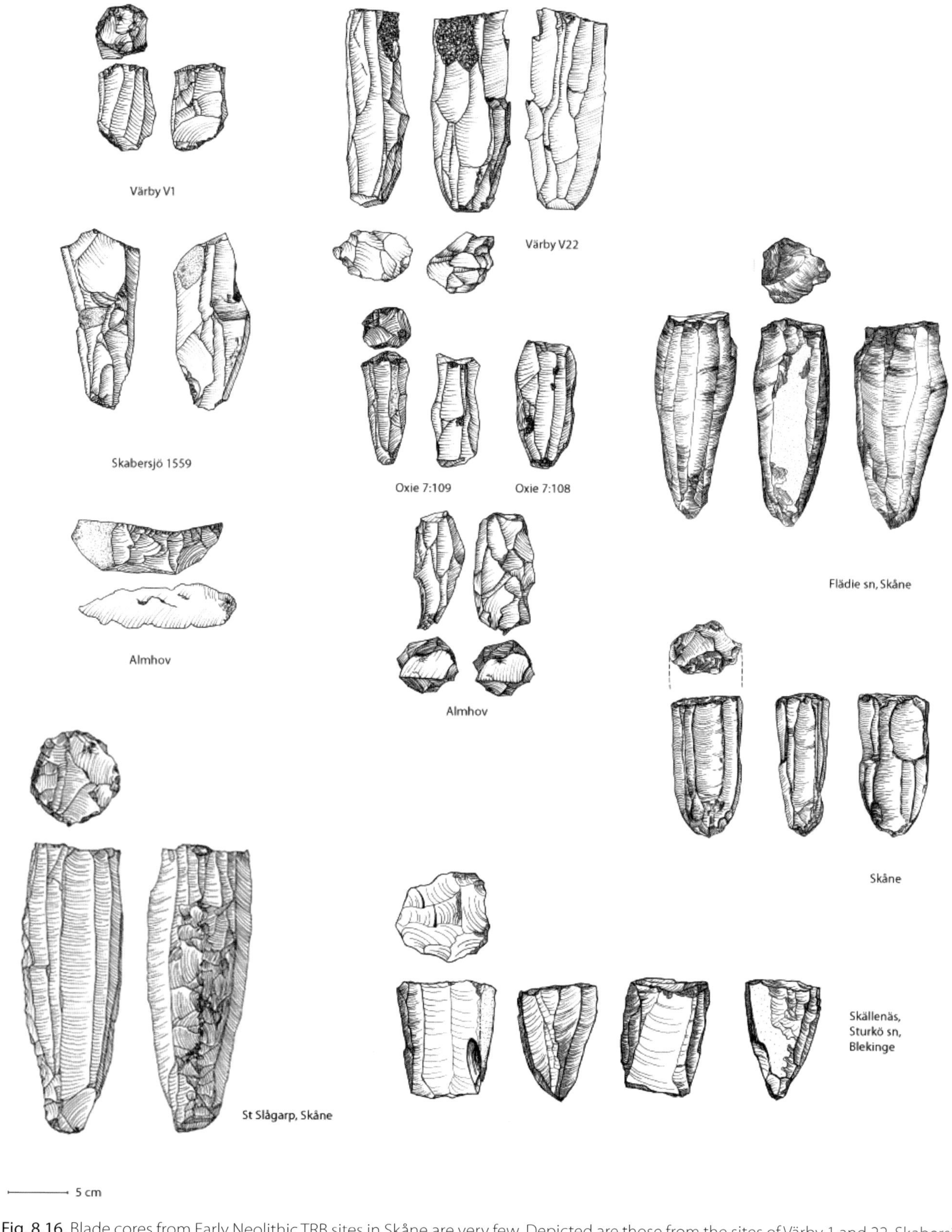

Fig. 8.16. Blade cores from Early Neolithic TRB sites in Skåne are very few. Depicted are those from the sites of Värby 1 and 22, Skabersjö and Oxie 7. In the recently excavated large site of Almhov only two cores connected to blade (microblade) production, which are presented here were found. The four blade cores found in Skåne are stored in the stray find collection in the National Historical Museum in Stockholm. The core from St Slågarp parish in Skåne is an exceptional find in Sweden and it lacks counterparts in the blades hitherto found (from Knutsson 1995,153).

production traces on the blades from these sites, especially the longer ones. Cores are thus, in contrast to the situation in the Mesolithic, unusual finds from EN settlements, and only one core used to produce long blades has been found in Sweden. Yet the sites constantly yield small numbers of long blades.

Only five cores, which would correspond in size and method to these blades, could be interpreted as remains from the blade production typical of the TRB or CWC were found in the stray find collections of the Historical Museums in Stockholm and Lund (Knutsson 1995, 153f; Callahan 1995, 236). Three of them are made of good-quality flint from the Baltic, and have a brownish patina, developed during a deposition in the acidic soils, most probably bogs. Interestingly enough, one of these cores substantially exceeds the size of all blades hitherto found in Sweden. It shows technological similarities to cores from the White Painted Pottery group in Poland an early part of the Polish Linearbandkeramik tradition (Migal 2006, 390, 394).

Some Southern Comparisons

In the farming communities of the Near East, southern and central Europe, the occurrence of blades seems to be closely connected with the appearance of farming techniques and domesticated species. The systematic exploitation of flint, chert, radiolarite and obsidian sources in the same areas may be attributed to the production of these blades. Large production sites have been found and long distance distribution has been mapped for blades, although not in Scandinavia (for example Balcer 1980; Dijkman 1980; Perlès 1992; Vencl 1967; Weiner 1992; Zimmermann 1981). This production and distribution is dated to the earliest Neolithic and continues (although not without interruptions) through the Bronze Age.

The blades themselves seem from the beginning to have been produced and used for the harvesting of crops. In the southern regions, they might have been of a great importance. The localities of their production show extensive amounts of waste and the blades all over Europe show a high degree of regularity. Perlès' (1992) analyses led her to propose that during the Early and Middle Neolithic the blades of southeastern Europe were produced by itinerant specialists carrying with them the raw material in the form of prepared cores from different sources quite far from the places where the blades were later deposited (see also Gronenborn 1992; 1994). In the regions closest to central Sweden, systematic blade production and distribution has been detected in and around the mines of Świeciechów (Fig. 8.17) in central Poland as early as from the Early Neolithic (EN) and down to the times of the Polish Middle Neolithic Funnelbeaker culture (TRB) (Balcer 1980; Migal 2006).

We know that since the beginning of blade production, some of the techniques were quickly brought to perfection. After that, the development or modifications yielded new production techniques, and new preferences concerning the form of the final products that changed in the very slow mode of fashions. Sometimes small products were required, sometimes they were to be large, and other times they were to be narrow and very long. There exist only a limited number of methods for making blades, but the techniques involved can be combined in different ways so it is possible to identify different cultural traditions or transformations of know-how and knowledge of production. Nevertheless, modern flintknappers point to the complexity of blade production and only very few are able to produce copies of some of these blades.

Not only are blades special products, but throughout Europe the depositional treatment of blades was also special. They tend to be found in intentional deposits such as burials and hoards. Formal deposition of blades and blade products seems to have become a systematic and very important feature of the burial customs. On the other hand, Corded Ware Culture settlement contexts contain very small amounts of flint and no blades at all. Exactly the same pattern occurs in other regions of central Europe as well as in Sweden (Knutsson 1995).

Some European grave finds contain large numbers of blades. Such finds have been made in contexts connected with the groups of Gatersleben, Rössen, Walternienburg (Funnelbeaker), the Corded Ware Culture, and other Neolithic archaeological cultures. In Pevestorf 19, Ldkr. Lüchow-Dannenberg, in Germany (Fig. 8.17), a Neolithic burial site dated to the Middle and Late Neolithic and Bernburger, Schönfelder and Globular Amphora Cultures, yielded

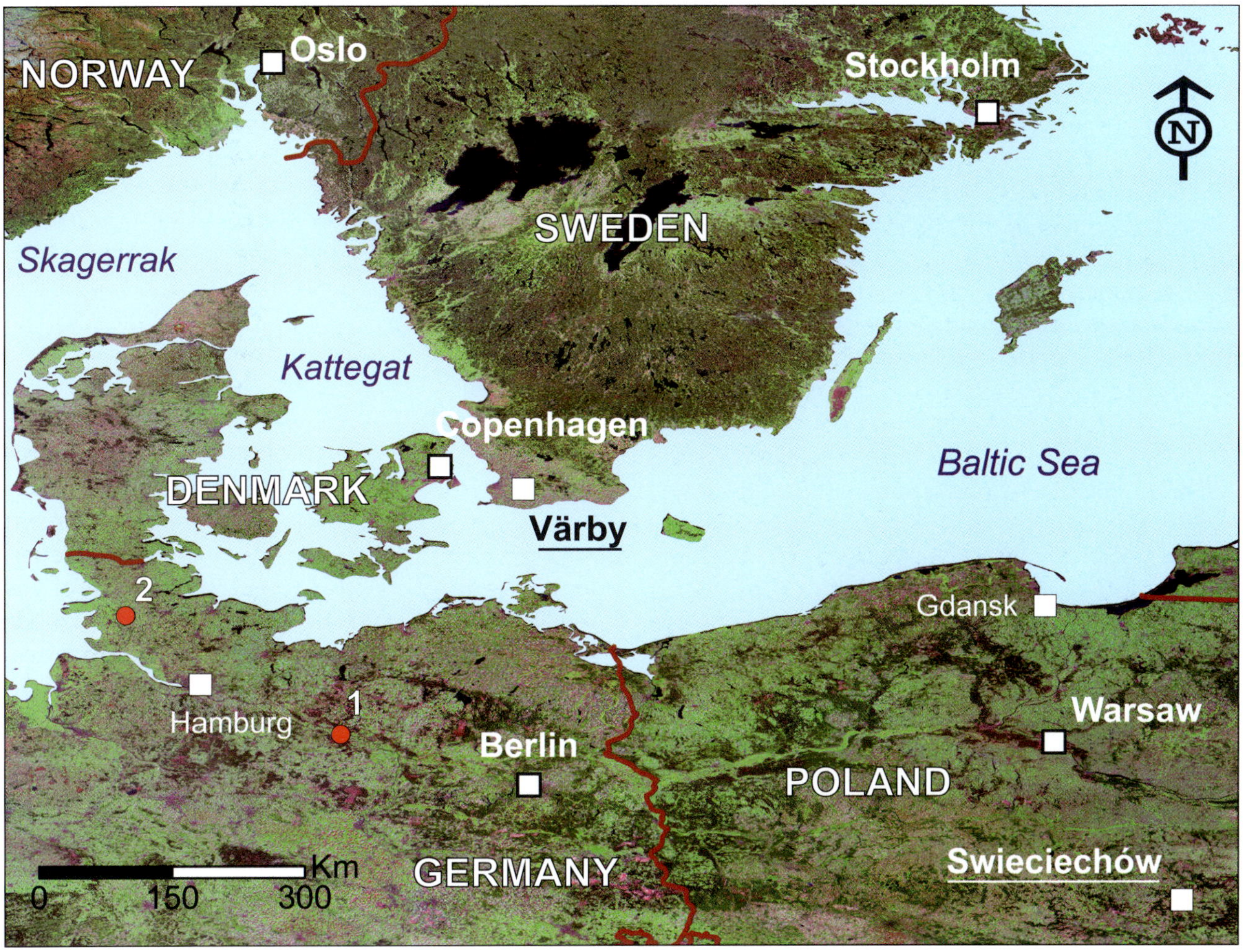

Fig. 8.17. Map of northern Europe with the towns of Świeciechów in Poland, Värby in Sweden, as well as the archaeological sites of 1) Pevestorf 19; 2) Schwienhusen, as mentioned in Chapter 8.5. Map: R. Lugon, J.-C. Loubier and A. Chevalier.

38 inhumations and 13 cremations. 26 of these burials were not disturbed and yielded burial gifts (Mayer 1993). Of these graves, nine had more than 8 blades each and in one there were 92. Several of the blades in each grave could be refitted, which means that they were struck from the same core, but these cores were missing. Other cores, occasionally found in the site, could not be associated with production of the blades in the graves. Also, as usual in these contexts, very few signs of a contemporary settlement were found. The blades occur in all four chronological groups of this cemetery (Mayer 1988, 75–85; 1993).

Another example is a cemetery, belonging to the Corded Ware Culture in Bohemia. There the blades were found in approximately half of the 177 graves. An analysis of the blades showed a similarity to production of blades in Swedish Corded Ware Culture burials, as well as to blades described from some sites in the Middle East (Knutsson 1995).

Purposeful depositions of blades in groups and bundles are known in central, northern and western Europe. Most of them are undated, but many of them have been interpreted as remains from the Neolithic. An example is the famous find from La Creusette near Le Grand Pressigny (Fig. 8.18), where one find of 122 and another of 133 blades was found together with a polished axe and a *livre de beurre* (pound of butter) core. This dates the finds into the very late Neolithic or the beginning of the Bronze Age (Geslin *et al.* 1980, 289ff). Other depositions of blades have been found in Poland, Germany, the Czech Republic and Slovakia, in France, Sardinia, Portugal (Vencl 1967, 183, lists some of these finds; Schardt 1980, 284f; Balcer 1980, 310ff; Pélegrin 2006, 37–68).

Several authors note that long blades (often more than 8 cm), appear in small amounts in the graves and in the settlements (Pélegrin 2006; Vencl 1967).

Fig. 8.18. Location of Le Grand Pressigny in France, with the archaeological site of La Creusette. Map: R. Lugon, J.-C. Loubier and A. Chevalier.

Finds of really long blades (over 20–25 cm) are rare and appear in some graves (for example at Varna, Bulgaria), in purposeful depositions (like La Creusette) and in production places close to the Neolithic mines. At least in Poland, the production and distribution of blades and the long blades in particular reaches its maximum during the Funnelbeaker (TRB) culture. In France, where long blades are most numerous, the maximum production and distribution is dated to the Late Neolithic and Bronze Age. In northern Germany, the long blades seem to have been produced during the Early Neolithic and deposited as stray finds. After that, a production of shorter, but still well-made blades struck from facetted platforms continued (in a workshop-like production system) and the blades, in some cases in considerable numbers, were still deposited in graves.

So far, I know of only one place with a large deposition of cores, in Schwienhusen in Schleswig-Holstein (Fig. 8.17). Around 40 blade cores have been uncovered on the shore of a bog, and no other flint was found near the cores. It has been interpreted as a workshop hiding-place or a votive deposit. The cores of the Schwienhusen deposit were prepared for blade production but used very little for this purpose. Their size matches the blades of the Corded Ware Culture rather than the older types, as the blades from the region dated to Early Neolithic were produced from larger cores. Such Early Neolithic blades of high quality are often found in small deposits, as well as in some production sites (for example Golisch 2007; Weiner 1992).

Conclusion: Interpreting the Scandinavian Blades

I presented an idea that the longer blades in south Scandinavia were among the symbols or metaphors closely connected to agricultural production and especially the harvesting of grasses, like cereals. This suggestion is based on the observations that new methods for blade-making were introduced in correlation with the introduction of farming during the Early Neolithic Funnelbeaker period. Use-wear evidence on many of these blades documents their use as harvesting implements. Further, the blades are spread evenly over a large area, in very small numbers (1–2/site), implying their special status. The manufacture of blades requires specialised skills, and we can assume that they were made by itinerant specialists bringing the materials with them, or imported from production sites near the flint sources, as for the Polish Funnelbeaker long blade production. In some cases, but not always, they seem to me to be made of flint imported to Skåne from elsewhere (*i.e.* from south and east Baltic regions). They show similarities with the flint blades from the later Middle Neolithic Corded Ware Culture burials. Like the polished thin-butted axes, the blades show a tendency to be longer in the sites situated far away from the mines and flint sources of the Baltic region (cf. Sundström 2003).

It is unusual to find cores that might be connected to the production of these longer (over 10 cm) blades. In one of the sites, Värby (Fig. 8.17), one such item is found. Three others were found as stray finds in southern Sweden, made of high quality flint, and a further one of a local flint type. One of these cores is of exceptional size and quality. It is comparable to other, Early Neolithic Polish finds. Few similar cores and core pre-forms have been collected in Denmark. They are described by Vang Petersen (1999, 56f), Callahan (1995) and Pélegrin (2007). Very long blades (20–28.5 cm) are also found in Denmark and northern Germany and related to Neolithic periods. One possible interpretation is that along with a diminishing production of short blades, longer to very long blades were produced and distributed in Scandinavia in small numbers during the Early and Middle Neolithic. The reasons for this production are to be understood in other than subsistence-economical terms, because for harvesting purposes only (and from a functional point of view) blades of special lengths would not be required. In the Corded Ware Culture the blades were made for burials. Since the Early Neolithic, blades are found in settlements and hoards, so the questions to pose would be if they were intended for everyday use and as such reshaped and broken for different purposes (*i.e.* not recognisable as blades) and the remains of their production were either destroyed during use, or deposited where they were used, outside the settlements), or if they had special purposes (ceremonial, commemorative or 'ritual') and received a special treatment.

Interpretation of the importance of large blades is based on their relative scarcity in some places and abundance in others, the appearance of large production localities, the subsequent distribution of products from such production centres and the special deposition patterns in the receiving regions. In these deposition patterns we see *few* finds of the end products and a complete lack of production debris on the deposition sites in central Scandinavia. These distribution and deposition patterns exist in Scandinavia from the Early Neolithic through at least the Middle Neolithic everywhere there is evidence for agriculture, like cereal seeds, or bones from domesticated animals. This pattern seems to be repeated from the earliest appearance of agriculture in Europe and cross-cuts several classificatory culture boundaries including the Linearbandkeramik (Linear Band Ware), Funnelbeaker, Corded Ware Culture and their Scandinavian appearances. Such a pattern of appearance has a meaning reaching far beyond a practical use of these sharp pieces of stone. Since an earlier blade production method seems to have been taken out of use in the same time as domesticates appeared, it is reasonable to assume that the new imported method had more meaning than being merely a technique of production. I suggest that the blades, including their production methods, were part of a 'farming manual' that in some respect guaranteed successful harvests.

8.6. CONCLUDING REMARKS

Inja Smerdel

In concluding this chapter on acquiring skills and the transmission of knowledge, we are aware (as is stated in the introduction) that our contributions tackle only some of the possible topics of this wide and meaningful subject of research, which is so substantial when we deal with agriculture. Our examples vary not only in theme and narration but are also diversified with respect to research methods. Yet they reveal how knowledge and skills are pivotal factors for assuring continuity and stability and as initiators of change. New skills and changing bodies of knowledge are not only a response of individuals or communities to the introduction of cultural changes such as new or improved tools, crops, or different techniques, they depend on other motives for acquiring skills and gaining knowledge besides subsistence.

For example, starting with the closing contribution of Helena Knutsson – with her interpretations substantiated by experimentally 'reinventing ' the skills of prehistoric knappers and harvesters – we learn that new methods of blade production and use were introduced in correlation with the introduction of farming skills. Her evidence from use-wear studies demonstrates that not only are many blades used as harvesting implements, but that some patterns of their appearance can also have meanings extending beyond the practical uses of these sharp pieces of stone. Thomas Schippers – who on the basis of ethnographic fieldwork addresses skills as identity markers – indicates with his examples that technical solutions and the necessary skills to perform them are often closely related to specific groups. Not only is the social setting of importance in the acquisition, exercise and transmission of skills, but such skills also can be used in marking cultural boundaries. Such boundaries may prevent the use of certain skills or tools, not because they are unknown, but because they are explicitly refused and considered inappropriate by certain groups. Inja Smerdel discusses whetstone holders, eloquent witnesses to the hay harvest and scythe mowing, one of the most valued and socially recognised of manly skills. She reveals how mowers, as both skilled labourers and courting males, took great pains to be distinct individuals, using their whetstone holders as the external tokens of a mower's consciousness of his own personality and of his wish to be recognised and valued as an individual. Learning the knowledge and acquiring the skills of a mower thus, among other things, signified learning to become an individual, a personality, a real man. Inja Smerdel also addresses the issue of the apprenticeship of children in daily work with oxen, a case study based mostly on ethnographic fieldwork. From this, we learn that parallels and analogies between the training of young oxen and 'little humans' were widely recognised; that children's apprenticeship in animal husbandry was a school of life, required by fundamental socio-economic necessity; and that parallelism and analogies between training oxen and children can be traced throughout the processes of knowledge transmission and acquisition of skills. The need to guarantee continuity and stability, while remaining flexible, as knowledge and skills are transmitted to the next generations, ensures that in their 'imitative learning' by 'observing and doing', children are indeed crucial actors in this socio-historical game.

No technology is just an artefact. All require human knowledge and skills to operate, and thus all are imbued with meaning, acquired, transmitted, and performed in a social context. It is no surprise that the intangible aspects of agriculture are as important to people as the material ones, and these cases provide but a small sampling of the infinite interweaving of all aspects of agricultural life.

Chapter Note

1 This article is a part of a long-term project on stone technology, transmission of knowledge and cultural choices by Kjel and Helena Knutsson. Kim Darmark and Lars Sundström joined the project with statistical and practical work. This article has been written with financial help from the Birgit and Gad Rausing Foundation.

Chapter Bibliography

Albreht, I. (1945) Umirajoči in izumrli običaji. In *Vodnikova pratika*. Vodnikova družba, Ljubljana.

Arden, T. and Hudson, S. R. (eds.) (2006) *The Social Experience of Childhood in Ancient Mesoamerica*. University Press of Colorado, Boulder.

Balcer, B. (1980) Zu den Problemen der Verbreitung der Feuersteine von Świeciechów und Krzemionki. In G. Weisgerber, R. Slotta and J. Weiner (eds.) *5000 Jahre Feuersteinsbergbau. Die Suche nach dem Stahl der Steinzeit.* Deutsches Bergbau-Museum, Bochum, pp. 310–317.

Ballin, T. B. (1999) *Kronologiske og regionale forhold I sydnorsk stenalder. En analyse med udgangspunkt i bopldserne ved Lundevågen (Farsundprosjektet)*. Unpublished PhD thesis, Aarhus University.

Bartels, A. (1900) *Der Bauer in der deutschen Vergangenheit.* Monographien zur deutschen Kulturgeschichte 4. Diederichs, Leipzig.

Barth, F. (1969) Introduction. In F. Barth (ed.) *Ethnic Groups and Boundaries, The social organisation of culture difference*. Little Brown & Co, Boston, pp. 9–28.

Baxter, J. E. (2005) *The archaeology of childhood: Children, Gender, and Material Culture*. AltaMira Press, Walnut Creek.

Bell, J. (1985) Illustrations of the Use of Oxen in Ireland. *Ulster Folklive* 31, 73–75.

Bergin, O. and Best, R. I. (1977) 'Tochmarc Étaín'. *Ériu* 12 (1934–1938, (1977)).

Bermann, M. (1881) *Maria Theresia und Kaiser Josef II. in ihrem Leben und Wirken*. U. Hartleben's Verlag, Wien/Pest/Leipzig.

Blixen, K. (2004) *Spomin na Afriko*. Delo d.d., Ljubljana.

Braudel, F. (1988) *Strukture vsakdanjega življenja: mogoče in nemogoče, Materialna civilizacija, ekonomija in kapitalizem, 15.-18. stoletje*, 1, 2. ŠKUC, Filozofska fakulteta, Ljubljana.

Bromberger, C., Dossetto, D. and Schippers, T. K. (1982–83) L'ethnocartographie en Europe: coups d'oeil retrospectives et questions ouvertes. In *L'ethnocartographie en Europe*. CEM, Aix en Provence, pp. 15–40.

Bruneel, C., Daelemans, F., Helin, E., Leboutte, R., Lehners, J.-P. and Vellut, J.-L. (1990) Les migrations à longue distance (1500–1900), Belgique et Grand-Duché de Luxembourg. In A. Eiras Roel (ed.) *Long Distance Migrations (1500–1900)*. 17. Congreso Internacional de Ciencias Historicas (Madrid, 26 et 30 août 1990). Commission Internationale de Démographie Historique, Paris, pp. 45–60.

Callahan, E. (1995) Blades from Middle Neolithic Battle Axe Culture graves in Sweden. A preliminary technological study. In H. Knutsson, *Slutvandrat. Aspekter på övergången från rörlig till bofast tillvaro*. Aun 20. Appendix 2, pp. 225–237.

Cameron, E. L. (1996) *Isn't she a doll? Play and ritual in African sculpture*. UCLA Fowler Museum of Cultural History, Los Angeles.

Campbell, B. (2005) On 'Loving Your Water Buffalo More Than Your Own Mother': Relationships of Animal and Human Care in Nepal. In J. Knight (ed.) *Animals in Person, Cultural Perspectives on Human-Animal Intimacies.* Berg, Oxford, New York, pp. 79–100.

Carrithers, M. (1992) *Why Humans have Cultures, Explaining anthropology and social diversity*. Oxford Univeristy Press, Oxford.

Cevc, T. (1979) Otroške živalske igračke – 'buše'. *Traditiones* 5–6 (1976–1977), 69–78.

Dalin, G. (ed.) (1999) *Les bœufs au travail.* Actes du colloque du Festival Animalier International de Rambouillet (Bergerie Nationale de Rambouillet, samedi 26 septembre 1998). Bergerie Nationale, Rambouillet.

Digard, J.-P. (1988) Jalons pour une anthropologie de la domestication animale, *L'Homme* 108 (28, no. 4), 27–58.

Digard, J.-P. (2005) Entre los hombres y los animales: Una conquista recíproca. In J. L. Mingote Calderon (ed.) *Animalario, Visiones humanas sobre mundos animales*. Museo del Traje, Centro de Investigación del Patrimonio Etnológico, Madrid, pp. 27–38.

Dijkman, W. (1980) Mouthiers 'Le Martins', Dép. Charente. In G. Weisgerber, R. Slotta, and J. Weiner (eds.) *5000 Jahre Feuersteinsbergbau. Die Suche nach dem Stahl der*

Steinzeit. Deutsches Bergbau-Museum, Bochum, p. 482.

Erjavec, F. (1883) Iz pótne torbe. In Fr. Levec (ed.), *Letopis Matice slovenske za leto 1882 in 1883.* Matica slovenska, Ljubljana, pp. 195–351.

Fél, E. and Hofer, T. (1974) *Geräte der Átányer Bauern.* Kommission der Königlich Dänischen Akademie der Wissenschaften zur Erforschung der Geschichte der Ackerbaugeräte und der Feldstrukturen (Publikation Nr. 2), Kopenhagen.

Fenton, A. (1976) *Scottish Country Life.* John Donald Publishers LTD, Edinburgh.

FN 1992, 1993. Smerdel, I., Field notes on hay harvest, Archives of the SEM (Slovene Ethnographic Museum).

FN 2007. Smerdel, I., Field notes from Marija Reka, Archives of the SEM.

Forni, G. (1990) *Gli albori dell'agricoltura. Origine ed evoluzione fino agli Etruschi ed Italici.* Reda, Roma.

García-Hoz Rosales, C. (2005) Los animales de juguete. Arquetipos y modelos. In J. L. Mingote Calderon (ed.) *Animalario. Visiones humanas sobre mundos animales.* Museo del Traje, Centro de Investigación del Patrimonio Etnológico, Madrid, pp. 163–172.

Geslin, M., Bastien, G. and Mallet, N. (1980) Das Klingendepot von La Creusette, Dé. Indre-et-Loire. In G. Weisgerber, R. Slotta, and J. Weiner (eds.) *5000 Jahre Feuersteinsbergbau. Die Suche nach dem Stahl der Steinzeit.* Deutsches Bergbau-Museum, Bochum, pp. 289–293.

Golisch, T. H. (2007) *Weddingsted-Nordgerlin, Kr. Dithmarschen. Untersuchungen zur Technologie und Chronologie eines neolithischen Schlagplatzes.* Veröffentlichungen des Dithmarschen archäologischen Vereins. Band 1. Heide.

Grieshofer, F., Pallestrang, K. and Witzmann, N. (eds.) (2004) *Ur- Ethnographie. Auf der Suche nach dem Elementaren in der Kultur: Die Sammlung Eugenie Goldstern.* ÖVerlag des Österreichischen Museums für Volkskunde, Wien.

Griffin-Kremer, C. (2001) Bovine Bodies and the Domestication of the Human Mind. In L. Del Giudice and G. Porter (eds.) *Imagined States.* State University Press, Logan Utah, pp. 167–192.

Griffin-Kremer, C. (2007) Du joug de tête au joug de garrot: Récit mythique et changement technique? In *Les bovins: de la domestication à l'élevage, Ethnozootechnie* 79, 51–67.

Gronenborn, D. (1992) *Silexartefakte der ältestband-keramischen Kultur.* PhD thesis, Universität am Main.

Gronenborn, D. (1994) Überlegungen zur Ausbreitung der bäuerlichen Wirtschaft in Mitteleuropa – Versuch einer kulturhistorischen Interpretation ältestbandkeramischer Silexinventare. *Praehistorische Zeitschrift* 70 (1), 137–151.

Hansen, W. (1984) *Kalenderminiaturen der Stundenbücher. Mittelalterliches Leben im Jahreslauf.* Callwey, München.

Haraway, D. J. (2008) *When Species Meet.* University of Minnesota Press, Minneapolis.

Harris, M. (1989) Mother Cow. In *Cows, Pigs, Wars and Witches. The Riddles of Culture.* Vintage Books, New York, pp. 11–32.

Ingold, T. (2000) *The Perception of the Environment. Essays in Livelihood, Dwelling and Skill.* Routledge, London and New York.

Inkret, A. (1982) Neprizanesljiva in presunljiva dokumentacija o otroštvu. In A. Puhar, *Prvotno besedilo življenja. Oris zgodovine otroštva na Slovenskem v 19. stoletju.* Globus, Zagreb, pp. 7–14.

Jochaud, P. (2002) *Une vie de paysan - Souvenirs de la traverse du siècle.* L'Ecomusée rural de Vigneux de Bretagne, Petit Patrimoine (extraits)

Kamp, K. A. (2001) Where Have All the Children Gone? *Journal of Archaeological Method and Theory* 8 (1), 1–34.

Kamp, K. A. (ed.) (2002) *Children in the Prehistoric Puebloan Southwest.* University of Utah Press, Salt Lake City.

Klein, E. (1920) *Stenåldersliv.* Stockholm.

Knutsson, H. (1995) *Slutvandrat. Aspekter på övergången från rörlig till bofast tillvaro.* Aun 20. Uppsala.

Knutsson, H. (2000) Rituella och funktionella flintor. In E. Hjärthner-Holdar and C. Risberg (eds.) *Hantverkets roll i samhället - produktion och reproduktion. Rapport från den tredje workshopen om hantverksspecialisering Uppsala 18-19 mars 1999.* Uppsala, pp. 17–33.

Knutsson, H. (2001) Technology, mythology and the travels of the agricultural package in Europe. *Documenta Praehistorica* 38, 117–132.

Knutsson, H. (2012) Using Worldwide Samples in Understanding Mesolithic-Neolithic Transition in Northern Europe. In A. Cannon (ed.) *Structured Worlds: The Archaeology of Hunter-Gatherer Thought and Action. Approaches to Anthropological Archaeology.* Sheffield, Equinox Publishing.

Knutsson, K. (1995) Funktionsanalys av flintspån från stridsyxegravar i Vikletice, Tjeckiska republiken. Appendix 1. In H. Knutsson (1995) *Slutvandrat. Aspekter på övergången från rörlig till bofast tillvaro.* Aun 20. Uppsala, pp. 221–224.

Kocbek, Fr. and I. Šašelj (1934) *Slovenski pregovori, reki in prilike.* Družba sv. Mohorja, Celje.

Koprivec, I. (1939) *Kmetje včeraj in danes. Narodopisne črtice iz Slovenskih goric.* Glavna hranilnica, Sv. Lenart v Slovenskih goricah.

Koštiál, I. (1939) O oselniku. *Etnolog* 10–11(1937–1939), 22–24.

Kuhar, B. (1972) *Odmirajoči stari svet vasi.* Prešernova družba, Ljubljana.

Kuret, N. (1989) *Praznično leto Slovencev. Starosvetne šege in navade od pomladi do zime,1.* Družina, Ljubljana.

Leach, Sir Edmund R. (1961/2004) *Rethinking Anthropology.* Berg, Oxford.

Lech, J. (1980) Das Grab eines Steinschmiedes von Michałowice bei Kraków und das Problem der Arbeitsspezialisierung unter der ersten Ackerbau betreibenden Bevölkerung Mitteleuropas. In G. Weisgerber, R. Slotta, and J. Weiner (eds.) *5000 Jahre*

Feuersteinsbergbau. Die Suche nach dem Stahl der Steinzeit. Bochum, pp. 272–278.

Lerche, G. (1994) *Ploughing Implements and Tillage Practices in Denmark from the Viking Period to About 1800.* Poul Kristensen, Herning.

Leroi-Gourhan, H. (1945/1973) *Milieu et techniques.* Albin Michel, Paris.

Lévi-Strauss, C. (1962) *La Pensée sauvage.* Plon, Paris.

Lillehammer, G. (1989) A child is born: the child's world in an archaeological perspective. *Norwegian Archaeological Review* 22(2), 89–105.

Llaty, C. (1997) Le fléau, la planche et le mulet, l'égrenage des céréales dans les Alpes du Sud. *Le Monde alpin et rhodanien*, 83–100.

Lucassen, J. (1990) Dutch migration, 1600–1900. In A. Eiras Roel (ed.) *Long Distance Migrations (1500–1900).* 17. Congreso Internacional de Ciencias Historicas (Madrid, 26 et 30 août 1990). Commission Internationale de Démographie Historique, Paris, pp. 19–44.

Majer, M. (1848) *Pravila kako izobraževati ilirsko narečje i u obče slavenski jezik.* Ljubljana.

Makarovič, G. (1973) *Cvetlice v ljudski umetnosti: motivi v oblikovanju za kmetije.* Vodnik po razstavi. Slovenski etnografski muzej, Ljubljana.

Makarovič, G. (1995) *Slovenci in čas. Odnos do časa kot okvir in sestavina vsakdanjega življenja.* Krtina (Knjižna zbirka Krt; 94), Ljubljana.

Makarovič, M. (1994) *Sele in Selani. Narodopisna podoba ljudi in krajev pod Košuto.* Celovec.

Mallet, N. (1980) Grand Pressigny, Touraine, Dép. Indre-& Loire. In G. Weisgerber, R. Slotta and J. Weiner (eds.) *5000 Jahre Feuersteinsbergbau. Die Suche nach dem Stahl der Steinzeit.* Deutsches Bergbau-Museum, Bochum, pp. 483–485.

Malmer, M. (1962) *Jungneolitische Studien.* Lund.

Matičič, I. (1943) *Petrinka.* Vodnikova družba, Ljubljana.

Mayer, M. (1988) Flintzusammenpassungen aus neolithischen Gräbern von Pevestorf, Ldkr. Lüchow-Dannenberg. *Die Kunde* N.F. 39, 75–85.

Mayer, M. (1993) *Pevestorf 19. Ein mehrperiodiger Fundplatz im Landkreis Lüchow-Dannenberg.* Veröffentlichungen der urgeschichtlichen Sammlungen des Landesmuseums zu Hannover, Band 41. Oldenburg.

Mencej, M. (2006) Nesnovna kulturna dediščina Kozjanskega in Obsotelja in njena uporaba v turizmu. In *Mesto in trg na meji / Grad i trg na granici.* 9. Vzporednice med slovensko in hrvaško etnologijo / 9. Hrvatsko-slovenske etnološke paralele. SED (Knjižnica Glasnika SED; 38), Ljubljana, pp. 55–80.

Migal, W. (2006) The macrolithic flint blades of the Neolithic times in Poland. In J. Apel and K. Knutsson (eds.) *Skilled production and social reproduction. Aspects of traditional stone-tool technologies.* SAU Stone Studies 2. Uppsala, pp. 387–398.

Mrkun, A. (1943) *Etnografija velikolaškega okraja, 1. zv., Kmetijstvo.* Videm - Dobrépolje.

Otte, M. and Behm-Blancke, M. R. (1992) Das lithische Inventar. Die Rekonstruktion technischer Verfahrensweisen. In M. R. Behm-Blancke (ed.) *Hassek Höyük. Naturwissenschaftliche Untersuchungen und lithische Industrie.* Istanbuler Forschungen, Band 38. Tübingen, pp. 165–176.

Parain, Ch. (1937/1979) Les anciens procédés de battage et de dépiquage en France. Reprinted in Parain, Ch. *Outils, ethnies et développement historique.* Éditions socials, Paris.

Pélegrin, J. (2006) Long blade technology in the Old World: and experimental approach and some archeological results. In J. Apel and K. Knutsson (eds.) *Skilled production and social reproduction. Aspects of traditional stone-tool technologies.* SAU Stone Studies 2. Uppsala, pp. 37–68.

Pélegrin, J. (2007) E-letter dated 2007-03-06.

Perlès, C. (1992) Systems of exchange and organization of production in Neolithic Greece. *Journal of Mediterranean Archaeology* 5 (2), 115–164.

Prežihov, V. (1969) Jirs in Bavh. In P. Voranc, *Izbrano delo*, 3. Mladinska knjiga, Ljubljana, pp. 51–61.

Puhar, A. (1982) *Prvotno besedilo življenja. Oris zgodovine otroštva na Slovenskem v 19. stoletju.* Globus, Zagreb.

Ramšak, M. (2007) Etnološko proučevanje otrok v slovenski etnologiji in folkloristiki. *Etnolog* 17, 31–41.

Rebula, A. (1957) *Klic v Sredozemlje.* Mohorjeva družba, Celje.

Ribstein, J. (2006) La transhumance bovine dans le massif vosgien et l'arc alpin: approche ethno-écologique. In *Ethnozootechnie* 7 (Hors série), 7–144.

Rinieri-Villain, F. (1997) Attelages de la valle des Merveilles. In F. Sigaut, J. M. Duplan and N. Bochet (eds.) *Les bœufs au travail.* Journée d'étude de la Société d'ethnozootechnie (17 octobre 1997). AFMA, Paris, pp. 3–10 (*Ethnozootechnie* 60).

Rognon, X. and Lefèvre, C. (eds.) (2007) *Les bovins: de la domestication à l'élevage.* Journées d'étude de la Société d'ethnozootechnie et L'Homme et l'Animal, Société de Recherche Interdisciplinaire (Paris, 16 et 17 novembre 2006). *Ethnozootechnie* 79.

Sand, G. (1962) Le labour. In G. Sand, *La Mare au Diable. François le Champi.* Éditions Garnier Frères, Paris, pp. 13–25.

Schardt, H. G. (1980) Das Klingendepot von Dorsheim, Kr. Bad Kreuznach. In G. Weisgerber, R. Slotta and J. Weiner (eds.) *5000 Jahre Feuersteinsbergbau. Die Suche nach dem Stahl der Steinzeit.* Bochum, pp. 284–288.

Schippers, T. K. (1993) L'égrenage des cereals dans la haute vallée de l'Ubaye. In J.-L. Miège (ed.) *Les cereals en Méditerranée.* CNRS Éditions, Paris, pp.171–178.

Schmidt, L. (1952) *Gestaltheiligkeit im bäuerlichen Arbeitsmythos. Studien zu den Erntschnittgeräten und ihrer Stellung im europäischen Volksglauben und Volksbrauch.* Verlag des Österreichischen Museums für Volkskunde, Wien.

Sigaut, F. (1998) Compagnie des animaux utiles et utilité des animaux de compagnie. In B. Cyrulnik (ed.) *Si les lions pouvaient parler: Essais sur la condition animale.* Gallimard, Paris, pp. 1078–1085.

Sigaut, F., Duplan, J. M. and Bochet, N. (eds.) (1997) *Les bœufs au travail.* Journée d'étude de la Société d'ethnozootechnie (17 octobre 1997). AFMA, Paris (*Ethnozootechnie* 60).

Smerdel, I. (1991) Prelomna in druga bistvena gospodarska dogajanja v zgodovini agrarnih panog v 19. stoletju na Slovenskem. *Slovenski etnograf* 33–34 (1988–1990), 25–60.

Smerdel, I. (1994) *Oselniki: Zbirka Slovenskega etnografskega muzeja. / Whetstone Holders: The Collection of the Slovene Ethnographic Museum.* SEM (Knjižnica Slovenskega etnografskega muzeja / Slovene Ethnographic Museum Library; 4), Ljubljana.

Smerdel, I. (2005) »Bol si pámetən ku člóvk,« je rekel volu? O razmerju človek – vol v vsakdanu in kulturi pivškega kmeta. In J. Boštjančič (ed.) *Slavenski zbornik.* Kulturno društvo, Slavina, pp. 341–379.

Smerdel, I. (2007) « Les gens ressentent plus de compassion à la mort d'un boeuf que lorsque meurt un homme... ». Sur les soins pour la santé des boeufs de travail en Slovénie (aux 18e, 19e et jusqu'au dernier quart du 20e siècle). In *Les bovins: de la domestication à l'élevage, Ethnozootechnie* 79, 141–154.

Smerdel, I. (2009) Le son des mots, la voix des sons: Sur la culture de communication avec les boeufs de travail en Slovénie. In *L'homme et l'animal - voix, sons, musique... Ethnozootechnie* 84, 49–71.

Soafer Deverenski, J. (ed.) 2000 *Children and Material Culture.* Routledge, London/New York.

Sørensen, M. (2006) Rethinking the lithic blade definition: towards a dynamic understanding. In J. Apel and K. Knutsson (eds.) *Skilled production and social reproduction. Aspects of traditional stone-tool technologies.* SAU Stone Studies 2. Uppsala, pp. 277–298.

Steensberg, A. (1993) *Fire-Clearance Husbandry. Traditional Techniques Throughout the World.* Poul Kristensen, Herning.

Sundström, L. (2003) *Det hotade kollektivet. Neolitiseringsprocessen ur ett östmellansvenskt perspektiv.* Coast to cast-books no. 6. Uppsala.

Šarf, F. (1974) Košnja na visokoležečih predplaninskih območjih. *Traditiones* 3, 133–146.

Štrekelj, K. (1908–1923) *Slovenske narodne pesmi, 4.* Slovenska matica, Ljubljana.

Tabet, P. (1998) *La construction sociale de l'inégalité des sexes, Des outils et des corps.* L'Harmattan, Bibliothèque du féminisme, Paris-Montréal.

Trdina, J. (1987) »Vsaka svinja naj si rije svoje korenje«, Peripatetikar. In S. Štabi and I. Kramberger (eds.) *Podobe prednikov. Zapiski Janeza Trdine iz obdobja 1870–1879, 2. knjiga.* Krt, Ljubljana.

Vacha, B., Pohl, W. and Vocelka, K. (1994) *Habsburžani. Zgodovina evropske rodbine.* Mladinska knjiga, Ljubljana.

Vang Petersen, P. (1999) *Flint fra Danmarks oldtid.* København.

Vencl, S. (1967) Depot čepelí z Borče, o. Litoměřice. *Archeologické rozhledy* 19, 180–186.

Versini, F. (1997) De la compétition des bœufs-tirants et des charetiers en Guadeloupe. In F. Sigaut, J. M. Duplan and N. Bochet (eds.) *Les bœufs au travail.* Journée d'étude de la Société d'ethnozootechnie (17 octobre 1997). AFMA, Paris, pp. 75–84 (*Ethnozootechnie* 60).

Vuk, S. (1998) O mornarju, ki se je vrnil pod Kras. In S. Vuk, *Pomlad pod Krasom. Izbrane pesmi, proza in pisma.* Obzorja (Iz slovenske kulturne zakladnice; 32), Maribor.

Zelenin, D. (1927) *Russische (Ostslavische) Volkskunde.* Berlin/Leipzig.

9 Religious and Legal Aspects of Agrarian Life

9.1. INTRODUCTION

Annelou van Gijn and José Luis Mingote Calderón

In the previous chapters, occasional reference has been made to the role religious beliefs or world view played in agricultural practices. In this chapter, we will focus on this issue a little further. It should be stressed that we do not intend to study religious beliefs or the supernatural per se, but instead how techniques and tools were influenced by and related to religious beliefs. In order to understand the influence of world view or religion in the pre-industrial agrarian world, we need to start by accepting an important premise: in these societies, religion is not a separate aspect of life, but, on the contrary, religion regulates life. It permeates every aspect of society and structures the interactions between the inhabitants of a community. People are usually largely unaware of the extent to which religion determines and regulates their life, because in many societies ritual and secular are not seen as dichotomous spheres (Brück 1999; Hansson and Heiss 2014; Heiss and Hansson 2014).

Because in this book we are dealing with prehistoric societies dating from around 8000 BCE all the way to the present, the way in which we use the term 'religion' needs some clarification. Obviously the term is usually connected with the world religions we know today, such as Christianity, Islam, Hinduism, Judaism and Buddhism. However, people generally referred to as being animists also have religion. While the modern world religions are well known systems of beliefs and ritual, earlier, animistic religions were predominated by natural spirits and ancestral beings and their rituals remain largely unknown. Still, the social and cultural life of their devotees was certainly regulated and determined just as much if not more by their beliefs. In this chapter we will therefore subsume terms as ideology, cosmology and world-view, terms which are usually employed to describe past belief systems and even present-day animistic religions, under the heading 'religion'. Defining religion is difficult and many books have addressed this issue (see, for example, Geertz 1968; Insoll 2004). However, religion always refers and tries to deal with those aspects of the human condition that are beyond human control, that is, the intangible and unexplainable. In this section we will use the term religion for all matters related to the interaction between humans in their daily life and the world beyond, but we will also use terms such as world view and belief system when referring to earlier periods not documented historically.

Another term that needs some clarification is the term ritual. We commonly associate rituals with religion, but there are many secular rituals as well, if only because repetition is central to ritual behaviour. Brück (1999) has made abundantly clear that the term ritual is actually a product of western rationalism, arguing instead for a better comprehension of other conceptions of causality and rationality. Often objects and material practices can shift from the profane to the ritual and back, from the domestic to the sacred, with distinctions being blurred and often not explicitly recognised by the actors themselves. Domestic practices that occur on a daily basis are often referred to as ritual because such practices share a lot of characteristics with rituals in the non-domestic sphere such as formalisation and repetition. Ritual is always formalised, done in 'the proper way', because one purpose of ritual is comunication, be it with other people or with the gods. This implies that rituals are performed in a standardised manner and because they are related to events that occur at a certain regularity, they are also repeated in the same way again and again. Such a formality ensures that the messages and meanings of these rituals are understood by those for whom they are meant. The standardised way of carrying out rituals results in patterned behaviour, the material remnants of which are often recognisable to archaeologists. For example, we find a recurrent association between plough (or ard) marks and burial monuments in Bronze Age Denmark, which can hardly be fortuitous (Bradley 2005, 25–27). The exact meanings of these archaeological patterns of 'special significance' will, however, elude us forever as 'rituals and symbols will always be easier to recognise than they are to interpret' (Bradley 2005, 6).

Yet another commonly used term is ritualisation, a process by which certain actions become 'special' through regular performances. Activities commonly carried out within the domestic sphere can become ritualised at certain intervals, including the tools involved in these actions. For example, sickles, clearly functional objects, are used to reap cereals in daily life, but they are often assigned symbolic importance at some stage in their cultural biography, as for instance shown by their deposition in small roadside chapels in Spain, or by their symbolism connecting agriculture and religious reverence in the Alpine Notburga myth (Fig. 9.1).

The permeation of religious and daily life appears as especially acute for agricultural societies. Obviously hunter-fisher-gatherers turned to the spirits of

Fig. 9.1. Fountain at the monastery of St. Georgenberg in Tyrol, Austria, depicting St. Notburga of Rottenburg/Rattenberg. Local Catholic lore has it that this 13th-century figure claimed her right to attend the vespers (evening prayer) after fieldwork by simply hanging her sickle into thin air, where it was held by a sunbeam. Notburga is still considered a patron saint of farmers in the eastern Alps. Image: A. G. Heiss.

nature when setting out to hunt (see Zvelebil 1997, among others). However, agriculture is something that at least for early agriculturalists ran somewhat contrary to nature: in order for the fields to be laid out and cultivation to proceed, the life-giving forest first needed to be destroyed by means of axes and fire. Hence, death and rebirth are closely intertwined here. Subsequently, the fields had to be prepared and the crop sown at the right time. Then came a long period of waiting and tending, a period which was perceived as precarious because a good harvest was by no means assured. As was shown in Chapter 3, this period of waiting was surrounded with rituals such as offerings to the gods to ensure good weather and a prosperous yield (Mingote Calderón, Chapter 3). The time of harvest was a time of relief and was of old surrounded by festivities. Clearly the agricultural cycle extends over a large portion of the year, encompassing several seasons. Timing is thus crucial in the agricultural cycle and it is not strange to see that various ways of controlling ('domesticating') time or at least of monitoring seasons increased after the introduction of agriculture. It can hardly be coincidental that the alignments of many Neolithic monuments seem to mark the occurrence of winter and summer solstices (Lucas 2005). Stonehenge is of course a famous example (Parker Pearson 2008).

The interaction between agriculturalists and nature is thus a very different one from hunter-gatherers: whereas the latter considered themselves very much part of nature, agricultural people in a way attempted to control nature. However, such an attempt at domination was bound to fail to a large extent and farmers today are just as much subject to the vagaries of nature as the farmers of the past. It is not surprising therefore that agriculture and religion are closely linked. In many societies, divinities are the 'providers' of harvests. People believe that carrying out the different agrarian tasks properly was not the only factor that ensured a good harvest. The influence of the gods and spirits was of capital importance as they ultimately would decide on the quality and quantity of the yield. Therefore, rituals were necessary to ensure that the divine powers were positively inclined to provide a good harvest. In return, people needed to conduct an adequate or 'acceptable' (from a moral point of view) way of life as otherwise the divinity would punish the offender with harvest losses or by sending plagues, storms or droughts. This reciprocity between the people and their gods is crucial in agricultural societies. In Roman religion it was known as the principle of *do ut des* ('I give so that you may give') (Warrior 2006). Gifts were supposed to commit the divinity to giving something else in return, which again obliged the devotee to offer another gift of gratitude (Rüpke 2007).

Another aspect to stress is the fact that agricultural activities are a communal affair. The clearing of the fields, the harvest and the threshing are all tasks that are more effectively done with several people. The way people participate in these tasks needs to be regulated to some extent in order for everyone to participate and put in their fair share of work. In historical contexts, we know that agricultural tasks were linked with religion and it is likely that this was the case in the past as well. Religion draws on the authority of supernatural beings and is therefore impersonal and (generally) impartial. If communal tasks are put in a religious 'straightjacket' they do not need to be reinforced by members of the community; the gods and spirits will ensure that everyone does their duty and, inversely, will also benefit fairly from their work input.

This leads us to the relationship between religion and legal systems. In many European countries a strict social control over individual behaviour has existed since early times. This is clearly reflected in the numerous laws and regulations that were imposed on many practices, behaviours and ways of doing things, including agriculture. For many researchers, the continuous renewal of these local laws is a sign of the little impact they seem to have had on the population. But this assertion corresponds to a modern perception of society where control is exerted by a specific body of authority beyond the local farmers. On the contrary, in pre-industrial societies, although guards existed and were paid by the community, farmers could also be the keepers of the proper functioning of things. Hence, continuous repetition of laws and local regulations serves to reinforce their validity throughout time. The interplay between religion and society results in a great diversity of behaviours in pre-industrial agricultural societies which are radically different from those of industrial societies. Religion is part and parcel of daily life and the relationships between individual people and the supernatural world are reflected in offerings of all kinds of objects to the gods and spirits.

This chapter thus illustrates the multitude of ways in which the agricultural process is intertwined with religion and how agricultural practices are surrounded by 'special behaviour'. In the first contribution, Knutsson argues that the long transition process towards an agricultural Neolithic way of life in Central Sweden should be seen as a period of experimentation and obtaining the necessary practical knowledge to allow a shift to a new subsistence. She interprets the special 'foreign' material culture as mnemotechnical devices and proposes a practical explanation for some of the features other archaeologists may refer to as ritual. The next contribution by Van Gijn addresses the various agricultural tools encountered in Early and Middle Neolithic sites in the present-day Netherlands. She shows that utilitarian items, with a clear 'functional' purpose in domestic context could be ritualised at some point in their cultural biography as they underwent a special treatment upon deposition. In the next paper, Beckmann argues that the frescoes in Minoan palaces on Crete are not merely beautiful decorations, but are actually an agricultural 'calendar'. She states that the plants depicted are symbols of life and death. In the following contribution, Beckmann provides a few salient examples of the symbolic connotations of agricultural tools depicted and described in ancient and medieval texts and representations. This contribution is in support of the notion that the secular and the sacred are closely entwined. The contribution by Schippers shows that laws and regulations continue to determine the technical solutions chosen by farmers, even after they have lost their apparent relevance. The last paper by Mingote Calderón illustrates how legal systems are entwined, justified and endorsed by religion, using historical images of agricultural practices.

9.2. THE COMPLEX ART OF CHANGING LIFESTYLES ON THE VERGE OF THE NEOLITHIC

Helena Knutsson

Introduction[1]

Recent archaeological research papers and excavation reports point to a quite complex subsistence situation in Scandinavia during the period when different agricultures reached the region. Experimentation with alternative choices about how to make best use of wild and cultivated resources lasted for several centuries or even millennia. Although people seem to have been aware of farming practices, and perhaps even knowledgeable about them, they preferred to keep to the economy of hunting, fishing and gathering as well. Nonetheless, the agricultural expansion into northern Scandinavia was a part of colonisation, as seen from the most recent genetic evidence (Skoglund *et al.* 2012). This means that we have to look at this process with slightly different eyes.

In archaeological material, rituality has been defined as odd, perhaps unexpected and formalised behaviour, often possible to trace across different units (Hodder 1992, 222f). According to Bradley, ritual is 'a specialised form of behaviour which emphasises some of the concerns of daily life through a kind of performance' (Bradley 2005, xiii, 30f). As such, it is assumed to have played an important role in the transition to farming economies. The aim of this paper is to discuss the prehistoric religious *reality* in concordance with this view. Another, parallel line of archaeological thought highlights the role of economic and adaptive responses to changes in the climate and resource situation (for example Rowley-Conwy 2004; Louwe Kooijmans 2004; Binford 2002, 400). The discussion continues and adjusts to new data from excavated sites and assemblages. An increasing body of archaeological data from central Sweden makes it possible to contribute to this discussion. We cannot avoid the importance of seeing humans from an evolutionary standpoint, and, consequently, we need to try to establish a fruitful combination of the different observations the two important branches of archaeology offer to interpretations of prehistoric remains. Thus, the practical perspective of the process also has to be taken into consideration. This contribution argues for moderation in the interpretation of agricultural remains as evidence for ritual behavior. In doing that, I will confront Bradley's view of some parts of central Swedish sites as remains of ritual or votive actions.

The Complexity of the Early Neolithic Sites in the Region

Domesticated species were brought to Scandinavia around 4500–3500 cal BCE, a period that coincided with the post-glacial climate warming, but also with the beginning phase of climatic regression which might have changed availability of some wild resources (Rowley-Conwy 1984; 2004). Farming would have been possible throughout most of Scandinavia at that time. The few seed imprints in pottery show the use of different kinds of cereals (mostly wheats and barley) and fruit (among them grapes) from the beginning (Welinder 1998, 56f; Hallgren and Sundström 2008, 199f; Hallgren 2008, 115f). Tool kits usually referred to as the Funnelbeaker Culture (TRB or TBK from German Trichter(-rand-)becherkultur) were introduced into

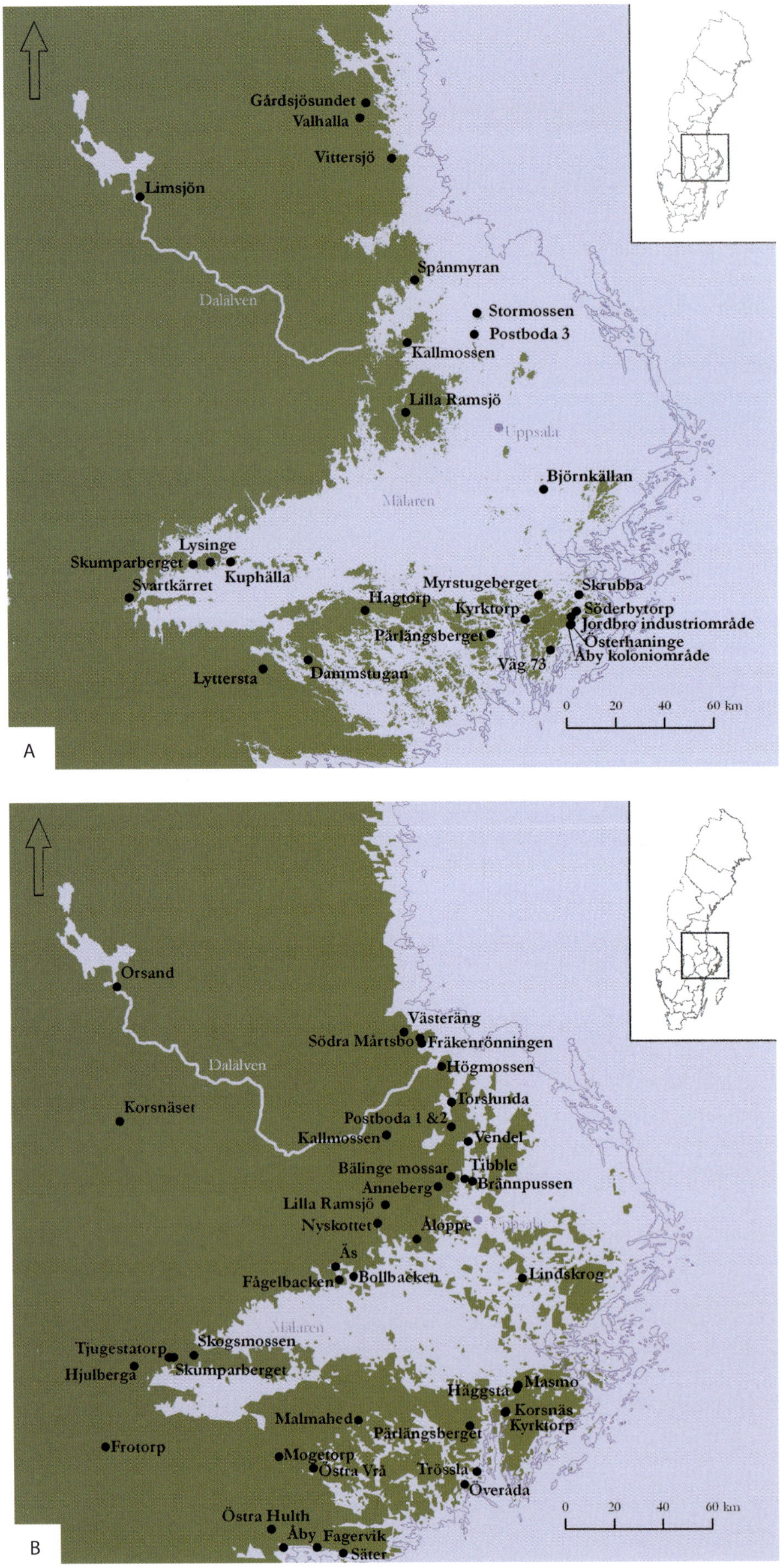

Fig. 9.2. Map of the region presented (the faint blue line illustrates the modern coastline): A: Mesolithic (*ca.* 6,000 BCE), and B: Neolithic (*ca.* 3,500 BCE) coastlines and sites excavated and analysed (Stenbäck 2008: 16–17).

southern and central Scandinavia at this time. The settlement sites in the region show clear differences in their internal structure. Three of these sites will be presented below.

We still have a few sites where the actual production of food was carried out or food remains deposited. The scarcity of domesticated species in the sites has brought into question archaeological interpretations of the TRB in the Scandinavian Early Neolithic as a farming culture (Johansson 2003; Rowley-Conwy 2004). In contrast, hunting is represented in all central Swedish TRB sites, as evidenced by fossil fauna of large and small game, fish and sea mammals. Gathering has hitherto only been assumed, by association with historical and modern analogies, as it even today makes a substantial contribution to nourishment in many farming communities (Ertug 2006; Welinder 2002).

According to radiometric dates, the Funnelbeaker material culture and some domesticated species arrived within a very short period of time, simultaneously over a large area, from modern northern Germany to central Sweden, a distance of more than 1000 km. The statistical corrections of the radiocarbon dates do not allow for a precise statement of the speed of the movement. A modern parallel is the well-documented rapid spread of English farming culture over the whole continent of

Structures, material/type of site	Inland sites 'Farmsteads'	Coast bound sites 'Gathering places'	Coast bound sites Hunting/fishing camps
Settlement areas > 20 000 m^2	x		
Settlement areas < 10 000 m^2		x	x
Settlements divided into distinct task areas	x		
Houses	x		
Huts small structures		x	(x)
Graves/scattered human bones		x	
Remains from dolmens/long barrows (assumed)	(x)		
Depositions/offerings	x		
Bones from domesticated animals	x	(x)	
Bones from wild animals and fish		x	(x)
Domesticated plants – seeds or imprints in pottery	x		
Pottery of TRB style	x	x	
Polished axes < 25 cm, fragments of them	x		(x)
Polished flint axes > 25 cm		x	
Axe production remains	x		
Quartz tools and debris	x	x	x
Grinding stone depositions, often fragmentary	x	x	
Grindings stones production or repair remains	x		
Battle axes	x		

Fig. 9.3. The information in this table is gathered from a comparably small sample of finds. It is, as already mentioned, supported by a number of similar finds in southern Scandinavia (for example information in Gidlöv *et al.* 2006; Madsen 1997; Nielsen 1993; Stafford 1999; Larsson 1984; Hallgren and Sundström 2008; Hallgren 2008). It is possible to interpret these different patterns as remains from two different cultures, economies, traditions, etc. Since there is a certain overlap in the material culture found on the sites, variations in site use are possible interpretations as well. The pattern can be interpreted in terms of adaptation (Hallgren 2008). Then the coast-bound sites would be remains of seasonal visits at times when game was abundant. This interpretation is of course a practical, non-speculative depiction based on economic theories. Sundström's (2004) suggestion which involves historicity (i.e. importance of and embeddedness in traditions) and frustrations as driving forces is supported by the occurrence of burial structures and burnt human bones on these sites. This offers a possibility to interpret the remains as two aspects of one and the same, probably hesitating, human group.

Australia between 1788 and 1850, something made possible despite older communication technologies and a partly hostile environment. Radiometric Funnelbeaker dates in the region overlap with dates of assemblages lacking remains from farming activities and some 'mixed' sites, furnished with some TRB objects and traces of structures with burned human bones. In central Sweden, a group of sites is interpreted as a regional cluster, based on similarity of material culture and ^{14}C-dates: four inland sites with strong evidence for farming, two river bank sites and six coastal sites, the two latter types with a Mesolithic settlement structure dominated by wild plant and animal remains (Fig. 9.3).

This combination of contemporaneous sites of different character offers a starting point for the exploration of the complex processes of transformation from hunter to farmer in Scandinavia and in much of northwestern Europe, where similar find situations are documented.

Archaeological investigations in Central Sweden support the interpretation of TRB as a culture which at the beginning exhibits a strong survival of the habits bound to a mobile lifestyle. Sundström (2003) has reconstructed a settlement pattern that consisted of farmsteads with strong similarities to one another in buildings and use of space. In the same region at the same time, there were scattered settlements with completely different use of space and material culture more akin to hunter-gatherers. A regional study in Denmark (Nielsen 1993; Sundström 2003, 152f) shows a similar pattern. The most important differences between the sites are illustrated in Fig. 9.3.

The tools and items imported to central Scandinavia during the 5th to 4th millennia have only vague connections to a supposed farming subsistence, but they have a history and an origin clearly related to the new economy elsewhere. Five 'new' types of artifacts appear in the region: fragments of polished flint axes, battle axes, flint blades, saddle-shaped quern stones and pottery (Knutsson and Knutsson 2001; Lidström Holmberg 2004; Hallgren *et al.* 1995). The changes in material culture seem to have been introduced in a Mesolithic cultural and spatial substrate. The changing site structure and material culture may be linked to a number of changes in social habits and behaviours. They may have been carried by migrating humans together with the new material culture and the domesticates, or could have occurred as responses to the changes in the everyday lives of these northern foragers and collectors/hunters, fishermen and gatherers. According to ethnographic sources, any of these scenarios is possible.

'The Mesolithic Substrate' and Subsequent Changes

The dominant types of Mesolithic sites in the region are small to large coastal settlements containing quartz and quartzite tools and production debris, waste from axe manufacture, and imported objects of southern as well as northern origins (microblades made of flint, volcanic rocks and occasionally of distant/non-regional varieties of quartzite). The ecofacts from these sites consist normally of small amounts of burned bones of fish, large game such as elk, deer, and seal, and small game like otter, beaver, hedgehog, hare, and occasionally fowl.

Adzes[2] made of raw materials like diorite, diabase and amphibolite are usually found on the sites. Among imported raw materials, flint is the most visible, although not the most usual. Flint, quartz and volcanic rocks, were knapped into small handle-cores to produce microblades or small blades and blade-like flakes. This type of technology is spread over large areas covering almost all of Scandinavia during the Late Mesolithic. According to use-wear analyses, the small tool kits seem to have consisted of scrapers, knives, planers and various composite tools. Traces from cutting plants were found on the straight edges of tools made of different raw materials. Several tool categories were found among the unaltered flakes, for example, corner knives used to tear and cut meat and rawhide, or wide short flakes with very large platforms and convex sharp distal edges used as knives (Fig. 9.4). These types of tools are also represented in the later assemblages dated to the transition time to Scandinavian Early Neolithic (Guinard and Groop 2008, 207f; Falkenström and Lindberg 2008, 227, 267; Knutsson *et al.* 2003).

Another raw material present is slate, which was cut and polished into knives, points and arrows, and sometimes into figures of animals or humans. This

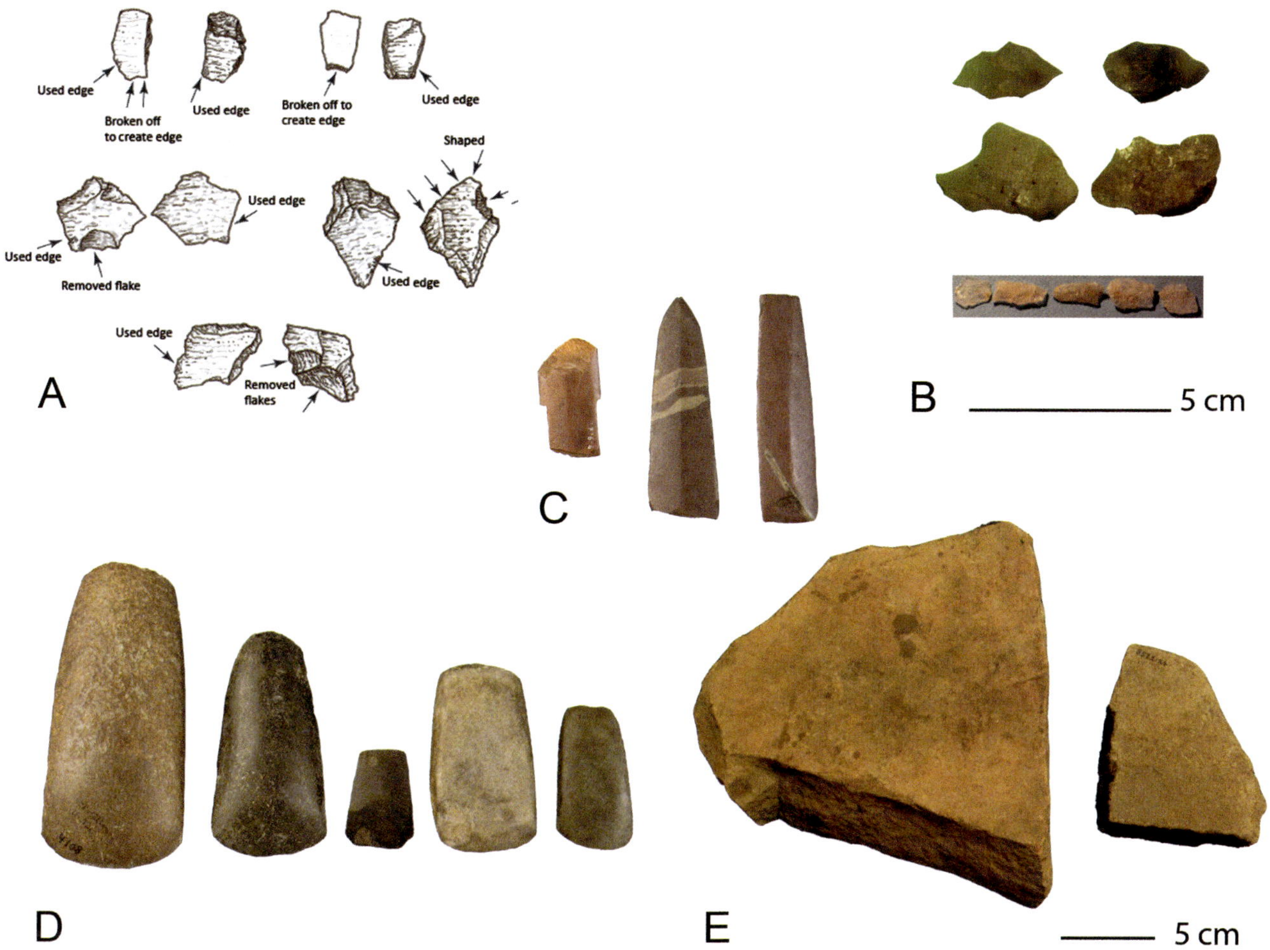

Fig. 9.4. Typical tools from Mesolithic sites in central Sweden: A: tools made of quartz flakes recognised by use-wear analysis, B: large wide flakes and series of inserts made of quartz, C: slate objects, D: axes of greenstone and E: fragments of grinding stones.

raw material is mostly confined to the mountain ranges of western and northern Scandinavia (Taffinder 1998), although the items made of it are found in Late Mesolithic and Funnelbeaker central Swedish sites, probably as objects of exchange. The assemblages also contain some broken and some complete grinding stones for axe manufacture.

In the Neolithic sites, simple quartz tool production continues parallel to the imported items. The Scandinavian Early Neolithic remains are usually deposited in the immediate vicinity of or mixed with Mesolithic finds. In fact, the Early Neolithic in these sites has often been recognised via domesticates or radiocarbon dates of burned seeds or hazelnut shells, and not via the 'new' material culture. A new feature is burning of stone and flint items, nodules and chunks. The actual physical remains from farming and stock-breeding activities are quite faint, but present. Burned seeds and bones of domesticates are clearly dated to the Early Neolithic. In the sites dated to that period, the imported amounts of flint increase but unevenly. From at least three sites in the region, large flint blades are found, produced by the same methods as for the southern Scandinavian Funnelbeaker blades (Fig. 9.5).

The flint axes from settlement sites are seldom complete, and when found they are short (less than 25 cm and resharpened; Fig. 9.5). Hoards of long (25–45 cm), unused, polished axes are also found in the region (Hallgren and Sundström 2008; Karsten 1994). For some reason the Early Neolithic peoples of central Sweden deposited unused axes of exceptional quality outside their farming settlements. Burned items of the same shape are found in the settlements. They used some of them as sources for production of flakes for other tools (Knutsson 1984; Sundström 2004). On several sites, blades and flakes with the remains of polish on their dorsal aspects are struck off the thin-butted

Fig. 9.5. A) thin-butted axes from central Sweden; B) polygonal axes; C) types of pottery; D) a quern; and E) flint blades from some central Swedish sites.

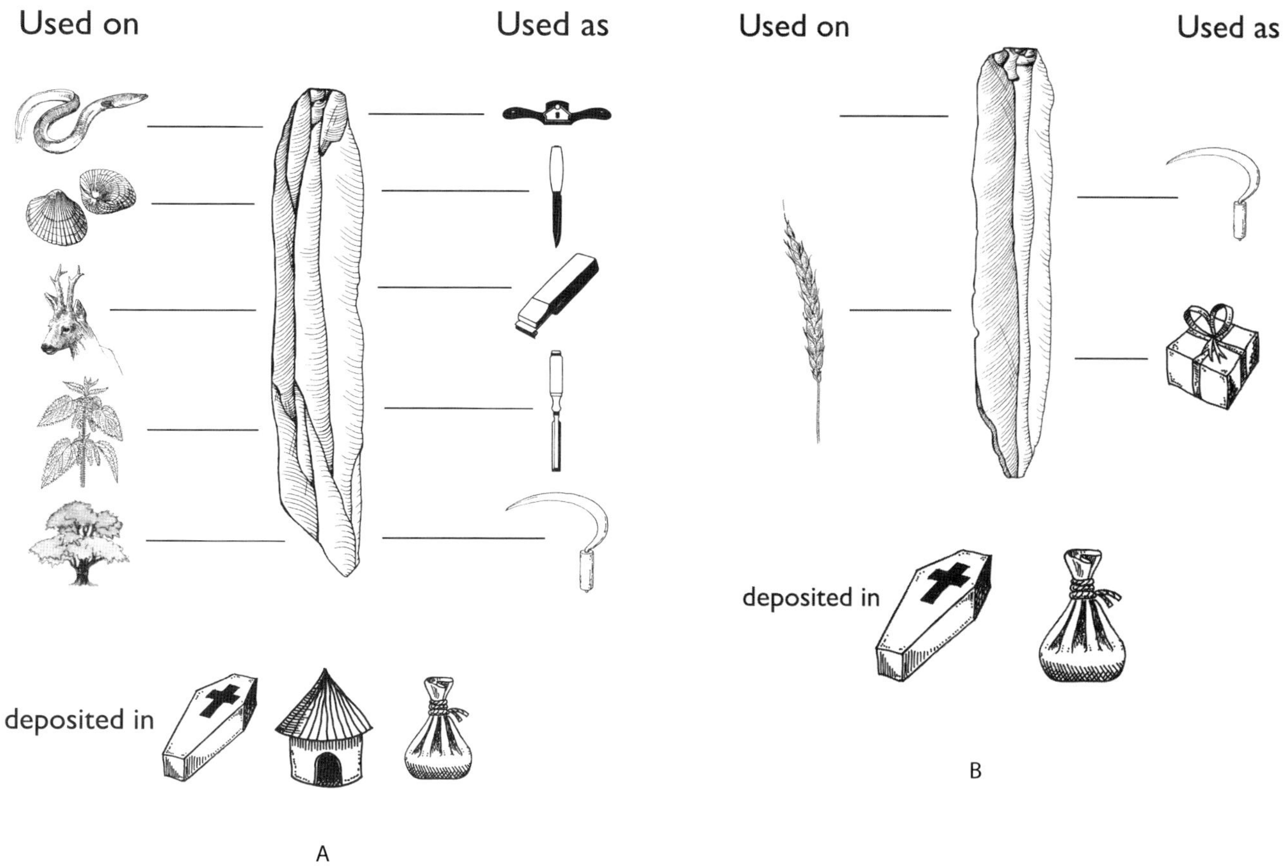

Fig. 9.6. A summary of uses of blades during the Mesolithic (A) and Neolithic (B) periods in Sweden, shown by use-wear analysis. During the Mesolithic, blades were used for many different tasks and deposited as personal possessions in graves, but also thrown away, or lost in settlement sites. In the Neolithic of Sweden, the use of blades was restricted to harvesting purposes. They were deposited mainly in graves, used, or even not used at all (i.e. newly produced). The symbols refer to contact materials: meat, wood, bone/antler and plants; tool types (as compared to modern tools): shaver/scarper, knife, etc.) and different social situations of use: as burial gifts, as everyday tools and as hoards or other purposeful depositions.

axes. In the flint-rich areas along the southern Swedish coast, these patterns are not so obvious, but burning of flint on the sites is as usual as in flint-poor central Sweden (Hallgren 2008, 233f; Apel *et al.* 1995, 113f; 1997, 7ff). This seems to be a general practice on Neolithic sites in Scandinavia and of course has been given ritual connotations. Polygonal battle-axes dated to the Early Neolithic found on the sites are usually fragmented, either intentionally or, accidentally, by use (Fig. 9.5). Outside of the settlements, not only preforms but also finished axes are found (*ca.* 200 from Sweden and Norway). 'Battle-axes' are a common form in European Neolithic contexts (Zapotocký 1992). Hallgren (2008, 215f) shows that at least two types belong to the Swedish 'region' of the Funnelbeaker culture. It is also possible to trace this type of regional artifact tradition within pottery production as well as within the spatial patterns of the polished adzes made of porphyrite, dolerite, amphibolite, etc. The import of Funnelbeaker tools from elsewhere is accompanied by production and use of regional varieties of common European types. The principles of treatment such as burning and use of some of these implements, as well as general outlines of the farming sites, seems to be similar over the whole area.

Blades were skillfully produced and distributed around Scandinavia from the Early Mesolithic. With the onset of the Early Neolithic, a change occurred in their production method and deposition patterns. The varied everyday use of the Mesolithic blades in multiple tasks has been documented and discussed by Juel Jensen (1986) and myself. During the Neolithic, the blades were only used as harvesting knives or sickles (Knutsson 1995; 1999; 2000; 2001; Juel Jensen 1994). The full extent of the

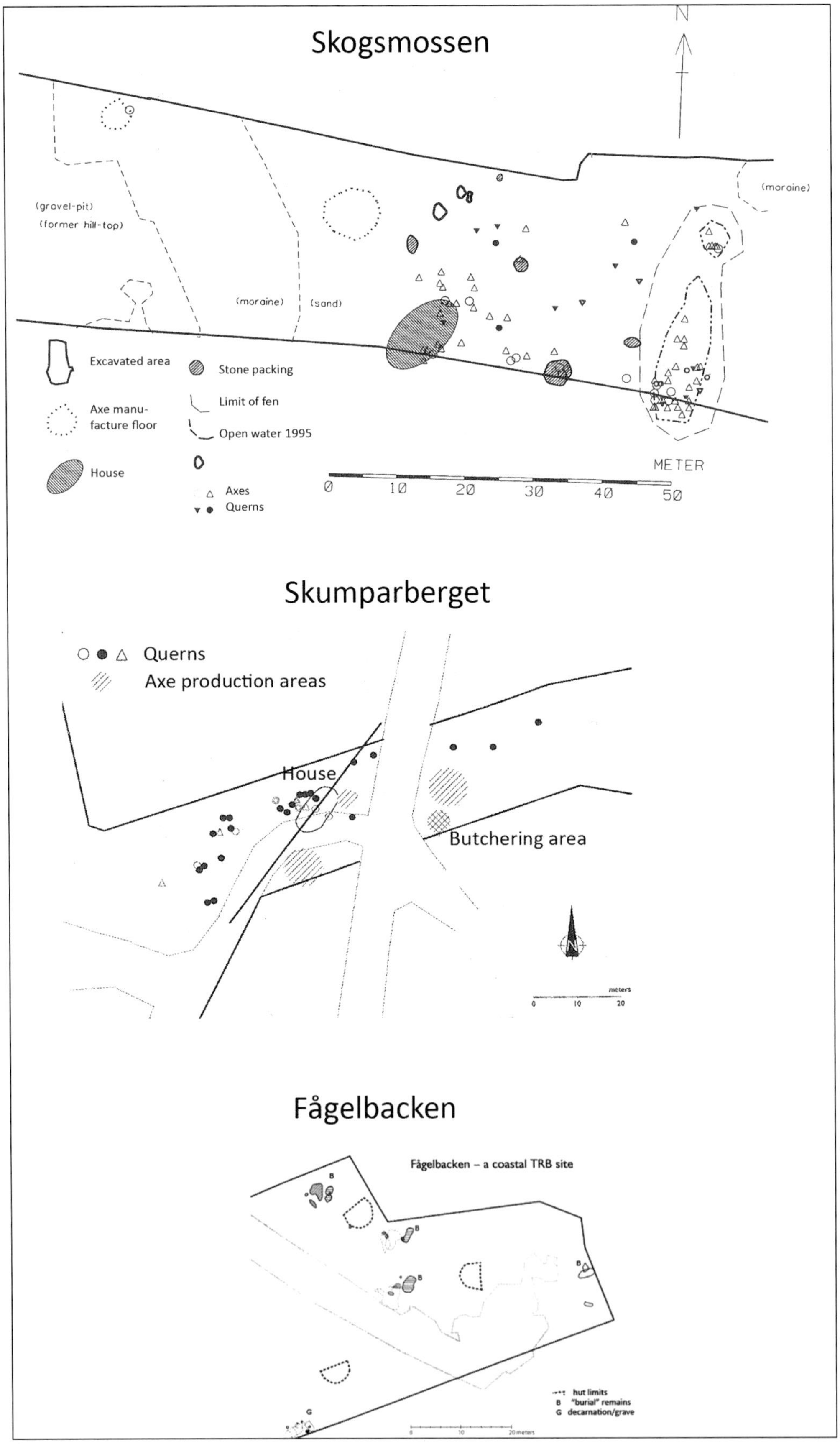

Fig. 9.7. Drawings of activity and artifact distributions in the three sites described of Skogsmossen, Skumparberget and Fågelbacken.

difference in use as well as in deposition is visible only in comparisons between the material from Mesolithic sites (burials and settlements in the same places) and the Late Middle Neolithic graves (no blades from contemporaneous settlements have yet been found). The differences between the two approaches to blades are summarised in Fig. 9.6. Early Neolithic blades have been found in small numbers in settlements. Their production seems to copy Neolithic continental methods or was actually carried out by continental (*i.e.* non-Scandinavian) knappers. The use-wear patterns have not been studied yet, but the depositions to some extent resemble or precede the depositions of the Middle Neolithic Corded Ware Culture (CWC) settings.

The Early Neolithic blades are found in large refuse pits together with large numbers of burned flint lumps, burned bones and pottery. There they are few in number, many of them complete and most of them unburned. Considering the scarcity of flint in central Scandinavia, this pattern is even stronger here. Stray finds of 264 blades have been made in the region. Some of them led archaeologists to finds of new sites, and from excavations there are now eight Early Neolithic Funnelbeaker sites known with one to ten blades, in some cases even secondary products of blades (scrapers, 'knives'). Blade production remains are not known from the region: a few cores have been found buried in bogs in southern Sweden. Although practical tools, one possible interpretation is that a necessary tool-kit was produced or imported for 'proper treatment' of the domestic species harvested in the first period of agricultural experiments in southern Scandinavia.

The patterns which we interpret as ritual seem to be the same over the whole area, even if some of the 'ritual' contexts differ between south and north. In central Sweden, the sites still show a relatively strong presence of material remains which can be connected to previous hunting-gathering lifestyles in the region.

Site Structures

As an illustration of the Early Neolithic variation in Funnelbeaker site organisation in central Sweden: three sites are presented here: Skogsmossen, Skumparberget and Fågelbacken.

Skogsmossen

The site of Skogsmossen (Fig. 9.7) has been interpreted by Hallgren (2008) as a ritually structured farm. The site contained faint remains of a house, three hearths and a cooking pit at 10–15 metres distance from the house. Axes were produced and left at the site and also on a spot 120 metres from it (possibly belonging to another settlement). Grinding stones seem to have been at least repaired at the site. A small (40 × 10 metres) fen was found *ca.* 30 metres from the house. The fen contained two concentrations of potsherds (90 kg), daub (53 kg), flint and porphyrite axes (more than 35 pieces), one polygonal axe in fragments, a slate knife, grinding stones (fragments of four saddle stones and 18 loaf stones), flint artefacts (scrapers, knives and arrow points), waste from stone tool production (flakes and cores of quartz, flakes of greenstones and flint) and a sandstone piece with ochre traces. Very few ecofacts were found in the bog, only a few bones in bad condition, and some cereal grains (13 whole and 16 fragments). Remains of two constructions were correlated with the distribution of the daub. The excavators interpreted the postholes as belonging to some kind of platforms from which the items were placed in the fen. Burned stones in the fen are suggested to have come from hearths placed on the probably clay-plastered platforms. The fen was interpreted as a place of sacrifice. The ritual behaviour on the site is not only represented by the bog findings, but also in the remains of the everyday behaviour in the settlement part. The main argument is the strict organisation of space into clearly delimited knapping areas, and a 'kitchen area' with finds of deliberately burned and/or crushed tools and pots (Hallgren *et al.* 1997). No remains of graves or human bones have been found.

Skumparberget

The site Skumparberget 2 (Fig. 9.7), *ca.* 10 km west of Skogsmossen, shows a similar setting. The site has a clear division of space, but no fen or other sacrificial area has been found, although this may have to do with the size of the excavation. The wattle and daub house was burnt, as were many of the artefacts found on the site. Large numbers of flint arefacts were recovered and the excavators suggested there was a small-scale production of transverse arrowheads on the site, as well as greenstone adze-knapping and polishing on a larger scale (Apel *et al.* 1997). A

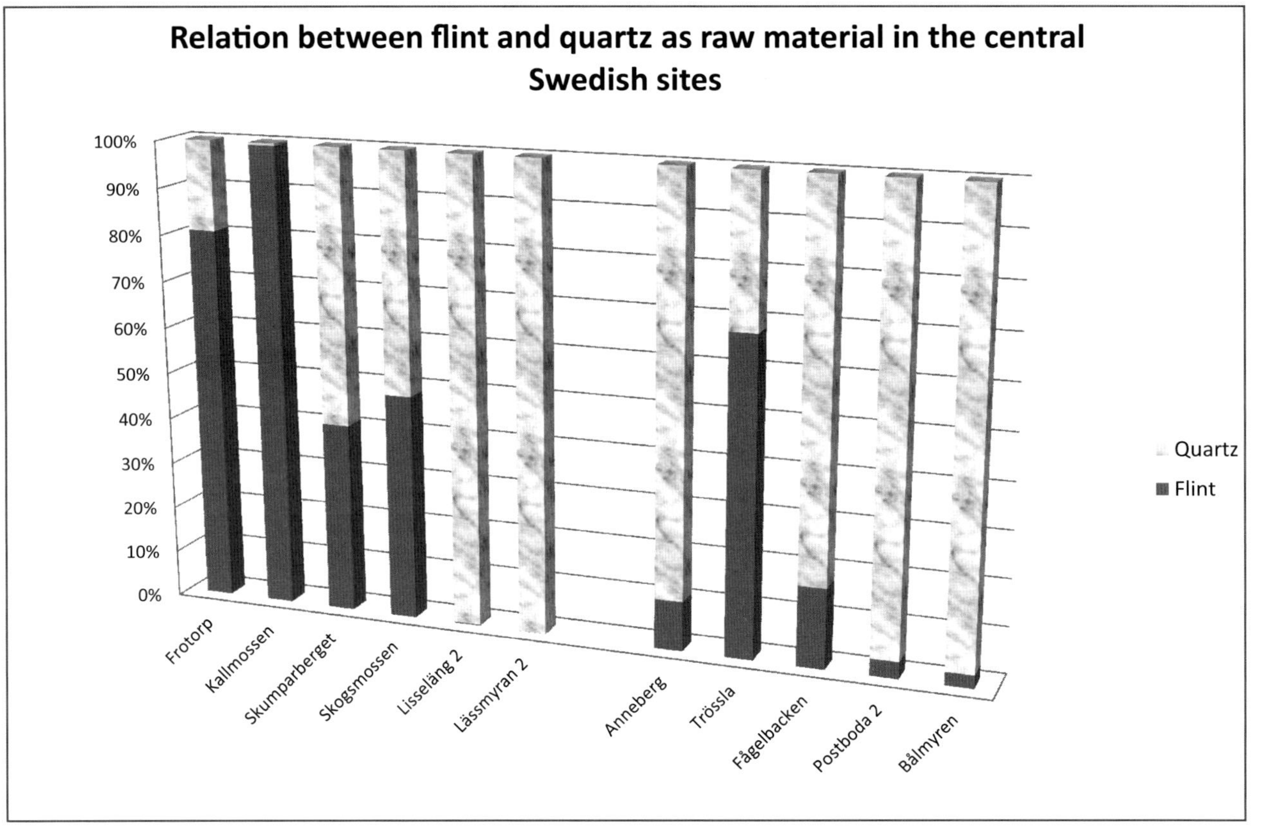

A

Site	Flint	Quartz
Frotorp	223	53
Kallmossen	219	1
Skumparberget	504	728
Skogsmossen	1 650	1 750
Lisseläng 2	8	3 477
Lässmyran 2	11	30 183
Anneberg	1 765	14 668
Trössla	36	17
Fågelbacken	173	834
Postboda 2	17	449
Bålmyren	11	375

B

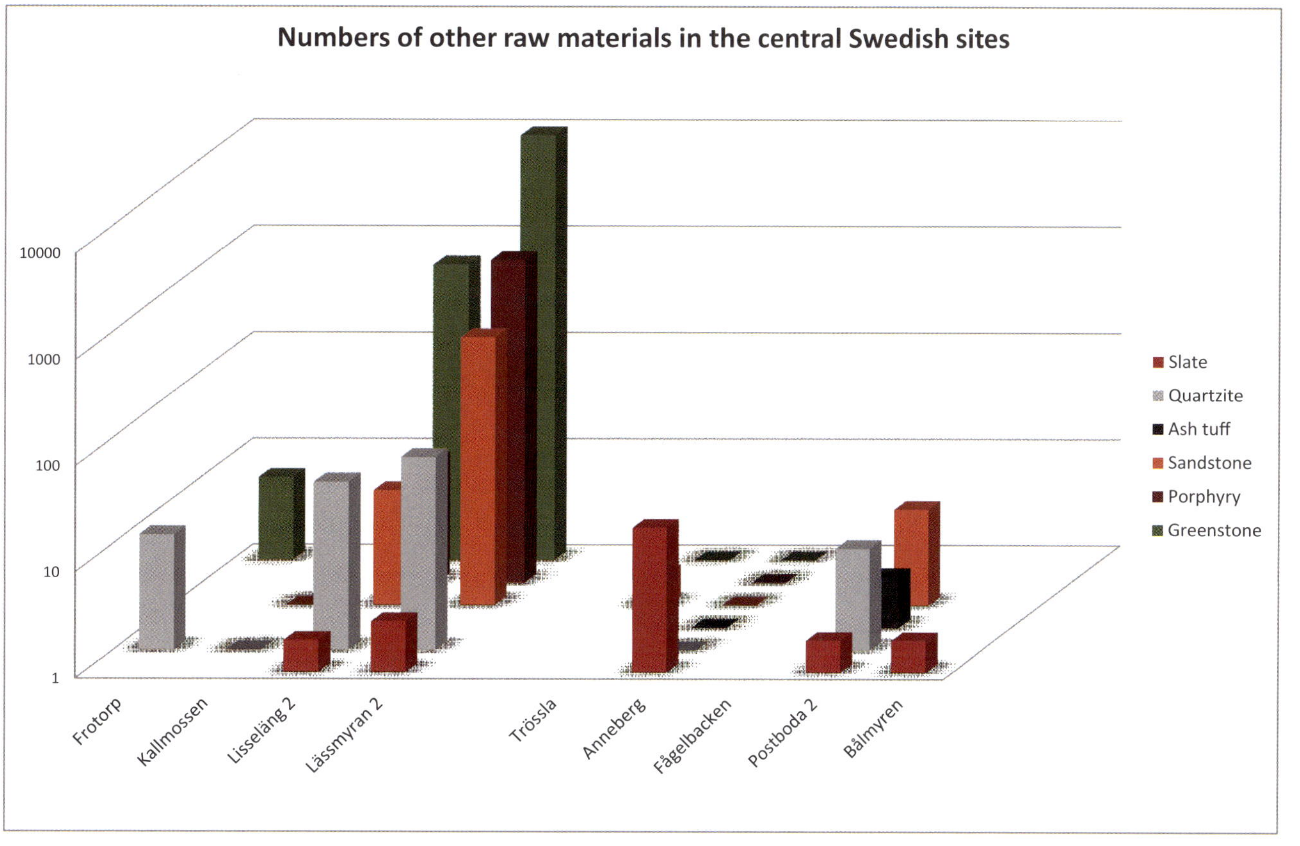

C

Site	Slate	Quartzite	Ash tuff	Sandstone	Porphyry	Greenstone
Frotorp		12				6
Kallmossen		1		1		
Lisseläng 2	2	38		12	14	601
Lässmyran 2	3	65		329	1,080	9,925
Trössla				2		1
Anneberg	23	1	1	1	1	1
Fågelbacken						
Postboda 2	2	9	3	8		
Bålmyren	2					

D

Fig. 9.8. A: The finding sites for the materials indicated in the diagram, split into inland sites ('introvert'; bars to the left) which were situated away from the Neolithic coast, and coastal sites (bars to the right). There is a slight overrepresentation of sites with flint in the inland sites. To see the proportions within the smaller materials, the two largest numbers have been limited to 5000 pieces. Nevertheless, they are presented in a logarithmic scale. Although larger numbers of flint have been found in the inland sites, the materials are so different in sizes that it is not possible to present a valid tendency. B: The actual find numbers behind the diagram. C: In addition to these two raw material categories, other imports as well as local stones are found in the sites. The axes were produced in a number of volcanic or metamorphous rocks collectively named green-stone. Knives and points were produced of different varieties of slate, small numbers of microblade flakes of flint-like raw materials of different colors have reached the places. The largest stone products were the grinding stones and the querns. They were made of sandstones and coarse grained erratic or metamorphous rocks. Pounders and rounded knapping stones of granite, porphyry were usually chosen from glacifluvial remains or from the beach material. D: Find numbers to Fig. 9.8 C.

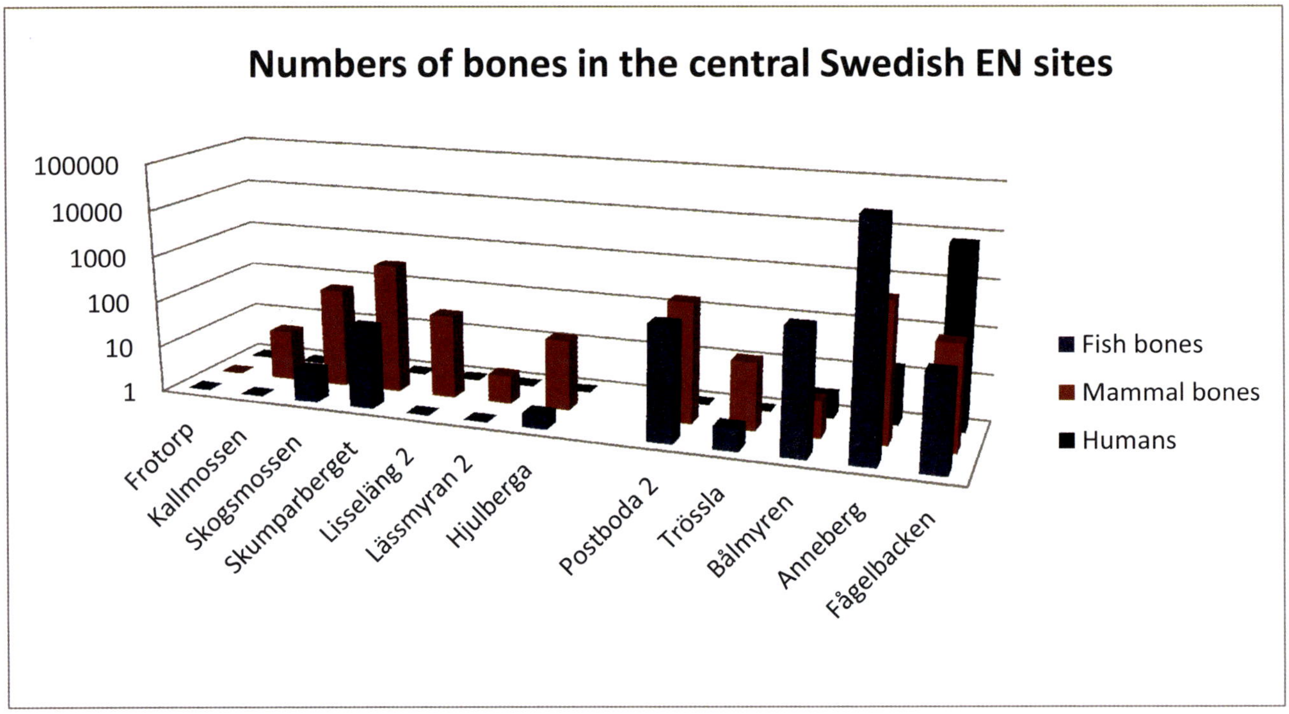

A

B

Site	Fish bones	Mammal bones	Humans
Frotorp			
Kallmossen	1	12	
Skogsmossen	5	134	
Skumparberget	57	601	
Lisseläng 2	1	62	
Lässmyran 2		4	
Hjulberga	2	32	
Postboda 2	272	368	1
Trössla	3	28	
Bålmyren	437	6	3
Anneberg	74 446	909	15
Fågelbacken	104	168	7 401

Fig. 9.9. A: The real proportions of the numbers of bones in the sites are illustrated in the diagram. The enormous sample of fish bones in the site of Anneberg was possible due to exceptional preservation circumstances and a meticulous use of small-meshed sieves in a scientific project. The bone materials are also quite varying in numbers, depending on predepositional treatment, postdepositional and other taphonomic (how plant and animal remains accumulate and differentially preserve within archaeological sites) processes, the excavation strategies and the fragmentation of the material. The category wild/domesticated is an illustration to this, as it comprises pig bones, which are inseparable and those large game bones which were not possible to separate into categories such as cow, deer and elk. Many of the places presented also contained material from the Bronze Age and Iron Age, and some of the domesticated bone remains were dated to these periods. The association with the Stone Age is often stratigraphical and consequently sometimes uncertain. To be able to see the relation between the mammal bones, fish remains and human bones, the numbers are presented in a logarithmic scale. The sites are divided into inland sites (bars to the left) and coastal sites (bars to the right). Human bones were found only on three sites situated at the coast. B: The data behind Fig. 9.9A.

comparison between the sites of Skogsmossen and Skumparberget offers an illustration of similarities and differences in structure in the archaeological materials hitherto uncovered. Since then, new, similarly structured sites with ample evidence for domesticates have been found, but all with slightly different organisation of space (Fig. 9.3).

Fågelbacken

A completely different pattern was found in the contemporaneous site Fågelbacken, situated on the Stone Age coast *ca.* 70 km northeast of the former sites (Fig. 9.2). Instead of houses, D-shaped huts of *ca.* 6 × 4 metres were identified by the excavators (Fig. 9.7). Hearths found on the site are not connected to the huts. At several places within the site, burnt human bones were found in different structures, among others in the supportive rectangular foundations for a number of large posts surrounded by thin sticks, and in other pits. The amount of human bones on the site was unusually large, considering that they were burnt (5.53 kg, *ca.* 95% of all bones). They were found together with animal bones of both wild and domesticated species and 'secondary burned' potsherds of Funnelbeaker type. A partially destroyed rectangular structure 6m × 6 m was excavated at the site with remains of more than 60 Funnelbeaker vessels (some having large parts preserved), burnt animal bones, a greenstone axe, a quern, grinding stone fragments and fragments of quartz and flint (Apel *et al.* 1995, 82f). The structure has been interpreted as a timbered equivalent to a megalithic grave (Hallgren 2008, 107f) or as a charnel or bone house (Sundström 2003, 130) Its connection to the Funnelbeaker settlement is demonstrated by the ^{14}C-dates and the pottery styles. The site as a whole has been interpreted as a ritual burial site (Hallgren *ibid.*) and as a social gathering site (Sundström *ibid.*) for the Early Neolithic groups on the farmsteads in the inland settings. The burnt animal bones on this Funnelbeaker site were all from wild species.

Discussion

Before considering Funnelbeaker ritual behavior based on this evidence, an analysis of prehistoric farming needs to consider problems connected to farming practice. Many practical things had to function if people were to profit from such a manipulation of nature. Diseases threatening both the domesticates and people had to be prevented or treated, climate had to be predicted from sowing to harvest, and animals protected until the next generation of stock was secured. The seeds had to be preserved without being used or destroyed by moisture, rats, birds, insects, or neighbours. When farming and stock-breeding moved to another part of the continent as was the case here, they had to be adjusted to this new context, which may have involved new types of threats and obstacles. For example, salt, not available everywhere, was a necessary food complement for stock once breeding started in Europe (Chapman *et al.* 2001; Chapman and Gaydarska 2003).

Ethno-historical and ethnographic reports in the 18th and 19th century interpreted many practical behaviours in 'primitive' societies as ritual and ceremonial. Not all did so, and a number of 19th-century ethnographies (like those written about Mbuti by Schebesta and Turnbull) describe the practical and down-to-earth thinking and doings of the people involved (Knutsson 1995; 2012). But the former have influenced archaeologists to interpret many traits in excavated materials, not least those of Mesolithic and Neolithic origin, in ritual terms. Anomalies in the excavated sites are often thought to represent ritual features and behaviours. Examples of this can be read for example in Bradley (2005 and literature cited). Ceremonies and rituals were part of prehistoric peoples' lives (as they are in modern society), but I mean that they were formed to support the pragmatic aspects of everyday doings. A person endowed with so-called common sense would never rely solely on God while trying to survive. Only elites with a supporting organisation that tends to their survival can interpret reality in such strange terms. The thirty to fifty generations of people who during the Early Neolithic came into contact with agriculture once it was available in the form of stories, seeds and animals, had to retain a more practical attitude towards life, to be able to gain access to this new economy, as also shown in the exhaustive report on the large Funnelbeaker site Almhov in Scania, southern Sweden (Rudebeck 2010, 117f).

Consequently, for reasons of survival, farming needed to be well-tested before it could be an accepted subsistence practice. Otherwise it could create quite a shaky basis for making a living.

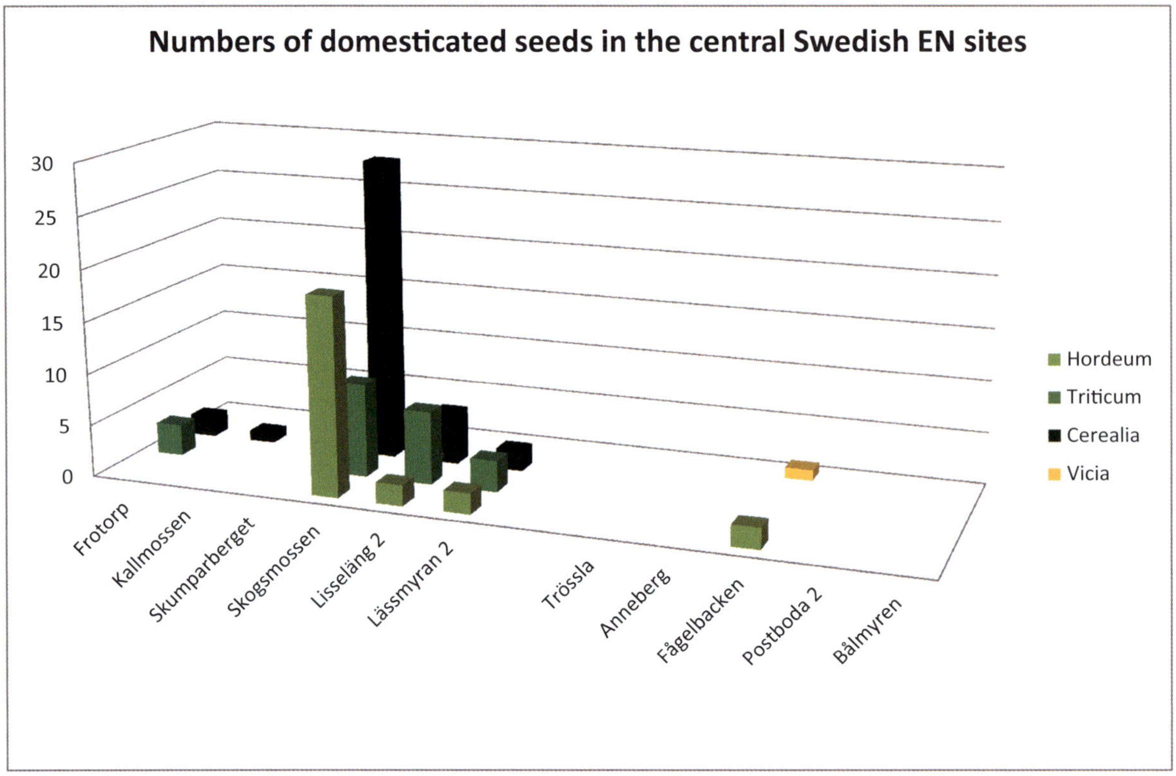

A

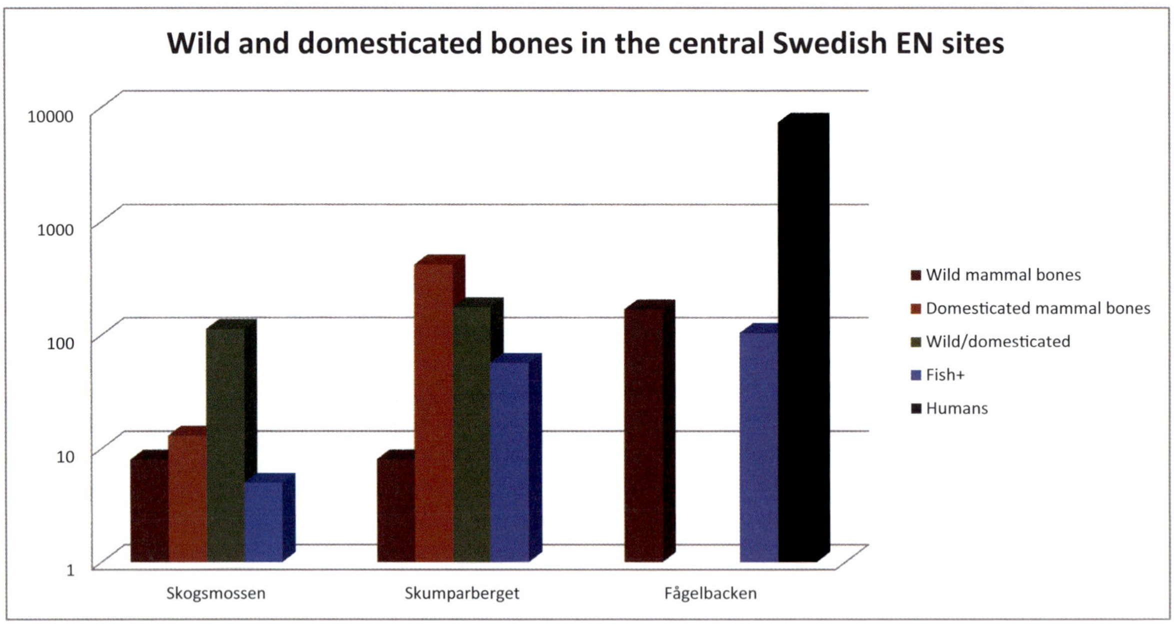

B

Fig. 9.10. A: The seed remains from cereals were very few (scale 1–50 items), and a difference between the coastal and the inland sites is quite clear. On the site of Skumparberget some imprints from cereal seeds in the potsherds are mentioned, but no analysis of plant macrofossil remains was done because of a budgetary shortage. From the site of Fågelbacken only two seeds are reported. These proportions (together with proportions in wild/domesticated bones and burial remains) form the base for the interpretations of the sites as different economies, niches, or social practices. The differences are further completed by different site structures. B and C: Occurrence of human bones in the Early Neolithic sites is a negative reflection of the occurrence of seeds.

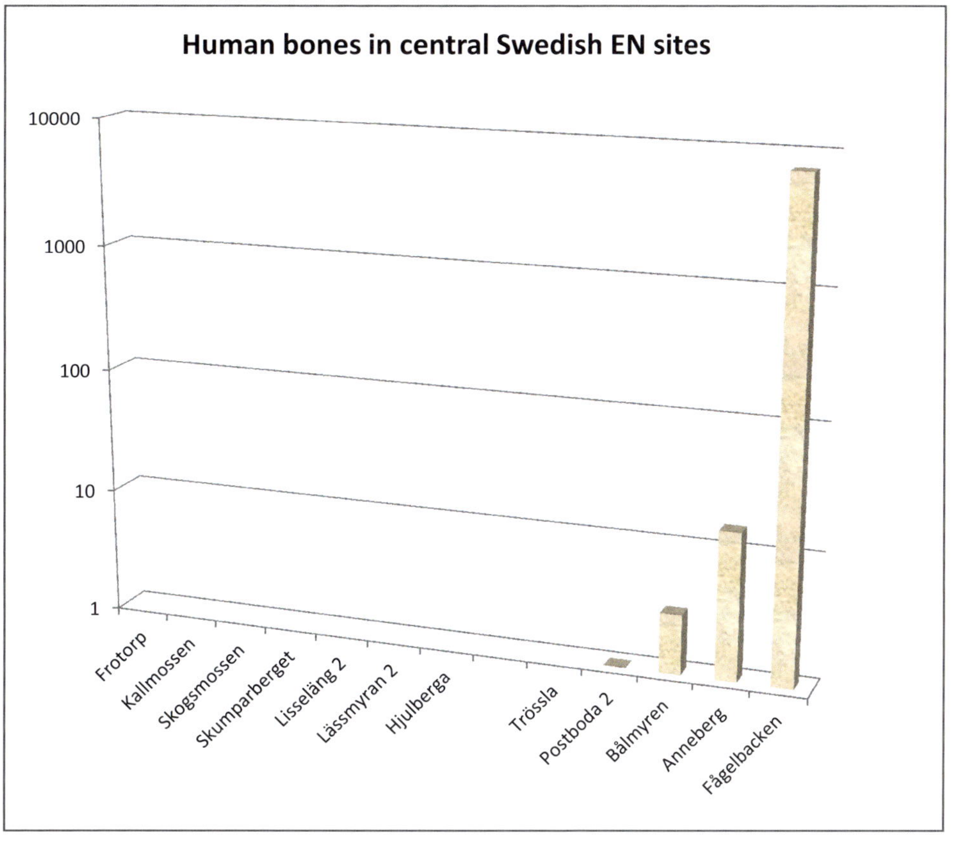

C

Due to such circumstances, it is understandable that the process of transition must have involved some, or even a lot of hesitation by the hunter-gatherers. One way to confront such hesitation and maintain interest for the idea from those who craved farming life for reasons of enthusiasm, lust for power or for economic gain, was to try to sell the idea together with different useful manuals for treatment and handling of the system. These manuals could consist of systems of commemorative ceremonies (involving relationships with different ghosts/gods, which was impractical, as only in 'lucky' moments could the receivers see any results), farmers' almanacs for seasonal needs, rules to keep the domesticates alive and practical sets of tools and containers for storage. The behaviour or operational sequences had to be remembered. In due course, they were adapted to local conditions and changed according to them, and the historical traditions of the particular group of people involved. In Scandinavia, the long period of such hesitation is visible in the archaeological material. Hunting and gathering remained important for a long period and were from time to time even the most important subsistence system used throughout most of the Funnelbeaker period.

Ritual Remains and Ritual Actions – A Summary

During the period of introduction of domesticated species into the central belt of Scandinavia, new material culture was added to an original Mesolithic set of tools and utensils. Remains of these new tool-kits are found in several types of settings, sometimes together with domesticates, sometimes together with remains of hunting, fishing and gathering. Finds of broken and burned flint axes and crushed and burned flint chunks and nodules are more numerous in south Scandinavia, but small echoes of such handling are found in the central region as well. Another deliberately broken category concerns the querns (and grinding stones used for polishing of tools). In some sites, the pots were burned a second time, after they had been used, together with the bodies of dead humans and animals, and the remains put in pits on the sites. Pottery is normally found crushed, but almost complete vessels have been found in some sites. The interpretation must be that at least some were deposited or *left behind* whole.

Based on ethnographical observations, it can be proposed that the introduction of farming practices could be a tricky business and that some kind of theory and practice supporting predictions about results, considerations of environmental suitability, and rules for the preservation of the harvest and survival of animals must have been important. I suggest that the new material culture must be seen not only as valuable exchange goods. It might have served as a memory-enforcing kit, a kind of physical manual with attached narratives, which ensured the success of the enterprise. The uniformity of treatment of the utensils and the tools over the entire Funnelbeaker region, together with traces of re-organisations of *some* of the sites, gives support to such an assumption. It is possible that a lot of the remains from agricultural settings from later times are the result of strictly defined and rigid rituals as well, where the performers had little knowledge of their initial meanings. I doubt that this was the case in the earliest period of farming for two reasons. The first reason is that this ritualised kit was introduced to groups which were completely aware of the relation between the outcome of work and survival. The second is that the introduction phase must be seen as an experimental phase, comparable to modern science. All results in the new environment were impossible to predict, as well as the preservation of the breeding pools. In this light, it would be better to take a higher opinion of the capacity and understanding of the earliest agricultural experimenters and understand their material remains rather as mnemotechnical resources or even as fashion. Burning of goods and remains could also be seen in a light other than that of religious-ritual. Staying in one place, perhaps together with animals and preserved seeds, required quite different hygiene measures than a mobile way of life (where risks of diseases were eliminated by simply moving on). Pottery remains have already been suggested to have served as by-products of imports for festive activities (Sherratt 1991; Sundström 2004; Fischer 2002). There are remains of systematic behaviour which we, for lack of a better explanation or perhaps slightly coloured by remains of social-evolutionist views, interpret as ritual. If we need to use such a category as ritual, it is important to acknowledge it as a classificatory term, and not as an interpretation. Such interpretation possibly does not allow for a fair depiction of the earliest farmers.

If we look at the earliest farming from this point of view, it seems necessary to enlarge our knowledge of how the first farmers, in all the corners of the world where farming slowly became the way of life, managed all the obstacles and endured, developed and improved the techniques and processes until they could depend on production of food from entirely *non-endemic* species. All later catastrophes and pests show that this was not an easy task. Even today, on local levels, food production sometimes fails, and still today, even in Sweden, complements to food production are gathered with great lust from the 'wild'.

9.3. THE RITUALISATION OF AGRICULTURAL TOOLS DURING THE NEOLITHIC AND THE EARLY BRONZE AGE

Annelou van Gijn

Introduction

Ritual behaviour is usually highly structured and repetitive and therefore results in archaeologically recognisable patterns. However, even though the detection of such patterns may be possible, attributing meaning to them is often impossible (Bradley 2005). Some clues about the possible ritualisation of objects can be obtained if we look at their entire life history, not just their use or their find location. In this endeavor, the concept of cultural biography (Kopytoff 1986) has turned out to be instrumental. Objects, because of their place in daily domestic life, not only symbolise the social configuration, but also structure and reinforce the interactions of prehistoric agents. In order to understand this role we need to study the entire life-cycle of an object: from the selection of raw material, through the manufacture, the use and the final discard or deposition and the treatments it may have undergone during this life cycle (Van Gijn 2010).

Such an approach, including detailed microscopic use-wear and residue analyses, has shown that typical agricultural tools from various periods of the Dutch Neolithic and Bronze Age display a special cultural biography, one that is not easily explained from a strictly functionalistic or utilitarian perspective. These agricultural tools, which had a distinct utilitarian role in domestic daily life, changed their 'meaning' at the end of their use life and underwent a special treatment. This ritualisation of utilitarian objects will be illustrated with four examples deriving from Early, Middle and Late Neolithic contexts, as well as from the Bronze Age. The examples show, I would argue, that cultivation was surrounded by ritual throughout this period.

'Killing' the Querns: An Example from the Dutch Early Neolithic Linearbandkeramik Culture

The first example comes from the farmers of the Linearbandkeramik (LBK) culture (5300–4900 cal BCE). Around 5300 cal BCE, LBK farmers settled on the loess zones of the very southeastern corner of the Netherlands. They cleared fields in the woods on the plateaus, probably by means of the stone adzes that are frequently found. Other agricultural tools retrieved from LBK contexts include sickle blades and querns. The sickle blades probably formed part of composite sickles akin to the famous Karanovo type (see Chapter 10, Fig. 10.4 and Fig. 10.5). Querns, adzes and sickle blades were found in graves such as those in the cemetery of Elsloo (Modderman 1970; Van de Velde 1979). I would argue that it is highly relevant that these three types of tools, all involved in cultivation, were considered appropriate burial goods. This indicates the close relationship between death and agricultural practices shown again and again in prehistory (see for examples Bradley 2005). Another indication for the special significance of querns involves the hoards of such items found in Belgium and northern France (Jadin 2003; Hamon 2008). More surprising is the finding that also the querns found in settlement context display evidence for a special biography. This will be illustrated with an example from the settlement of Geleen-Janskamperveld (Verbaas and Van Gijn 2007).

These querns were made of sandstone cobbles of medium-grain size that could be easily collected in the nearby riverbeds of the river Meuse. These cobbles were shaped by means of hard hammer percussion after which the grinding surface itself was pecked. This pecking was probably done by means of conically shaped exhausted flint cores, many of which have been retrieved from Linearbandkeramik settlement sites (see Chapter 7, Fig. 7.8). The resulting querns had an elongated shape and were quite flat. Most of the querns display evidence of having been used to process cereals (Verbaas and Van Gijn 2007). During their use life, the active surface was recurrently refurbished by means of pecking, resulting in an increasingly concave shape (see Chapter 7, Fig. 7.15).

The querns found in settlements are for the most part broken. Remarkably enough however, the querns were rarely broken in the middle only, at the place which would be thinnest through recurrent resharpening. Instead, many querns were fragmented in several pieces, showing up to five fracture planes (Fig. 9.11A). These fractures were 3–4 cm thick, too thick to have been the result of accidental breakage during refurbishing. These observations led us to propose that these querns were fractured intentionally (Verbaas and Van Gijn 2007). It is remarkable that none of the quern fragments could be refitted, something one would have expected if these fractures had occurred accidentally. Instead, they seem to have been dispersed, most likely outside of the settlement. Geleen-Janskamperveld has been excavated almost completely, so we can only conclude that these quern fragments were taken out of the settlement and deposited elsewhere. Another feature pointing to the ritualisation of these querns is that the fractures of the broken querns also display traces of a red residue. This residue was analysed by means of X-ray diffraction analysis and was shown to be ochre (Fig. 9.11B). Although it was long believed that these querns may have been employed to grind the haematite nodules that were also retrieved from the Linearbandkeramik sites, there is no evidence for them having been used as such, as use-wear traces from this activity have not been found. Instead, it is likely that the querns were, before being completely exhausted and unsuitable for further use, taken out of circulation and intentionally destroyed and dispersed before being deposited. Their fractures were rubbed with ochre, a material repeatedly associated with ritual behaviour in past and present (McBryde 1997; Taçon 2004; Cristiani, Pedrotti *et al.* 2009).

Destroying Sickles: The Case of the Middle Neolithic Hazendonk Sites in the Wetlands

The next example comes from the Middle Neolithic Hazendonk group (3750–3400 cal BCE), from the wetlands of the Rhine/Meuse delta where the transition from hunting-fishing-gathering to agriculture took well over a thousand years (Louwe Kooijmans 2007). The wetland sites of the Hazendonk group have provided evidence for cereal growing and harvesting, despite the fact that their location on relatively small dunes surrounded by marshland cannot be considered ideal for crop growing (Kubiak-Martens 2006). The fields were likely to have been small, probably not providing sufficiently for the emmer and barley to have contributed in bulk to the subsistence economy. It conerned minor crops, possibly intended more as special food, to be consumed on festive occasions (Fisher 2002).

A

B

Fig. 9.11. A) Broken quern from the Linearbandkeramik settlement of Geleen-Janskamperveld displaying multiple fragments; B) ochre on the broken side. Image: Laboratory for Artefact Studies, Leiden University.

Fig. 9.12. The Dutch archaeological sites of 1) Heiloo; 2) Schipluiden; 3) Wateringen; 4) Ypenburg; and 5) Geleen-Janskamperveld Map: R. Lugon, J.-C. Loubier and A. Chevalier.

The three Hazendonk sites that were studied for the presence of traces of wear and residue include Ypenburg, Schipluiden and Wateringen 4 (see Fig. 9.12; Raemaekers, Bakels *et al.* 1997; Van Gijn, Van Betuw *et al.* 2006; Van Gijn and Verbaas 2008). All have produced flint sickles for harvesting crops, albeit in small numbers. Interestingly, at Schipluiden and Ypenburg, these sickles bear evidence for a special cultural biography. First of all, the sickles were made of high quality Belgian flint. The lack of associated production waste in the coastal sites indicates they were imported as finished products. Their morphology is closely related to that displayed by Michelsberg flint assemblages from the Rhineland and Belgium. They were used for harvesting the cereals, probably considered to be a special activity as cereals were a 'novelty' in the western wetlands in this period. After use, their functional edge was destroyed by flaking. Signs of burning are visible on the destroyed edge, suggesting that the sickles were subsequently burned, after which they were deposited or left behind across the settlement (Fig. 9.13). On some sickles, a red residue was found that may be ochre (Van Gijn 2010, 71–74). All in all, it is very clear that these sickles were not just any flint tool, but had a very particular cultural biography.

Placing Agricultural Implements with the Dead: The Case of the Funnelbeaker Megaliths

The Funnelbeaker (TRB) culture in the Netherlands (3400–2900 cal BCE) has produced another example of the ritualisation of agricultural implements. During the TRB period, megaliths were constructed which served as communal burial places for local kinship groups. Flint objects abound in these burial chambers, largely consisting of unretouched material, cores and core fragments. In addition, a selection of domestic implements were deposited, all displaying intensive traces of use. These implements

Fig. 9.13. 'Destroyed' sickle from the Middle Neolithic site of Ypenburg, displaying signs of burning and fragmentation. Image: Q. Bourgeois, Leiden (from Van Gijn 2010, fig. 7.3a).

include polished axes, sickle blades and strike-a-lights (Figs. 9.14A and B). The worn out axes, which also display hafting traces, were used for a very long period and displayed several resharpening phases (Fig. 9.14C). Prior to deposition in the burial chambers, they were resharpened, obliterating most of the traces of past use. However, in the micro-scars, traces of polish from chopping wood are still visible (Van Gijn 2010, Fig. 6.8). It seems the axes were given a 'new life' in order to accompany the dead. The sickle blades display a typical 'cereal polish', with shallow striations indicating the direction of motion. The blades were hafted parallel to the handle, similarly to the way sickle blades were hafted in the Funnelbeaker north group (Juel Jensen 1994). The strike-a-lights are elongated pieces of flint with a sturdy point. Their use as strike-a-light is demonstrated by the fact that the point is very rounded and displays a multitude of stacked impact scars and linearly distributed patches of matt and rough polish (Fig. 9.14A).

What is so special about these observations of the Funnelbeaker funerary outfit? First, it should be stressed that used and worn out axes, sickle blades and strike-a-lights are hardly ever found in contemporaneous settlement assemblages. Moreover, they are heavily used implements which have one thing in common: they are all related to agricultural activities. The axes were used to chop wood, clearing the forest for laying out the fields. The strike-a-lights ignited the fires which burnt down the undergrowth, as we know the TRB people practised slash-and-burn agriculture (Bakels and Zeiler 2005). Last, the sickles served to reap the cereal stems. Agriculture was most likely a communal affair and the deposition of agricultural tools in the communal burial chamber indicates that these items were closely tied in with the continued existence of the local group (Van Gijn 2010, 135–136).

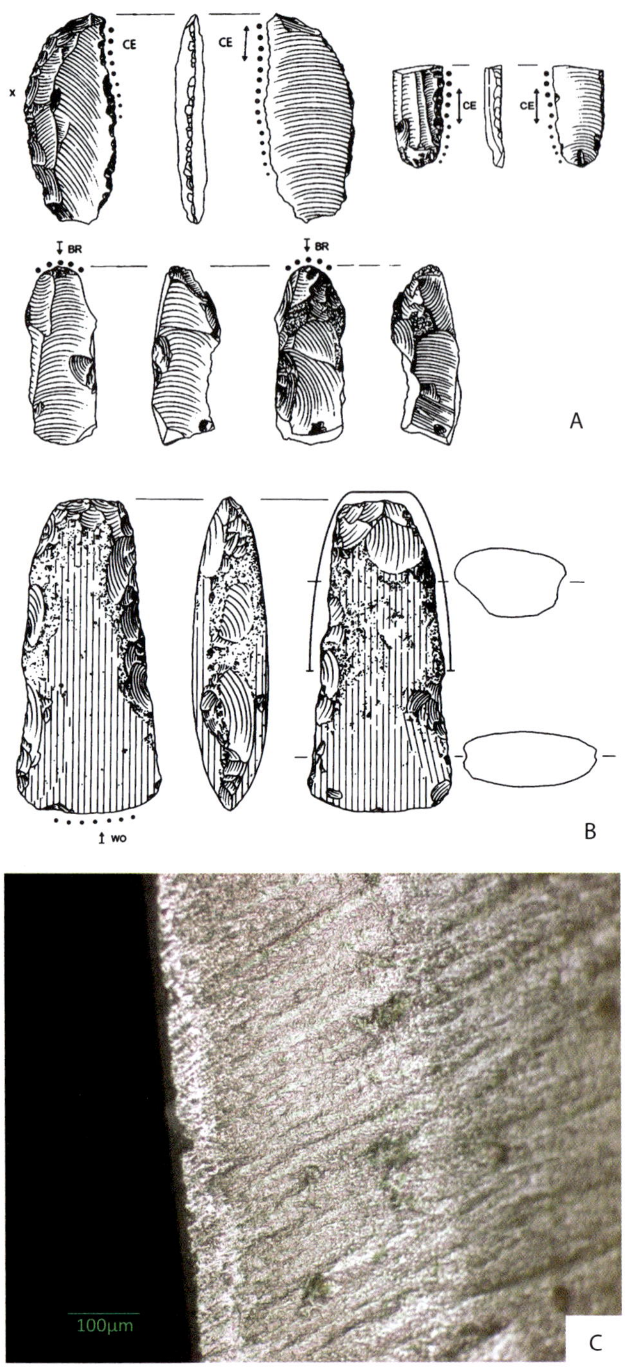

Fig. 9.14. A: Sickle blades and strike-a-light from the megalith of Mander (scale: 1:2). Drawing: R. Timmermans, Leiden (from Van Gijn 2010, figs. 4.7 and 6.7). B: Heavily worn axe from megalith D19 in Drouwen (scale: 2:3). Drawing: R. Timmermans. C: Resharpening traces observed in B (Photograph: Laboratory for Artefacts Studies, Leiden University (from Van Gijn 2010, fig. 6.8).

Offering Crescent-Shaped Sickles in the Late Bronze Age?

By the Late Bronze Age, flint tools had largely given way to metal counterparts (Van Gijn 2008; 2010). Nevertheless, during this time skilled flintknappers in southern Scandinavia and northern Germany mass-produced bifacially worked crescent-shaped tools of exceptional craftsmanship usually referred to as sickles (Fig. 9.15). These objects, made of fine-grained Scandinavian flint, were imported into the northern parts of the present-day Netherlands, especially West Friesland (Fig. 9.16). They were

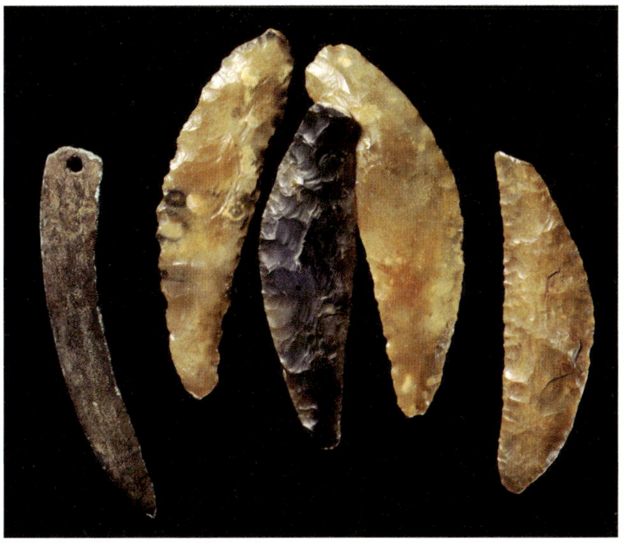

Fig. 9.15. Crescent-shaped 'sickles' and a bronze sickle from the hoard of Heiloo, West-Friesland. Image: Dutch National Museum of Antiquities, Leiden.

brought as finished products, because no production waste has been encountered in contemporary settlement sites in the northern Netherlands and high-quality flint suitable for the production of these implements was not available. Moreover, a large hoard of more than forty 'sickles' was found in Trendelbusch in Oldenburg, supporting the idea that they were trade objects (Jacob-Friesen 1955, 86–88).

Use-wear analysis of Danish specimens has shown them indeed to have been used as harvesting implements, supporting their classification as 'sickles' and indicating that they were intended to be used as such in their land of origin (Juel Jensen

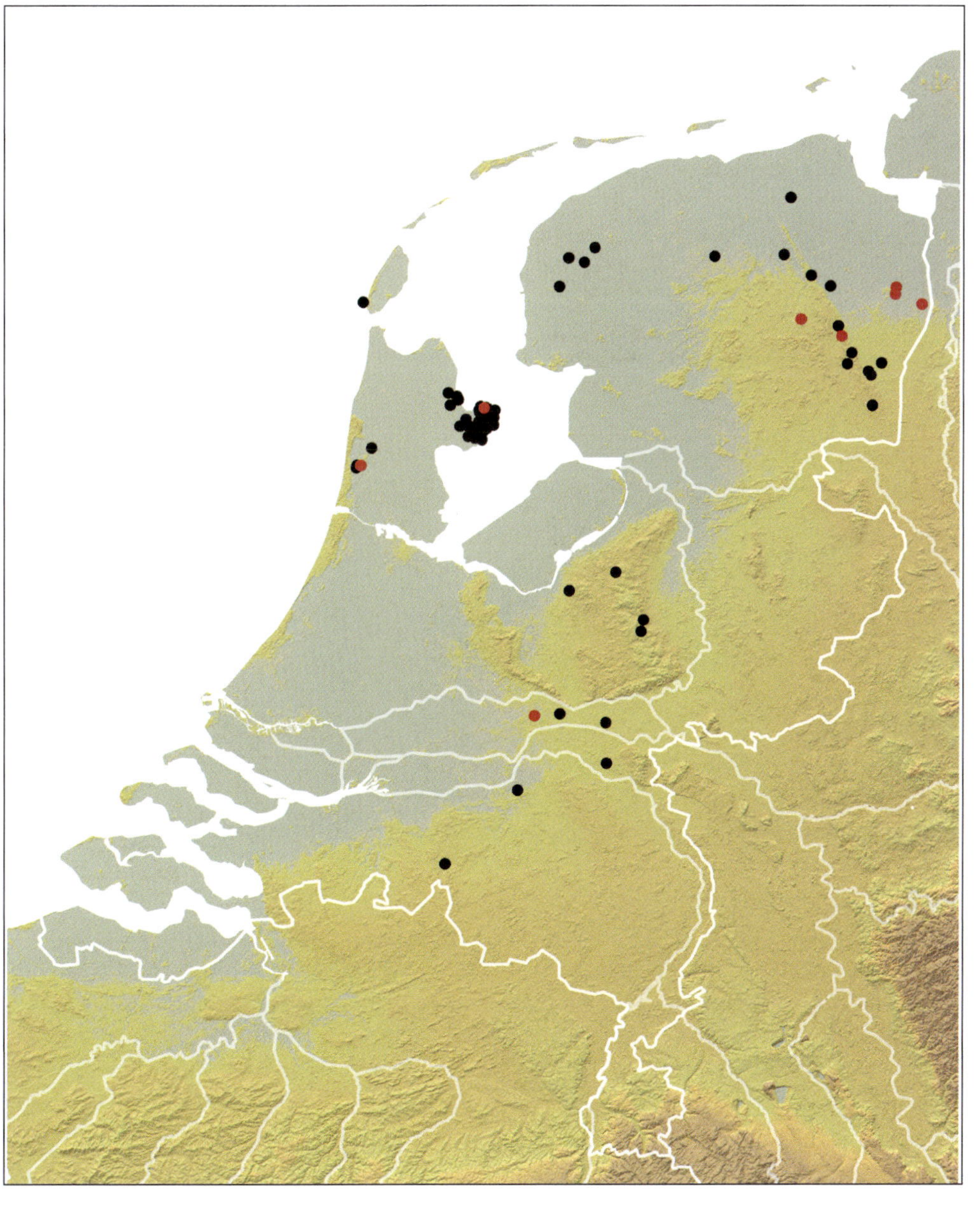

Fig. 9.16. Distribution map of the crescent-shaped 'sickles'. The red dots indicate the multiple object hoards, located mostly in the north-east of the country, the black dots the single finds. Drawing: K. Wentink (from Van Gijn 2010, fig. 7.17).

pers. com.). On the basis of their crescent shape and extensive gloss, these implements were also referred to as sickles in Dutch archaeological literature (Brongers and Woltering 1978). However, use-wear analysis of the 'Dutch sickles' showed them to have an extremely blunt edge, sometimes even 3–4 mm thick (Fig. 9.17A). The gloss was not just present on the edge, but extended across the entire surface of the tool, covering both lateral edges (Fig. 9.17B). There was no evidence for the presence of a haft, but instead the polish gradually faded out and was only absent on the very distal end of the object. The polish was rough and matt and displayed a large number of deep striations of varying width and length, with a random directionality.

Experiments with using replicas of such crescent-shaped tools showed them to be effective harvesting tools. However, even long-term use did not result in the rounding seen on the archaeological implements and the polish was always limited to a band *ca.* 1 cm wide along the concave cutting edge. The experimental polish was a typical 'cereal polish': smooth, highly reflective and displaying narrow, long and fine striations. Moreover, it turned out to be necessary to recurrently re-sharpen the edge in order for it to continue to be an effective harvesting tool. Such re-sharpening would have removed any existing traces of wear and prevented the occurrence of the extremely blunt edges seen on the archaeological specimens.

It seems therefore highly unlikely that these 'sickles' were indeed sickles in the sense that we know them: cereal harvesting implements. Instead the wear traces bear close similarities to the type of wear resulting from contact with soil. Use as a coulter has been proposed (Bruyn 1984), but was deemed unlikely due to the presence of polish towards the butt end: if only the very end of the 'sickle' would have been fitted into the ard, the leverage would have caused the tool to snap very quickly. Experiments were therefore conducted using these objects in a handheld fashion to cut through the grass and roots of sods, a task for which they proved to be effective (Van Gijn 1999). This activity produced traces very akin to the ones on the archaeological tools and a blunt edge was not experienced as making the tool less efficient. Sod-cutting seems a very mundane task for a highly crafted import object. However, we should realise that this part of the territory of the Netherlands is largely lacking in both trees and stones, leaving little choice in building materials. Turves provided the building material for both the erection of barrows and for the construction of dwellings, a practice that continued into historical times.

Unfortunately, the exact find context of many of these crescent-shaped tools is known in only a few cases. For the most part they are isolated finds, made in the course of digging activities for the re-organisation of the landscape. The complete ones are almost always found outside settlements; if we do encounter them in settlement context it always concerns fragments which have been transformed into scrapers or other domestic tools. There is one exception, a 'sickle' found near a granary (*spieker*) attributed to the earliest habitation phase of the *terp* of Middelstum, where simple constructions made of sods were recovered (Boersma 1988). The proximity to the wall of the structure makes it tempting to suggest that it concerned a house offering, but unfortunately the find circumstances do not allow such an inference. Still, considering the large number of isolated finds (a cursory inventory led to 116 complete specimens (Van Gijn 2010, 193)), it is unlikely that so many of these beautifully crafted import items were accidentally lost outside of the settlements. Rather, I suggest that they were left in those places where the sods were cut. Whether this was done for future use or whether they were left as a kind of 'offering' is difficult to ascertain.

In addition to the isolated finds, eight multiple depositions are currently known, composed of 3–5 objects. Some of the hoards in the northeastern part of the Netherlands, like the hoard of Norg, probably concerned a trader's deposit, as most of the 'sickles' did not display any traces of use, and because this location is 'on the way' from the centres of production in southern Scandinavia and the area of West Friesland where most of these objects were found. The hoard of Heiloo, in West-Friesland, is worth mentioning because it consists of four flint sickles and one bronze one (Van Gijn 2010, Fig. 8.5) (Brunsting 1962) (Fig. 9.15). The sickles were put in the ground in an upright position. This indicates that they were not buried for later use, but must have constituted a special deposition. To summarise, it thus seems that the crescent-shaped tools, in the south Scandinavian context used as harvesting tools, were traded south, probably in an unused state. They were re-interpreted in the region of the present-day Netherlands and fitted into the local technological system where they were seen as appropriate tools to collect building material. It may well be that it is exactly their agricultural significance and connotation that made them appropriate for such an important task: the construction of the houses of the living and the dead.

Conclusion

The four examples of tool use discussed in this contribution all concern implements involved in or associated with the agricultural process, more specifically cultivation. They display special cultural biographies, indicating the ritualisation of these objects during their life trajectory. Obviously attributing precise meanings to these different biographies is impossible, but the fact that all concern agricultural tools is, I believe, significant. The example from the Early Neolithic Bandkeramik shows that agricultural tools, that is sickles, querns and adzes, consituted key burial gifts ending their use life in funerary contexts. This illustrates the close association between death and cultivation and shows the importance of agriculture in the world view of these early farmers. Perhaps more surprisingly, it was also shown that querns found in settlement context were intentionally destroyed, dispersed and treated with ochre before deposition. Querns in ethnographic context frequently are the personal possessions of women and are closely entwined with their owner's lives. Sometimes they constitute heirlooms which are passed on to the descendants, but there are also instances in which these querns were believed to pose a threat to the living because they embodied the soul of the deceased. This could only be resolved by 'killing' these querns as well and dispersing them across the landscape.

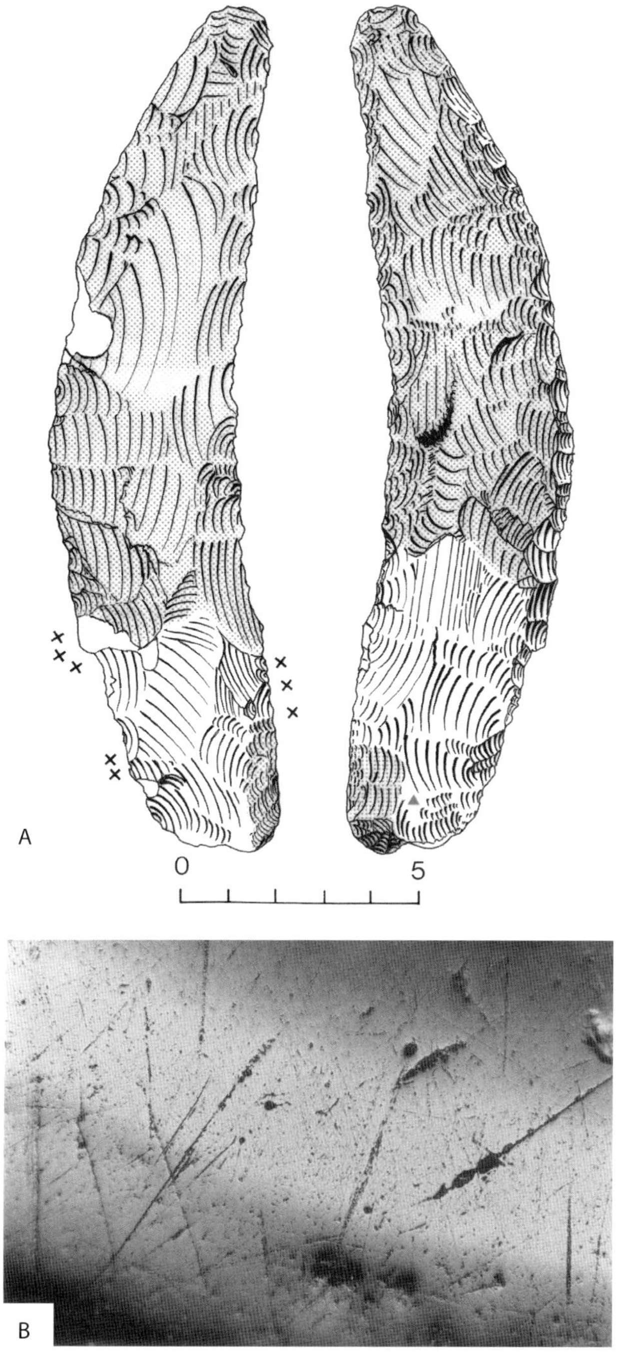

Fig. 9.17. A: Sickle from Andijk, West Friesland, showing the extent of the polish from cutting turfs and the area where the sickle was held (scale 1:1). B: Extensive gloss and rounding of the edge due to contact with soil (original magnification 100x). Images: Van Gijn (2010, fig. 7.16).

The special biographies of agricultural tools described above can be ascribed to the ambiguity of the agricultural process: new life can only grow after death. Cultivation begins with the destruction of the ancestral forest, the forest which formerly provided the necessary food and raw materials for survival. This awareness of the importance and perhaps even mythical significance of the forest, especially to communities which only recently incorporated cultivation in their subsistence system is, I would argue, demonstrated by the early farmers of the Middle Neolithic sites of Schipluiden and Ypenburg who still relied to a large extent on wild resources. It can be suggested that they destroyed their sickles because crop growing was perceived as a threat to the balance of nature. The third example, the deposition of agricultural implements with the dead by the Funnelbeaker people, can be interpreted as a way to ensure the continued existence and prosperity of the local group. This example again illustrates the close relationship between funerary practices and cultivation that persisted throughout much of the Neolithic and Bronze Age. Likewise, it may be no coincidence that during the Late Bronze and Early Iron Age the crescent-shaped sickles from southern Scandinavia were utilised to cut turves, a building material not only used to construct the dwellings in the treeless surroundings of the western and northern parts of the present-day Netherlands, but which was also frequently used to raise the barrows for the dead, something that required the stripping of extensive tracts of land. It was thus an agricultural tool that was re-interpreted and used for a very important other task, cutting turves, a task that was closely linked with funerary practices. In conclusion, these examples all show that agricultural practices, and especially crop cultivation, were as closely entwined with the social and religious life of the Neolithic and Bronze Age peoples in northwestern Europe as was the case in later pre-industrial agricultural societies.

9.4. AN 'AGRICULTURAL CALENDAR' FROM THE BRONZE AGE?

Sabine Beckmann

Introduction

Agricultural principles stood at the centre of the symbolic aspects of prehistoric communities' lives. One of the most important farmers' 'tools' must have been their orientation in the agricultural year-cycle which helped them sow and harvest at the right moments. Naturally with changing times and beliefs, older 'calendars' would not have been available without oral transmission from generation to generation, so possibly some of those known from later times may have originated much earlier.

Knowledge of when which work had to be done and how to recognise the correct moment in the agricultural year must have been most vital. Crucial agricultural intervals would be controlled by religious authorities who organised events or festivities, as for example in the ancient Greek festival year of the Classical period that is rather well documented (Isager and Skydsgaard 1992, 163.) A similar event in modern Greece is today's Assumption of Mary, a feast of transition from death to everlasting life in heaven, the most important religious festival in summer, that also used to be the end of the threshing period.

Certain star constellations were used for orientation in time as well as in space, most probably as early as the Bronze Age (Beckmann 2006a). The Pleiades' appearing on and vanishing from the early morning sky for instance was an important sign for recognising the times of sowing and harvest (Hesiod, 383,384). Their setting was seen as a sign for the beginning of the new agricultural year (in October/November). Even though a frieze from (probably) Hellenistic times is usually seen as the oldest available pictorial representation of a Greek calendar (Isager and Skydsgaard 1992, 168) there may have been earlier ones going back to an older or even Bronze Age tradition, as the following paragraphs will show.

Flowers and Seasons

> *'Soft crocuses mingled with irises and hyacinths, and rose-blooms and lilies,*
>
> *marvellous to see, and the narcissus which the wide earth caused to grow yellow as crocus'.*
>
> Hom. Hmyn to Demeter, 425.

Homer tells us here about Persephone's abduction by Hades while she is picking flowers. The daughter

Fig. 9.18. The Blue Bird Fresco from the House of Frescoes, Knossos, Crete – The part of the fresco exhibited in the Heraklion archaeological museum.

Fig. 9.19. Map of the Aegean Sea with the archaeological site of Knossos. Map: R. Lugon, J.-C. Loubier and A. Chevalier.

of the goddess of earth and fertility, Demeter obviously does not do this on a normal meadow, as none of these flowers ever appears in nature in bloom together with the other. So while she collects her bridal bouquet for the '*hieros gamos*' (the holy marriage) with the god of the underworld she can be understood to take with her from the surface to the realm of sleeping fertility, tokens of all phases of the flowering year, so that when she is gone nature will lie barren until she returns to the surface, bringing with her new fertility. Her mother's search for her daughter and her return to the world were also the myth on which the most important ancient Greek agriculturally inspired religious ritual/mystery cult, that of the Eleusinian Mysteries, was based. As will be seen below, this passage of the Homeric hymn may, *in nuce*, contain the basic elements of a pictorial agricultural 'calendar' known in the Aegean since the Bronze Age.

While Minoan wall paintings were often seen as nothing more than a pretty decoration (Evans 1964 II, 446), they should rather be understood as religious images, similar to other frescoes of the Bronze Age all over the eastern Mediterranean (Marinatos 1985, 221; Schaefer 1977, 12). The so-called 'Blue Bird Fresco'(Fig. 9.18), excavated in 1923 by Evans in the 'House of Frescoes' in Knossos (Fig. 9.19; Evans 1964 II, 446) can be read – just like the above passage of Homer – as an agricultural calendar when looked at in the right way. The part of the fresco exhibited in the Iraklion museum is only a fragment of a much larger frieze (Cameron 1968, fig. 13, coloured in Evely 1999, 247) (Fig. 9.20) that once decorated a room in a small building containing also other objects interpreted as belonging to a ritual context (Evans 1964). This adds emphasis to the interpretation of the fresco as the depiction of a 'religious landscape' (the term is from Frankfort 1948, 154; cf. Beckmann 2006b; Chapin 2004, 54; Immerwahr 1990, 50; Marinatos 1985, 221) and thus the House of Frescoes should be seen rather as an agricultural shrine than a 'home of a small burgher' (Evans 1964 II, 406). Whereas Marinatos regards the image as not realistic due to the different terrains shown in a very condensed fashion on a small image (thus a depiction of 'ideal spring') (Marinatos 1985, 221), this may be rather too simplistic an interpretation (Chapin 2004, 58).

The main feature of the originally long, frieze-like image giving it calendric value are the different plants and flowers shown in various areas of the fresco, usually named the Blue Bird Fresco for the central motif, although even looking only at the part exhibited (*ca.* 1/4 of the whole frieze, cf. Cameron 1968) gives a good impression, which is

Fig.9.20. Cameron's reconstruction of the whole frieze, including the now exhibited part on the right (from Evely, 1999, 247).

why I shall concentrate here only on this part of the whole image. Other plants, like the white lily and the olive, shown on those parts of the fresco which are not exhibited also play a prominent role in the annual cycle (Cameron 1968).What makes these plants understandable as calendar-images is the fact that they are all shown in bloom at the same time, an event that could never happen in a real garden as these phases are spread out over the year. By studying the plants surrounding the 'Blue Bird' (there were several birds on the whole fresco) one by one, the calendar can be 'read'.

Some basic knowledge of the Cretan year and the seasons are useful for this. While summer is the driest season of the year, in November, that is after the first rains (also the beginning of the later ancient Greek agricultural year, see above), the island becomes green within a few days – from a northern hemisphere perspective this should be called spring. It is the right time for farmers to start working their fields by ploughing and sowing. The flower on the fresco characterising this phase is the saffron crocus (Fig. 9.21).

Fig. 9.21. Saffron/crocus. A: Saffron crocus depicted in the House of Frescoes, Knossos (Heraklion archaeological museum). B: One kind of crocus (*Crocus laevigatus*) growing in Crete, in November.

Saffron

When the rain has been intense enough to spread humidity well into the soil, the first saffron-flowers appear on the still rather bare ground. Thus their flowering must have been seen not just as a pretty event but as the important sign of nature for farmers to start the new year's work by ploughing their fields and sowing cereals and pulses. With the beginning of the fertile time of the year (the return of Persephone from the underworld in later times), saffron-crocus can be collected, supplying a remedy for obstetrics and gynecology and a yellow-orange dye for clothes. Only high-status humans would wear this precious colour. Gods were more often described as saffron-clad in ancient sources such as Hesiod and Homer. Interestingly, a cast of priests in the Eleusinian Mysteries named *krokonidai* tied a saffron-dyed ribbon (*krokes*) around the initiate's right hand and left foot. The use of the plant in this famous agriculturally inspired ritual gives us another clue to its importance as a Minoan agricultural symbol of the return of life after the long parched summer.

Iris

The most intense part of the year happens in spring. When the sun warms up the earth after winter's darkness, plants grow and bloom within a few weeks, and even the most desolate parts of thorny *phrygana* (the scrubland ecoregions also called *garrigue* around the Mediterranean basin) are decorated by the blooms of *Iris cretica* (Fig. 9.22). Their rainbow-coloured petals open by March during the crucial time of growth for cereals. Only by taking a close look at iris blooms would the magic of the three-fold symmetry dominant in Minoan plant-iconography become clear. This three-fold symmetry is a feature that flowers like crocus and lily share to a degree, which is why they are sometimes difficult to tell apart in Minoan art. Naturally in this symmetry no part can be seen as more important than the other,

Fig. 9.22. Iris. A: Iris from the House of Frescoes (Heraklion Archaeological Museum). B: The small *Iris cretica* in nature, February.

just as in the case of the three-fold Cretan epochs: winter, spring, summer. Other features in Minoan art illustrate the importance of this arrangement for Minoan religion, too: the only clearly religious buildings, the large shrines, are recognisable as such mainly by their tripartite architecture (but throne-constructions are also set on tripartite platforms (cf. Marinatos 1985).[3]

Another element in iris-lore sheds light on the flower's possible symbolic function for the agricultural year. Since the 19th century, scholars have maintained that the flower called *hyakinthos* (see also the Homeric example cited above where it is translated as *hyacinth*) should be identified with some kind of small iris – long before the excavation of Knossos made *Iris cretica* a well-known feature of Minoan art. This identification enlightens the myth of the youth named Hyakinthos who lost his life when Apollo's discus accidentally hit him. Where his blood touched the earth a flower sprang up – the hyakinthos/iris. Here we see the light-god Apollo's sun-disc that makes flowers wither and die in late spring (Roscher 1886–1890).[4] In Crete, this may happen from April when days with hot wind – the first touch of devastating summer's heat – can kill sensitive flowers like *Iris cretica*, but may also erase all the farmers' work by parching his fields before his crop has produced seed. So, while crocus constitutes the flower of the transition from dry to humid (from death to life), iris/hyakinthos stands for the end phase of humid spring when death again lurks close to farmers' fields and the fertile season runs dry.

Even though at first glance there seems to be one plant for each central point in a season, the general idea of the image was not so simple. Other plant symbols stand for such concepts as the overcoming of summer and death by reaching out from spring to winter. One example is the pomegranate.

Pomegranate

When pomegranates begin to bloom, the dry part of the year has begun. But even though all green nature around seems to die, pomegranates bloom on far into summer while their first tiny green fruit begin to develop and cereals ripen towards harvest. Bright orange flowers with five to eight petals bloom between leaves with a tendency to groupings in three (Fig. 9.23). While the farmers' crops become ready for harvest, pomegranates' tiny green fruit still hold on to the withered but nevertheless bright orange remains of their flowers, attempting to be both flowering and fruiting at the same time despite the surrounding dryness.

Looking again at the Blue Bird Fresco it becomes obvious that what are often called roses are actually pomegranates (Fig. 9.23A): apart from their (slightly bleached by age) orange colour they show six petals – the Minoans must have known roses have five only. Also the three-fold shape of the leaves fits the pomegranate more than the rose. The oblong green spheres with orange tops called 'rosebuds' (Möbius 1933, 11) are obviously tiny pomegranates still carrying their dried flowers. Several features

Fig. 9.23. Pomegranate. A: Pomegranate from the House of Frescoes (Heraklion Archaeological Museum). B: Cretan pomegranate (*Punica granatum*), flowering in May.

make the pomegranate a symbol of fruitfulness: Its ability to thrive with very little water, its power to overcome certain death as (like the vine) it grows fruit all summer, its breast-shaped fruit with an abundance of blood-red kernels ripening from September onwards.

The pomegranate was associated with Aphrodite, Hera and especially Persephone. Persephone's eating a few pomegranate-seeds in the underworld was the reason why she had to stay there for months. And it is still one of the ingredients of *kolliva*, the traditional mixture of cooked cereals and pomegranate perfumed with mint (see below) distributed at obsequies or funerary rites in Greece.

The symbolic meaning of the plant contains a death-life transition by connecting two fertile seasons and overcoming the death of summer. Farmers would have finished harvest[5] by the end of the pomegranates' flowering and would be looking to its ripening fruits when meditating about when the next green phase of the year was to be expected (see also Hansson and Heiss 2014).

Mint

The plant next to the Blue Bird interpreted as 'wild peas or vetches' by Evans (1964 I, 454; Möbius, 1923) can by closer scrutiny be recognised as mint (Fig. 9.24), typically growing close to water (in the fresco a blue line). Mint blooms during the height of summer when all other greens have dried up. To know where mint grows means to know where water is, and farmers would have done well to choose each spot with mint as a place for a garden or small field for providing their family with fresh vegetables. Mint has always been an important healing plant.

Like the pomegranate, mint is a plant of transition as it defies dry summer but vanishes into the earth in winter when the rest of nature is green. Greek myth tells of the nymph Mentha, trod into the ground by jealous Persephone: she is sent into the 'underworld', possibly taking Persephone's place with Hades, while the goddess brings fresh growth up into the winter world. The Minoans gave mint the highest position of sacred plants in the Iris Fresco of Amnissos (Heraklion Museum): it is shown to rise (flanked by high *Iris germanica*) from a tripartite platform usually occupied only by the 'goddess' (Marinatos 1990), thus symbolising herself or her powers. Mint must have been her emblem, as it is presiding over the room from the goddess' *dais.* In a similar way the goddess in her human form is seen sitting on the centre of a tripartite platform on a fresco from Thera where her sacred saffron is presented to her (Marinatos 1990). All in all, mint was not only a practical indicator of the precious summer humidity, but also a symbol of overcoming death as the plant vanishes in green winter and triumphs over parching summer by 'knowing' the water. No wonder then that it was also holy to Persephone.

Fig. 9.24. Mint. A: Mints, depicted in the House of Frescoes (Heraklion Archaeological Museum). B: A typical mint (spearmint, *Mentha spicata*) variations of which grow wild in Crete.

Conclusion

On the whole the Blue Bird Fresco seems to provide onlookers (perhaps originally in a ritual context) with a synoptic view of important phases of the Minoan agricultural calendar by illustrating a symbolic year cycle that does not have beginning nor end but moves from phase to phase, including life and death. Thus we can understand why all these plants were still seen as symbols of life *and* death in later antiquity: only by a continuum of growth and decay can farming thrive. In practical terms, there can be no humus without previous decay of other plants, one of the principal meanings of fallow, compost and green manure. The ritual activity for which the fresco may have been the background must have included this kind of knowledge. It does not just show a simple 'farming calendar' but rather an 'epochologio' (from the Greek word for season: *epochi*, cf. Beckmann 2006b) that includes in symbols each phase of the agricultural year, an everlasting calendar of the mysteries of agriculture.[6]

9.5. THE SYMBOLIC CONNOTATIONS OF AGRICULTURAL TOOLS IN ANTIQUITY

Sabine Beckmann

Ancient authors tell us a lot about the symbolic meanings of all kinds of agricultural tools in recounting mythology or describing actual ritual use. For example, in Hesiod's *Theogony* (line 161), the world's first scythe is described as 'of a metal that was gray and very hard' the first mention of steel in the Greek language. This was not a human tool at all, but an invention of the Earth-goddess Gaia as a weapon of revenge to be used by her children against their father Uranos, who treated them badly. Kronos the Titan used the sickle/scythe to castrate and overthrow his father and has since remained a well-known figure: to the Romans, Saturn, and to later Europeans as Father Time, the old man with his scythe representing the image of death. This myth shows the intricate connection of fertility, death and continuity of life (from father to son) in the tool's ritual use.

In a Roman example, Pliny tells us about the religion of native Celts in Roman Britain and provides the example that is probably best-known to the general public for the use of a harvesting tool in religious context: The druids cutting mistletoe with a golden sickle (Pliny, *Natural History* 16, 95). As the ritual was accompanied by the slaughtering of two bulls, we understand that the sickle was also seen as bringing fruitfulness to the fields through the death of two animals known as typical examples of powerful fertility.

With the rise of Christianity, many of the old rituals were adjusted to this new religion. Since medieval times new ways were also found to connect agricultural tools to religious beliefs. One context providing us with a wealth of images is the 'Sunday-

Fig. 9.25. Sunday Christ (*Feiertagschristus*). Village church of St. Kanzian zu Saak, Carinthia, Austria, on the southern outer wall. Dated 1465, rediscovered in the early 80s. The lower part of the painting shows a man transporting wine on his horse, a sower and a man pulling a harrow. The tools surrounding Christ include weaver's shuttle, tongs, barrel, awl, shovel, scythe, axe, pitchfork, dungfork, flax comb, plow etc.

Christ' or 'Christ of the Trades'. From the 14th to the 16th century, Christ was often depicted surrounded by agricultural tools and tools of the trades (Fig. 9.25). This recalls the well known 'Arma Christi' images of Christ with the tools of his martyrdom (see below). Only about fifty images of this type are known all over Europe (Wildhaber 1981), conveying the same intention: to warn viewers to respect the Third Commandment (Remember the Sabbath and keep it holy). Tools typical of specific trades or professions are shown very close to or even touching the Man of Sorrows, symbolising that working on a holy day is forbidden and doing so actually 'hurts' Christ. These images draw on the subject of the Arma Christi images, but the tools of the Passion are replaced by 'tools of the trade' and agriculture. Here the singular event of Christ's Passion has become an everlasting, repeatable event (Zimmermann 1997, p. 30).

Images like these give us several very nice depictions of medieval tools, sometimes in the context of their use. Even though the images of 'Christ of the Trades' were meant to be sinister admonitions not to sin, they represent for us today a rare glimpse of the tools of the past.

9.6. TECHNICAL SOLUTIONS BETWEEN HABITS AND REGULATIONS

Thomas K. Schippers

It is often implicitly presumed by both archaeologists and ethnographers that rural societies have adopted and adapted the most suitable technological solutions in regard to their aims of maintaining and reproducing themselves in a given ecological and historical context. Apparent similarities of implements and technical solutions have also been academically used to outline 'cultural' entities or even define (successive) *époques*, especially when written sources were thought to be non-existent or of little value. But these Spencerian-like, deterministic views on technological choices (or rather their absence) have more recently become less clear and even been blurred by new types of sources and data. For example European folklorists have discovered in recent decades that many of the orally collected narratives in Europe in fact derived from earlier written texts, while specialists interested in agricultural technology have become increasingly aware of the role played by early written regulations and legislation regarding the use or the prohibition of particular tools or technical solutions.

A small but enlightening example of this is the following short passage from a charter of February 20th, 1448, quoted in a text about the history of the southern French Monastery of the Lérins Islands[7] (Antier 1975/1988, 67–68):

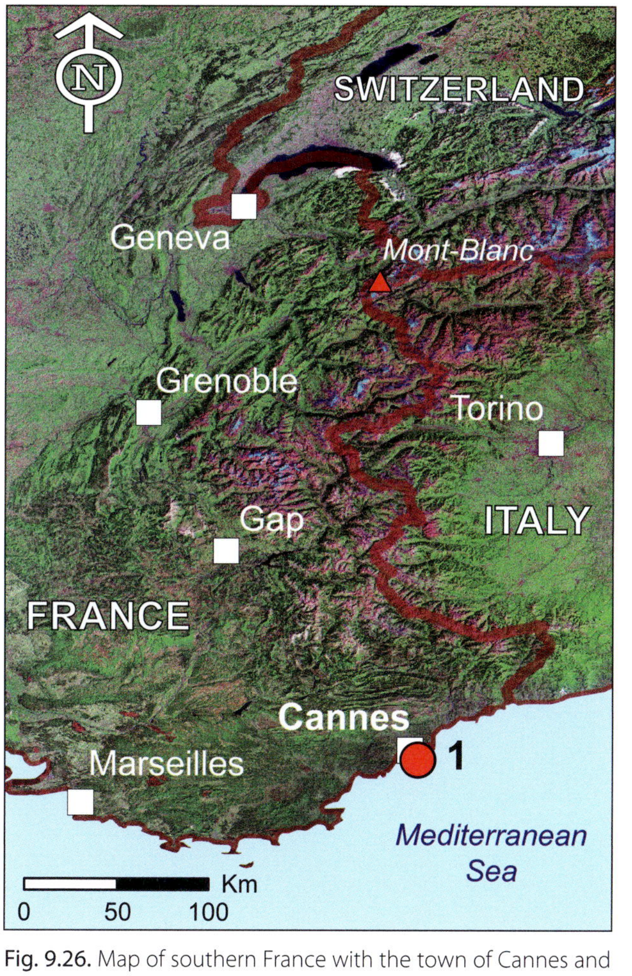

Fig. 9.26. Map of southern France with the town of Cannes and the site of 1) Lérins. Map: R. Lugon, J.-C. Loubier and A. Chevalier.

> '....*Le seigneur-abbé est tenu au temps de la moisson au terroir de Cannes, avoir et tenir jumens à suffisance pour fouler les bleds des habitants, les plus proches de la mer et en grand danger étant foulés en premiers. Au temps des moissons il est permis de batter avec des fleaux deux charges de bled, si on ne peut avoir de jumens.*'

(Translation: 'During the cereal harvest on the territory of Cannes (Fig. 9.26), the lord-abbot has to have and to provide enough mares to trample the cereals of the inhabitants, with those nearest to the sea and [so] in the greatest danger, having their wheat trampled first. During the cereal harvest, it is allowed to thresh [only] two loads of cereals with flails in case one cannot obtain mares.')

This rather neutral remark acquires a totally unsuspected significance for those familiar with threshing technologies in (southern) France and more generally the Mediterranean coastal regions. As cultural geographers and ethnographers like Charles Parain (1938) have often suggested, (most) 'Mediterraneans' do not use flails but animal trampling for cereal threshing; this includes the Cannes region in southern France. So the mention of the possibility of flail threshing appears as a complete surprise because ethnographic data have exclusively attested to threshing by trampling in this area since at least the 18th century.

The main interest of this short historical quotation is to open up new analytical perspectives on some of the technological 'habits' still frequently associated with pre-industrial agriculture: the role of regulations and legislation imposed by various authorities in regard to the use or prohibition of particular technical solutions. In the case of threshing by animal trampling in specific outdoor areas, it seems made possible a much more efficient (quantitative) surveillance by the feudal authorities than more individual flail threshing, which could take place in private, more hidden areas. Accordingly, all factual evidence of certain technical solutions should, when possible, be contextually reframed in terms of social and legal allowance, tolerance or prohibition. Especially in complex societies, agricultural tools and their uses are not only subject to criteria of efficiency, but also to a social consensus in terms of acceptability. Today, this dimension seems to become more and more prevalent in order to maintain sustainable agricultural practices for the future.

9.7. INFLUENCE OF LEGISLATION ON AGRARIAN TECHNIQUES: MEDIEVAL AND MODERN LOCAL LAWS IN THE IBERIAN PENINSULA

José Luis Mingote Calderón

Introduction

The information provided by local legislation (*ordenanzas*) shows that these local laws were strongly influenced by the community. These laws showed a society in which the control exerted by neighbours was very strong, enormously reducing individual freedom. In the Christian kingdoms, this type of legislation was deeply rooted in the so-called *fueros largos* (*fueros* are the group of laws which regulate the settlement in a new area) which regulated different aspects of the individuals' life, including those which nowadays could be referred to as civil issues, together with others related to the moral or religious spheres. Agricultural practices and the behaviour of farmers toward their neighbours' crops and fields were an important aspect of bodies of local law.

The local laws come from the neighbours, the married men living in the particular village where the laws are established. These laws need to be approved by the lord or the king and follow the more general laws from the kingdom (Fernández del Pozo 1988, 73). Therefore there is a dialectical relationship between law and custom. There are many examples of written laws based on traditional practices or customs (Rubio Pérez 1993, 225, 410 and 431).

During the 19th century, this corpus of laws started to collapse and lose its dominant power. However, in many areas these laws have been preserved until the last century and they have been accepted as cultural models with a long history behind them (Fernández del Pozo 1988, 35). In many cases there are appeals to the help of God or Mary in imposing laws. In addition there are references to the way people legislate, for example, it is common to say that the law is at the service of God (Rubio Pérez 1993, 260, 265, 269, 369, 379, 392, 400 and 401). This behaviour shows what is thought to be a stable ideological construction which aims to bring the divine order to earth. It is not possible to create laws against God or the church.

The extreme controlling role of these laws is very clear, and becomes evident when we look at the way this control is actually imposed. Punishment is applied right after the field guardians or neighbours denounce an irregularity, and it should be remembered that denouncing is an obligation (Rubio Pérez 1993, 236, 324 and 458). In many cases these tasks of control were carried out by the neighbours in rigorous turns and the accuser could get a percentage of the fine imposed on the offender.

Techniques and Ideology

When discussing agricultural tasks one should not forget that in many cases agrarian work was completely dependent upon religious ideology. This is reflected in many aspects as, for example, the impossibility of carrying out certain tasks during festivities or when rituals were being performed. These traditions have a long history which predates medieval times and Christian contexts (Mingote Calderón 2003; Philippart 1990). Their real meaning can be found in some of the laws imposed in the Catalan village of Ager (1278) referring to the need to respect Sundays and festivities, enforcing full participation by all members in the rituals needed for the community's prosperity. It is forbidden to make sheaves, cut hay, transport wood, stones or firewood, harvest saffron, spread cereals in the threshing yard, work on temporal tasks, harvest grapes and process them, make bread, wash clothes, hang clothes and so forth. In some laws, we find explicit orders which forbid work from sunset on Saturday until Monday morning (Serra-Rafols 1949, 236–237 and 245).

There are also rules regarding the prohibition of certain tasks during funerals. It is forbidden to yoke animals during the funeral period since everybody should participate in the funeral ritual. All neighbours have to participate in rituals related to prayers for fruit or water, or rituals related to animal welfare. Thanksgiving rituals are compulsory and considered a social commitment.

Regulation of Technical Aspects

There are numerous laws regarding technical aspects of agriculture. One of the most common ones refers to the fixing of dates to start or finish certain works. There is information on the role of the local communities in issues related to the sowing, or harvesting, the cutting of hay, the transportation of sheaves (or other products such as wood), the cleaning of threshing yards, the burning of stubble or the harvest of grapes. For example, sowing cereal in areas of *dehesas*[8] has to be done on the 29th September (Saint Michael-San Miguel) in La Alberca, Las Hurdes and Las Batuecas (Salamanca and Cáceres; see Fig. 9.27), and this is regulated in the local legislation of 1515, copied in 1668 (Berrogain 1930, 392–393). In Valencia, the mowing of meadows cannot be done after some of the festivities of July (Santa Marina on the 18th or Santiago-St Jacques on the 25th). This applies to the villages of Villaoliba and Sotillo de Boedo in 1587 and 1675 (Fig. 9.27; Díez Merino 1985, 309; Primo Ruiz 1963, 79). Laws related to transporting products in León Province can be found in my recent book (Mingote Calderón 2008, 67–92). The cleaning of threshing yards needs to be finished by San Lucas (18th October) as mentioned in the local laws of de Sotillo de Boedo (Primo Ruiz 1963, 76). In Jorquera (Albacete) the local laws of 1759 mention the prohibition of burning stubble before the 15th August and after the 1st of February (Almendros Toledo 1989, 127). The grape harvest is commonly related to St. Michael's day. In the local legislation related to grape harvesting of Robledillo (Madrid), we find mention of 2 'vineyard mayors' who had

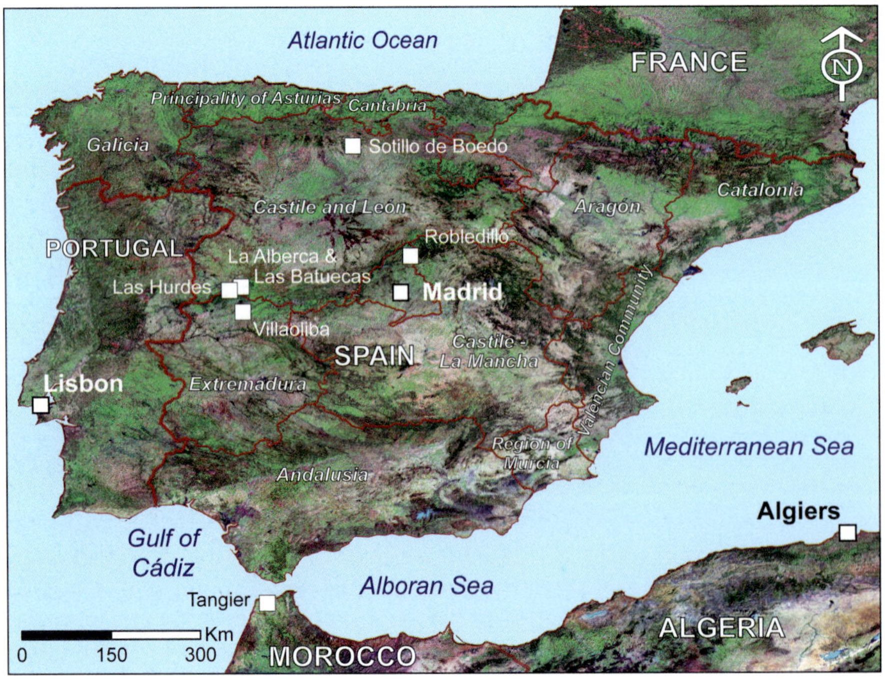

Fig. 9.27. Localisation of the Spanish towns of La Alberca, Las Batuecas, Las Hurdes, Robledillo, Sotillo de Boedo and Villaoliba, as well as the political communities of the Principality of Asturias, Cantabria, Castile and León, and of the Valencian Community. Map: R. Lugon, J.-C. Loubier and A. Chevalier.

several tasks under their control. They had to choose two 'honest' persons who would check the growth of the vines and would fix the starting date for the harvest (Fernández García 1966, 189).

There is also a long list of laws which regulate the passage of carts, pack animals, flocks and people across the agrarian landscape. This legislation shows the range of problems caused by the road system. The spatial organisation of land ownership and property throughout a region was also a factor influencing the regulation of the transit of people and animals. In fact, the tendency to use shortcuts to access property, and therefore cross other peoples' lands, including sown fields, led to new laws prohibiting the crossing of vineyards and cereal fields (see for examples Mingote Calderón 2008).

There are also many laws regarding the way certain tasks should be done. Many of them are deeply rooted in pre-industrial ideas of the existence of standardised local ways of doing things 'correctly.' Such ideas are reflected in many agrarian contracts and *fueros* (a body of laws that regulate the settlement in a new area) in which we find sentences like: 'working according to the customs of the good farmer' or 'working according to the customs of the area'. Clear examples come from León province where control is applied to the way the watering of the communal land is done, or more specifically, where the laws regulate the way waterlogged fields should be ploughed to avoid damage (Rubio Pérez 1993, 393 and 447; Martínez Martínez 2003, 72). Many examples of laws reveal rich technical information. For example in Calamocos (León province), the local laws written in 1683 established that threshed cereal should not be winnowed if this could disturb other people doing the flailing in the nearby area (Rubio Pérez 1993, 252–253).

Cutting wood in the forest is another issue very much related to the sphere of agrarian techniques. Local laws mention many uses of the forests, including cutting firewood and felling timber for house building. However, the aspects related to the cutting of wood for agrarian implements are particularly interesting. The prohibition against removing wood from the forest when its final use is not well specified is quite common in the local laws. This is the case revealed by the local laws (*Ordenanzas de Villa y Tierra)* of Buitrago in 1583 which specified that it is forbidden to remove wood from the forest, whether it be in segments already prepared for tool manufacturing, or as unprepared stems of wood. In addition, the law refers to prohibitions on collecting wood for beds, plough stilts or complete ploughs (Fernández García 1966, 28 and 31). In contrast, the law allows cutting oak wood to fix carts, provided that this use can be proven; in other words, the wood can be removed once the different pieces have been worked. It is also forbidden to cut wood for selling outside the village, and there are specific references to different parts of cart wheels (Mingote Calderón 2008). However, despite these laws, many people, particularly the poor, did not respect these rules and this is pointed out in documents from León and Cantabria at the end of the Ancient Regime.

Lastly, I will refer to some rules regarding the lending of animals for specific agrarian works. Examples refer to the need for asking permission of the council in order to have animals for four days during the ploughing period. This is well documented in 1638 in Toralino (León) (Rubio Pérez 1993, 458). It is likely that such a law is very much related to the use of communal land by people not having access to these pasture lands. Control over the number of heads allowed per neighbour may have led people to hide their excess animals by lending some of these animals to another neighbour (Fernández del Pozo 1988, 23).

Conclusion

The number of edited texts as well as their wide chronological span is barely reflected in these examples, but it is clear that there are many similar situations across a large territory and throughout a long period. One of the main conclusions is that, although minor changes may occur, laws are generally maintained through time without major transformations, preserving their original spirit and intent. In addition, when a change is made, there is a clear statement that the rule has to be followed. The permanent character of the laws and their repeated application made them a way of creating stable cultural patterns associated with agrarian work. This mechanism allowed the continuity and stability of a social model which, if not regulated by laws, would otherwise result in tense relationships amongst neighbours.

Chapter Notes

1 It is very easy to find all kinds of interpretations of rituals in the literature on the Early Neolithic in Sweden. At any rate, it is important to include a cautionary note on these interpretations. The following terms are used and abbreviated: M (Mesolithic: 10,000-6,000 BCE, see *e.g.* Larsson 1990); EN (Early Neolithic: 4,000–3,000 BCE); MN (Middle Neolithic: 3,000–2,000 BCE); LN (Late Neolithic: 1,800–1,500 BCE); TRB (*Trichterbecherkultur* = Funnel Beaker Culture, EN and MN); CWC (Corded Ware Culture, Late MN); PWC (Pitted Ware Culture, Early MN). The dates are related to Scandinavian chronology. In the comparative part of the article, the chronology is related to classifications in continental Europe. This article was written with financial help from the Birgit och Gad Rausings Stiftelse för Humanistik Forskning (Birgit and Gad Rausing Foundation), Lund.

2 An adze is a tool used for smoothing or carving wood in hand woodworking.

3 The so-called Iris-Fresco from Amnissos (see also here under 'mint') shows the larger *Iris germanica*, used at least since Middle Minoan times for the production of orris-perfume (Tzedakis and Martlew 1999). Together with the central mint, the iris in this fresco stands on a three-partite dais, emphasising its symbolical importance.

4 Lemma 'Hyakinthos' and 'Apollo'.

5 The central mystery of harvest probably belonged to the symbolism of the lily not mentioned in further detail here. The flower was present in a part of the fresco not exhibited (Cameron 1968).

6 Material presented here was already partly discussed in Beckmann (2006b) and in more detail especially for its calendrical aspects recently, in Beckmann (2012).

7 It was founded by Honorat (later beatified) in the early 5th century as the first monastery in western Europe. It later possessed large properties on the Provençal mainland, among them the village of Cannes.

8 A *dehesa* denominates a mixed agrosylvopastoral system in Spain – sunny, open oak forests which serve as a source of various goods. Apart from timber and firewood taken from the trees, the *dehesas* are sometimes used for cultivated crops, sometimes for hog feeding, and are also a source for game.

Chapter Bibliography

Ahlbäck, M. and Isaksson, M. (2007) *Riksväg 73 Slutundersökningar. RAÄ 661,664,665,666, Ösmo sn, Södermanland.* Rapporter från Arkeologikonsult AB 2037. Upplans Väsby.

Almendros, T. and Manuel, J. (1989) *Ordenanzas municipales de la Ribera del Júcar. Villa de Ves (1589) y Jorquera (1721)* Instituto de Estudios Albacetenses de la Excma. Diputación de Albacete, Albacete.

Andersen, N. H. (1997) *Sarup vol. 1. The Sarup enclosers. The Funnel Beaker Culture of the Sarup site including two causwayed camps compared to the contemporary settlements in the area and other European enclosures.* Jutland Archaeological Society Publications XXXIII:1. Aarhus: Aarhus University Press.

Andersen, N. H. (2000) *Kult og ritualer i den ældre bondestenalder.* Kuml 2000: pp. 13–58.

Antier, J. J. (1975–1988) *Les grandes heures des Îles de Lérions.* Éditions de may, Cannes (orig Librairie académique Perrin, Paris).

Apel, J. (1996) En rumslig tolkning av den tidigneolitiska boplatsen Fågelbacken In: P. Lekberg (ed.) *Fågelbacken ett fornlämningskomplex i östra Västmanland. Slutundersökningsrapport.* Tryckta rapporter från Arkeologikonsult AB nr 14. Upplands Väsby.

Apel, J., Bäckström, Y., Hallgren, F., Knutsson, K., Lekberg, P., Olsson, E., Steineke, M. and Sundström, L. (1995) Fågelbacken och trattbägarsamhället. Samhällsorganisation och rituella samlingsplatser vid övergången till bofast tillvaro i östra Mellansverige. *Tor* 27 (1), 47–132. Uppsala: Uppsala University.

Apel, J., Hadevik, C. and Sundström, L. (1997) Burning down the house. The transformational use of fire and other aspects of an Early Neolithic TRB site in eastern central Sweden. *Tor* 29, 5–47. Uppsala: Uppsala University.

Bakels, C. C. and Zeiler, J. (2005) The fruits of the land. Neolithic subsistence. In L. P. Louwe Kooijmans, P. W. Van den Broeke, H. Fokkens and A. L. van Gijn (eds.)

The prehistory of the Netherlands. Amsterdam University Press, Amsterdam, pp. 311–336.

Beckmann, S. (2006a) The mystery of Taurus - knowing the way. Possibilities for astronomical knowledge in the Bronze Age. *Journal of the Washington Academy of Sciences*, 91 (1), 55–68.

Beckmann, S. (2006b) Conscious of time - Minoan 'calendar'- symbolism in the 'Blue Bird Fresco'. *Proceedings of the 9th International Congress of Cretan Studies*. Elounda, Crete 2001, vol. I, 3. The Society of Cretan Historical Studies, Herakleio, pp. 65–82.

Beckmann, S. (2012) Beyond the Moon. Minoan 'Calendar'-Symbolism in the 'Blue Bird Fresco'. In J. Ben-Dov, W. Horowitz and J. M. Steele (eds.) *Living the Lunar Calendar*. Oxbow Books, Oxford, pp. 63–78.

Berrogain, G. (1930) Ordenanzas de La Alberca y sus términos: Las Hurdes y Las Batuecas. *Anuario de Historia del Derecho Español* VII, 381–441.

Blomqvist, M., Fagerlund, D., Eriksson, T. and Rosborg, B. (2006) *Bönderna i Frotorp - spar av aktiviteter från trattbägartid och bronsålder-äldre järnålder E20, Närke, Viby socken, Frotorp 2:6. RAÄ 220*. UV Gal Rapport 2006: 2. Uppsala.

Boersma, J. W. (1988). De datering van een vuurstenen sikkel uit Middelstum-Boerdamsterweg. In M. Bierma, A. T. Clason, K. E. and G. J. de Langen (eds.) *Terpen en wierden in het Fries-Groningse kustgebied*. Wolters-Noordhof, Groningen, pp. 31–35.

Bradley, R. (2005) *Ritual and domestic life in prehistoric Europe*. Routledge, London and New York.

Brongers, J. A. and Woltering, P. J. (1978) *De prehistorie van Nederland: economisch - technologisch*. Fibula – Van Dishoek, Haarlem.

Browall, H. (1986) *Alvastra pålbyggnad. Social och ekonomisk bas*. Theses and paperes in North European Archaeology 15. Stockholm University, Stockholm.

Brück, J. (1999) Ritual and rationality. Some problems of interpretaton in European archaeology. *European Journal of Archaeology* 2 (3), 313–344.

Brunsting, H. (1962) De sikkels van Heiloo. *Oudheidkundige Mededelingen uit het Rijksmuseum van Oudheden te Leiden* 43, 107–115.

Bruyn, A. (1984) Een vuurstenen 'sikkel' uit Medemblik. *Jaarverslag Rijksdienst voor Oudheidkundig Bodemonderzoek*, 89–94.

Cameron, M. (1968) Unpublished paintings from the 'House of the Frescoes' at Knossos. *Journal of the British School at Athens* 63, 1–31.

Chapin, A. (2004) Power, privilege, and landscape in Minoan Art. In Ann Chapin (ed.) *CHARIS: Essays in Honor of Sara A. Immerwahr*. Hesperia Supplement 33. The American School of Classical Studies, Princeton, pp. 47–64.

Chapman, J. and Gaydarska, B. (2003) The provision of salt to Tripolye mega-sites. In O. G. Korvin-Piotrovsky, V. O. Kruts and S. M. Kruts (eds.) *Tripolian settlements-giants*. Institute of Archaeology, Kiev, pp. 203–211.

Chapman, J., Monah, D. and Monah, F. (2001) *The exploitation of salt in the prehistory of Moldavia, Romania*. Archaeological Reports for 1999/2000. Durham and Newcastle, pp. 10–20.

Cristiani, E., Pedrotti, A. and Gialanella, S. (2009) Tradition and innovation between the Mesolithic and Early Neolithic in the Adige valley (Northwest Italy). New data from a functional and residues analyses of trapezes from Gaban rockshelter. *Documenta Praehistorica* XXXVI (36)(12), pp. 191–205.

Díez Merino, L. (1985) Ordenanzas locales de los pueblos de la peña palentina en el Siglo XV [*sic*]. *Publicaciones de la Institución Tello Téllez de Meneses* 53, 297–319.

Ertuğ, Z. Füsun (ed.) (2006) *Ethnobotany: at the junction of the continents and the disciplines. Proceedings of the IVth International Congress of Ethnobotany (ICEB, 2005), 21–26 August 2005, Istanbul, Turkey*. Yeditepe Üniversitesi, Yayinlari.

Evans, A. (1964) *The Palace of Minos at Knossos* (4 vols.) Biblo and Tannen, New York.

Evely, D. (1999) *Fresco: A Passport into the Past. Minoan Crete through the eyes of Mark Cameron*, British School at Athens, Athens.

Falkenström, P. and Lindberg, K.-F. (2008) Brott i kvarts. In N. Stenbäck (ed.) *Stenåldern I Uppland. Uppdragsarkeologi och eftertanke*. Arkeologi E4 I Uppland vol 1. pp. 227–266. Uppsala: Societas Archaeologica Upsaliensis, Riksantikvarieämbetet & Upplandsmuseet.

Fernández del Pozo, J. M. (1988) *Economía y vida popular en los concejos leoneses. Ordenanzas Municipales de la Ribera del Órbigo. Tres textos inéditos del S. XVI*. Ediciones Leonesas, León.

Fernández García, M. (1966) *Fuentes para la historia de Buitrago y su tierra. Volumen I. Ordenanzas*, Madrid.

Fischer, A. (2002) Food for feasting? An evaluation of explanations of the neolithisation of Denmark and southern Sweden. In A. Fischer and K. Kristiansen (eds.) *The Neolithisation of Denmark. 150 Years debate*. J. R. Collins Publications, Sheffield, pp. 343–393.

Frankfort, H. (1948) *Ancient Egyptian Religio: An Interpretation*. Columbia University Press, New York.

Geertz, C. (1968) *Islam observed*. Yale University Press, New Haven.

Gidlöf, K., Hammarstrand Dehman, K. and Johansson, T. (2006) *Almhov - delområde 1. Citytunnelprojektet*. Malmö kulturmiljö Rappert nr. 39. Malmö.

Van Gijn, A. L. (1999) The interpretation of 'sickles'. A cautionary tale. In P. C. Anderson (ed.) *The prehistory of agriculture*. Institute of Archaeology Monograph 40. University of California, Los Angeles, pp. 363–372.

Van Gijn, A. L. (2008) Not at all obsolete! The social significance of flint for Bronze Age communities, case studies from the Netherlands. In B. Eriksen (ed.) *Lithic technology in metal using societies*. Jutland

Archaeological Society publications Vol. 67, Aarhus, pp. 45–60.

Van Gijn, A. L. (2010) *Flint in focus. Lithic biographies in the Neolithic and Bronze Age.* Sidestone Press, Leiden.

Van Gijn, A. L., Van Betuw, V., Verbaas, A. and Wentink, K. (2006) Flint. Procurement and use. In L. P. Louwe Kooijmans and P. F. B. Jongste (eds.) *Schipluiden a Neolithic site on the Dutch north sea coast (3800–3500 BC)*, Analecta Praehistorica Leidensia 37/38, 129–166.

Van Gijn, A. L. and Verbaas, A. (2008) Het technologische systeem van Ypenburg. Een gebruikssporenanalyse van diverse werktuigtypen. In J. M. Koot, L. Bruning and R. A. Houkes (eds.) *Ypenburg-locatie 4, een nederzetting met grafveld uit het Midden-Neolithicum in het West-Nederlandse kustgebied.* Station Drukwerk Leiden, pp. 289–314.

Gronenborn, D. (1994) Überlegungen zur Ausbreitung der bäuerlichen Wirtschaft in Mitteleuropa – Versuch einer kulturhistorischen Interpretation ältestbandkeramischer Silexinventare. *Praehistorische Zeitschrift* 70 (1), 137–151.

Guinard, M. and Groop, N. (2008) Handtagskärnor och tvärpilar. In N. Stenbäck (ed.) *Stenåldern I Uppland. Uppdragsarkeologi och eftertanke.* Arkeologi E4 I Uppland vol 1: pp. 207–226. Uppsala: Societas Archaeologica Upsaliensis, Riksantikvarieämbetet & Upplandsmuseet.

Hallgren, F. (ed.) (2004) *Trössla, efterundersökning av en grustagen trattbägarboplatsi östra Södermanland.* SAU Rapport 2004:7. Uppsala.

Hallgren, F. (2008) *Identitet i praktik. Lokala, regionala och överregionala sociala sammanhang inom nordlig trattbägarkultur.* Coast to coast-book 17. Uppsala.

Hallgren, F., Bergström, Å. and Larsson, Å. (1995) *Pärlängsberget en kustboplats från övergången mellan senmesolitikum och tidigneolitikum.* Slutundersökningsrapport. Tryckta rapporter från Arkeologikonsult AB nr 13. Upplans Väsby.

Hallgren, F., Djerw, U., af Geijerstam, M. and Steineke, M. (1997) Skogsmossen, an Early neolithic settlement site and sacrificial fen in the northern borderland of the Funnel-beaker Culture. *Tor* 29, 49–112. Uppsala: Uppsala University.

Hallgren, F. and Possnert, G. (1997) Pottery design and time. The pottery from the TRB site Skogsmossen, in view of the AMS-datings of organic remains on potsherds. *Tor* 29, 113–136. Uppsala: Uppsala University.

Hamon, C. (2008) The symbolic value of grindingstones hoards: technical properties of Neolithic examples. In C. Hamon and B. Quilliec (eds.) *Hoards from the Neolithic to the Metal Ages. Technical and codified practices. Session of the XIth Annual Meeting of the European Association of Archaeologists*, Archaeopress, Oxford, pp. 19–28.

Hansson, A.-M. and Heiss, A. G. (2014) Plants used in Ritual Offerings and in Festive Contexts: Introduction. In A. Chevalier, E. Marinova and L. Peña-Chocarro (eds.) *Plants and People: Choices and Diversity through Time.* EARTH Monograph 1. Oxbow Books, Oxford, pp. 311–334.

Heiss, A. G. and Hansson, A.-M. (2014) Plants used in Ritual Offerings and in Festive Contexts: Conclusions. In A. Chevalier, E. Marinova and L. Peña-Chocarro (eds.) *Plants and People: Choices and Diversity through Time.* EARTH Monograph 1. Oxbow Books, Oxford, p. 374.

Hesiod, Works and Days. (2006) translated by Catherine M. Schlegel and Henry Weinfeld in: *Theogony and Works and Days.* The University of Michigan Press, Ann Arbor.

Immerwahr, S. A. (1990) *Aegean Painting In The Bronze Age.* Pennsylvania State University Press, University Park and London.

Insoll, T. (2004) *Archaeology, ritual, religion.* Routledge, New York.

Isager, S. and Skydsgaard, J. E. (1992) *Ancient Greek Agriculture, An Introduction.* Routledge, London.

Jacob-Friesen, K. H. (1955) *Einführung in Niedersachsens Urgeschichte, Teil I, Steinzeit.* Hildesheim.

Jadin, I. (2003) *Trois petits tours et puis s'en vont... La fin de la presence danubienne en Moyenne Belgique.* Etudes et Recherches Archéologiques de l'Université de Liège, Liège.

Johansson, P. (2003) *The lure of origins: an enquiry into human-environmental relations, focused on 'Neolithisation' of Sweden.* Coast to coast-book 8. Uppsala: Uppsala University.

Juel Jensen, H. (1994) *Flint tools and plant working. Hidden Traces of Stone Age Technology. A use wear study of some Danish Mesolithic and TRB implements.* Aarhus, Aarhus University Press.

Karsten, P. (1994) *Kasta yxan i sjön.* Acta Archaeologica Lundensia 8 (23). Lund, Lunds University.

Knutsson, H. (1995) Slutvandrat. Aspekter på övergången från rörlig till bofast tillvaro. *Aun* 20. Uppsala: Uppsala University.

Knutsson, H. (1999) Rituella och funktionella flintor. In E. Hjärthner-Holdar and C. Risberg (eds.) *Hantverkets roll i samhället – produktion och reproduktion. Rapport från den tredje workshopen om hantverksspecialisering Uppsala 18–19 mars 1999.* Uppsala, pp. 17–33.

Knutsson, H. (2001) *Technology, mythology and the travels of the agricultural package in Europe.* Documenta Praehistorica XXXVIII. Ljubljana, pp. 117–132.

Knutsson, H. (ed.) (2004) *Coast to coast – Arrival. Results and reflections.* Coast to coast-book 10. Uppsala: Uppsala University.

Knutsson, H. (2008) Spån och tidigt jordbruk. In N. Stenbäck (ed.) *Stenåldern I Uppland. Uppdragsarkeologi och eftertanke.* Arkeologi E4 I Uppland vol. 1, pp. 319–365. Uppsala: Societas Archaeologica Upsaliensis, Riksantikvarieämbetet & Upplandsmuseet.

Knutsson, H. (2012) Using Worldwide Samples in Understanding Mesolithic-Neolithic Transition in

Northern Europe. In A. Cannon (ed.) *Structured Worlds: The Archaeology of Hunter-Gatherer Thought and Action.* Approaches to Anthropological Archaeology. Sheffield, Equinox Publishing.

Knutsson, K., Falkenström, P. and Lidberg, K. F. (2003) Appropriation of the past. Neolithisation in the northern Scandinavian perspective. In L. Larsson, H. Kindgren, K. Knutsson, D. Loeffler and A. Åkerlund (eds.) *Mesolitic on the Move. Papers presented at the Sixth International Conference on the Mesolithic in Europe, Stockholm 2000.* Oxbow Books, Oxford, pp. 414–430.

Knutsson, K., Knutsson, H. and Apel, J. (in prep.) *En bok om stenar och människor. Om klassifikation av bearbetade stenartefakter.* Uppsala.

Koch, E. (1998) *Neolithic bog pots from Zealand, Møn, Lolland and Falster.* Nordiske Fortidsminder Serie B Volume 16. København.

Kopytoff, I. (1986) The cultural biography of things. Commoditization as process. In A. Appadurai. *The social life of things. Commodities in cultural perspective.* Cambridge University Press, Cambridge, pp. 64–91.

Kubiak-Martens, L. (2006) Botanical remains and plant food subsistence. In L. P. Louwe Kooijmans and P. F. B. Jongste (eds.) *Schipluiden. A Neolithic settlement on the Dutch North Sea coast c. 3500 cal BC,* Analecta Praehistorica Leidensia 37/38, Leiden, pp. 317–338.

Ladero, Q., Ángel, M. and Galán Parra, I. (1982) Las ordenanzas locales en la Corona de Castilla como fuente histórica y tema de investigación (siglos XIII al XVIII). *Anales de la Universidad de Alicante. Historia Medieval* 1, 21–243.

Ladero, Q. and Ángel, M. (1998) Las ordenanzas locales. Siglos XIII-XVIII. *La España Medieval* 21, 293–337.

Larsson, M. (1984) *Tidigneolitikum i Sydvästskåne. Kronologi och bosättningsmönster.* Acta Archaeologica Lundensia Series in 4°. No 17. Lund.

Larsson, L. (1999) The Mesolithic of Southern Scandinavia. Journal of World Prehistory 4 (3), 257–309.

Lidström Holmberg, C. (2004) Saddle Querns and gendered dynamics of the Early Neolithic in mid central Sweden. In H. Knutsson (ed.) *Coast to coast - Arrival. Results and reflections.* Coast to coast-book 10. Uppsala: Uppsala University.

Lindberg, K.-F. (2008) Mesolitiska och neolitiska verktyg. In N. Stenbäck (ed.) *Stenåldern I Uppland. Uppdragsarkeologi och eftertanke.* Arkeologi E4 I Uppland vol. 1, pp. 267–288. Uppsala: Societas Archaeologica Upsaliensis, Riksantikvarieämbetet & Upplandsmuseet.

Louwe Kooijmans, L. P. (2004) Comment. In P. Rowley-Conwy, How the west was lost. A reconsideration of agricultural origins in Britain, Ireland and southern Scandinavia. *Current Anthropology* 45 (Supplement August-October), 102–103.

Louwe Kooijmans, L. P. (2007) The gradual transition to farming in the Lower Rhine Basin. In A. Whittle and V. Cummings. *Going Over. The Mesolithic-Neolithic transition in north west Europe,* Oxford University Press, Oxford, pp. 287–309.

Lucas, G. (2005) *The archaeology of time.* Routledge, New York.

Madsen, T. (1997) Ideology and social structure in earlier neolithic of south Scandinavia. A view from the sources. In A. L. van Gijn and M. Zvelebil (eds.) *Ideology and social structure of Stone Age communities in Europe.* Analecta Praehistoria Leidensia 29. Leiden University, Leiden.

Malmer, M. P. (2002) *The neolithic of South Sweden, TRB, GRK and STR.* Alqvist & Wiksell International, Stockholm.

Marinatos, N. (1985) The function and interpretation of the Theran frescoes. *Bulletin de Correspondance Hellénique,* 11(1), 219–230.

Marinatos, N. (1990) Minoan-Cycladic Syncretism. In D. Hardy, C. G. Doumas, J. A. Sakellarakis and P. M. Warren (eds.) *Thera and the Aegean World III.1.* The Thera Foundation, London, pp. 370–377.

Martínez Martínez, M. (2003) Ordenanzas de Murias de Rechivaldo. Año 1872. *Tierra de León* 41, (117), 59–80.

McBryde, I. (1997) The cultural landscapes of Aboriginal long distance exchange systems. Can they be confined within our heritage registers? *Historic Environment* 13 (3/4), 6–14.

Mingote Calderón, J. L. (1995) *No todo es trabajo: técnicas agrícolas tradicionales.* Centro de Cultura Tradicional. Diputación de Salamanca, Salamanca.

Mingote Calderón, J. L. (2003) Des implications idéologiques de l'outil agricole dans la société médiévale et moderne hispanique. In G. Comet (ed.) *L'Outillage Agricole dans l'Europe médiévale et moderne. Actes des XXIIIe Journées Internationales d'Histoire de l'Abbaye de Flaran. 7, 8, 9 septembre 2001.* Études réunies par. Presses Universitaires du Mirail, Toulouse, pp. 95–146.

Mingote Calderón, J. L. (2008) *Forcados y carros.* El Diario de León. Edilesa, León.

Möbius, H. (1933) Pflanzenbilder der minoischen Kunst in botanischer Betrachtung. *Jahrbuch des Deutschen Archäologischen Instituts* 48, 1–39.

Modderman, P. J. R. (1970) Linearbandkeramik aus Elsloo und Stein. *Analecta Praehistorica Leidensia* 3, Leiden.

Nielsen, P. O. (1993) Settlement. In S. Hvass and B. Storgaard (eds.) *Digging into the past. 25 years of archaeology in Denmark.* The Royal Society of Northern Antiquities & Jutland Archaeological Society, Aarhus.

Parain, Ch. (1938) Les ancienprocédés de battage et de dépiquage en France.*Travaux du Premier Congrès international de Folklore,* Tours, pp. 89–90

Parker Pearson, M. (2008) New discoveries at Stonehenge. *C.J.C. Reuvenslezing* 20, Erfgoed Nederland, Amsterdam.

Philippart, G. (1990) Temps sacré temps chômé. Jours chômes en Occident de Caton l'Ancien à Louis le Pieux l'Occident médiéval. In J. Hamesse and C. Muraille-Samaran (eds.) *Le travail au Moyen Âge. Une approche interdisciplinaire. Actes du Colloque international de Louvain-la-Neuve. 21–23 mai 1987.* Université Catholique de Louvain, Louvain-la-Neuve, pp. 23–34.

Pliny the Elder. In John Bostock, M.D., F.R.S. H.T. Riley, Esq., B.A. London. *The Natural History.* Accessible on the Internet at: http://www.perseus.tufts.edu/hopper/t?doc=Perseus%3Atext%3A1999.02.0137%3Abook%3D16%3Achapter%3D95 (Accessed last 12/1/2011)

Primo Ruiz, J. (1963) Ordenanzas antiguas del Concejo de Sotillo de Boedo. *Publicaciones de la Institución Tello Téllez de Meneses* 23, 65–87.

Raemaekers, D., Bakels, C. C., Beerenhout, B., Van Gijn, A. L., Hanninen, K., Molenaar, S., Paalman, D., Verbruggen, M. and Vermeeren, C. (1997) Wateringen 4. A settlement of the Middle Neolithic Hazendonk 3 group in the Dutch coastal area, *Analecta Praehistorica Leidensia* 29, 143–192.

Rankama, T. (2002) Analyses of the Quartz Assemblages of Houses 34 and 35 at Kauvonkangas in tervola. In H. Ranta (ed.) *Huts and Houses. Stone Age and Early Metal Age buildings in Finland.* Helsinki, pp. 79–108.

Roscher (1886–1890) *Lexikon der Griechischen und Römischen Mythologie*, Leipzig.

Rowley-Conwy, P. (1984) *How the west was lost. A reconsideration of agricultural origins in Britain, Ireland and southern Scandinavia.* Current Anthropology 45 (Supplement August–October), 83–113.

Rubio Pérez, L. M. (1993) *El sistema político concejil en la provincia de León.* Universidad de León, León.

Rudebeck, E. (2010) I trädstodernas skugga – monumentala möten i neolitiseringens tid. In B. Nilson and E. Rudebeck (eds.) *Arkeologiska och förhistoriska världar: fält erfarenheter och stenåldersboplatser i sydvästra Skåne.* Malmöfynd 19. Malmö Museer, Arkeologienheten, Malmö.

Rüpke, J. (2007) *Religion of the Romans.* (Translation of 2001 edition by Richard Gordon). Polity Press, Cambridge, UK/Malden, MA.

Schäfer, J. (1977) Zur kunstgeschichtlichen Interpretation altägäischer Wandmalerei. *Jahrbuch des Deutschen Archäologischen Instituts* 92, 1–23.

Segerberg, A. (1999) Bälinge mossar. Kustbor i Uppland under yngre stenålder. *Aun* 26. Uppsala: Uppsala University.

Serra-Rafols, I. de C. (1949) Legislación medieval en un valle pirenaico. Las 'Ordinacions de la vila d'Ager' en el año 1278. *Pirineos* V, 11–12, 219–251.

Sherratt, A. (1991) Sacred and profane substances: the ritual use of narcotics in later Neolithic Europe. In P. Garwood, D. Jennings, R. Skeates and J. Toms (eds.) *Sacred and profane: Proceedings of a conference on archaeology, ritual and religion, Oxford 1989.* Oxford University, Oxford, 50–64.

Skoglund, P., Malmström, H., Raghavan, M., Storå, J., Hall, P., Willerslev, E., Thomas, M., Gilbert, P., Götherström, A. and Jakobsson, M. (2012) Origins and Genetic Legacy of Neolithic Farmers and Hunter-Gatherers in Europe. Science 336 (6080), 466–469.

Stafford, M. (1999) From forager to farmer in flint. A lithic analysis of the prehistoric transition to agriculture in Southern Scandinavia. Aarhus: Aarhus University Press.

Stenbäck, N. (ed.) *Stenåldern I Uppland. Uppdragsarkeologi och eftertanke.* Arkeologi E4 I Uppland vol 1. Uppsala: Societas Archaeologica Upsaliensis, Riksantikvarieämbetet & Upplandsmuseet.

Sundström, L. (2003) *Det hotade kollektivet. Neolitiseringsprocessen ur ett östmellansvenskt perspektiv.* Coast to coast-books no. 6. Uppsala: Uppsala University.

Sundström, L. and Darmark, K. (2006) *Bålmyren. En familjebaserad tidigneolitisk kustboplats i Uppland.* SAU Skrifter 7. Uppsala.

Sundström, L., Darmark, K. and Stenbäck, N. (eds.) (2006) *Postboda 2 och 1. Sässongsboplatser med gropkeramik från övergången tidigneolitikum-mellanneolitikum i norra Uppland.* SAU Skrifter 10. Uppsala.

Sundström, L. and Hallgren, F. (2008) *Tidigneolitisk trattbägarkultur i Uppland.* In E. Hjärthner-Holdar, H. Ranheden and A. Seiler (eds.) *Land och samhälle i förändring. Uppländska bygder i ett långtidsperspektiv.* Arkeologi E4 I Uppland vol 1, pp. 199–228. Uppsala: Societas Archaeologica Upsaliensis, Riksantikvarieämbetet & Upplandsmuseet.

Svensson, M. (1993) *Hindby offerkärr – en ovanlig och komplicerad fyndplats. Fynd Nr. 2, 1993.* Göteborg, pp. 5–11.

Taçon, P. S. C. (2004) Ochre, clay, stone and art. In N. Boivin and M. A. Owic (eds.) *Soils, stones and symbols. Cultural perceptions of the mineral world.* UCL Press, London, pp. 31–42.

Taffinder, J. (1998) The allure of the exotic. The social use of non local raw materials during the Stone Age in Sweden. Aun 25. Uppsala: Uppsala University.

Thomas, J. (1988) Neolithic explanations revisited: the Mesolithic-neolithic transition in britain and south Scandinavia. *Proceedings of the Prehistoric Society* 54, 59–66.

Thomas, J. (1991) Rethinking the Neolithic. Cambridge University Press, Cambridge.

Thomas, J. (1999) *Understanding the Neolithic.* Routledge, London.

Thomas, J. (2004) Comment. In P. Rowley-Conwy, How the west was lost. A reconsideration of agricultural origins in Britain, ireland and southern Scandinavia. Current Anthropology 45 (Supplement August–October), 105–106.

Tzedakis, Y. and Martlew, H. (eds.) (1999) *Minoans and Mycenaeans: Flavours of their Time.* National Archaeological Museum/Kapon, Athens.

Van de Velde, P. (1979) *On bandkeramik social structure.* PhD thesis, Leiden University, Analecta Praehistorica Leidensia 12.

Verbaas, A. and Van Gijn, A. L. (2007) Querns and other

hard stone tools from Geleen-Janskamperveld. In P. Van de Velde, *Geleen-Janskamperveld*, Analecta Praehistorica Leidensia 39, Leiden, pp. 191–204.

Warrior, V. M. (2006) *Roman Religion, Cambridge Introduction to Roman Civilization Series*. Cambridge University Press, Cambridge.

Welinder, S. (1974) Neoliticum - bronsålder 3900–500 F. Kr. In S. Welinder, E. A. Pedersen, and M. Widgren (eds.) *Jordbrukets första femtusen år. 4000 f. Kr. - 1000 e. Kr.* Det svenska jordbrukets historia del 1. Natur och Kultur/LTs förlag, Stockholm.

Wildhaber, R. (1981) Feiertagschristus. In. *Reallexikon zur Deutschen Kunstgeschichte*. Bd. VI: beg. von O. Schmitt, fortgeführt von E. Gall, L. H. Heydenreich, H. M. von Erffa und K. A. Wirth, hg. vom Zentralinstitut für Kunstgeschichte München. pp. 1002–1010. Online at (accessed last on 12/1/2011): http://rdk.zikg.net/gsdl/cgi-bin/library.exe (lemma Feiertagschristus).

Zápotocký, M. (1992) Streitäxte des mitteleuropäischen Äneolithikums. VCH Verlagsgesellschaft, Weinheim.

Zimmermann, A. (1997) *Jesus Christus als 'Schmerzensmann' in hoch- und spätmittelalterlichen Darstellungen der bildenden Kunst: eine Analyse ihres Sinngehalts*. PhD Dissertation Halle 1997 - (Accessed last 12/1/2011) online at: http://sundoc.bibliothek.uni-halle.de/diss-online/97/98H110/index.htm

Zvelebil, M. (1997) Hunter-gatherer ritual landscapes: spatial organisation, social structure and ideology among hunter-gatherers of Northern Europe and Western Siberia. In A. L. van Gijn and M. Zvelebil (eds.) *Ideology and social structure of Stone Age communities in Europe*. Analecta Praehistorica Leidensia 29, Leiden, pp. 3–12.

10 Agricultural Practices: Change and Stability

10.1. INTRODUCTION

Thomas K. Schippers

Agriculture in general and agricultural practices in particular have frequently been considered proverbial examples of stability and routine until the dawn of industrialisation. These stereotypical views have often been based on imprecise technological and ecological data and models that advocated continuity and even technical stagnation in rural societies. But more recent, detailed information on bio-climatic changes, as well the quantitative and qualitative enrichment of precise historical and ethnographic sources, tend to shatter these stereotypes. This chapter will illustrate both stability and change with some examples of the great complexity of agricultural practices as dynamic results of not only ecological and technical 'constraints' but also of fluctuating socio-economic and cultural contexts. Agricultural practices, as means of subsistence literally 'grounded' in place and time, not only embody the necessary expertise to deal with current situations, but also accumulate and preserve collective, trans-generational memories and local knowledge in order to be resilient when challenged by internal or external changes or crises.

Agricultural practices can be defined as specific ensembles of knowledge and skills brought into play to domesticate certain plants and/or animals in order to satisfy human nutritional, material or immaterial needs. As such, agricultural practices rely on the social transmission of environmental/ecological knowledge and the concrete implementation of learned skills, gestures and very often, specific tools. As pioneers in the field of the study of agricultural practices like André-Georges Haudricourt have pointed out, the cognitive and social dimensions of agricultural practices most often prevail over the material aspects. Many ethnographic and also historical examples emphasise the importance of executing agricultural practices in a 'proper way' according to local social, cultural and sometimes religious standards. This social dimension offers an insight into the force of their stability over time. For example, in all agricultural societies bio-climatic cycles are domesticated with the help of perceived environmental markers and socialised with the help of calendar prescriptions and rituals (see Chapter 9). In more complex agricultural societies, collective rules in regard to the rotation of cultivating and fallowing or the cycles of irrigation have durably formatted agricultural practices, often with the help of very strict oral or written rules. In feudal agricultural societies, landowners and local authorities, always keen to strictly control peasants and harvests, usually regulated both labour calendars and 'permitted' techniques in very precise ways. These legal dimensions of agricultural practices have of course reinforced the stability of socio-cultural norms related to bio-climatic and agro-technical constraints, producing circular patterns of causality.

But if these circular causalities at work in agricultural practices have more generally contributed to great (systemic) stability over time, they are not necessarily synonymous with rigidity. The inseparable twins of human perception and action at work in all processes of domestication (Fig. 10.1), allow – within limits – adjustments and even important changes when confronted with the modification of internal and/or external factors. The temporal scales of these adjustments in agricultural practices can range from a few months to several centuries according to the triggering factors: seasonal weather conditions (droughts, floods), demographic changes (epidemics, migrations), bio-climatic changes, technical improvements (innovations), political events (wars, conquests) and so forth. So, in order to deal with changes, agricultural societies can either draw on local knowledge and skills or on borrowed innovations, as some examples in this chapter will show. Adjustments relying on locally accumulated expertise, available technologies, and social structures may either be successful by assuring the durability of the rural society when systemic changes occur, or they may prove insufficient, resulting in decline and even disappearance. Archaeologists, historians and ethnographers have attempted to document these changes with increasing success. More diversified and precise sources of information such as bioclimatology, palynology, microwear analysis, ethnobotany, quantitative ethnography, technological analyses and even experimental reconstructions enable us to outline more systemic connections and correlations between various factors and parameters.

The qualitative and quantitative improvements and increased diversity of data about past and present pre-industrial agricultural practices have also led to an enrichment of scholarly analysis and understanding of underlying processes. Old theoretical issues about technological choices, borrowings or apparent refusals are today the object of more nuanced, multi-factorial answers. Historically or ethnographically documented case-studies offer stimulating models for the interpretation of archaeological data regarding changing practices. For example, the precise measurements and experiments by the French ethnographer-anthropologist Maurice Godelier in the 1960s regarding the recent replacement of stone axes by imported steel ones in the Baruya slash-and-burn society of New Guinea, help to understand how this technological improvement not only increased the size of the cultivated plots, but also male domination over the women (who cultivate them) and the societal consequences of this (Godelier and Garanger 1973; Godelier 1982). The hypotheses formulated by André Leroi-Gourhan in the 1940s stressing the importance of considering techniques, skills and practices together as complex entities (*milieu technique*), still offer, with today's enriched, diversified data, a fruitful way to understand both change and stability (Leroi-Gourhan 1945/1973, 303–401). For example, his idea that an innovation needs to encounter a *milieu favorable* where it can fill a potential need in order to be adopted (if compatible with the available material and immaterial means), seems especially stimulating when examining cases of technological refusal or stagnation. Many well documented historical and ethnographic examples have shown how social factors can slow down or even block technical changes that do not fit into current local practices and values. On the other hand, especially in the field of tastes and cuisine, which often contribute to specific group identities, singular practices can be maintained for long periods within changing technological and economic contexts. Finally, new technologies and/or practices can sometimes fully replace former solutions which may be abandoned as obsolete and disappear quite suddenly, as for example, what happened to most of the plants cultivated in Europe for dyeing. Modern mechanisation as well as colonisation and trade have in recent times contributed to the decline and disappearance of many local skills and practices and related tools and crops, but also to a growing awareness of the importance of the preservation of local cultivars and knowledge associated with them.

The contributions in this chapter offer some glimpses into the issues of change and stability in agricultural practices at various spatio-temporal scales. Maria Gurova presents some Bulgarian examples of the appearance in the 4th millennium BCE of denticulated blades as sickle inserts, as well as the more puzzling case of the dearth of archaeological evidence of threshing sledges before the historically and ethnographically documented

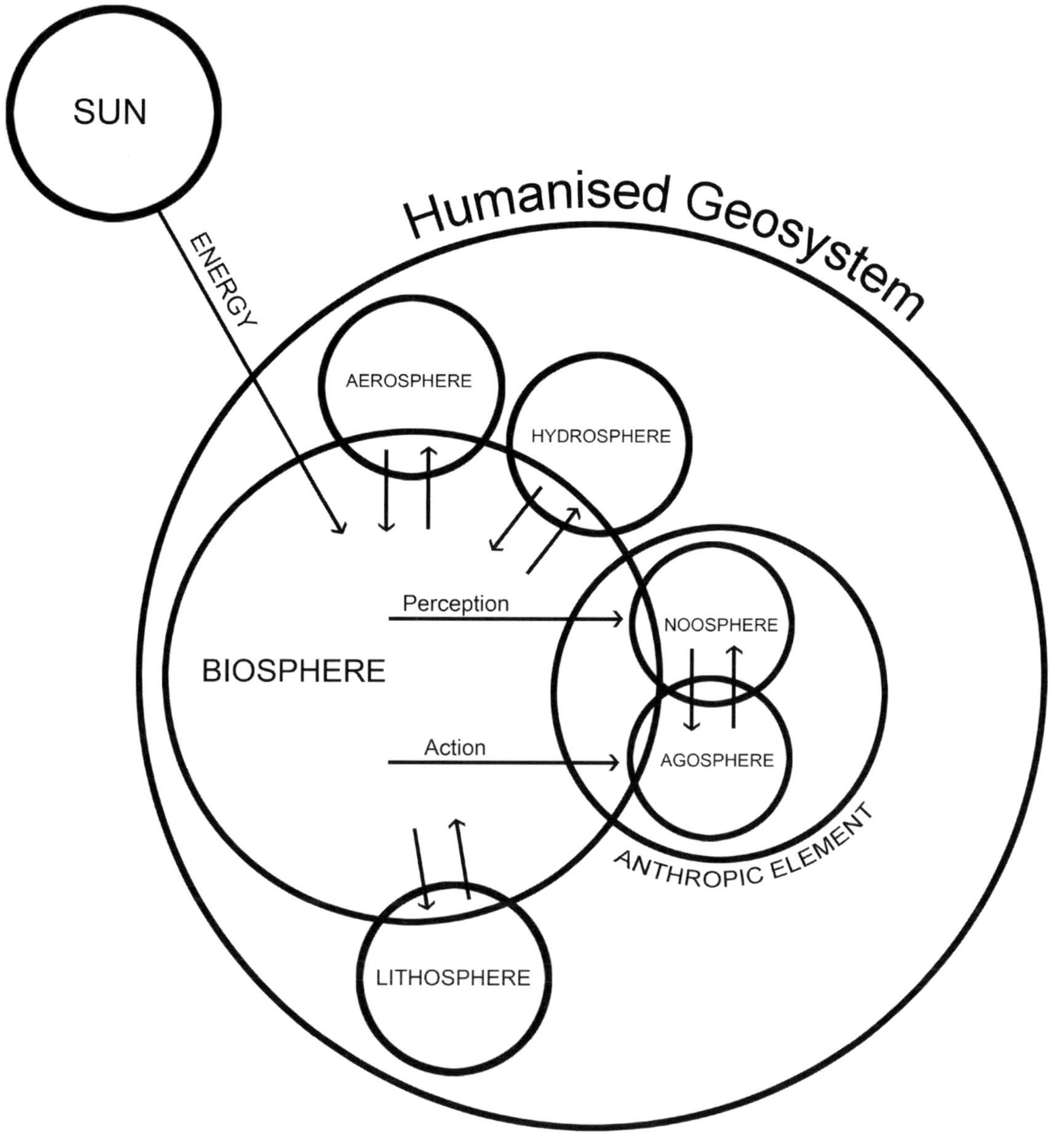

Fig 10.1. Human perception and action as the anthropic interface within a geo-system.

ones from Roman, Byzantine and later periods. Yorgos Brokalakis questions the apparent stability of agricultural tools during the centuries-long Byzantine era, which may be the result of missing archaeological data and the predominance of imprecise textual and pictorial sources. John C. Whittaker briefly reflects on some of the conditions surrounding the decline of the threshing sledge in Cyprus and Turkey in the 20th century. In the last contribution of this chapter, Thomas K. Schippers presents an ethno-historic case study of the complexity of change in a small rural society in the southern Alps during the last two centuries, analysing the various feedback mechanisms at work in the local society and in its agricultural practices when the context slowly changes.

10.2. SOME EXAMPLES OF TECHNOLOGICAL CHANGE IN PREHISTORIC BULGARIA

Maria Gurova

Introduction

The evidence of prehistoric flint assemblages from Bulgaria shows a remarkable long-standing development and continuity from the beginning of the Neolithic to the end of the Chalcolithic, a time span of two millennia. Although the study of numerous flint toolkits reveals some changes in technology and of raw material supply, no signs of any drastic crisis are attested. For centuries our prehistoric predecessors apparently successfully managed with different revival strategies leading to different changes in social organisation, adaptation and function. These changes seem to appear in the *habitus* arrangement, a presumed (but not clearly revealed) gender division of labour, an increasing craft standardisation, a sophistication of ritual practices and certainly an evolution in lithic production, directly related to the agricultural practices and subsistence activities.

The flint toolkit considered here is an intrinsic part of the prehistoric agricultural technical equipment which consists of two main tool types: the well known and elaborate sickles from Early Neolithic Karanovo I culture (VI mill. BCE) and the *tribulum* (threshing sledge), the use of which use can be dated with confidence from the Drama-Merdžumekja site (Chalcolithic period Karanovo V and VI–V mill. cal BCE). Both sickles and threshing sledges have their own 'biography' of evolution within the context of non-industrial agriculture. Due to the vagaries of specialised studies and physical evidence, their 'biographies' are difficult to follow during the many centuries from late prehistory until the complex late Medieval Ages. The abundant ethnographical data from modern times show that both sickles (made of metal) and *tribula* (mostly with flint inserts) show different degrees of resemblance to their prehistoric predecessors and were still in active use in the beginning of the 20th century in Bulgaria.

Prehistoric Sickles

Harvesting tools (sickles) represent the most characteristic and peculiar tool categories from the very beginning of the Neolithic in the Balkans (end of the seventh millennium cal BCE; see the sites in Fig. 10.2 as mentioned below). They are quite conservative and are among the most long-lasting items in the tool repertoire. Despite their stylistic variability, sickles can be reliably identified by handle fragments or typical micro use-wear of their flint inserts (Fig. 10.3).

The flint industry is often considered to be one of the most conservative elements of material culture. For a prehistoric population, it was common practice to keep, maintain and reproduce a range of efficient tools for household and subsistence activities over a long time. In Bulgarian prehistory, the Karanovo type of sickle, successfully used for more than two millennia, represents a good example of such a long-lasting tool. Here it is worth remembering that from the beginning of the Neolithic period in Bulgaria, a complete 'Neolithic package' was present in terms of pottery production, agricultural toolkit, domesticated animals and crop cultigens (Colledge and Conolly 2007; Marinova 2007). The sickles from Tell Karanovo and other Thracian Neolithic

Fig. 10.2. Bulgarian sites as mentioned in Chapter 10.2. 1) Ada Tepe; 2) Aul Kaja; 3) Drama – Merdžumekja; 4) Kamenska Čuka; 5) Karanovo (with Tell Karanovo); 6) Yunatsite. Map: R. Lugon, J.-C. Loubier and A. Chevalier.

tells, with obliquely inserted flint segments in slightly curved antler handles, are well known and often used as a reference for the reconstruction of prehistoric harvesting practices (Figs. 10.4 and 10.5). It should be stressed that almost all sickle inserts identified among the numerous assemblages from the Neolithic and Chalcolithic periods in Bulgaria have a diagonal polish that indicates the angular fixation in arched handles. A few examples with a narrow lateral polish suggest parallel hafting or even use without a handle. Within the framework of the studied assemblages, these examples cannot be interpreted as preceding the Karanovo sickle type, but rather as 'reaping knives', evidence of which has been rather sporadic and contemporary with the typical sickle (Gurova 1997). Detailed studies of the origin, spread, spatial and temporary distribution, and function of Neolithic sickles in Bulgaria have been published in recent years (Gurova 2004a; 2008a).

In the beginning of the Bronze Age (4th millennium BCE), a shift in sickle morphology and usage can be recognised. Traditional sickles were first partially and later entirely replaced by a new form in the flint tool repertoire: denticulated blades. Denticulated implements appear sporadically as an innovation in the flint toolkit during the late Chalcolithic, but it was only during the Bronze Age that they became a diagnostic feature of the flint industry for this period. Their appearance, functional identification and more general interpretation have already been discussed in specialised literature on Bulgarian prehistory (Gurova 2001a; 2004b; 2005; 2007; 2008b). Until quite recently there was not sufficient evidence about material culture during the Bronze Age in Bulgaria, but this scarcity of archaeological records has been successfully made up for through extensive contemporary surveys and wide-scale rescue excavations.

As mentioned, the denticulated implements represent an innovation. They consist mainly of blades with different dimensions and more rarely of blade-like flakes. The most characteristic feature of these artefacts is the intentional denticulated retouch varying from fine and delicate to pronounced bifacial notches on the working edges. This retouch undoubtedly was intended to improve the efficiency of the cutting tools. The microwear traces on these denticulated implements show a typical 'cereal polish', oriented parallel or at a slight angle to the edges. Morpho-metrical parameters of the artefacts suggest at least two options regarding their use: insertion in a composite sickle, or simply working without a handle. Among the denticulated implements from Bulgarian assemblages that have been studied, the most interesting and representative of these tools are several items from the following sites (see Fig. 10.2): Lepitsa (northern Bulgaria, Early Bronze Age, excavation in progress), the multi-layer Tell Yunatsite (southern Bulgaria, Late Bronze Age, exploration finished), the Late Bronze Age site at Kamenska Čuka (south Bulgaria), Bronze and Iron Age sites in Svilengrad district, Bronze Age sites Aul Kaja and Ada Tepe[1] in the eastern Rhodopes (Gurova 2006, 2008b) (Figs. 10.6 and 10.7).

What caused such a radical change in these flint tools so important to the subsistence economy? A convincing answer has come from palaeo-botanical studies, which do not show dramatic changes in

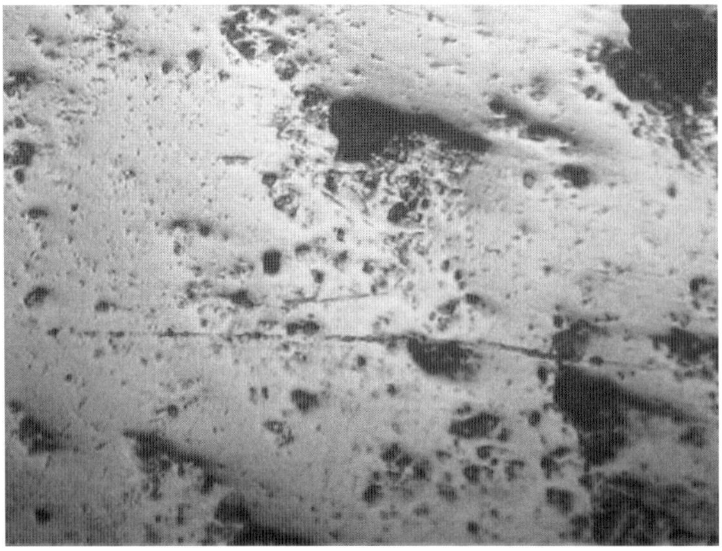

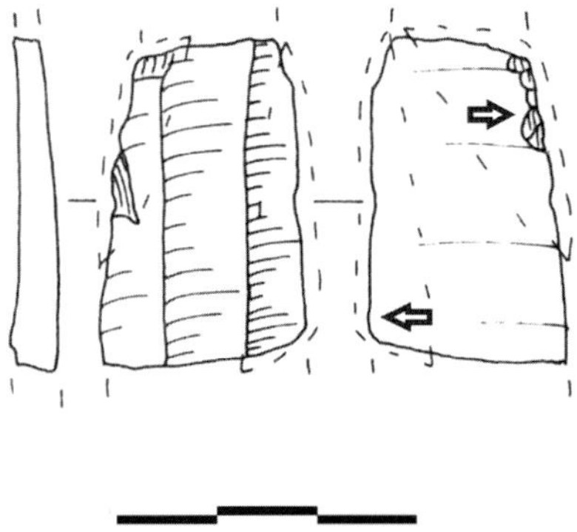

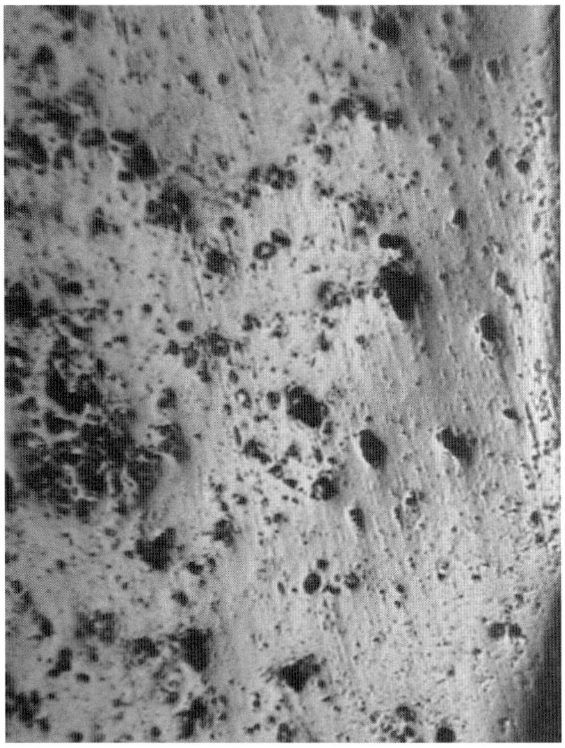

Fig. 10.3. Sickle insert from Chalcolithic site of Drama – Merdžumekja site. Drawing and photomicrograph of cereal-polish (x100): M. Gurova.

the composition of the cereal taxa, but rather a misbalanced ratio between them. Einkorn *(Triticum monococcum)* predominates in combination with barley (*Hordeum vulgare* var. *vulgare*). During the LBA, an important increase in millet (*Panicum miliaceum*) cultivation is attested among the cereal remains (Popova 1995). Millet is more resistant to climatic and environmental changes due to its stronger stems. This fact, as well as the increased growth of weeds, such as *Setaria* sp., *Polygonum/Persicaria* spp., *Agrostemma githago, Chenopodium* type and *Carthamus* sp. (Stefanovich and Bankoff 1998, 275) is quite indicative. The stronger stems of millet, plus an increase in tough weeds, could be an explanation of this change in the harvesting toolkit to a more effective denticulated sickle.

My study of the flint assemblages from the Bronze Age site of Troy in Turkey (Çanakkale district) allowed me to identify a type of tool, made mostly of blades or thin blade-like pieces of tabular flint with cortex on both faces (Fig. 10.8). The typical features of these tools are their impressive dimensions and intentionally denticulated working

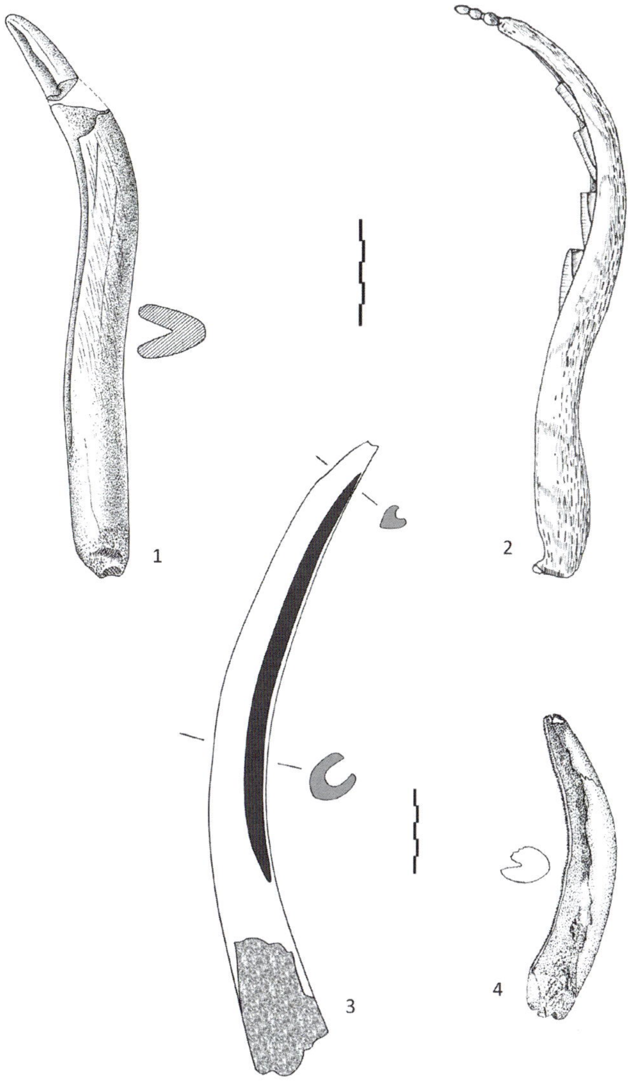

Fig. 10.4. Prehistoric sickles: 1) site of Čavdar (according to K. Kančev, unpublished monographs, used with his permission); 2) Tell Karanovo according to G. Georgiev (1958, 371, fig. 2); 3) Kovačevo (according to Sidera 1998, 225, fig. 9); 4) Samovodene (according to Stanev 2002, 287, fig. 96).

edges with detectable use-wear traces of cereal-processing preserved particularly on the teeth of the denticulation (Fig. 10.9). Apart from this particular category of denticulated implements, some quite massive blades as well as blade-like flakes with partial and variably pronounced denticulated retouch and parallel bilateral cereal polishing are among the late assemblages from Troy. A third category of sickle artefacts consists of truncated and simply retouched blades, backed pieces (including geometric segments) with parallel to lightly diagonally oriented polishes that suggest that they were inserted in a curved handle (Gurova forthcoming). It is worth stressing here that the palaeobotanical data from Late Bronze Age Troy do not differ noticeably from the above mentioned data from the Bulgarian Bronze Age (Riehl 1999, 60).

The formal typological similarity between some of the denticulated tools from Troy and the flint implements from the Bulgarian Bronze age discussed above are obvious, but in a detailed comparison – including use-wear and raw material analysis – some particularities and differences appear, which require further and adequate interpretation based on wider experimentation and parallel case studies. A good example of a thorough approach to the problematics of the sickles was proposed by S. Rosen (1997).

Later stages in the evolution of the sickle can be seen in a glance at some metal sickles, common around the Mediterranean in traditional farming households of the 19th and 20th century and still

Fig. 10.5. Preserved prehistoric sickle from tell Karanovo: 1) from G. Georgiev' excavation; 2) V. Mikov excavation. Pictures of the NAIM exhibition objects.

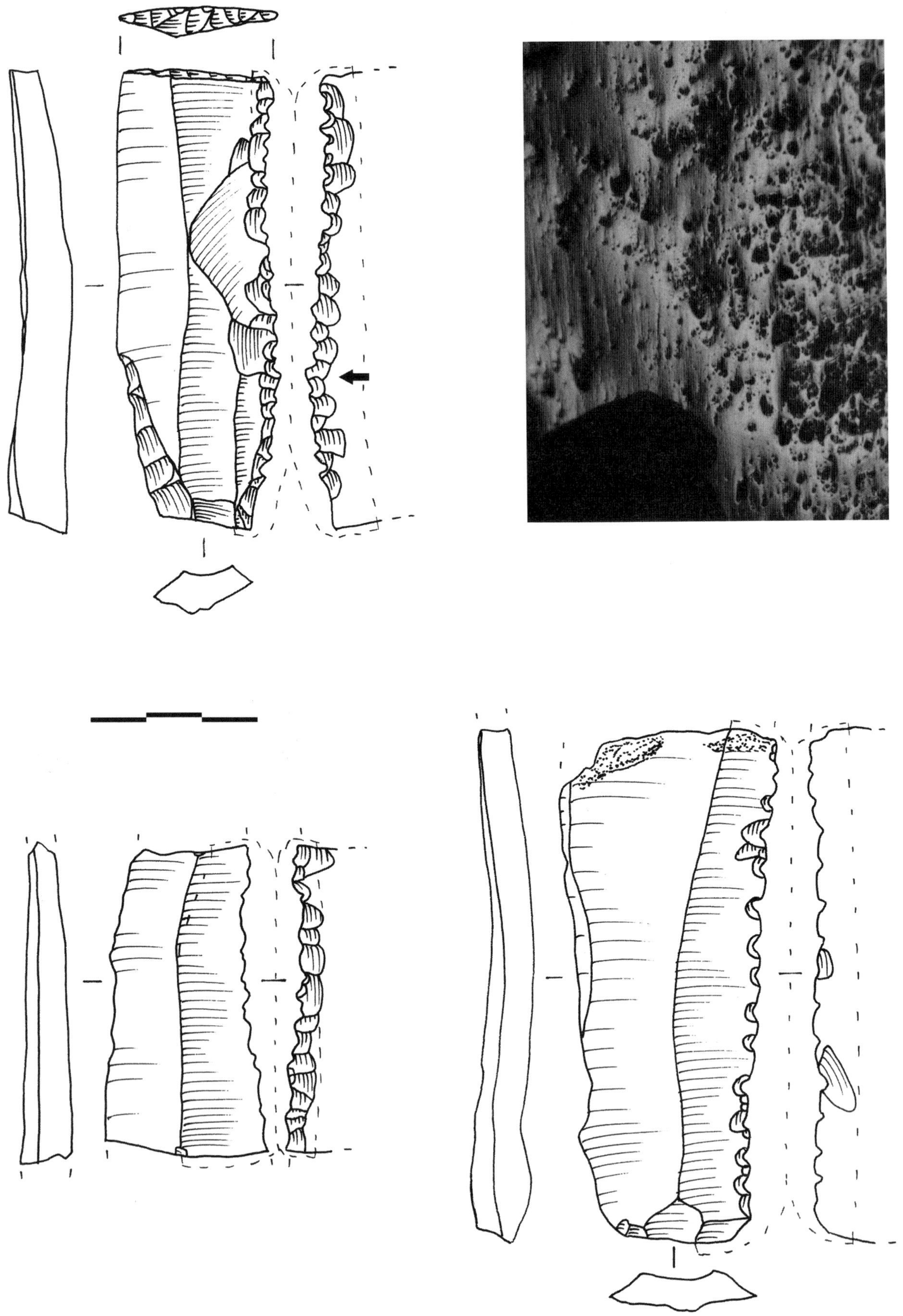

Fig. 10.6. Denticulated sickle blades from Bronze Age site of Aul Kaja (Rhodopes). Drawing and photomicrograph (x100): M. Gurova.

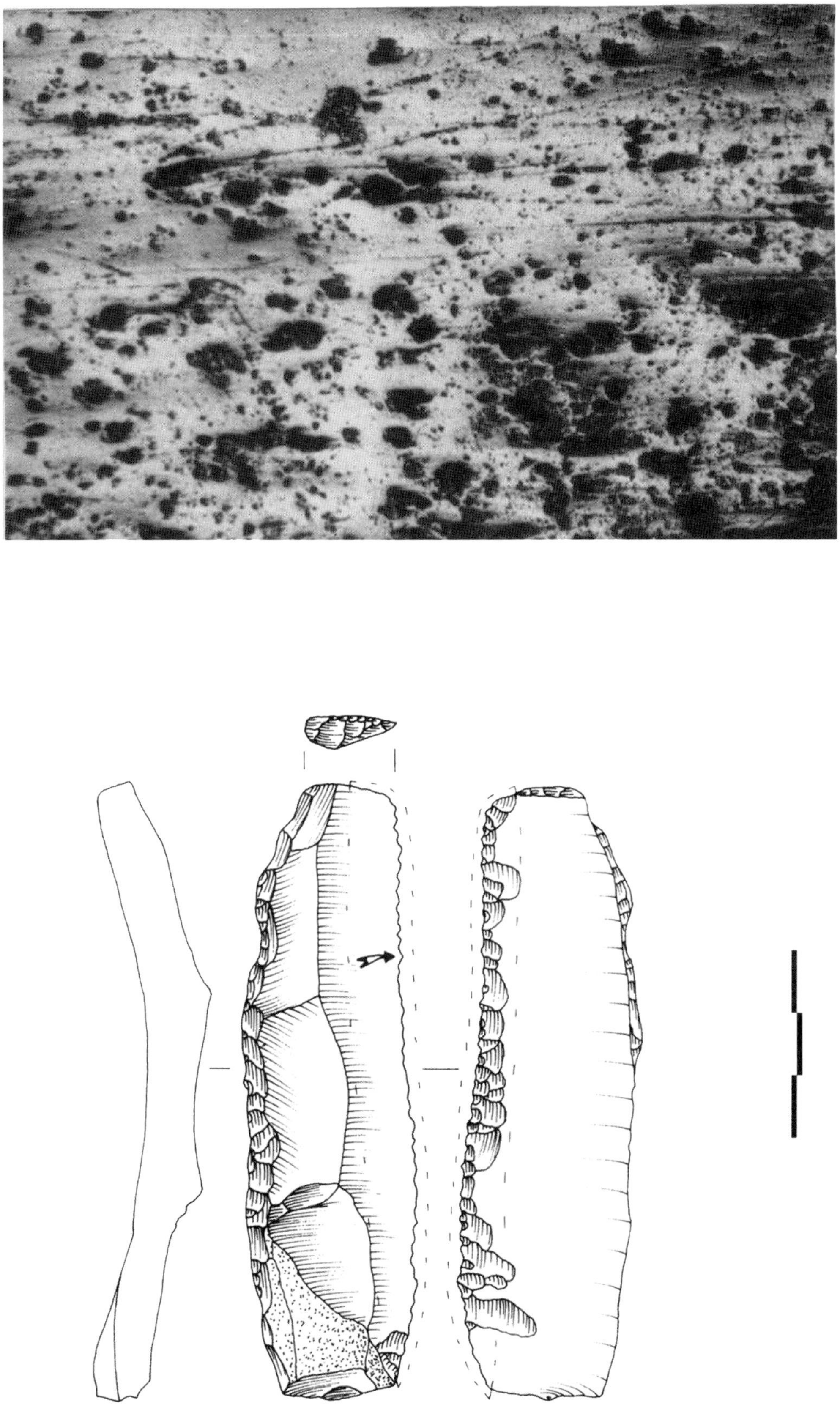

Fig. 10.7. Denticulated sickle blade from Late Bronze Age site of Kamenska Čuka. Drawing and photomicrograph of cereal polish (x 100): M. Gurova.

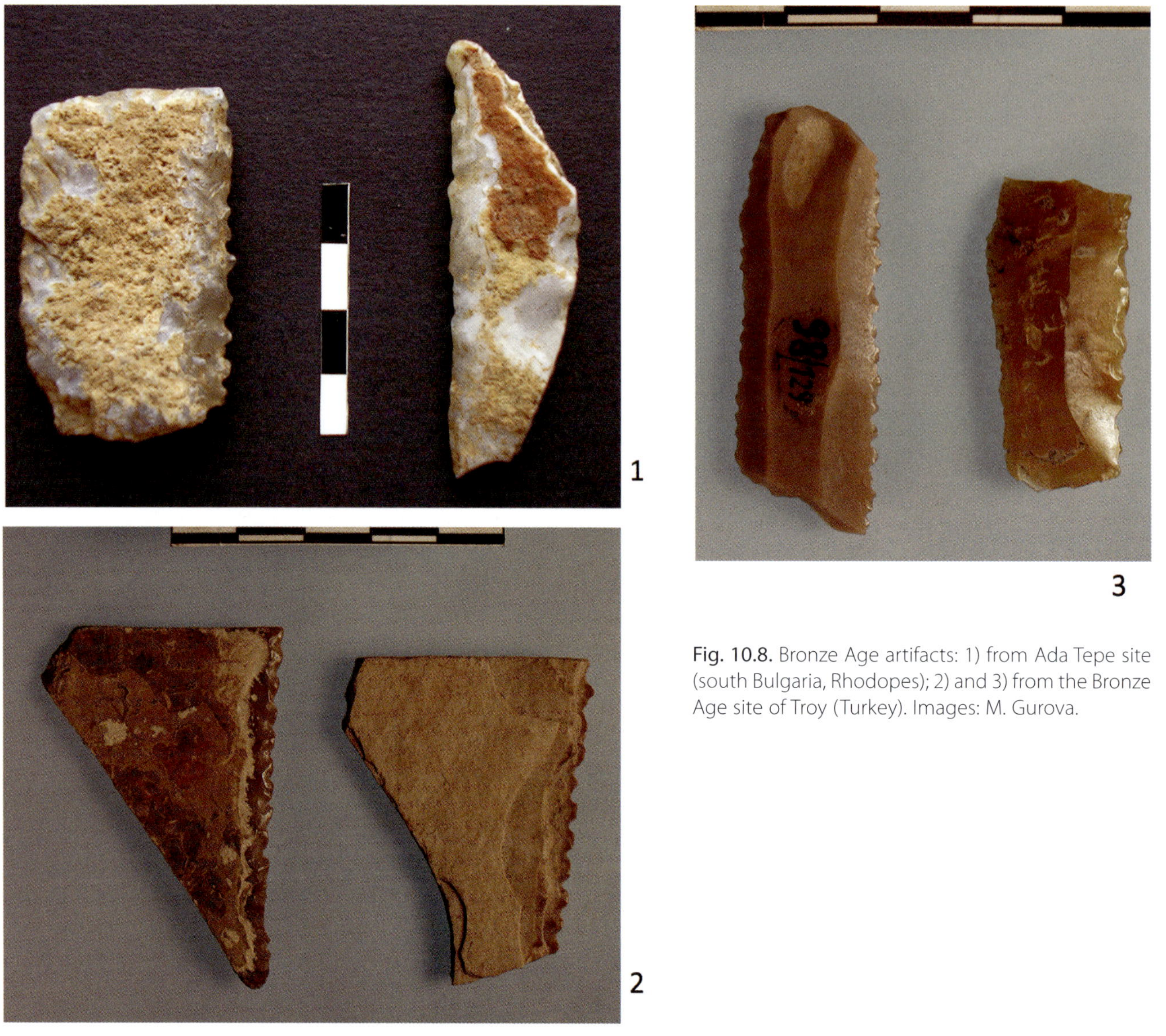

Fig. 10.8. Bronze Age artifacts: 1) from Ada Tepe site (south Bulgaria, Rhodopes); 2) and 3) from the Bronze Age site of Troy (Turkey). Images: M. Gurova.

in use today. These sickles have a strongly curved arched blade and working edge with a short wooden handle (Fig. 10.10). They have a formal ancestry in the first metal sickles, known mainly from several hoards dating from the Bronze and Iron Age, whose shape is basically an optimisation of the efficient prehistoric Karanovo I flint-toothed sickle.

The Appearance and Use of the *Tribulum*

The threshing sledge (*tribulum*) is one of the most ancient and enduring agricultural implements and it has been widely presented in the specialised literature (see, for example, Anderson 2003; Skakun 1999; 2006; as well as Chapter 5). The documented geographic distribution of this instrument from prehistoric times to the ethnographic present covered a vast zone around the Mediterranean, from the Near East to the Iberian peninsula and from the Caucasus to North Africa. Nevertheless, the latest scientific reconstructions suggest a significant difference between prehistoric *tribulum* construction and use in the Mediterranean area and those from northern Mesopotamia. Significant evidence from ethnographic *tribula*, and particularly their inserts, from the southeastern Balkans are presented and described (see Chapter 5.5) as a referential corpus in regard to a series of archaeological *tribulum* inserts identified by myself and coming from a Late

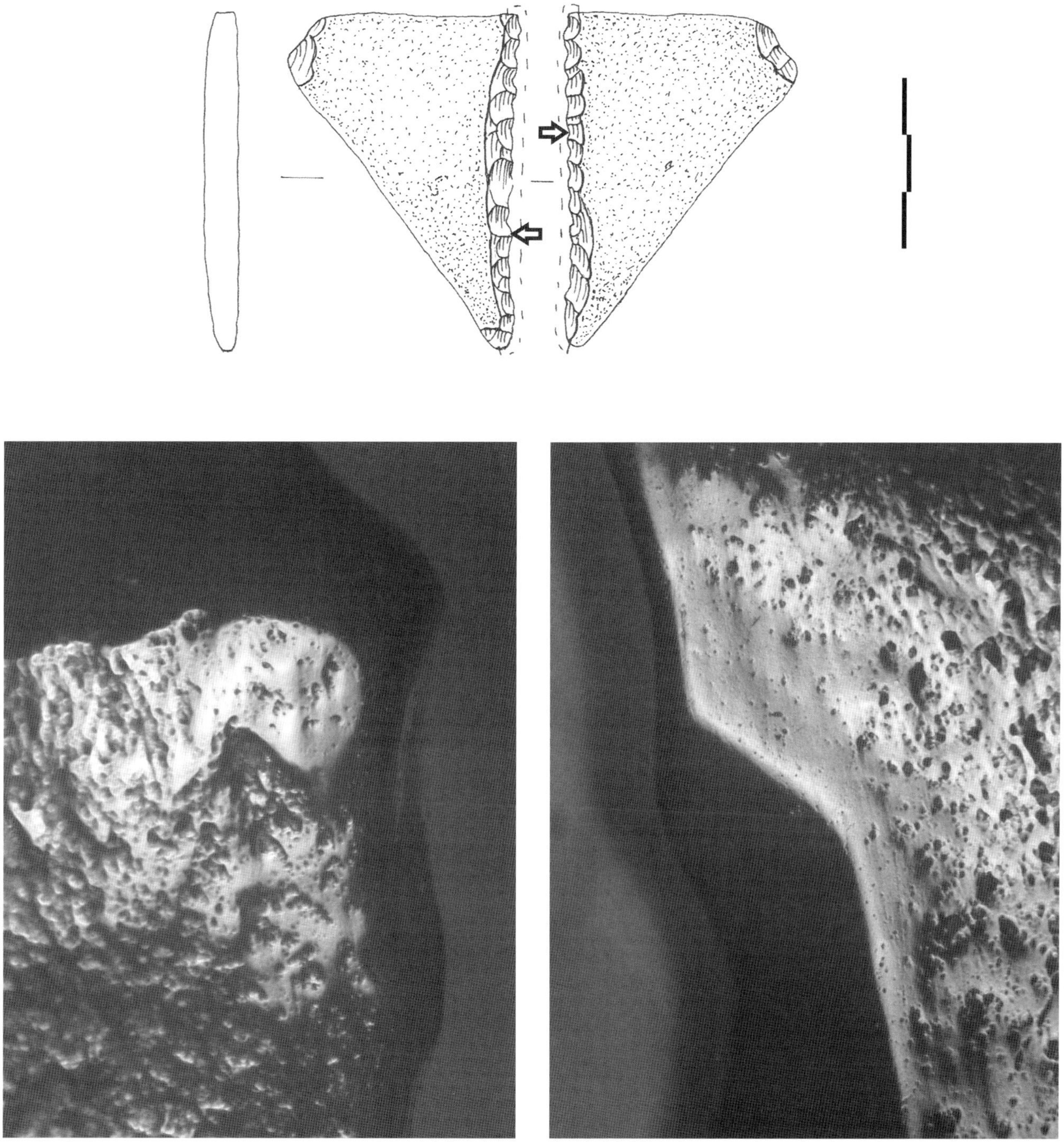

Fig. 10.9. Denticulated sickle blade from Late Bronze Age site of Troy (phase VI–VII). Drawing and photomicrograph of cereal polish (x 100): M. Gurova.

Chalcolithic site in Bulgarian Thrace (see Chapter 5.9).

The comparison of ethnographic reference collections with archaeological flint artefacts led me to conclude that the oldest unequivocal evidence for the use of the *tribulum* in Bulgaria dates to the Chalcolithic strata (Karanovo V and VI period – first half of the 5th millennium BCE). A very small number of such pieces comes from Late Neolithic contexts (Gurova 2001b) but should be considered anecdotal rather than clear evidence of *tribulum* presence. The stratigraphically uncertain data published by Skakun (1993a) on *tribulum* elements in

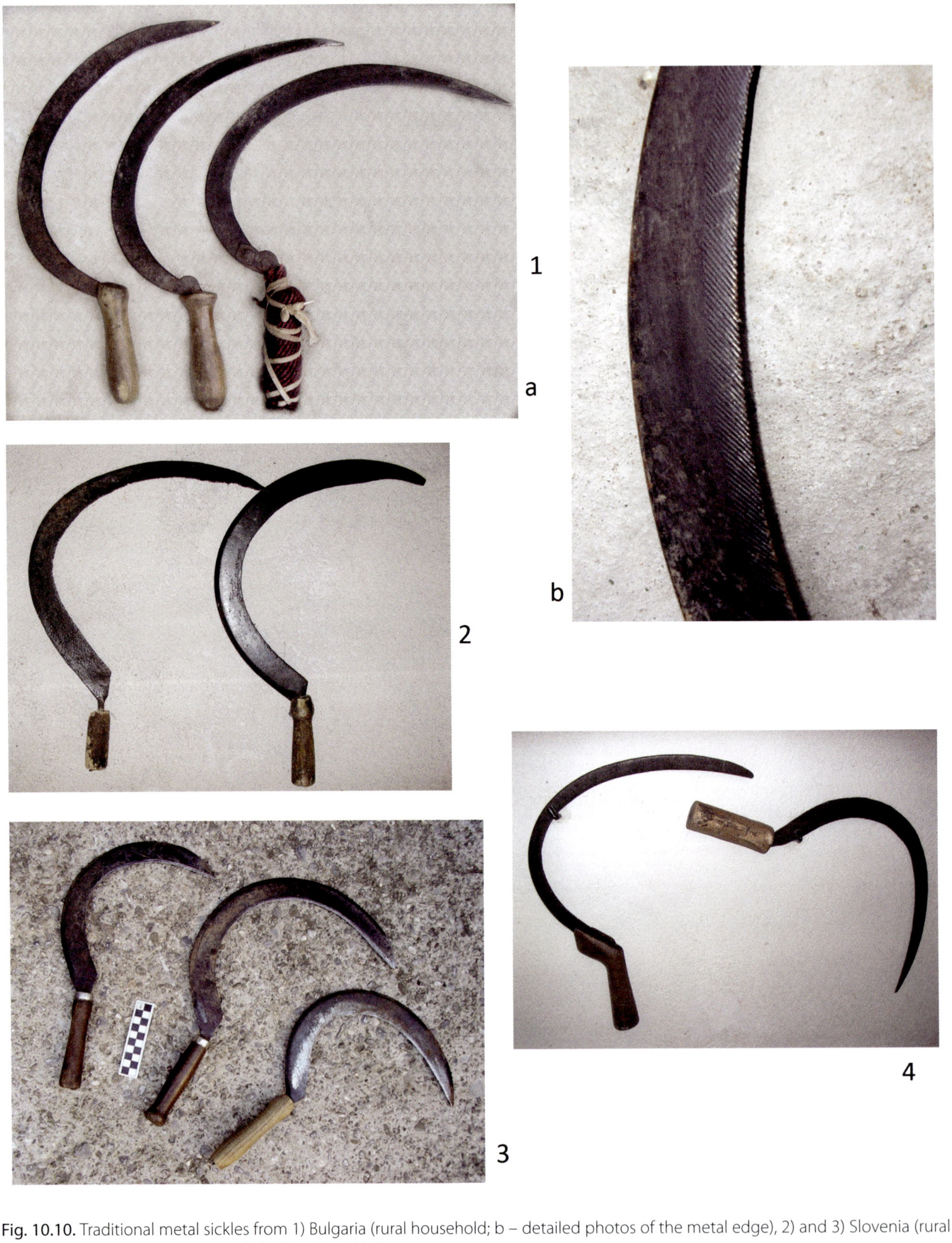

Fig. 10.10. Traditional metal sickles from 1) Bulgaria (rural household; b – detailed photos of the metal edge), 2) and 3) Slovenia (rural household), 4) Melos (museum exhibit). Images: M. Gurova.

western Bulgarian late Neolithic assemblages should be regarded with caution. Apart from a series from the Drama-Merdžumekja site (see Chapter 5.9) and generalised conclusions in Skakun's work (Skakun 1992; 1993 a, b; 1994), only some very rare examples of (possible) *tribulum* inserts are attested among the scarce Bronze Age flint assemblages (Gurova 2006, forthcoming).[2] From later historical times in this region, the Roman, Byzantine, Bulgarian Kingdom and Ottoman periods, archaeological evidence for the agricultural use of *tribula* is so far lacking. On the other hand, a Late Medieval 'cultic' use of *tribula* in funerary contexts is attested. However, in contrast, a considerable ethnographic 'corpus' of threshing sledges is available: those instruments used by traditional, pre-industrial Bulgarian farmers until the 1950s.

To conclude, from the two examples presented, the following seems to become obvious: both sickle and *tribulum*, attested as agricultural tools in prehistoric Bulgarian contexts, reveal the significant temporal persistance embodied by optimal and rather stable decision-making capacities of the prehistoric farmers while dealing with palaeo-environmental changes and social constraints and restrictions.

10.3. THE STABILITY OF BYZANTINE TOOLS

Yorgos Brokalakis

It is generally argued that the Mediterranean farmer used the same basic equipment from Classical Antiquity till contemporary times (Amouretti 1986, 108–110; see also Nouschi 2006, 18–19). The Byzantine farmer seems also to follow this generalisation, and Byzantine agricultural tools are thought by modern scholars to have remained the same despite the fact that the Empire, during its thousand years of existence, encountered transitional situations and crises that affected agriculture to at least some degree.[3]

Fig. 10.11. Agricultural implements depicted in the manuscript of Hesiod's *Works and Days* kept in Venice (Marc. gr. Z.464) and dated to the 14th century. (Bryer 2002, fig. 2)

In pre-industrial societies, many agricultural tools indeed had a very long life span. Since they were conceived to perform certain functions, and since the techniques for which they were employed remained unchanged, there apparently was little need for the tools to be modified (Asdrachas 2007, 188–189 and 194–195). Furthermore, the stability of implements in Byzantium has been interpreted as evidence of how successfully the tools were adapted to the environmental conditions (Lefort 2002, 235).

This notion of stability, as well as our general knowledge of Byzantine agricultural implements, is based on research carried out mostly by historians or art historians and derives mainly from pictorial evidence such as representations of tools on frescoes or in manuscripts (see briefly Λιβέρη 2000 and Bryer 2002). For example, the illuminated manuscripts of Hesiod's *Works and Days* (Fig. 10.11), dating back to the 10th century and later, are considered to give some insight into the equipment of Byzantine farmers and to provide evidence for the continuity of agricultural traditions (Derenzini and Maccagni 1970; Bryer 1986 and Derenzini 1995).

On the other hand, little attention has been paid to the archaeological material. Actually, the implements found so far in archaeological excavations are not plentiful, at least in Greece and Asia Minor. This can of course be related to the lack of concern and attention on behalf of the archaeologists in this region. Here, in contrast to the northern Balkans where medieval archaeology developed an early and

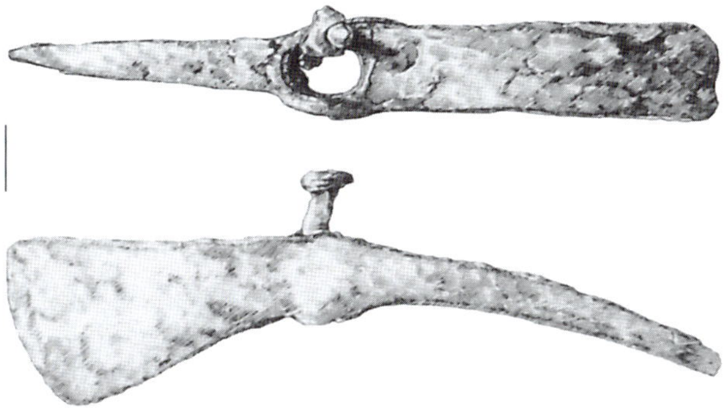

Fig. 10.12. Cross-hoes from Early Byzantine Olympia (a. Völling 2002, fig. 11)

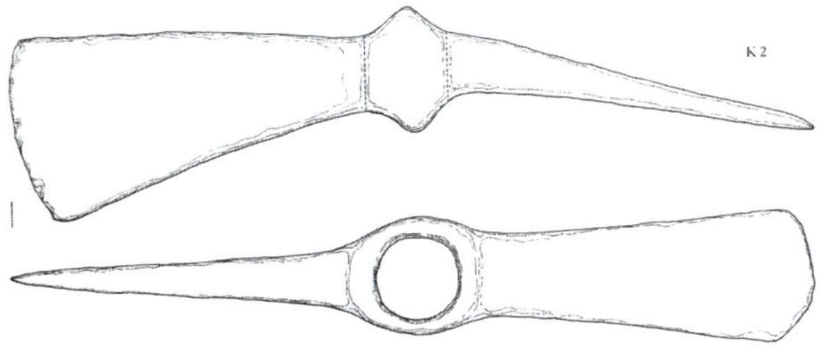

Fig. 10.13. Cross-hoes from Late Byzantine (?) Pergamon (Gaitzsch 2005, Tab. 23, K2)

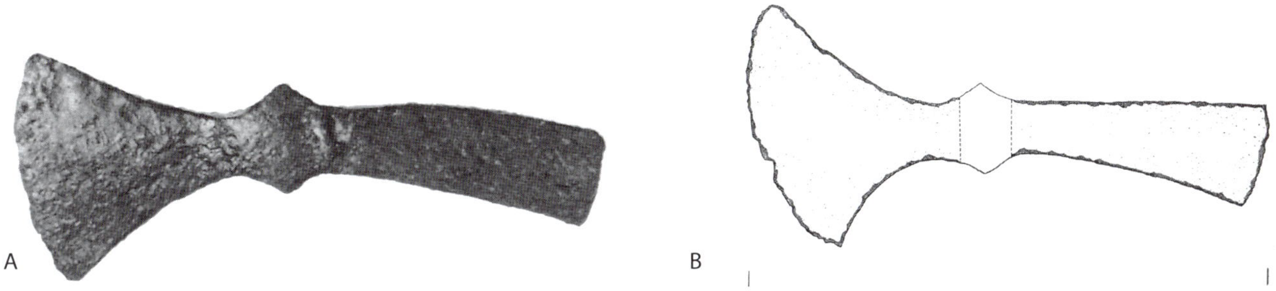

Fig. 10.14. A: Double-axe from Early Byzantine Nemea (Miller 1988, Pl. 7c); B) Double-axe from Late Byzantine Pergamon (Gaitzsch 2005, Tab. 5, BA4).

strong interest in exploring the aspects of material culture,[4] archaeological investigations have only recently started to produce evidence concerning agricultural technologies (for recently published artefacts from Greece, see Völling 2002, and from Asia Minor, see Gaitzsch 2005, 103–113).

Despite the limited evidence that has come to light, one could speak of some kind of stability concerning agricultural tools. Furthermore, the persistence of some forms and shapes are sometimes impressive, as, for example, of the cross-hoe (Figs. 10.12 and 10.13) or the double-axe (Fig. 10.14) that are found

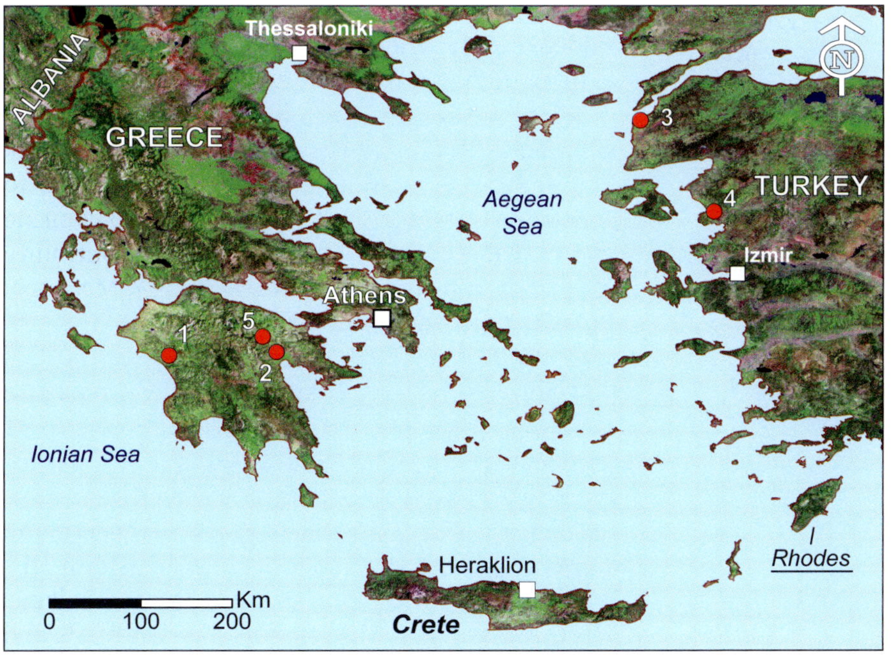

Fig. 10.15. Map of the Aegean with the archaeological sites of 1) Olympia; 2) Pyrgouthi; 3) Troy; 4) Pergamon; 5) Nemea; and the island of Rhodes mentioned in Chapters 10.2 and 10.3. Map: R. Lugon, J.-C. Loubier and A. Chevalier.

in Early and Late Byzantine sites (4th–7th century and 13th–15th century, respectively).[5]

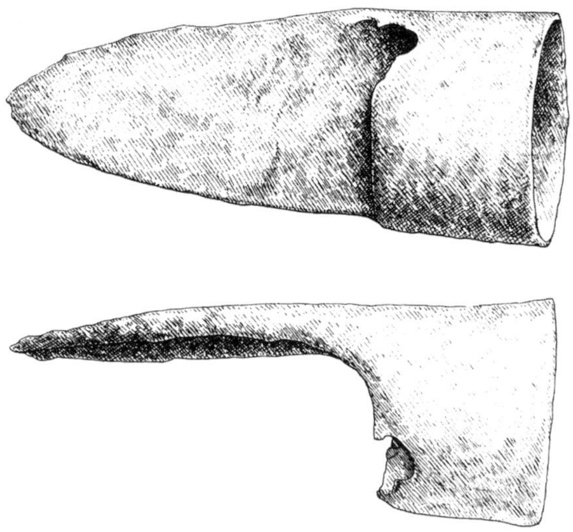

Fig. 10.16. Ploughshares from Early Byzantine Olympia (Völling 2002, fig. 4)

Fig. 10.17. Ploughshares from Early Byzantine Pyrgouthi (Hjohlman 2005, fig. 93)

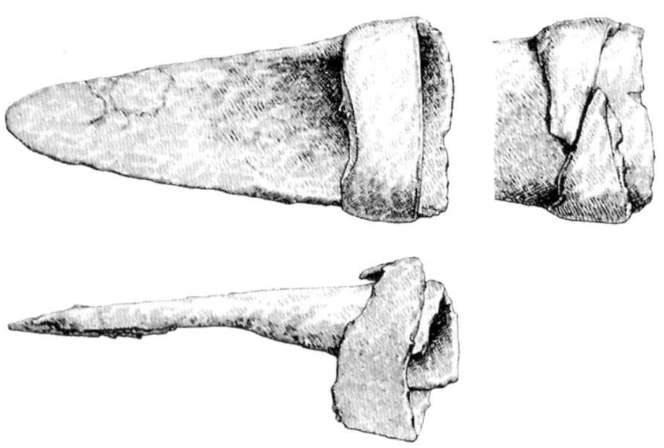

Fig. 10.18. Ploughshares from Early Byzantine Olympia (Völling 2002, fig. 3)

The archaeological data has surely enriched the variety of tool forms known, but their meaning has not been properly studied yet nor have comparisons been possible between the implements of different periods within a specific geographical region, given our current knowledge.[6] So, considering the paucity of the existing evidence, some remarks must be made about change or improvement in the agricultural equipment.

In Greece, for example, the archaeological finds show that the larger balanced sickle was not used in ancient times.[7] This sickle is curently known mainly from Early Byzantine contexts (one on Rhodes, Papanikola-Bakirtzi 2002, 126, item no. 123, and the other one in Olympia, Völling 2002, 200, fig. 13; See also Fig. 10.15). Such an innovative implement, lighter and more efficient than earlier sickles, did not have a long tradition in Greece and might have been introduced by the Romans who had a more advanced agricultural technology.[8] Of course this point of view is based on a limited number of surviving examples and it is possible that the result of future excavations will oblige us to reconsider such an assumption.

Furthermore, the ploughshares from Early Byzantine Olympia (Völling 2002, 196) and Pyrgouthi (Hjohlman 2005, 243) on the Peloponnese (Figs. 10.16 and. 10.17) could be related to some kind of improvement (sites see Fig. 10.15). For attaching the leaf-shaped shares to the plough-sole, it was common to use a separate ring (Fig. 10.18). In these cases however the ring was manufactured simultaneously with the share out of a single piece of iron. This represents a new solution that increases stability during ploughing and demonstrates, according to Völling, 'an effort to achieve a lasting and permanent connection between sole and share' (Völling 2002, 196). Whatever the circumstances that favoured such an improvement, these artefacts from the Peloponnese make us think about the potentialities of accumulating more archaeological evidence and how this could enrich our knowledge of Byzantine agricultural equipment.

As the number of archaeologically documented artefacts continues to grow, we expect that future statements about the stability of the Byzantine agricultural tool assemblage will be assessed from a more extensive archaeological knowledge base, still keeping in mind that also changes or improvements may continue to be discovered.

10.4. SOME PRINCIPLES OF TECHNOLOGICAL DECLINE: THE CASE OF THE *TRIBULUM*

John C. Whittaker

The threshing sledge in Cyprus (and elsewhere) provides a good example of common traits of technological decline. If we do not carry the analogy too far, it is sometimes useful to think of technologies as analogous to organisms: they occupy a particular functional niche within their social and technical environment, interacting and sometimes competing with other technologies.

A technology survives as long as it maintains a competitive edge – technical, economic, or even social – over technologies with similar functions in the system. In some regions, the manufacture of stone blades to arm threshing sledges continued long after other late stone tool technologies like gun-flints and strike-a-lights, because it apparently out-competed alternatives. The raw material, usually flint or *chert*, was relatively cheap and common, and thousands of blades could be made in a day by a single worker with minimal equipment. The metal blades used on *tribula* in some areas in modern times, especially as stone knapping declined, may be more durable, but they must be individually forged or cut, requiring much more time and equipment.

The *tribulum* itself competed in some areas with other threshing techniques such as stone or wooden rollers, or flails, or as in Cyprus with trampling by animals alone. Although absolute measures are lacking, the *tribulum* had advantages of efficiency in threshing large quantities, and in producing finely chopped chaff, which was a valuable resource in Cyprus, especially for animal fodder. In Cyprus after World War II, the *tribulum* was rapidly replaced by mechanised threshers, which were much more efficient. They represented a much larger investment, but that could be offset socially by cooperative use and ownership, or by systems of leasing or hiring.

Technologies are just as interdependent as communities of organisms where the loss of one member may upset the equilibrium of all. As the *tribulum* became obsolete, other parts of the system were affected. The knappers and carpenters who made the flints and the sledges declined in number, and eventually moved on to other pursuits or retired. Blacksmiths increasingly competed against mass-produced imports of most of their repertoire of tools. As farm equipment modernised, blacksmiths lost more customers when they no longer needed to make the metal fittings and harnesses used with the *tribulum*, or the tools used by flint knappers. The need for animal traction and all the trades supporting it declined too. As fewer farmers used donkeys or oxen to pull threshing sledges and transport crops, the chaff used as fodder became less necessary, reducing one of the advantages of the threshing sledge in comparison to other ways of threshing.

On the land, the threshing floors that every household had to have were given over to other uses or obliterated by fields or new construction. In the last few decades in Cyprus, the tractor-powered thresher in its turn has been losing ground to larger combine harvesters. As a result, the small individually owned traditional field systems are increasingly being bought up and united into larger farms, with old boundary walls removed, especially on the coastal plains. In the more mountainous

Fig. 10.19. Threshing sledges in Turkey survived a little longer than in Cyprus but suffered the same processes of decline in the 1980s. Here a threshing sledge advertises the menus of a restaurant in the tourist center of Goreme, Capadokia.

districts, it is difficult to use large machines on the smaller, steeper holdings, and the tractor-powered threshers bought in the 1950s to the 1980s continue in use.

Cultural forms and social behaviour associated with traditional threshing also have been lost. Village families helped each other thresh and winnow, and labourers were fed by the family they helped until the threshing was done. Our old informants in Cyprus had variable memories of threshing time (Whittaker 2000). Some remembered enjoying the cooperation and the social activities that accompanied threshing, while others recalled the hard work and the tedium of driving the sledge around and around.

As a technology declines, it becomes limited in distribution, and fewer people use it or remember the necessary skills to use it efficiently. The technology may be simplified or modified to fit circumstances, for instance by replacing lost flints in a *tribulum* with metal blades as flint-knappers become fewer and their products harder to buy. At the end of its span, a technology may only survive for symbolic, nostalgic or even emblematic reasons. One of the last makers of Cypriot threshing sledges produced miniatures for the tourist market. Today, antique dealers sell them as exotic decorations to tourists, and restaurants hang them on their walls to suggest that their cuisine is authentically traditional (Fig. 10.19).

10.5. TRANSFORMATION AND RESILIENCE IN A LOCAL AGRO-SYSTEM: THE CANTON OF COMPS IN THE SOUTHERN FRENCH ALPS 1850–1990

Thomas K. Schippers

Introduction

This paragraph will present a case study of crisis, change and resilience in a small region (ten villages) in the southern French Alps. The study is based on multidisciplinary sources: history, agronomy, ethnography and social anthropology (Dufour and Schippers 1987, 78–123; 1993, 169–187). It describes the transformation of the local agro-pastoral economy during the last two centuries due to both internal causes (over-population followed by subsistence crisis and emigration) and external ones (enlargement of economic circuits, introduction of potatoes as a new crop). The aim here will be to exemplify how social, economic and technical forms of adjustment and change can interact. This provides an interesting view of the resilience of a small population confronted with a series of successive crises of diverse natures. It has been mainly by choosing among traditional subsistence, poly-cultural, agro-pastoral techniques and subsequent specialisations that the inhabitants have managed to maintain themselves in a more and more open socio-economic context. But this persistent human presence has gone hand in hand with important changes in the landscape as well as in social and economic relations. The landscape, the archives, and the farmers' collective memories have preserved here, like a *palimpsest*, the visible and invisible traces of successive transformations, rendered legible and comprehensible by our multidisciplinary approach.

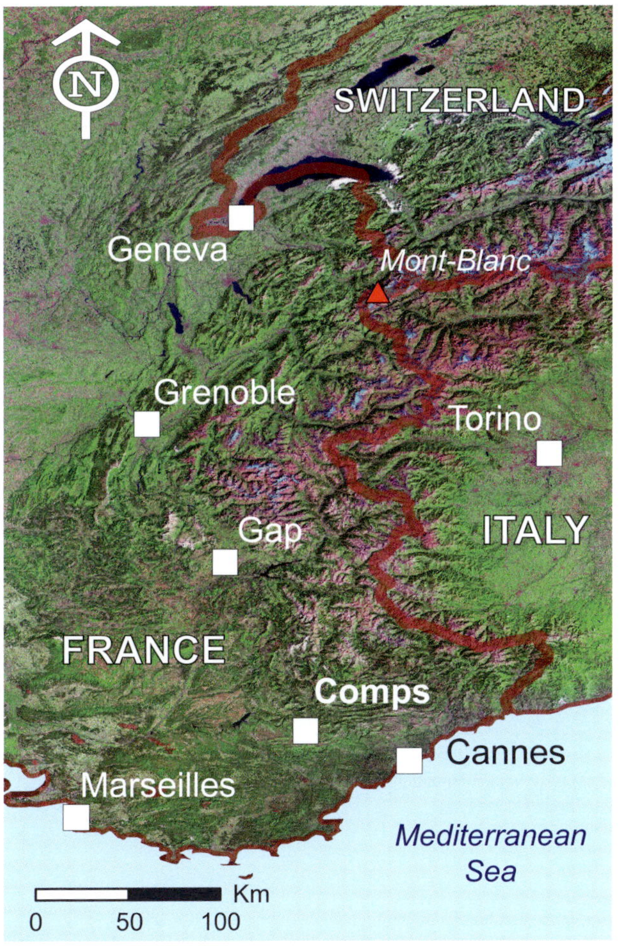

Fig. 10.20. Map of southern France with the town of Comps. Map: R. Lugon, J.-C. Loubier and A. Chevalier.

The canton of Comps (originally about 300 km²) is situated in the transition zone between the Mediterranean lower Provence region and the southwestern part of the Alps (Fig. 10.20). Even the canton[9] itself is divided in two parts: a southern area of dry, hilly high-plateaux with an average height around 900 m, and a northern zone composed of small valleys and some steep mountains with altitudes up to around 1600 m. Climatically speaking, the canton is subject to the negative aspects of both Mediterranean and Alpine influences: summer drought and rather long cold winters.

The habitat is composed of tiny hill-top villages (Fig. 10.21) grouped around the remains of small medieval Templar castles, as well as isolated farms and hamlets (Fig. 10.22). Historically, the inhabitants have always been considered poor and peripheral in the wider regional context of the Provence. Self-sufficient, small-scale farming based on the cultivation of cereals and some flocks (sheep, goats, cows, donkeys and mules) seems to have been predominant in this remote area. It was considered for centuries by administrators to be peripheral, poor and 'without event' in comparison to the often quite politically agitated more southern populations. According to the few available historical sources (Blanchard 1945, 469; Achard 1787 I, 107 and 1788 II, 521), a considerable part (up to 40%) of the

Fig. 10.21. Hill-top village of Trigance in the Canton of Comps.

Fig. 10.22. Isolated farm in the Canton of Comps.

inhabitants of the canton seems to migrate every winter to the southern coastal regions in order to work as farm labourers, servants or even simply to survive by beggary in urban centres at least until the end of the 18th century.

A Subsistence Crisis in the Early 1800s

An agro-historical study of the canton (de los Llanos *et al.* 1981, 21–44) has shown that due to demographic pressures, the maximum extension of cultivated land (22% of the canton) was reached around 1830. The widest expansion of cultivated terraces on hill and mountain slopes is from that period, and their remains today silently bear witness to a 'subsistence crisis'. In the early 19th century, the available arable land is estimated to have been 1.6 ha/inhabitant, but because it was farmed under an archaic bi-annual crop rotation system, the usable area in any year was halved to about 0.8 ha/inhabitant with an average cereal yield of 5 to 7 quintals (500–700 kg) per hectare (de los Llanoss *et al.* 1981, 25).

This subsistence crisis in the 1830s resulted in the start of a demographic decline, with the permanent emigration of many of the younger heirs in the following period. And one observes that between 1872 and 1892 almost 17% of the farms (and related households) disappear (Fig. 10.23), while at that date about 40% of the heads of the remaining farm households are over 60 years of age. But the permanent departure of part of the younger generations also allowed the remaining households to increase the size of their cultivated plots. During the second part of the 19th century, two other factors also helped to resolve the earlier subsistence crisis: the introduction of a tri-annual crop rotation, including fodder crops like trefoil, *sainfoin* (*Onobrychis viciifolia*) and *luzerne* (alfalfa), and the introduction of field cultivation of potatoes. So by the end of the 19th century, the subsistence crisis had been resolved as about 1.1 ha of cereals and 0.5 ha of potatoes per inhabitant were being cultivated; part of the potatoes were sold in the coastal towns and some fodder was also being grown to feed working and transport horses and oxen in the South. As a consequence of these agronomic and economic changes during the second part of the 19th century, the local economy began to become integrated into a wider regional one, a change also made possible by the construction and improvement of many roads during the Second Empire (1851–1870) period.

Year	Population	Family farms
1836	3549	
1846	3546	
1851	3441	
1861	3246	683
1876	2878	662
1891	2565	550
1901	2238	468
1911	1954	375
1921	1528	337
1931	1268	279
1945	1026	
1954	899	147
1968	788	117
1981	734	41
1985	788	
(1990)	(928)	
(1999)	(1109)	(44)

Fig. 10.23. The population and the number of farm households in the 19th and 20th centuries.

Economic and Social Crisis Around 1900

But this integration of the farmers of the canton into a regional and even national economy not only improved their situation but also made them more dependent on wider markets. So when changes in overseas production of cereals, potatoes, and cattle caused a series of declines in agricultural prices between the 1890s and 1910, more than 30% of the farm households of the canton disappeared. In contrast to the preceding period of demographic decline, the remaining farmers were this time not able to increase the size of their farmlands, because their workforce was limited to their household members. So some cultivated lands, especially the colder valley bottoms, were abandoned and became

Fig. 10.24. A landscape of fields overgrown by pine-trees.

Fig. 10.25. Former cultivated fields used as pastures and grasslands.

overgrown by pine trees, still predominant in the landscape today (Fig. 10.24). The First World War strongly accelerated this demographic decline as a proportionally high number of young men were killed, most of them enrolled in the elite frontline regiments of Alpine troops. This also resulted in an important emigration of the women of the same generation.

Interestingly, much of the resilience of the surviving

agricultural economy came from farmers increasing their flocks of sheep in the southern part of the canton and stocks of cows, horses and mules in the northern part. This increase of the herds did not require an increase in the workforce, and the abandoned fields offered convenient pastures (Fig. 10.25). So during the inter-bellum period, an increase and an intensification of animal breeding allowed an accumulation of capital in the form of livestock, in spite of a continuous demographic decline. During this period, traditional secondary sources of income were maintained, like the seasonal migrations of youngsters to the South during the wine and olive harvest. Some new sources of income even developed, like the gathering and 'export' of fresh, wild mushrooms. During the season, a merchant with a motor-lorry would collect 1–2 tons of mushrooms in the canton every evening. Also considerable quantities of hay were sold to more southern vine-growing farmers before the introduction of tractors after the second World War.

Fig. 10.26. Modern mechanical potato harvesting.

Specialisation in the Mid-20th Century

After the end of the Second World War, the sales of livestock allowed the remaining farmers of the canton to slowly modernise and mechanise, apparently without too much financial difficulty . The demographic decline, especially on the more arid southern plateaux, made the farmers there specialise in extensive (non-transhumant) sheep breeding that was able to compete successfully with overseas (New Zealand) meat prices. But in the 1970s most of the sheep farmers were expelled from this area as one of Europe's most extensive military training camps was established here and on neighbouring cantons, the Canjuers Military Camp of over 35,000 ha.

In the northern part of the canton, the remaining farmers have specialised in modern potato production with the help of mechanisation and irrigation (Fig. 10.26). Due to the efforts of some innovative leaders, a small cooperative of producers was created in order to allow direct sales in regional urban and touristic centres. In order to protect and valorise the local potato production a label was created, reversing the former negative stereotype of this inhospitable region into a positive one of pristine mountain purity (Fig. 10.27). From the

Fig. 10.27. The label on the local potato cooperative with a (slightly exaggerated) mountain landscape as an emblem of good quality and a natural environment.

1970s on, a few post-1968 urban 'neo-rurals' also started to settle in the canton and to take over or rent farms. They have largely contributed to a slow demographic rebound as new sources of income developed, linked with tourism and vacation houses.

The canton is near the world famous Verdon canyon and is nowadays not far from the densely populated French Riviera. During the summer months, the landscape now hosts numerous practitioners of open-air activities and 'adventure sports' ranging from hiking to free-style mountain climbing, from rafting to para-gliding, among others. The development of this rural tourism also allows local farmers to directly sell some of their produce, such as potatoes, blueberries and cheese to tourists and second-home owners.

Conclusion

This brief evocation of the agro-economic history of the canton of Comps (studied in detail in Dufour and Schippers 1987, 78–123 and 1993, 169–187) is aimed at showing its many-sided character. While scholarly analysis generally tends to separate according to disciplinary boundaries, and to simplify various factors and relations in order to gain a better understanding of processes at work, we would like to stress the benefits of articulating various types of sources produced by different disciplines. Here agronomic, demographic and historical data provided the factual foundations on which further ethnography-based interviews and observations could build to gain better insights into the way transformations have been carried out and perceived by the local population. For example, collecting family histories and stories provided an enriched understanding of agronomic choices and strategies. Here the area of the cultivated lands, the crop yields, the rotation system, and the level of mechanisation, as well as the average size and age of the households could be correlated in successive complex 'local systems'. A critical comparison of, on one hand, various factual data provided by archives and statistics, and on the other hand the 'grounded' perceptions and discourses of the local inhabitants, proved here to enlighten our understanding of local resilience at work during the past two centuries.

Chapter Notes

1 I would like to thank my colleague G. Nehrizov, who gave me access to the extremely interesting materials from their BA sites excavations.

2 The Bronze Age flint assemblage from the Durankulak site contains a securely stratified *tribulum* insert: study in progress for further publication.

3 See Kazhdan and Epstein 1985, 27: 'Byzantine technology was extremely conservative. Byzantine agricultural implements remained virtually unchanged from Roman times'. Additionally, Lefort (2002, 234) comments 'how little the peasantry's tools changed through the ages and how rudimentary they were'.

4 Henning (1987), for example, studying the passage from Antiquity to the Middle Ages in the northern Balkans, made a great use of the enormous quantity of existing archaeological records concerning agricultural equipment.

5 For cross-hoes, compare the ones from Early Byzantine Olympia (Völling 1995, 437–438, no. 10–13, pl. 95a–d, and Völling 2002, 199, fig. 11) with those from Late Byzantine Pergamon (Gaitzsch 2005, 105–107 and 178–179, K1–14, pl. 23–25). It is worth noting that Gaitzsch emphasises the difficulty of dating such artefacts. He proposes for some an Early Byzantine date because of the similarities with those from Olympia. For double-axes, see the example from Early Byzantine Nemea (Miller 1988, 5, pl. 7 c), and the one from Late Byzantine Pergamon (Gaitzsch 2005, 72 and 166, BA4, pl. 5).

6 See, on the contrary, Henning 1982, in which he explores the changes in the use of ploughshares in the Lower Danube region.

7 For the form and shape of the sickles in Ancient Greece see Kron (1998).

8 Isager and Skydsgaard 1992, 45. See also Henning (1987, 87), who remarks that the balanced sickle was introduced in the northern Balkans in the Roman Period.

9 This administrative entity was created in the aftermath of the 1789 Revolution like everywhere in the country and includes several former (feudal) parishes.

Chapter Bibliography

Achard, C. (1787 and 1788) *Description historique, géographique et topographique des villes, bourgs, villages et hameaux de la Provence ancienne et moderne, du Comtat venaissin, de la Principauté d'Orange et du Comté de Nice*, 3 volumes. Imprimerie Pierre-Joseph Calmen, Aix en Provence.

Amouretti, M.-C. (1986) *Le pain et l'huile dans la Grèce antique*. Annales littéraires de l'Université de Besançon 328, Centre de recherche d'histoire ancienne 67. Belles Lettres, Paris.

Anderson, P. C. (2003) Observation on the threshing sledge and its products in ancient and present-day Mesopotamia. In P. C. Anderson, L. C. Cummings, T. K. Schippers and B. Simonel (eds.) *Le traitemet des récoltes: un regard sur la diversité du Néolithique au present*. CNRS Editions, Antibes, pp. 417–438.

Asdrachas, S. I. (2007) The constant of agrarian techniques. In S. I. Asdrachas *et al.*, *Greek economic history, 15th-19th Centuries* 1. Piraeus Group Bank Cultural Foundation, Athens, pp. 188–207.

Ataman, K. (1999) Threshing Sledge and Archaeology. In P. C. Anderson (ed.) *Prehistory of Agriculture. New Experimental and Ethnographic Approaches*. Institute of Archaeology Monographs 40. University of California, Los Angeles, 211–222.

Blanchard, R. (1945) *Les Alpes occidentales*, Volume IV: *les Préalpes du Sud*, Arthaud, Grenoble.

Bryer, A. (1986) Byzantine agricultural implements: The evidence of medieval illustrations of Hesiod's *Works and Days*. *The Annual of the British School at Athens* 81, 45–80.

Bryer, A. (2002) The means of agricultural production: Muscle and tools. In A. E. Laiou (ed.) *The economic history of Byzantium: From the seventh through the fifteenth century*. Dumbarton Oaks Studies 39. Dumbarton Oaks Research Library and Collection, Washington, pp. 101–113.

Colledge, S. and Conolly, J. (2007) The Neolithisation of the Balkans: A Review of the Archaeological Evidence. In M. Spataro and P. Biagi (eds.) *A Short Walk Through the Balkans: The First Farmers of the Carpathian Bassin and Adjacent Regions*. volume 12, Trieste, 25–38.

Derenzini, G. (1995) I disegni nei manoscritti delle *Opere e i Giorni* di Esiodo: Problemi della tradizione iconografica e testuale. In A. Iacobini and E. Zanini (eds.) *Arte profana e arte sacra a Bisanzio. Atti del Convegno internazionale di studi, Roma 22-23 novembre 1990*. Milion 3. Argos, Rome, pp. 447–461

Derenzini, G. and Maccagni, C. (1970) Per la storia degli attrezzi agricoli. Una tradizione iconografica nei codici esiodei? *Le Machine* 6–7, 1–29.

Gaitzsch, W. (2005) Eisenfunde aus Pergamon, Geräte, Werkzeuge and Waffen. *Pergamenische Forschungen* 14. Walter de Gruyter, Berlin/New York.

Gurova, M. (1997) Gebrauchsspurenanalyse des neolithischen Feuersteininventars. In S. Hiller, and V. Nikolov (Hrsg.), Karanovo, I, *Die Ausgrabungen im Südsektor 1984-1992*, Salzburg/Sofia, pp. 363–375.

Gurova, M. (2001a) Kremachna kolektsija ot praistoriceskoto selishte Mihalic (razkopki 1998–1999). *Godishnik na Arheologiceskija institut s muzei* I, 192–203 (in Bulgarian).

Gurova, M. (2001b) Eléments de tribulum de la Bulgarie – références ethnographiques et contexte préhistorique. *Archaeologia Bulgarica* 5 (1), 1–19.

Gurova, M. (2004a) Outils agricoles des assemblages néolithiques en silex de la Bulgarie – un contexte plus large. *BAR Internatianal Series* 1303, 37–51.

Gurova, M. (2004b) Die Feuerstenartefakte im Kontext der diagnostischen Funde. In V. Nikolov and K. Bacvarov (eds.) *Von Domica bis Drama. Gedenkschrift für Jan Lichardus*, Sofia, pp. 181–189.

Gurova, M. (2005) Kremachnite artefkti v konteksta na diagnostichnite nahodki. *Godishnik na Departament Aheologia*, NBU, VI, 88–103 (in Bulgarian).

Gurova, M. (2006) Kremachen ansambal ot iamno svetilishte ot zheliaznata epoha i selishte ot rannata bronzova epoha pri Svilengrad. In V. Nikolov, G. Nehrizov and Iu. Tsvetkova (eds.) *Arheologicheski razkopki po traseto na zhelezopatnata linija Plovdiv-Svilengrad prez* 2004. V. Tarnovo, 2006, pp. 512–517 (in Bulgarian).

Gurova, M. (2007) Kremachna kolektsia ot Perperokon: Simvolni I utilitarni vnushenia. *Godishnik na Natsionalnia Arheologicheski Musei* 11, 7–15 (in Bulgarian).

Gurova, M. (2008a) Prehistoric agricultural tools – indicators of the process of Neolithization. In M. Gurova (ed.) *Prehistoric research in Bulgaria: New challenges*, Sofia, pp. 39–55 (in Bulgarian).

Gurova, M. (2008b) Kremachni artefakti ot Svilengrad. In V. Nikolov, G. Nehrizov and Iu. Tsvetkova (eds.) *Spasitelni arheologicheski razkopki po traseto na zhelezopatnata linija Plovdiv-Svilengrad prez 2005*. V. Tarnovo, pp. 527–533 (in Bulgarian).

Gurova, M. (forthcoming) Troy I–V chipped stone assemblages: functional connotations. Troia Publication vol. 2.

Henning, J. (1982) Die Entwicklung der Landwirtschaftstechnik und gesellschaftliche Veränderungen im Übergang von der Antike zum Mittelalter im unteren Donaugebiet. In J. Herrmann and I. Sellnow (eds.) *Produktivkräfte und Gesellschaftsformationen in Vorkapitalistischer Zeit*. Akademie Verlag, Berlin, pp. 525–547.

Henning, J. (1987) Südosteuropa zwischen Antike und Mittelalter, *Archäologische Beiträge zur Landwirtschaft des 1. Jahrtausends u.Z. Schriften zur Ur- und Frühgeschichte*, 42, Akademie Verlag, Berlin.

Hjohlman, J. (2005) Pyrgouthi in Late Antiquity. In J. Hjohlman, A. Penttinen and B. Wells (eds.) Pyrgouthi, A rural site in the Berbati valley, From the Early Iron Age to Late Antiquity, Excavations by the Swedish Institute at Athens, 1995 and 1997, *Acta Instituti Atheniensis Regni Sueciae*, Series in 4o, 52. Paul Åströms, Stockholm.

Isager, S. and Skydsgaard, J. E. (1992) *Ancient Greek agriculture, An introduction*. Routledge, London/New York.

Kazhdan, A. P. and Epstein, A. W. (1985) *Change in Byzantine culture in the eleventh and twelfth centuries. The transformation of the classical heritage* 7. University of California Press, Berkley.

Kron, U. (1998) Sickles in Greek sanctuaries: Votives and cultic instruments. In R. Hägg (ed.) Ancient Greek cult practice from the archaeological evidence, Proceedings of the fourth international seminar on Ancient Greek cult, organised by the Swedish Institute at Athens, 22–24 October 1993. *Acta Instituti Atheniensis Regni Sueciae*, Series in 8o, 15. Paul Åströms, Stockholm, pp. 187–215.

Lefort, J. (2002) The Rural Economy, Seventh-Twelfth Centuries. In A. E. Laiou (ed.) *The economic history of Byzantium: From the seventh through the fifteenth century*. Dumbarton Oaks Studies 39. Dumbarton Oaks Research Library and Collection, Washington, pp. 231–310.

Λιβέρη, A. (2000) Βυζαντινά γεωργικά εργαλεία και μηχανές. *Δελτίον της Χριστιανικής Αρχαιολογικής Εταιρείας* 21, 275–285.

Llanos C. de los, *et al.* (1981) *Évolution et transformation de l'utilisation du territoire dans le Canton de Comps sur Artuby*, unpublished thesis, Institut National de Recherches agronomique (INRA), Paris-Gringon.

Marinova, E. (2007) Archaeobotanical data from the early Neolithic of Bulgaria. In S. Colledge and J. Conolly (eds.) *The Origin, Spread and Use of Domestic Plants in Neolithic Southwest Asia and Europe*. Left Coast Press, Walnut Creek, pp. 93–109.

Miller, S. G. (1988) Excavations at Nemea, 1984–1986. *Hesperia* 57, 1–20.

Nouschi, A. (2006) *Il Mediterraneo contemporaneo, Il XX secolo*. Besa, Nardò.

Papanikola-Bakirtzi, D. (ed.) (2002) *Everyday life in Byzantium*, Thessaloniki, White Tower, October 2001–

January 2002 (catalogue of the exhibition 'Byzantine Hours, Works and days in Byzantium'), Hellenic Ministry of Culture, Kapon Editions, Athens.

Popova, T. (1995) Plant Remains from Bulgarian Prehistory (7000–2000 BC). In D. Bayley and I. Panayatov (eds.) *Prehistoric Bulgaria. Monographs in World Archaeology*, 22. Madison Wiskonsin, pp. 193–208.

Riehl, S. (1999) *Bronze Age Environment and Economy in the Troad. The Archaeobotany of Kumtepe and Troy*. Mo-Vince-Verlag, Tübingen.

Rosen, S. (1997) *Lithics After the Stone Age: A Handbook of Stone Tools from the Levant*. Altamira Press, Walnut Creek, London/New Delhi.

Skakun, N. (1993b) Agricultural implements in the Neolithic and Eneolithic cultures of Bulgaria. In P. C. Anderson, S. Beyries, M. Otte and H. Plisson (eds.) Traces et fonctions: les gestes retrouvés. *ERAUL* 50/2, Liege, 361–368.

Skakun, N. (1992) Evolution des techniques agricoles in Bulgarie chalcolithique (d'après les analyses tracéologiques). In P. Anderson (ed.) *Préhistoire de l'agriculture: nouvelles approches expérimentales et ethnographiques*. Monographie de CRA, 6. Paris, pp. 289–303.

Skakun, N. (1993a) Results of traceological Examination of flint implements from neolithic settlements in western Bulgaria. In Gatsov, I. *Neolithic Chipped Stone Industries in Western Bulgaria*. Jagellonian University Press, Kraków, pp. 52–54.

Skakun, N. (1994 (1997)) Agricultural implements and the problem of spreading of agriculture in southeastern Europe. In A. L. van Gijn (ed.) Enigmatic wear traces on stone implements: evidence for handicraft in prehistory. *Helinium*, XXXIV/2, 294–305.

Skakun, N. (1999) Evolution of Agricultural Techniques in Eneolithic (Chalkolithic) Bulgaria. Data from Use-wear Analysis. In P. C. Anderson (ed.) *Prehistory of Agriculture. New Experimental and Ethnographic Approaches*. University of California Monograph 40, Los Angeles, pp. 199–210.

Skakun, N. (2006) *Orudia truda i hoziaistvo drevnezemledelcheskih plemen Iugo-vostochnoi Evropy v epohu eneolita*. Nestor-Istoria, St. Petersburg, (in Russian).

Stefanovich, M. and Bankoff, H. (1998) Kamenska Čuka 1993–1995. Preliminary report. In M. Stefanovich, H. Todorova and H. Hauptman (Hrsh.) *James Harvey Gaul – in memoriam*. The James Harvey Gaul Foundation, Sofia, pp. 255–338.

Völling, T. (1995) Ein frühbyzantinischer Hortfund aus Olympia. *Mitteillungen des Deutschen Archäologischen Instituts, Athenische Abteilung* 110, 425–459.

Völling, T. (2002) Early Byzantine agricultural implements from Olympia (5th/6th Centuries AD). In Π. Θέμελης και Β. Κοντή (eds.) *Πρωτοβυζαντινή Μεσσήνη και Ολυμπία, Αστικός και αγροτικός χώρος στη Δυτική Πελοπόννησο, Πρακτικά του Διεθνούς Συμποσίου*, Αθήνα 29–30 Μαΐου 1998. Εταιρεία Μεσσηνιακών Αρχαιολογικών Σπουδών, Ινστιτούτο Βυζαντινών Ερευνών/ΕΙΕ, Athens, pp. 195–207.

Whittaker, J. (2000) Alonia and Dhoukanes: The Ethnoarchaeology of Threshing in Cyprus. *Near Eastern Archaeology* 63 (2), 62–69.

Annexes

CONTRIBUTORS

Dr Patricia C. Anderson
Université Nice Sophia Antipolis
CEPAM (Cultures and Environments, Prehistory, Antiquity, Middle Ages), UMR 7264 and GDR 2517, CNRS (Centre for National Scientific Research)
St. Jean d'Angély 3
06357 Nice Cedex 4 • France
Email: *patricia.anderson@cepam.cnrs.fr*

Dr Álvaro R. Arizaga-Castro
Escola Superior de Conservación e Restauración de Bens Culturais de Galicia
Rúa Xeneral Martitegui, s/n
36002 Pontevedra • Spain
Email: *alvaroarizagacastro@hotmail.com*

Professor Corrie Bakels
Universiteit Leiden
Instituut voor Prehistorie
PO Box 9515
2300 RA Leiden • The Netherlands
Tel: +31 71 527 2382
Fax: +31 71 527 2429
Email: *c.bakels@arch.leidenuniv.nl*

Dr Sabine Beckmann
Kroustas Forest Historical Landscape Park
721 00 Agios Nikolaos, Crete • Greece
Email: *sabine7@otenet.gr*

Dr Yorgos Brokalakis
Leoforos Ikarou 57
713 07 Poros, Herakleion, Crete • Greece
Email: *ybrokalakis@yahoo.gr*

Dr Carolina Carpinschi
42 Avenue des Gobelins
75013 Paris • France
Email: *carpcaro@gmail.com*

Dr Juan Francisco Gibaja
Consejo Superior de Investigaciones Científicas, Institució Milà i Fontanals (CSIC-IMF)
Calle Egipciaques 15
08001 Barcelona • Spain
Email: *jfgibaja@imf.csic.es*

Professor Annelou van Gijn
Universiteit Leiden
Faculteit Archeologie
PO Box 9514
Leiden 2300 RA • The Netherlands
Email: *a.l.van.gijn@arch.leidenuniv.nl*

Professor Jesús Emilio González Urquijo
Instituto Internacional de Investigaciones Prehistóricas de Cantabria (IIIC)
Avenida de los Castros, s/n
39005 Santander • Spain
Email: *jesuse.gonzalez@unican.es*

Dr Maria Gurova
Bulgarian Academy of Sciences
National Institute of Archaeology with Museum
2, Saborna
1000 Sofia • Bulgaria
Email: *gurovam@yahoo.fr*

Dr Juan José Ibáñez Estévez
Consejo Superior de Investigaciones Científicas, Institució Milà i Fontanals (CSIC-IMF)
Calle Egipciaques 15
08001 Barcelona • Spain
Email: *ibanezjj@imf.csic.es*

Dr Ouafaa Kadim
Lebanon
Email: *kadim_ouafaa@yahoo.fr*

Dr Helena Knutsson
Societas Archaeologica Upsaliensis
Kyrkogårdsgatan 8A
753 12 Uppsala • Sweden
Email: stonesslab@gmail.com

Dr Grith Lerche
Den Kongelige Veterinær og Landbohøjskole
Institut for fødevarevidenskab
Bülowsvej 17
1870 Frederiksberg C • Denmark
Email: *gle@life.ku.dk*

Monder Mhamdi
Université Nice Sophia Antipolis
CEPAM (Cultures and Environments, Prehistory, Antiquity, Middle Ages), UMR 7264 and GDR 2517, CNRS (Centre for National Scientific Research)
St. Jean d'Angély 3
06357 Nice Cedex 4 • France
Email: *mhamdi@cepam.cnrs.fr*

Dr José Luis Mingote Calderón
Museo Nacional de Antropología
Centro de Investigación del Patrimonio Etnológico
Calle Alfonso XII, 68
28014 Madrid • Spain
Email: *jluis.mingote@mcu.es*

Dr Marta Moreno García
Consejo Superior de Investigaciones Científicas (CSIC)
Instituto de Historia
Albasanz 26-28
28037 Madrid • Spain
Email: *marta.moreno@cchs.csic.es*

Dr Leonor Peña-Chocarro
Consejo Superior de Investigaciones Científicas (CSIC)
Escuela Española de Historia y Arqueología en Roma
Via di Torre Argentina 18
00186 Roma • Italy
Email: *leonor.chocarro@csic.it*

Dr Hara Procopiou
Université Paris I
Archéologie, préhistoire et Antiquité (ARSCAN)
3 rue Michelet
75006 Paris • France
Email: *haris.procopiou@univ-paris1.fr*

Dr Isabelle Rodet-Belarbi
INRAP
CEPAM (Cultures and Environments, Prehistory, Antiquity, Middle Ages), UMR 7264 and GDR 2517, CNRS (Centre for National Scientific Research)
St. Jean d'Angély 3
06357 Nice Cedex 4 • France
Email: *isabelle.rodet-belarbi@inrap.fr*

Dr Marie Russel
Institut National de la Recherche Agronomique (INRA)
Unité Collège de direction (CODIR)
147, rue de l'Université
75338 Paris Cedex 07
Email: *marie.russel@paris.inra.fr*

Dr Thomas K. Schippers
Université de Provence, Aix-Marseille I
Maison Méditerranéenne des Sciences de l'Homme (MMSH)
5 rue du Château de l'Horloge / BP 647
13094 Aix-en-Provence cedex 2 • France
Email: *t-schip@libello.com*

Dr François Sigaut († 2012)
Ecole des Hautes Etudes en Sciences Sociales (EHESS)
Centre de Recherches Historiques
54, Boulevard Raspail
Paris Cedex 6 75270 • France

Bernard Simonel
Université Nice Sophia Antipolis
CEPAM (Cultures and Environments, Prehistory, Antiquity, Middle Ages), UMR 7264 and GDR 2517, CNRS (Centre for National Scientific Research)
St. Jean d'Angély 3
06357 Nice Cedex 4 • France
Email: *simonel@laposte.net*

Dr Inja Smerdel
Slovenski etnografski muzej
Metelkova 2
1000 Ljubljana • Slovenia
Email: *inja.smerdel@etno-muzej.si*

Pascal Verdin
INRAP
CEPAM (Cultures and Environments, Prehistory, Antiquity, Middle Ages), UMR 7264 and GDR 2517, CNRS (Centre for National Scientific Research)
St. Jean d'Angély 3
06357 Nice Cedex 4 • France
Email: *pascal.verdin@inrap.fr*

Professor John C. Whittaker
Grinnell College
Department of Anthropology
1118 Park Street
Grinnell IA 50112 • USA
Email: *whittake@grinnell.edu*

THE EARTH PROGRAMME

Steering Committee, 2004–2009

Dr Patricia C. Anderson (Co-Chair)
CNRS, CEPAM Nice • France

Professor Corrie Bakels
Universiteit Leiden • The Netherlands

Professor Miquel Barceló (2004–2006)
Universitat Autonoma de Barcelona • Spain

Professor Sofus Christiansen († 2007)
Københavns Universitet • Denmark

Professor Gabriel Cooney (from 2008)
University College Dublin, College of Arts & Celtic Studies • Ireland

Professor Mareile Flitsch
Völkerkundemuseum der Universität Zürich • Switzerland

Dr Karl-Erik Frandsen
Københavns Universitet • Denmark

Dr Michael Given (Co-Chair 2004–2006)
University of Glasgow • United Kingdom

Dr Eric Huysecom
Université de Genève • Switzerland

Professor Urve Miller
Stockholms universitet • Sweden

Dr Leonor Peña Chocarro (Co-Chair from 2006)
Consejo Superior de Investigaciones Científicas (CSIC), Rome • Italy / Madrid • Spain

Professor Georges Raepsaet (2004–2005)
Université libre de Bruxelles • Belgium

Professor Bjarne Rogan
Universitetet i Oslo • Norway

Professor Erik Thoen
Universiteit Gent • Belgium

Professor Marijke van der Veen (from 2007)
University of Leicester • United Kingdom

Dr Irmeli Vuorela († 2012)
Helsingin yliopisto • Finland

Coordinators

Dr Marie Russel (Scientific Coordinator 2004–2009) INRA (Institut National de la Recherche Agronomique) • France

Dr Andreas G. Heiss (Publication Coordinator from 2010) Vienna Institute for Archaeological Science (VIAS)/Universität für Bodenkultur Wien (BOKU) • Austria

Members of Team 1, 'Crop Choices and Diversity'

Dr Laurent Bouby
CNRS, Université Montpellier II • France

Dr Bui Thi Mai
CNRS, Université Nice Sophia Antipolis • France

Dr Alexandre Chevalier (Team Leader from 2008)
Institut royal des Sciences Naturelles de Belgique / Koninklijk Belgisch Instituut voor Natuurwetenschappen • Belgium

Dr Linda Scott Cummings
Paleoresearch Institute, Golden, CO • USA

Dr Füsun Ertuğ (Team Leader 2004–2008)
İznik • Turkey

Dr Cozette Griffin-Kremer
Université de Bretagne Occidentale, Brest • France

Dr Mária Hajnalová
Univerzita Konštantína Filozofa, Nitra • Slovakia

Dr Ann-Marie Hansson
Stockholms Universitet • Slovakia

Dr Andreas G. Heiss
Vienna Institute for Archaeological Science (VIAS)/Universität für Bodenkultur Wien (BOKU) • Austria

Dr Stefanie Jacomet
Universität Basel • Switzerland

Dr Elena Marinova
Katholieke Universiteit Leuven • Belgium

Dr Jacob Morales
Universidad de Las Palmas de Gran Ganaria • Spain

Dr Carol Palmer (Team Leader 2006–2008)
University of Leicester • United Kingdom

Dr Leonor Peña-Chocarro (Team Leader 2004–2006)
Consejo Superior de Investigaciones Científicas (CSIC), Rome • Italy / Madrid • Spain

Dr Mauro Rottoli
Musei Civici di Como • Italy

Dr François Sigaut († 2012)
Ecole des Hautes Etudes en Sciences Sociales (EHESS), Paris • France

Dr Julian Wiethold
Institut National de Recherche Archéologique Préventive (INRAP), Metz • France

Dr Lydia Zapata
Euskal Herriko Unibertsitatea, Vitoria-Gasteiz • Spain

Members of Team 2, 'Skills, Processes and Tools'

Dr Álvaro R. Arizaga Castro
Universidade de Santiago de Compostela • Spain

Dr Sabine Beckmann
University of Crete • Greece

Yorgos Brokalakis
University of Crete • Greece

Dr Carolina Carpinschi (Team Leader 2004–2008)
CDHT-CNAM-EHESS, Paris • France

Professor Annelou van Gijn (Team Leader from 2005)
Universiteit Leiden • The Netherlands

Professor Detlef Gronenborn (Team Leader 2004-2005)
Universität Mainz • Germany

Dr Maria Gurova
Bulgarian Academy of Sciences, Sofia • Bulgaria

Dr Ouafaa Kadim
Lebanon

Dr Helena Knutsson
Uppsala universitet • Sweden

Dr Grith Lerche
Den Kongelige Veterinær og Landbohøjskole, København • Denmark

Dr José Luis Mingote Calderón
Museo Nacional de Antropología, Madrid • Spain

Dr Hara Procopiou
Université Paris I • France

Dr Thomas K. Schippers
Université de Provence, Aix-Marseille I • France

Bernard Simonel (Team Leader 2008)
CNRS, Université Nice Sophia Antipolis • France

Dr Inja Smerdel
Slovenski etnografski muzej • Slovenia

Pascal Verdin
Institut National de Recherche Archéologique Préventive (INRAP), Nice • France

Professor John C. Whittaker
Grinnell College • USA, Iowa

Members of Team 3, 'Agricultural Landscapes'

Dr Uzi Avner
Ben-Gurion University at Eilat • Israel

Dr Andrea L. Balbo
Consejo Superior de Investigaciones Científicas (CSIC), Barcelona • Spain

Dr Karoline Daugstad
Norges teknisk-naturvitenskapelige universitet, Trondheim • Norway

Dr Althea Davies
University of Stirling • United Kingdom

Dr Dagmar Dreslerová
Archeologický ústav AV ČR, Praha • Czech Republic

Dr Aline Durand
Université de Provence, Aix-Marseille I • France

Dr Christian Keller
Universitetet I Oslo • Norway

Dr Valter Lang
Tartu Ülikool • Estonia

Dr Esa Mikkola
Museovirasto, Helsinki • Finland

Dr Per Grau Møller
Syddansk Universitet, Odense • Denmark

Dr Fèlix Retamero (Team Leader from 2004)
Universitat Autònoma de Barcelona • Spain

Dr Inge R. Schjellerup (Team Leader from 2004)
Nationalmuseet København • Denmark

Dr Luc Vrydaghs
Universiteit Gent • Belgium

SCIENTIFIC NETWORKING WORKSHOPS CONTRIBUTING TO THE CONTENTS OF THIS BOOK

Plenary Meetings

February 2007: 'Stability and balance'
Organiser: Leonor Peña-Chocarro, Centro de Ciencias Humanas y Sociales (CSIC), Madrid • Spain
Venue: Hotel Alixares, Granada • Spain
Additional funding: Unicaja Granada

February 2006: 'Agricultural decision-making'
Organiser: Annelou van Gijn, Universiteit Leiden • The Netherlands
Venue: Universiteit Leiden • The Netherlands

March 2005: 'The dynamics of non-industrial agriculture: a human-centred approach'
Organisers: Michael Given, University of Glasgow, G. Sprott, D. Dornan and E. Edwards, Museum of Scottish Country Life, East Kilbride, Scotland • United Kingdom
Venue: Museum of Scottish Country Life, East Kilbride, Scotland • United Kingdom

Scientific Workshops

Team 1

July 2007: 'Understanding local diversity: from routine practice through to times of crisis and transfer'
Organiser: Stephanie Jacomet, IPNA, Basel University, Basel • Switzerland
Venue: Oberwalliser Mittelschule (OMS) St. Ursula, Brig, Valais • Switzerland
Additional funding: Fonds zur Förderung von Lehre und Forschung

May 2006: 'The role of wild plants within an agricultural system: stability and balance'
Organiser: Füsun Ertuğ, İstanbul Üniversitesi, Prehistorya ve Arkeoloji Bölümü, Istanbul • Turkey
Venue: İlçe Halk Kütüphanesi (Library of Buldan), Buldan • Turkey
Additional support: Buldan Nature Preservation Association

June 2005: 'Assessing and interpreting crop diversity'
Organiser: Leonor Peña-Chocarro, Centro de Ciencias Humanas y Sociales (CSIC), Madrid • Spain
Venue: Casa de Cultura, Proaza • Spain
Additional funding: CSIC

Team 2

June 2008: 'Skills, Processes, and Tools'
Organiser: I. Smerdel, Slovenski etnografski muzej, Ljubljana • Slovenia
Venues: Slovenski etnografski muzej, Ljubljana, and Pliskovica konferenčna dvorana, Lipica • Slovenia
Additional funding: Slovenski etnografski muzej, Republika Slovenija, ministrstvo za kulturo

September 2007: 'Innovation, resilience and revival'
Organiser: Hara Procopiou, Université de Paris I • France
Venue: Municipality of Milos, Plaka, and Milos Conference Centre George Eliopoulos, Adamas • Greece
Additional funding: Costopoulos Foundation and the municipality of Milos

May 2006: 'Local knowledge and technological stability'
Organiser: B. Simonel, Université Nice Sophia Antipolis • France
Venue: Maison d'Hôtes Kasbah Tizourgane, Chouka Ait Baha, and Hôtel Les Amandiers, Tafraoute • Morocco

September 2005: 'The social and cultural dimensions of tool use: perspectives on choice'
Organiser: Carolina Carpinschi, CDHT-CNAM-EHESS, Paris
Venues: Ecomusée Rural du Pays Nantais, Vigneux-de-Bretagne, Musée du Pays de Retz, Bourgneuf-en-Retz and Musée du Vignoble Nantais, Le Pallet • France
Additional support: René Bourrigaud

Team 3

April 2008: 'Agricultural landscapes: synthesis'
Organiser: Aline Durand, Maison méditerranéenne des Sciences de l'Homme, Aix-en-Provence • France
Venue: Maison méditerranéenne des Sciences de l'Homme, Aix-en-Provence • France

May 2007: 'Agricultural landscapes: crisis and change'
Organiser: Valter Lang, Tartu Ülikool, Tartu • Estonia
Venue: Estonian Open Air Museum and Hotel in Sagadi • Estonia

June 2006: 'Agriculture, environment and society in equilibrium: managing slopes; livestock and pastoralism'
Organiser: Inje Schjellerup, Nationalmuseet, København • Denmark
Venue: Nationalmuseet, København • Denmark

October 2005: 'The changing patterns of the agricultural landscapes'
Organiser: Felix Retamero, Universitat Autònoma de Barcelona • Spain
Venue: Hotel Agamenon, Es Castell, Menorca • Spain

Editorial meetings

September 2006 and September 2011: Organiser: Patricia C. Anderson, CNRS, CEPAM, Nice • France
Venue: St. Vallier de Thiey • France

March 2009 and May 2010 : Team 1
Organiser: Alexandre Chevalier, Institut Royal des Sciences Naturelles de Belgique, Brussels • Belgium
Venue: Institut Royal des Sciences Naturelles de Belgique, Brussels • Belgium

February 2009: Team 3
Organiser: Dragomir Dreslerová, Archeologický ústav AV ČR, Praha • Czech Republic
Venue: Archeologický ústav AV ČR, Praha • Czech Republic

February/March 2009: Team 2
Organiser: Pascal Verdin, CEPAM, INRAP, CNRS, Sophia Antipolis • France
Venue: Palais des Congrès, Juan-les-Pins • France

August 2008 : Team 1
Organiser: Leonor Peña-Chocarro, Centro de Ciencias Humanas y Sociales (CSIC), Madrid • Spain
Venue: Centro de Ciencias Humanas y Sociales (CSIC), Madrid • Spain

Summer schools taught by members of the Earth Programme

September 2006 and August 2008: Summer school 'Researching non-industrial agriculture' (Proaza, Spain)[1]
Organiser: L. Peña-Chocarro, Centro de Ciencias Humanas y Sociales (CSIC), Madrid • Spain
Venue: Casa de Cultura, Proaza • Spain
Additional funding: CSIC, Caja Astur

Note

1 We want to thank Valentina Palacios, Pepe Álvarez, Margarita Fernández Mier, Laura Panera and Juaco López Álvarez for giving the EARTH Team and the invited students so much insight into traditional agrarian practices in the Asturias region of Spain.

Index

Page numbers in italics are figures.